Innovative Waste Management Technologies for Sustainable Development

Rouf Ahmad Bhat
Sri Pratap College, India

Humaira Qadri
Sri Pratap College, India & Cluster University Srinagar, India

Khursheed Ahmad Wani
Government Degree College Bijbehara, India

Gowhar Hamid Dar
Sri Pratap College, India & Cluster University Srinagar, India

Mohammad Aneesul Mehmood
Sri Pratap College, India & Cluster University Srinagar, India

IGI Global
DISSEMINATOR OF KNOWLEDGE

A volume in the Practice, Progress, and Proficiency in Sustainability (PPPS) Book Series

Published in the United States of America by
IGI Global
Engineering Science Reference (an imprint of IGI Global)
701 E. Chocolate Avenue
Hershey PA, USA 17033
Tel: 717-533-8845
Fax: 717-533-8661
E-mail: cust@igi-global.com
Web site: http://www.igi-global.com

Library of Congress Cataloging-in-Publication Data

Names: Bhat, Rouf Ahmad, 1983- editor.
Title: Innovative waste management technologies for sustainable development / Rouf Ahmad Bhat, Humaira Qadri, Khursheed Ahmad Wani, Gowhar Hamid Dar, and Mohammad Aneesul Mehmood, editors.
Description: Hershey, Pa : Engineering Science Reference, [2019] | Includes bibliographical references.
Identifiers: LCCN 2019015488| ISBN 9781799800316 (h/c) | ISBN 9781799800330 (eISBN) | ISBN 9781799800323 (s/c)
Subjects: LCSH: Refuse and refuse disposal. | Recycling (Waste, etc.)--Environmental aspects. | Sustainable engineering. | Biomass energy.
Classification: LCC TD785 .I48 2019 | DDC 628.4--dc23 LC record available at https://lccn.loc.gov/2019015488

This book is published in the IGI Global book series Practice, Progress, and Proficiency in Sustainability (PPPS) (ISSN: 2330-3271; eISSN: 2330-328X)

British Cataloguing in Publication Data
A Cataloguing in Publication record for this book is available from the British Library.

For electronic access to this publication, please contact: eresources@igi-global.com.

Practice, Progress, and Proficiency in Sustainability (PPPS) Book Series

Ayman Batisha
International Sustainability Institute, Egypt

ISSN:2330-3271
EISSN:2330-328X

Mission

In a world where traditional business practices are reconsidered and economic activity is performed in a global context, new areas of economic developments are recognized as the key enablers of wealth and income production. This knowledge of information technologies provides infrastructures, systems, and services towards sustainable development.

The **Practices, Progress, and Proficiency in Sustainability (PPPS) Book Series** focuses on the local and global challenges, business opportunities, and societal needs surrounding international collaboration and sustainable development of technology. This series brings together academics, researchers, entrepreneurs, policy makers and government officers aiming to contribute to the progress and proficiency in sustainability.

Coverage

- Innovation Networks
- Eco-Innovation
- Technological learning
- Global Content and Knowledge Repositories
- Environmental informatics
- ICT and knowledge for development
- Intellectual Capital
- Outsourcing
- Sustainable Development
- E-Development

IGI Global is currently accepting manuscripts for publication within this series. To submit a proposal for a volume in this series, please contact our Acquisition Editors at acquisitions@igi-global.com or visit: https://www.igi-global.com/publish/.

The Practice, Progress, and Proficiency in Sustainability (PPPS) Book Series (ISSN 2330-3271) is published by IGI Global, 701 E. Chocolate Avenue, Hershey, PA 17033-1240, USA, www.igi-global.com. This series is composed of titles available for purchase individually; each title is edited to be contextually exclusive from any other title within the series. For pricing and ordering information please visit https://www.igi-global.com/book-series/practice-progress-proficiency-sustainability/73810. Postmaster: Send all address changes to above address.

Titles in this Series

For a list of additional titles in this series, please visit: www.igi-global.com/book-series

Driving the Development, Management, and Sustainability of Cognitiv Cities
Kiran Ahuja (DAV Institute of Engineering and Technology, India) and Arun Khosla (Dr. B.R. Ambedkar National Institute of Technology, India)
Engineering Science Reference • ©2019 • 337pp • H/C (ISBN: 9781522580850) • US $225.00

Technology-Driven Innovation in Gulf Cooperation Council (GCC) Countries Emerging Research and Opportunities
Anna Visvizi (Deree – The American College of Greece, Greece & Effat University, Saudi Arabia) Saad Bakry (King Saud University, Saudi Arabia) Miltiadis D. Lytras (Deree – The American College of Greece, Greece & Effat University, Saudi Arabia) and Wadee Alhalabi (Effat University, Saudi rabia)
Business Science Reference • ©2019 • 180pp • H/C (ISBN: 9781522590125) • US $195.00

Novel Technologies and Systems for Food Preservation
Pedro Dinis Gaspar (University of Beira Interior, Portugal) and Pedro Dinho da Silva (University of Beira Interior, Portugal)
Engineering Science Reference • ©2019 • 416pp • H/C (ISBN: 9781522578949) • US $345.00

Bioeconomical Solutions and Investments in Sustainable City Development
José G. Vargas-Hernández (University of Guadalajara, Mexico) and Justyna Anna Zdunek-Wielgołaska (Warsaw University of Technology, Poland)
Engineering Science Reference • ©2019 • 306pp • H/C (ISBN: 9781522579588) • US $195.00

Optimizing the Use of Farm Waste and Non-Farm Waste to Increase Productivity and Food Security Emerging Research and Opportunities
Leighton Naraine (Clarence Fitzroy Bryant College, St. Kitts-Nevis)
Engineering Science Reference • ©2019 • 223pp • H/C (ISBN: 9781522579342) • US $155.00

Handbook of Research on International Collaboration, Economic Development, and Sustainability in the Arctic
Vasilii Erokhin (Harbin Engineering University, China) Tianming Gao (Harbin Engineering University, China) and Xiuhua Zhang (Harbin Engineering University, China)
Business Science Reference • ©2019 • 703pp • H/C (ISBN: 9781522569541) • US $235.00

Handbook of Research on Rural Sociology and Community Mobilization for Sustainable Growth
Qaiser Rafique Yasser (Centre for Rural Economy, Pakistan)
Information Science Reference • ©2019 • 393pp • H/C (ISBN: 9781522571582) • US $245.00

IGI Global
DISSEMINATOR OF KNOWLEDGE

701 East Chocolate Avenue, Hershey, PA 17033, USA
Tel: 717-533-8845 x100 • Fax: 717-533-8661
E-Mail: cust@igi-global.com • www.igi-global.com

Editorial Advisory Board

Table of Contents

Detailed Table of Contents

A social dilemma is used to chronicle problems with a particular group of people living in that area. It is always the outcome of assorted factors looming beyond the distinctive control and when it comes to waste management, it encompasses all the execution process involved in managing the waste material. This chapter is an attempt to highlight the social problems associated with waste management. People, who are just helpless to place and pay for sanitation and spruceness services, tend to bundle their succulent and solid waste in the open. Open dumps along the roadsides are primarily in practice, which becomes the main reason for blockage of drains, spreading contagious infection, spreading of flies, and stray dog menace. The situation is worse in rural areas where the town area committees are not that active. Adverse breach of calamitous waste casting can be seen everywhere. Due to the influx in the population, the method used is inadequate, undervalued, and paucity of congressional exercises related to waste management are not functioning generously.

Soil, a natural medium for plant growth, provides the anchorage to the plants and supplies various nutrients required for the plant growth. It is an important ecosystem that not only provides the shelter to various organisms but also participates directly or indirectly in various biogeochemical cycles. However, in recent years, the earth's soil has been stripped away, rendered sterile, and contaminated with toxic chemicals due to various anthropogenic activities. This increasing wide spread pollution has caused vast areas of land to become non-arable and hazardous for both wildlife and human populations. Unlike many other organic pollutants, which are degraded in the soils, some of the hazardous compounds like heavy metals,

Electronic waste (e-waste) is one of the swift waste streams and comprises of end of life electronic products. The Western countries as alternative destinations for disposal ship the wastes to underdeveloped and developing countries where labor cost is reasonably meager and environmental laws are feebly implemented. When not recycled, the e-waste is either incinerated or landfilled. These methods involve not only wasting valuable metals but also creating potential risk for the environment. These substances are detrimental to nervous system, kidneys, bones, reproductive system, and endocrine system, and some of them are even carcinogenic and neurotoxic. Thus, extensive research is needed to evolve sophisticated technology which may help to curb environmental pollution and contribute towards sustainable development in terms of recycling of precious metals.

Biomedical wastes (BMWs) are generated by the treatment of living organism as well as during research activities. With the increase of population in developing countries, generation of BMWs is increasing, but with inappropriate disposal management technologies. BMWs pose threats to the humans and also degrade the quality of environment. Handling of the BMWs by inexperienced and untrained staff increases the risk of accumulation of waste at the source as well as increases the chances of the spread of the dreadful diseases. The implementation of BMWs regulations is inadequate and unsatisfactory in developing countries: a major concern. Safe and reliable methods for management of BMWs are of utmost importance. The proper management of BMWs can be helpful in maintaining the overall quality of the environment. In order to get good results, the need is to increase the involvement of the ecofriendly techniques as well as increase awareness among the citizens, staff, and the workers directly involved in the management of the BMWs.

Foreword

Generation of waste materials is one of the greatest concerns of the modern world. Globally, various strategies are implemented to overcome and manage the ever-increasing waste material. However, there are various issues associated with the applied waste management strategies, including the unscientific treatment of waste material, improper collection of waste and ethical issues. The derisory management of waste disposal has despoiled the quality of natural resources globally. Hence, the demand for the improvement of sustainable, efficient, and low-cost technologies to monitor and properly manage the huge quantities of wastes in a scientific manner is continuously growing. Hence, considering the currently available solutions, there is a global call for sustainable, cost-effective (and risk-based), as well as integrated approaches, to better manage risks and recover resources from waste.

The Current book entitled 'Innovative Waste Management Technologies for Sustainable Development' edited by Bhat et al., comprising of 15 chapters on various topics of the innovative technological methodologies in solving the issues of waste management in a sustainable manner, is a welcome addition of the ongoing research activities and devising policy matters on some innovative technological advancements in managing the waste materials. This book will serve an essential reference source discussing the management of diverse types of wastes and provides pertinent theoretical frameworks about new waste management technologies for control of air, water and soil pollution. The connotations of improper management of different wastes have been discussed in an elaborate manner.

I hope this book will generate consciousness towards waste and its impact on the environment, human health and other ecosystems along with some of the better management strategies. I am sure that this book will help the researchers to review the negative impacts of waste and they will start searching out different ways by involving different approaches.

Khalid Ul Rehman Hakeem
King Abdulaziz University, Saudi Arabia

Preface

As the world accelerates towards its urban future, the rate of waste generation is faster than the rate of development. Waste is inextricably linked to urbanization and economic development. The ever increasing standard of living results in a corresponding increase in the amount of waste generated. Unmanaged or crude dumping of any kind of waste has huge short-term as well as long-term potential adverse impacts on every component of the environment. The generation and release of green house gases, toxic leachates, lethal dioxins, noxious gases and heavy metals are some of the issues pertaining to inefficient disposal of wastes. The generated wastes release different categories of pollutants into the environment and even become the home for deadly microbial diseases. In addition to that, some of these toxicants released from wastes tend to accumulate in food chains, posing major risks to humans and wildlife.

The major problems affecting waste management are unscientific treatment, improper collection of waste and ethical issues. This in turn leads to hazards like environmental degradation, water, soil and air pollution. The inadequate management of waste disposal has degraded the quality of the natural resources on regional, state, country basis and even globally.

Therefore, there is an existent demand for the improvement of sustainable, efficient, and low-cost technologies to monitor and properly manage the huge quantities of wastes in scientific manner and convert these wastes into the energy sources. When considering solutions, no matter how advanced the solution is, unless and until it is applied to the ground zero level, it will not be effective. In essence, the current philosophy calls for sustainable, cost effective (and risk based), as well as integrated approaches to better manage risks and recover resources from waste.

In this backdrop, the present book titled *Innovative Waste Management Technologies for Sustainable Development* is an essential reference source that discusses the management of different types of wastes and provides relevant theoretical frameworks about new waste management technologies for control of air, water and soil pollution. The implications of improper management of different wastes have been discussed in an elaborate manner. The present book explores the innovative concepts of recycling, recovery, regeneration, waste to energy, and waste conversion to economically valued products, bio-conversion and vermi-technology that further paves way to a zero waste society. This book is ideally designed for environmentalists, policy makers, professionals, researchers, scientists, industrialists and environmental agencies. Chapter 1 argues that is an attempt to highlight the social problems associated with waste management in the form of a case study from Kashmir. People are just helpless to place and pay for sanitation and spruceness services (which are not provided by the government to the fullest), tend to bundle their succulent and solid waste in plastic wraps and cast them in the society. Open dumps along the road sides are primarily in practice which becomes the main reason for blockage of drains, spreading contagious infection, spreading of flies, and these plunks also led the menace of stray dogs.

The situation is worse in Far-flung sectors of the Valley where the Municipal Committees and Town Area Committees are not that active in checking the menace. Adverse breach of calamitous waste casting can be seen everywhere in the Valley especially you can see a municipal dump in the heart of the city near Poloview. Due to the influx in the population, the method used is inadequate, undervalued and paucity of congressional exercises related to waste management are not functioning generously.

Chapter 2 states that soil, a natural medium for plant growth, consists of layers of inorganic and organic constituents of inconsistent thicknesses varying from its parent rock in the morphological, physical, biological, chemical and mineralogical characteristics. It provides the anchorage to the plants and supplies various nutrients required for the plant growth. Apart from this, it is an important ecosystem that not only provides the shelter to various organisms but participates directly or indirectly in various biogeochemical cycles. Unlike many other organic pollutants, which are anthropogenic and often degraded in the soils, some of the hazardous chemicals/compounds like heavy metals, pesticides, polycyclic aromatic compounds and phthalates are of special concern as most of these are conserved i.e. either non degradable and degrade very slowly.

Chapter 3 advocates that management of solid wastes nowadays becomes a grave concern because of its associated significant negative impacts on quality of the environs. Composting an environmentally sound and sustainable approach to manage biodegradable fractions of solid waste. It has received considerable attention in the last few decades because of its potential of redressing the environmental pollution concerns associated with other waste disposal methods. Further, the different composting technologies and their effects on the physical and chemical properties of the soil, nutrient cycling and crop nutrition are impressed upon. The superiority of using composting technology in maintaining the soil health over the chemical fertilizer application will also be reviewed.

Chapter 4 states that Plastic derived from the petrochemical industry having high molecular weight constitute about 9-13% of total solid waste. After industrial revolution, the use of plastic has been increased manifold without improving its adequate management as a waste. Most of the plastic waste produced in the world is mainly from packaging industry followed by building and construction. Plastic is a non-degradable deadly pollutant to degrade environmental quality and are known to remain in water and soil for years without going any change in their structure. Due to enormous generation, open burning of plastic is also preferred due to the lack of resource in the developing countries thus releasing toxic gases thereby causing air pollution.

Chapter 5 indicates that the western countries as alternative destinations for disposal by shipping the wastes to underdeveloped and developing countries where labor cost is reasonably meager and environmental laws are feebly implemented. When not recycled; the E-waste is either incinerated or land-filled. These methods involve not only wasting valuable metals but also create potential risk for the environment. These substances are detrimental to nervous system, kidneys, bones, reproductive system and endocrine system; and some of them are even carcinogenic and neurotoxic. Thus, extensive research is needed to evolve sophisticated technology which may help to curb environmental pollution and contribute towards sustainable development in terms of recycling of precious metals.

Chapter 6 addresses that the implementation of BMWs regulations is inadequate and unsatisfactory in developing countries, which is a major concern nowadays. Safe and reliable methods for management of BMWs are of utmost importance. The awareness about generation and management of BMWs need to be spread among qualified medical personnel, including hospital administrators, private/governmental institutes and laboratory bearers. The proper management of BMWs can be helpful in maintaining the overall quality of the environment. In order to get good result, the need is to increase the involvement of the ecofriendly techniques as well as increase awareness among the citizens, staff and the workers directly involved in the management of the BMWs.

Chapter 7 reviews the different types of waste that is produced from sericulture and can be used form value added services. In order to ensure a profitable activity, it is necessary to process these secondary and waste products to obtain biologically active substances with important uses in various industries. This waste if do not managed properly has environmental and health implications. However, on the other hand the quality the waste from sericulture units can be converted into value added products. The conversion of waste into vermicompost is seen as an opportunity by different workers that may be advantageous in term of improving the structure and physical conditions of the soil, increase the water retention capacity and availability of phosphorous to the growing plants, prevention of leaching loss of soluble inorganic nitrogen and addition of nutrients/ essential micro nutrients. It is observed that the vermicompost obtained from seriwaste has better nutrient value as compared to conventional fertilizers. The success of sericulture industry in future will depend on sustainability of organic fertilizers both in terms of nutrient management and environmental protection. Similarly the waste can be used for biogas production, as activated carbon, feed for the livestock etc are the opportunities of converting waste into value added products.

Chapter 8 addresses the Zero Waste Strategies that may be applied to companies, communities, industrial sectors, schools and homes. Since it include many stakeholders, not only those of the environment, but also technological aspects. Sustainability is also strongly supported by environmental protection, cost reduction and additional jobs when it comes to waste management and handling back into the industrial cycle. Lowering global resource requirements force us to consider resource management and product management. The management of zero waste is therefore a holistic view of the sustainable avoidance and management of waste and resources.

Chapter 9 states that the waste management is undertaken to recycle the wastes so as to reduce the ill effects of wastes on environment, health and aesthetics. There are several kinds of wastes produced such as agricultural wastes, municipal wastes, industrial waste, mining waste. Some wastes are more hazardous such as medical wastes and nuclear wastes. Various techniques are used for the management of wastes which includes landfilling, incineration, anaerobic digestion, pyrolysis, plasma gasification, recycling, composting. Landfilling and incineration pollute the atmosphere by releasing toxic gases and greenhouse gases into air. Anaerobic digestion produces biofuel in the form of biogas. Plasma gasification results in the generation of electricity from wastes. Recycling of wastes involves the collection, sorting and reprocessing of wastes into new products.

Chapter 10 advocates that rapid generation and accumulation of waste in developing countries is due to the increase in development, urbanization, industrialization, poor government policies and population explosion. Various kinds of waste are produced in the developing countries but due to non availability of ecofriendly environmental techniques, lack of resources or the technical expertise, it is becoming difficult to treat the waste at the disposal sites. Most of the waste produced in developing countries is directly disposed off to the landfills without any proper sorting and segregation, where it produces green house gases thus results in global warming. Recycling and Composting is helpful in reducing the volume of the waste and producing valuable products which can have multiple applications.

Chapter 11 argues that there are plenty of techniques available (incineration, pyrolysis, biogas generation based plants, gasification, mechanical biological treatment, anaerobic digestion etc.,), which are used to covert wastes into energy or other useful products. Besides, microbial fuel cells are directly used to generate electricity from the oxidation of dissolved organic matter. These valuable energy converting techniques if properly implemented will be the asset for economic development and importantly will manage the wastes adequately.

Chapter 12 believes that inadequate management of these solid waste constituents could lead to drastic change in physico-chemical properties of soils. Howbeit, the waste product, which is discarded into the environment, is loaded with valuable compounds. They are new, innate and monetary sources of colorants, protein, dietary fiber, flavoring, antimicrobials, and antioxidants, which can be utilized in the food industry as a basis of natural food additives. Thus, a novel aspects concerning to the use of these by-products for advance utilization on the production of food additives with high nutritional value have gained immense interest because of their characterization among high-value products. Therefore, their recovery and improvement may be economically attractive and beneficial and in a sense will be environmental friendly approach.

The major objective of chapter 13 is to identify the main aspects presented in three companies of energy generation from biogas produced by landfills in three cities of the country: Benlesa in Salinas Victoria, Nuevo León; Ylem Energy Limited in Aguascalientes, Aguascalientes; and Biogas Juárez in Ciudad Juárez, Chihuahua. Analyzing these companies, their way of operating and business environment, it is possible to detect the existence of business opportunities in this area. It is concluded that there is a great potential of wasted businesses that works through market mechanisms in strategic alliance with public entities as a more viable option for the reduction of investment risk and the viability of the projects. The use of high technologies, composition of city waste and available infrastructure are key factors. The study of three companies in operation allows the identification of factors that can help in the implementation of more companies of energy generation based on urban solid waste in more areas of the country and elsewhere with successful and profitable results for both private capital and social interest.

Chapter 14 discusses the different technologies in India that are available to convert waste into energy apart from the technologies that have impact on the environment. The Nisargruna biogas technology which has been used to set up several biogas plants in the country is an excellent example of the technology. Scientists have developed a solid and liquid resource management model, also known as the Vellore model that has been adopted by many states in India. Researchers invented a technology that uses plastic waste to construct more durable roads. Several plastic roads have since been laid out in various cities in India with the help of this technology. Waste is used to run poultry and dairy farms in India are an admirable example of converting waste into a useful resource. These technologies are also generating employment in rural areas of India and have increased the standard of living of the waste handlers. However, there is a dire need to increase the efficiency of these environmental friendly technologies to a larger scale.

Chapter 15 studies that Vermicompost is humus like, finely granulated and stabilized material. To reintegrate the organic matter to the agricultural soils, it can be used as a soil conditioner. A potential raw material for "vermitechnology", could be non-toxic and organic industrial wastes. In the last two decades, "vermitechnology" has been applied for the management of industrial wastes and sludge to convert them into vermicompost for land restoration practices. The success of "vermitechnology" mainly depends upon several process parameters like quality of raw material, aeration, temperature, moisture, pH, type of vermicomposting system and the species of earthworm used. A suitable means for waste remediation and organic manure production is vermicomposting.

The book will generate awareness towards waste and its impact on environment, human health and other ecosystems along with some of the better management strategies. This publication will help the researchers to assess the negative impacts of waste and they will start searching out different ways by involving different approaches. The literature included in this publication will help to reduce the burden of waste and will indirectly reduce the load of diseases caused due to different types of waste. This may also help people to search avenues for the economic aspects of waste management.

Rouf Ahmad Bhat
Sri Pratap College, India & Cluster University Srinagar, India

Humaira Qadri
Sri Pratap College, India & Cluster University Srinagar, India

Khursheed Ahmad Wani
Government Degree College Bijbehara, India

Gowhar Hamid Dar
Sri Pratap College, India & Cluster University Srinagar, India

Mohammad Aneesul Mehmood
Sri Pratap College, India & Cluster University Srinagar, India

Acknowledgment

First and foremost we would like to place on record our gratitude and sincere thanks to Almighty Allah^{SWT} who always helped us surreptitiously while instilling passion of taking the stupendous jobs at hand. It is because of HIS divine intervention that we could finally complete this job in time.

We are at loss of words in extending heartfelt thanks to the anonymous learned reviewers who put every effort to make this project possible for us. We are thankful to the authorities at Cluster University SP College Srinagar and Government Degree College Bijbehara for providing facility of computers and internet during the course of editing. We are extremely thankful to Courtney Tychinski Development Editor, IGI Global for being so kind to us during the editing of book.

To our parents, we are speechless! We can barely find the words to express all the wisdom, love and support you've given us. To our children: you were kind serene during the editing of the book. Our love for you has increased and we hope you will bear with us in future as well.

To rest of the family members, we do love you guys very much, collectively as well as, individually for different reasons. You all have always made us feel as if we are proud members, in spite of all our imperfections. Finally, our thanks are due to all the people who have supported us in completing and compiling this voluminous book directly or indirectly.

Rouf Ahmad Bhat
Sri Pratap College, India & Cluster University Srinagar, India

Humaira Qadri
Sri Pratap College, India & Cluster University Srinagar, India

Khursheed Ahmad Wani
Government Degree College Bijbehara, India

Gowhar Hamid Dar
Sri Pratap College, India & Cluster University Srinagar, India

Mohammad Aneesul Mehmood
Sri Pratap College, India & Cluster University Srinagar, India

Chapter 1
Social Causatums of Improper Waste Management

G. Hassan Lone
University of Kashmir, India

Fatima Buchh
University of Kashmir, India

Naila Irshad Shah
University of Kashmir, India

ABSTRACT

A social dilemma is used to chronicle problems with a particular group of people living in that area. It is always the outcome of assorted factors looming beyond the distinctive control and when it comes to waste management, it encompasses all the execution process involved in managing the waste material. This chapter is an attempt to highlight the social problems associated with waste management. People, who are just helpless to place and pay for sanitation and spruceness services, tend to bundle their succulent and solid waste in the open. Open dumps along the roadsides are primarily in practice, which becomes the main reason for blockage of drains, spreading contagious infection, spreading of flies, and stray dog menace. The situation is worse in rural areas where the town area committees are not that active. Adverse breach of calamitous waste casting can be seen everywhere. Due to the influx in the population, the method used is inadequate, undervalued, and paucity of congressional exercises related to waste management are not functioning generously.

INTRODUCTION

Dilemma attributed to waste engenderment commenced from awful human acts, which actually needs diligent management. It has been found that the waste management becomes more intricate and multifarious in the more highly developed parts of the world (Kumar et al., 2013). Waste generation ratio is augmenting globally. The world's cities in 2016 generated 2.01 billion tons of garbage/ solid waste which is equal to 0.74 kilograms/person/day. With the increasing population, urbanization and indus-

DOI: 10.4018/978-1-7998-0031-6.ch001

trialization, generation of waste is likely to swell by 70% per year from 2016 levels to 3.40 billion tons in 2050 (World Bank report, 2018). In many developing and indigent countries 90% of people tend to bundle their succulent and solid waste in plastic wraps and cast them in the society. Open dumps along the road sides are primarily in practice (Bhat et al., 2018). Ahmed et al (2015) stated that Himalayans are striving for constructive and judicious plan for municipal waste management. In Western Himalayas, the rapidly growing City is Srinagar and municipal waste management is still a colossal challenge for government. Environmental Pollution Act (1986) characterizes waste as "*Any substance or object which the holder discards or tends to discard*". In other words waste can be defined as scrapped/ abandoned material like glass, sewage mud, ash, metal, plastic bags, and dead animals etc. Waste generation is not fixed to a particular quantity, it diverges from 2.75- 4.00 / capita/day in cost effective countries although in developing countries it is as low as 0.5 kg/capita/day. It has been found that waste generation increases with increasing population and increasing demand of land (Indris et al., 2004; Bhat et al., 2014).

Effective waste management is a global concern now. Even the developed countries are grappled with this issue. In most of the developing countries one can observe several problems associated with improper waste management. The environmental problems associated with improper waste management are known to everyone but there are also various social causatums of improper waste management which are rarely looked into by many environmentalists or the waste management experts. These social consequences include branding the locality living near the waste sites as 'dirty community' irrespective of their economic and social status which often leads to other associated problems. The other problem is the low land valuation of the area being in the vicinity of waste sites and temptation to migrate from the area. Before proceeding further let's enlist few facts which signifies the importance of proper waste management.

- Waste management reduces the chances of environmental pollution
- Waste management reduces spread of various diseases
- Waste management generates employment
- Waste management supports the normal societal functioning
- Waste management helps to conservation of energy
- Waste management reduces waste
- Waste management prevents misuse of open lands

CLASSIFICATION OF WASTE

Waste can be classified into various ways depending upon their type and source.

Classification Based on Type of Waste

1. **Solid Wastes**: Includes undesirable discarded material by people. These encompass urban waste, industrial waste, and agricultural, biomedical and radioactive wastes.
2. **Liquid Wastes**: Includes wastes generated from washing, flushing, brimming process of industries.
3. **Gaseous Wastes**: Includes wastes that are produced from the gases generated by vehicles and factories.

Classification Based on Sources of Waste

1. **Municipal Sources of Wastes**: Includes debris/garbage from family unit, workplaces, schools, souks and from other public places, items like plastic bags, plastic bottles, food trash, clothing, household broken thing and many more.
2. **Agriculture Source of Wastes:** Includes wastes produces as a result of various agricultural operations, may include fertilizers, pesticides, wastes from farm, poultry houses and slaughter houses etc.
3. **Medical Sources of Wastes**: Includes waste generated from hospitals, clinics, veterinary hospitals, laboratories, these mainly include surgical items, needles, blood, parts of body, pharmaceuticals, dressing material, syringes etc.
4. **Industrial Sources of Wastes**: Includes wastes discharged from manufacturing and processing units like food processing industries, cement factories, power plants, textile industries etc.
5. **Electronic Source of Wastes**: Also known as e- waste or e- scrap includes telephones, television, music player, DVDs, CDs, computers, mobiles etc., which are of no use and are discarded.
6. **Mining Source of Wastes**: Includes wastes produced from mine tailings, harmful gases released by blasting etc.
7. **Wastes From Demolition or Construction**: Includes waste generated from construction of roads, buildings and houses and demolition of old buildings and structures producing colossal construction debris.
8. **Radioactive Source of Wastes**: Includes wastes produced from nuclear explosion, atomic energy/nuclear power, digging of radioactive substances etc.

With rapid growth and development, industrialization and urbanization in Asia and Pacific region and its sub regions and increase in engrossing detrimental materials thus generating a huge quantity of perilous wastes. The consequence of diverse categories of wastes on air, water, soil and land in some selected countries of various sub regions is presented in Table 1. Municipal waste generation varies widely from country to country and it is found that economic development plays a lead role in its production. Developed countries like Australia, Hong Kong, Korea, Singapore and China produce 1.1-5 kg/capita/day (solid municipal waste), middle income countries like Malaysia, Indonesia and Thailand produce 0.25- 1 kg/capita/day. However low-income countries like India, Pakistan, Bangladesh produce 0.45-0.89 kg/capita/day of solid municipal waste. World Bank in 1999 reported that the Asian and Pacific region on the whole produces 1.5 million tons of solid municipal waste per day this is expected to more than double by 2025.

Asian and Pacific region ensphere extensive range of industrial solid materials which are deleterious to ecosystem. It mostly includes packaging material, paper, paint, food processing wastes, oil, ceramic, cloth, plastic etc. Agricultural production has increased which in turn resulted in the increase of animal's residue from farms, agriculture crop lees and agro industrial wastes (ESCAP, 1997). In the Asia and pacific region, Peoples Republic of China produces largest quantities of agriculture and crop residue followed by India. Almost 587 million tons of agricultural wastes are generated annually from Peoples Republic of China and in Pakistan about 56.22 million tons of crop residue are generated an-

Figure 1. Classification of waste

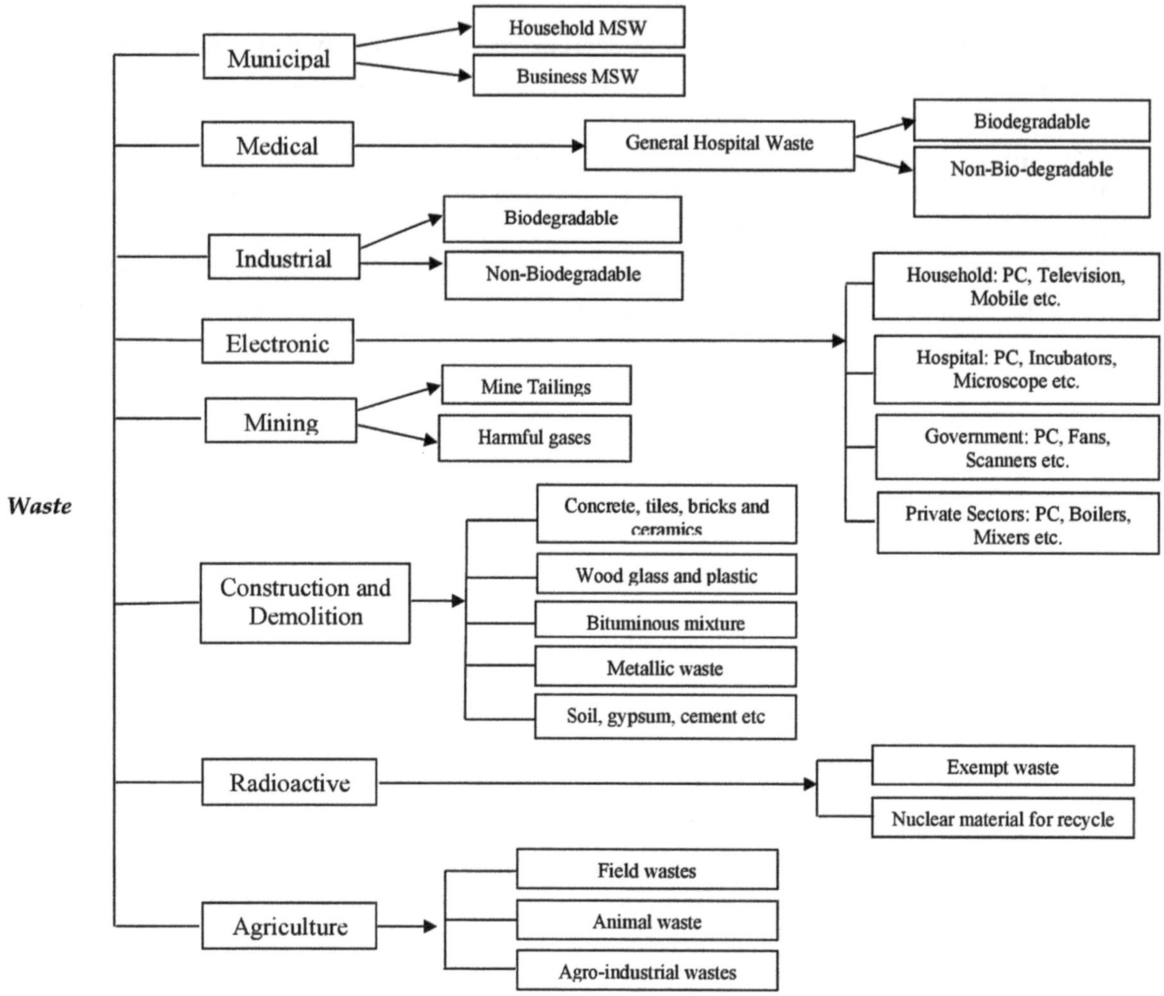

nually (ESCAP, 1997; UNEP, 1997). The majority of hazardous waste is consequence of widespread of agronomical, industrial, centric expansion, medical centers and health services. Significant proportion of industrial perilous waste are mainly produced from petroleum, petrochemicals, metals, pulp and paper, textile factories and hospitals etc.

WASTE GENERATION, TREATMENT AND COLLECTION SYSTEM

As per census 2011 the total population of India is 1.21 billion which accounted for almost 17.5% of the world's population. With the rapid growth of urbanization and industrialization, management of municipal waste persists to be one of the neglected areas of urban and rural development. People have tendency to throw the waste from distance to communal point that present a view of spilled over debris as can be seen in Figure 2.

Table 1. Impacts of various categories of wastes on water, soil and air in selected countries of different sub regions

Countries	Agricultural Wastes & Residues			Municipal Wastes			Industrial Wastes			Hazardous Wastes		
	Water Pollution	Land Pollution	Air Pollution	Water Pollution	Land Pollution	Air Pollution	Water Pollution	Land Pollution	Air Pollution	Water Pollution	Land Pollution	Air Pollution
Australia	⊕	⊕		⊕	⊕		⊕	⊕		⊕	⊕	
Bangladesh	•	⊕		•	⊕	⊕	•	⊕		•	⊕	
Brunei Darussalam				⊕								
Cambodia	•	⊕		•	⊕		•	⊕		•	⊕	
PR China	•	⊕	⊕	•	⊕		•	⊕		•	⊕	
Cooks Islands	⊕	⊕		⊕	⊕		⊕	⊕		⊕	⊕	
Fiji	⊕	⊕		•	⊕		•	⊕		•	⊕	
Hong Kong, China	⊕	⊕		⊕	⊕		⊕	⊕		⊕		
India	•	⊕		•	⊕		•	⊕	⊕	•	⊕	
Indonesia	•	⊕		•	•		•	•		•	⊕	
Japan	⊕			⊕			⊕	⊕		⊕	⊕	
Kazakhstan	⊕			•			⊕			⊕		
Kiribati	⊕			⊕			⊕			⊕		
Lao People's Democratic	•	⊕		•	•	⊕	⊕	⊕		⊕	⊕	
Republic												
Malaysia	•	⊕		•	⊕		•	•		•	•	
Maldives	⊕			⊕			⊕			⊕		
Marshall Islands	⊕			⊕	⊕		⊕	⊕		⊕		
Micronesia Federated	⊕			⊕	⊕		⊕			⊕	⊕	
States of												
Mongolia	⊕	⊕		•	⊕		⊕			⊕		
Myanmar	⊕	⊕		•	•		⊕	⊕		⊕	⊕	
Nepal	⊕	⊕		⊕	⊕		⊕	⊕		⊕	⊕	
New Zealand	⊕			⊕	⊕		⊕	⊕		⊕	⊕	
Pakistan	⊕	⊕		•	⊕		•	⊕	⊕	•	⊕	⊕
Papua New Guinea	⊕	⊕		•	⊕		⊕	⊕		⊕	⊕	
Philippines	•	⊕		•	•		•	•		•	•	
Rep. of Korea	⊕	⊕		⊕	⊕		⊕	⊕		⊕	⊕	
Samoa	⊕	⊕		⊕	⊕		⊕	⊕		⊕	⊕	
Singapore												
Solomon Islands	⊕	⊕		⊕	⊕		⊕	⊕		⊕	⊕	
Sri Lanka	⊕	⊕		•	⊕		•	⊕		•	⊕	
Tonga	⊕	⊕		⊕	⊕		⊕			⊕		
Tuvalu	⊕			⊕	⊕		⊕			⊕		
Thailand	•	⊕		•	•		•	•	⊕	•	•	⊕
Vanuatu				⊕	⊕		⊕	⊕		⊕	⊕	
Viet Nam	•	•		•	•		•	•	⊕	•	•	

Source: AnjelloandRanawana1996, ESCAP1997, Higham 1998, Hunt 1996, Kiser1998, KoeandAziz1995, World Bank 1995 and 1998, ENV 1997

Figure 2. View of spilled over debris

The waste management services in India are not up to the mark and there is a risk to environment quality in general and public health in particular (Supreme Court Committee Report, 1999, Vij, 2012; Bhat et al., 2012). Joshi R et al (2016) reported that about 12 million tons of waste collected from sweeping, construction and demolition and in landfill site occupy about 1/3rd of the total municipal solid waste. In India, due to indigent collection and insufficient transportation we can found solid municipal waste everywhere which is dumped openly along the roadsides. Unscientific dumping, collection mechanism is very low, open dumping, open burning and un-engineered sanitary land filling are the dumping means all over India (Bundela et al.,2010; Gidde et al., 2008; Sharholy et al., 2007) as can be seen in Figures 3 and 4. These methods of waste dumping/ discarding cause adverse effects on human health and ecological degradation throughout the country (Sahu. S.et al., 2014; Shazwin and Nakagoshi., 2010). Several researchers have reported improper management of solid waste in different cities of developing countries (Berkun et al., 2005; Chatterjee, 2010; Chung and Carols, 2008; Das and Bhattacharya, 2013; Imam et al, 2008; Jafari et al., 2010; Mohanty et al.,2014; Noorjahan et al., 2012; Bhat et al., 2013).

Most of the cosmopolitan cities in India like Delhi, Chennai, Kolkata and Mumbai compute about 42% of the total urban population which means that more solid municipal waste is produced in these parts of the country (Gosh & Khansal., 2014). Table 2 presents a detailed outline of Indian cities to assess and find out various issues associated with municipal solid waste management. Table 2 depicts that per capita production of solid municipal wastes is huge in some cities for example Delhi, Jammu, Kochi, Srinagar, Coimbatore, Agra, Ludhiana, Jaipur. This is probably high because of change in life style, high growth, and urbanization in cities. In many cities like, Imphal, Kavaratti, Rajkot, Gandhinagar, Nasik etc. biodegradable material (40%-60%) and inorganic waste (30-50%) is found to be the main source of solid municipal waste in urban areas of India (Bhat et al., 2014). Organic wastes were found to be escalating with the declining socio- economic status. Moreover, it has been found that recyclable

Table 2. Municipal solid waste generation rates in different cities of India

Sample No.	Name of City	Population (as per 2001census)	Area (Km²)	Waste quantity (TPD)	Waste Generation Rate (kg/ (cap · day))	Compostable (%)	Recyclables (%)	C/N ratio	HCV* (Kcal/ kg)	Moisture (%)
1.	Kavaratti	10119	4	3	0.30	46.01	27.20	18.04	2242	25
2.	Shillong	132867	10	45	0.34	62.54	17.27	28.86	2736	63
3.	Shimla	142555	20	39	0.27	43.02	36.64	23.76	2572	60
4.	Agartala	18998	63	77	0.40	58.57	13.68	30.02	2427	60
5.	Gandhinagar	195985	57	44	0.22	34.30	13.20	36.05	698	24
6.	Dhanbad	199258	24	77	0.39	46.93	16.16	18.22	591	50
7.	Pondicherry	220865	19	130	0.59	46.96	24.29	36.86	1846	54
8.	Imphal	221492	34	43	0.19	60.00	18.51	22.34	3766	40
9.	Aizwal	228280	117	57	0.25	54.24	20.97	27.45	3766	43
10.	Jammu	369959	102	215	0.58	51.51	21.08	26.79	1782	40
11.	Dehradun	424674	67	131	0.31	51.37	19.58	25.90	2445	60
12.	Asansol	475439	127	207	0.44	50.33	14.21	14.08	1156	54
13.	Kochi	595575	98	400	0.67	57.34	19.36	18.22	591	50
14.	Raipur	605747	56	184	0.30	51.40	16.31	223.50	1273	29
15.	Bhubaneswar	648032	135	234	0.36	49.81	12.69	20.57	742	59
16.	Tiruvanantapuram	744983	142	171	0.23	72.96	14.36	35.19	2378	60
17.	Chandigarh	808515	114	326	0.40	57.18	10.91	20.52	1408	64
18.	Guwahati	809895	218	166	0.20	53.69	23.28	17.71	1519	61
19.	Ranchi	847093	224	208	0.25	51.49	9.86	20.23	1060	49
20.	Vijaywada	851282	58	374	0.44	59.43	17.40	33.90	1910	46
21.	Srinagar	898440	341	428	0.48	61.77	17.76	22.46	1264	61
22.	Madurai	928868	52	275	0.30	55.32	17.25	32.69	1813	46
23.	Coimbatore	930882	107	530	0.57	50.06	15.52	45.83	2381	54
24.	Jabalpur	932484	134	216	0.23	58.07	16.61	28.22	2051	35
25.	Amritsar	966862	77	438	0.45	65.02	13.94	30.69	1836	61
26.	Rajkot	967476	105	207	0.21	41.50	11.20	52.56	687	17
27.	Allahabad	975393	71	509	0.52	35.49	19.22	19.00	1180	18
28.	Vishakhapatnam	982904	110	584	0.59	45.96	24.20	41.70	1602	53
29.	Faridabad	1055938	216	448	0.42	42.06	23.31	18.58	1319	34
30.	Meerut	1068772	142	490	0.46	54.54	10.96	19.24	1089	32
31.	Nashik	1077236	269	200	0.19	39.52	25.11	37.20	2762	62
32.	Varanasi	1091918	80	425	0.39	45.18	17.23	19.40	804	44
33.	Jamshedpur	1104713	64	338	0.31	43.36	15.69	19.69	1009	48
34.	Agra	1275135	140	654	0.51	46.38	15.79	21.56	520	28
35.	Vadodara	1306227	240	357	0.27	47.43	14.50	40.34	1781	25
36.	Patna	1366444	107	511	0.37	51.96	12.57	18.62	819	36
37.	Ludhiyana	1398467	159	735	0.53	49.80	19.32	52.17	2559	65
38.	Mumbai	1437354	286	574	0.40	52.44	22.33	21.58	1421	43
39.	Indore	1474968	130	557	0.38	48.97	12.57	29.30	1437	31
40.	Nagpur	2052066	218	504	0.25	47.41	15.53	26.37	2632	41
41.	Lucknow	2185927	310	475	0.22	47.41	15.53	21.41	1557	60

continued on the following page

Table 2. Continued

Sample No.	Name of City	Population (as per 2001census)	Area (Km²)	Waste quantity (TPD)	Waste Generation Rate (kg/ (cap · day))	Compostable (%)	Recyclables (%)	C/N ratio	HCV* (Kcal/ kg)	Moisture (%)
42.	Jaipur	2322575	518	904	0.39	45.50	12.10	43.29	834	21
42.	Surat	2433835	112	1000	0.41	56.87	11.21	42.16	990	51
44.	Pune	2538473	244	1175	0.46	62.44	16.66	35.54	2531	63
45.	Kanpur	2551337	267	1100	0.43	47.52	11.93	27.64	1571	46
46.	Ahmedabad	3520085	191	1302	0.37	40.81	11.65	29.64	1180	32
47.	Hyderabad	3843585	169	2187	0.57	54.20	21.60	25.90	1969	46
48.	Bangalore	4301326	226	1669	0.39	51.84	22.43	35.12	2386	55
49.	Chennai	4343645	174	3036	0.62	41.34	16.34	29.25	2594	47
50.	Kolkata	4572876	187	2653	0.58	50.56	11.48	31.81	1201	46
51.	Delhi	10306452	1483	5922	0.57	54.42	15.52	34.87	1802	49
52.	Greater Mumbai	11978450	437	5320	0.45	62.44	16.66	39.04	1786	54

Figure 3. Dumping site

material like glass, paper, and metal and plastic is absolutely less, reason being the rag pickers who segregate and collect material from dumping spot (Gupta et al., 2015).

In India waste compilation is very disorganized and haphazard. It is very unfortunate that most of the urban areas lack municipal solid waste depot. In many cities the collection bins are common for decomposable and non- decomposable wastes. The collection bins are neither properly designed, situated nor maintained, which are transpired into pathetic collecting ability. In India the collection ability of solid municipal waste is about 70% in cities (Saxena et al., 2010; Rathi, 2006; Siddique et al., 2006; Gupta et al., 1998; Maudgal, 1995; Khan 1994).

Figure 4. Un-engineered sanitary landfill

Pollution of air, water and soil is caused by delusive and inapt practice of solid waste management. Filthiness caused by dumping of debris in urban areas, solid waste blocks drains, effectuating stagnant water which pollutes the surface and ground water supplies. Improper burning of wastes in the dumps causes air pollution. Arboretum/ greenhouse gases produced from decomposition of organic waste in dumping ground pollutes nearby soil and water bodies. Rajput et al 2009 reported that 95% of dumping and disposal of municipal solid waste causes negative impact on the environment.

WASTE PROCESSING TECHNIQUES

Sanitary Land Filling

Kansal, 2002 stated that sanitary land filling is completely engineered dumping alternative, by which detrimental causatums of unrestrained dumping can be averted. Everything in the sanitary land fill is carefully monitored by engineers from centralizing to canopying the waste land. It discards the garbage/ debris by minimizing the surface water infiltrating into gas eluding from the debris (Kansal, 2002).

Compositing

It is a natural way of recycling, compositing biodegradable waste like municipal waste and agricultural waste and turns it to beneficial biotic compost. Compositing is a purely biological process which cannot be completed without restrained aerobic conditions (which requires oxygen). In this method organic matter break down into simpler substances, its efficacy is dependent on the ecological conditions i.e. oxygen, organic matter, material disturbance, temperature etc. Sharholy et al., 2008 reported that large scale composite plants with quantity of approximately 150- 300 tons/ day were found to be functional in Delhi, Jaipur, Baroda, Bangalore and Mumbai. Gupta et al 2007 and Srivastava (2005) stated that 9% of the solid waste is being treated by compositing.

Vermi-Composting

Is a kind of compost in which certain species of earthworms are used to inflate the process of converting the biodegradable waste/ organic waste into nutrient rich soil. The organic waste is used as a feed to the earthworms which when passes through their digestive system discharge out granular form of cocoons which is known as vermin- composting (Ansani, 2006). Vermi- compost is a dark brown, odor-free bio-fertilizer which is acquired by the process of vermin-composting (Manyuchi and Phiri, 2013a; Manyuchi et al.,2013; Abassietal., 2009; Aalok et al., 2008; Chauduri et al., 2000). Vermi-composting solutions are at the moment accessible in countries like Italy, Japan, USA and Canada (Aalok et al., 2008).

Incineration

It is a waste management process that entails the process of combusting of the organic matter/ biodegradable material. In incineration the combustible waste is burned at high temperature i.e. 900-1000 0C so that all the material gets turned into ash and the inflammable material is disposed in a dumping ground. Incineration decreases the quantity of waste by 75%- 95%. Incineration in Indian cities is restricted to hospital and other biological wastes (Zaman, 2009; DEFRA, 2007).

Pyrolysis

Is the thermal degradation of the organic/ biodegradable material in the absence of oxygen. Thermal decomposition of organic wastes begins at 350 -550 °C and can reach up to 700 - 800 °C in the paucity of air or oxygen. This is a method by which we can convert municipal waste into products which can be used for the production of heat, electricity and chemicals. The main products obtained by Pyrolysis of municipal waste are high calorific value gas (Sngas), bio-fuel (bio-oil) and soil residue (char).

Bio-Methane

It is a process of transformation of biodegradable/ organic matter into methane and production of compost by microbial activity in the absence of air through a method called anaerobic digestion. Bio- methane is most effective and practicable as it produces useful products like biogas and enriched compost, the solid waste from agro- based industries have high organic content which is useful for producing bio- methane. The bio-methanation is a two-stage process consisting of acidification and methanation (CPCB, 2007).

Recycling and Reuse

Demirbas (2010) reported that solid waste management delineates a big ecological and monetary concern all around the world. Recycling is actually the reusing of material into new product. It restrains conducive material for not being wasted; it reduces the use of raw material, consequently greenhouse gas dissemination. Moore (2012) and Al- Madeed (2012) affirmed that the management of solid waste entail new model arranged as 3R's or RRR in which first R stands for reduce, the next R stands for reuse and the last R for recycle. The 3R approach intends to bring forth the transition of solid wastes into new resources thereby reducing environmental pollution.

Non- Engineered Disposal

It is a widely used method in developing countries. In many Indian cities open, chaotic, unrestrained and poorly formulated dumping is mostly observed, conferring the growth of ecological degradation. Sharholy et al., 2008 reported that more than 90% of the solid wastes in cities and towns are openly discharged on land in a scanty manner. Kashmir Valley is failing in providing hale abstersion practices of municipal waste and is not conscientiously designed. Solid waste is discharged freely on the roadsides. Solid municipal wastes is being spilled in end ran, open spaces, around the street corners and partially open which is being burnt which result in environmental pollution, outbreak of diseases and aesthetically bad look. Collection of waste is fairly atrocious; waste agglomeration is confined to influential areas only. Largely caste aside areas are indigent societies which are being neglected as solid waste remains messy and piled. Social en-durability in municipal waste management can be characterized as procurement of prudent and healthy waste management services and to provide hygienic convenience to people. The poor hygienic conditions and waste management have negative impact on the health and wellbeing of people. It has been found that poor sanitation and improper waste management have direct and indirect connections to social, health and environmental repercussions. The unhygienic condition and tarnished ecology pass on negative message to other communities and disparagement by others as well. In some areas (especially Achan and in the heart of city poloview) of valley because of these pathetic conditions people don't prefer to marry their daughter and/ sons in that particular areas. Achan is a landfill wherein State Municipal Corporation used to open dump municipal wastes for about 23 years till 2008. People pile up solid waste in dustbins and plastic bags and strew it in dumping site or in vacant plots along the road sides. Sweepers are prearranged for sweeping roads and streets. Rag pickers gather the material which is recyclable. Open trucks and hand carts are used to collect municipal waste from diverse places to collection point. Mechanized and roofed carriers are used to collect municipal wastes from cities. Kumar (2013) stated that about 75%-80% of waste is being expelled in the embankment of river Ravi where the waste is being openly burnt.

HEALTH IMPACT OF WASTE MANAGEMENT

Imprudent management of waste and awful handling/carrying of waste has detrimental causatum on ecology and health of people in general. Disposal and improper handling practices have detrimental causatums resulting in air, water and soil pollution. Air pollution, environmental diminution and contamination of water are caused by unchecked mal-administering of waste.

Incineration and land filling are chiefly used to carry out last stage of waste disposal. Methane is released from trash dump by decomposition of biodegradable waste under anaerobic conditions, which plays the leading role in global warming, moreover open burning cause's brume and respiratory diseases (Sridevi et al., 2012). Odor is another serious problem particularly when the temperature exceeds 45oC especially in summers (Dasgupta et al., 2013). Solid municipal wastes turn into breeding place for mosquitoes; augmenting the peril of diseases like dengue, malaria, West Nile Virus (Sridevi et al., 2012). Central Pollution Control Board (2000) reported that the indigent waste management led to negative health causatums like nose and throat infections, dyspnea, respiratory problems, anemia, allergy, asthma, bacterial infection and other infections.

SOCIAL IMPACT OF WASTE MANAGEMENT

The improper waste management has many social causatums like branding the community living close to the waste site as 'dirty community' which in turn leads to decreased self esteem of the community members irrespective of their economic or social status in the larger society. The people often tend to migrate to other areas in order to protect themselves from the problems of unpleasant smell, smoke, noise of birds feeding on the waste and enormous flies living on the waste site. This migration is disadvantageous to the community in a manner that they sell off their land and residential/commercial assets at a very low cost, as the land around/near the waste site are highly devalued owing to its disadvantaged location. This leads to economic distress of the migrating families. However, it becomes a compulsion for those to live in the vicinity of the waste sites who do not possess enough resources to move out from the area. The residential areas near dumping sites being unhygienic and unaesthetic pave way for refusal of marriage proposal owing to the fact that these areas are considered to be more susceptible to communicable and infectious diseases thus creating problems for timely settlement of the family. The other social causatum includes the concentration of poor families/migrant people who indulge in rag picking/ isolation of plastic and metal items from the open waste.

CONCLUSION

In many developing countries proper collection and disposal facilities of municipal waste is not provided to the population. The present dilemma is that under developed countries relies on scanty and tenuous waste infrastructure, the communal section and waste evacuation. General public awareness, high-tech proficiency, sufficient resources, execution of acts is the need of the hour. Count of collection dumps/ collection sites and vehicles used for collection of debris is very less in relation to the size of population and the amount of waste generated. The current method of evacuation of municipal waste is in an unorganized manner which retrogrades the characteristics of underground water, air and environmental pollution by releasing of hazardous gases. Keeping in view predominantly the detrimental causatums related with patulous disposal and open burning, this routine should be expelled.

REFERENCES

Abbasi, T., Gajalakshmi, S., & Abbasi, S. A. (2009). Towards modelling and design of vermicomposting systems: Mechanisms of composting/vermicomposting and their implications. *Indian Journal of Biotechnology*, *8*, 177–182.

Ahmed, L., Bhat, F. A., & Sultan, S. (2015). Municipal Solid Waste Management in Srinagar City, Kashmir, India. *International Journal of Innovative Research in Science. Engineering and Technology*, *4*(7). doi:10.15680/IJIRSET.2015.0407148

Al-Maaded, M., Madi, N. M., Kahraman, R., Hodzic, A., & Ozerkan, N. G. (2012). An Overview of Solid Waste Management and Plastic Recycling in Qatar. *Journal of Polymers and the Environment*, *20*(1), 186–194. doi:10.100710924-011-0332-2

Anjello, R., & Ranawana, A. (1996). Death in Slow Motion- India has become the dumping ground for the West's toxic Waste. *Asiaweek*.

Arvind, K., & Surat, S. (2013). Domestic Solid Waste Generation– A Case Study of Semi Urban Area of Kathua District, Jammu, J & K, India. International Journal of Scientific and Research Publications, 3(5).

Asha Aalok, A. K. (2008). Vermi-composting: A Better Option for Organic Solid Waste Management. *Journal of Human Ecology (Delhi, India)*, *24*(1), 59–64. doi:10.1080/09709274.2008.11906100

Asnani, P. U. (2006). *Solid Waste Management*. India Infrastructure Report 160–189.

Berkun, M., Aras, E., & Nemlioglu, S. (2005). Disposal of solid waste in Istanbul and along the black sea coast of Turkey. *Waste Management (New York, N.Y.)*, *25*(8), 847–855. doi:10.1016/j.wasman.2005.04.004 PMID:15949934

Bhat RA, Dar GH, Jehangir A, Bhat BA, Yousuf AR (2012) Municipal solid waste generation and present scenario of waste management during Yatra season in Pahalgam: A tourist health resort of Kashmir valley. *Int. J. Cur. Res.*, *4*, 4-9.

Bhat, R. A., Dervash, M. A., Mehmood, M. A., & Hakeem, K. R. (2018). Municipal solid waste generation and its management, a growing threat to fragile ecosystem in Kashmir 756 Himalaya. *American Journal of Environmental Sciences*, *2*(11), 145–167.

Bhat, R. A., Kamili, A. N., & Bandh, S. A. (2013). Characterisation and composition of Municipal Solid Waste (MSW) generated in Yusmarg: A health resort of Kashmir valley: A Glance at the World. *Waste Management (New York, N.Y.)*, *33*(3), 774–777. doi:10.1016/j.wasman.2013.01.009

Bhat, R. A., Nazir, R., Ashraf, S., Ali, M., Bandh, S. A., & Kamili, A. N. (2014). Municipal solid waste generation rates and its management at Yusmarg forest ecosystem, a tourist resort in Kashmir. *Waste Management & Research*, *32*(2), 165–169. doi:10.1177/0734242X13518089 PMID:24519231

Bundela, P. S., Gautam, S. P., Pandey, A. K., Awasthi, M. K., & Sarsaiya, S. (2010). Municipal solid waste management in Indian cities. *Int. J. Environ. Sci.*, *1*(4), 591–605.

Census. (2011). Provisional Population Totals, India.

Central Pollution Control Board. (2005). *Management of Municipal Solid Waste*. Delhi, India: Ministry of Environment and Forests.

Central Pollution Control Board. (2007). *Bio-methanation potential of solid wastes from Agro- based Industries*. Delhi, India: Ministry of Environment and Forests.

Chatterjee, R. (2010). Municipal solid waste management in Kohima city— India. *Iranian Journal of Environmental Health Sciences & Engineering*, *7*(2), 173–180.

Chaudhuri, P. S., Pal, T. K., Bhattacharjee, G., & Dey, S. K. (2000). Chemical changes during vermin-composting (Perionyxexcavatus) of kitchen wastes. *Tropical Ecology*, *41*, 107–110.

Chung, S. S., & Lo Carlos, W. H. (2008). Local waste management constraints and waste administrators in China. *Waste Management (New York, N.Y.)*, *28*(2), 272–281. doi:10.1016/j.wasman.2006.11.013 PMID:17280826

CPCB. (2000a). *Status of Solid Waste Generation, Collection, Treatment and Disposal in Metro cities, Series: CUPS/46/1999–2000*. Author.

CPCB. (2000b). *Status of Municipal Solid waste Generation, Collection, Treatment and Disposal in Class I Cities, Series: ADSORBS/31/1999–2000*. Author.

CPCB. (2000c). *Management of municipal solid waste in Delhi*. Retrieved from http://www.cpcb.nic.in/divisionsofheadoffice/pcp/MSW_Report

CPCB. (2004a). *Management of Municipal Solid Waste*. New Delhi, India: Ministry of Environment and Forests.

CPCB. (2004b). Management of Municipal Solid Waste. Ministry of Environment and Forests.

CPCB. (2000d). *Management of municipal solid waste Delhi*. Retrieved from http://www.cpcb.nic.in/divisionsofheadoffice/pcp/ MSW_Report.pdf

Dasgupta, B., Yadav, V.L., & Mondal, M.K. (2013). Seasonal characterization and present status of municipal solid management in Varanasi, India. *Adv. Environ. Res.*, *2*, 51–60. doi:.1.051 doi:10.12989/aer.2013.2

DEFRA. (2007). Incineration of Municipal Solid Waste. Report of Department for Environment, Food & Rural Affairs.

Demirbas, A. (2010). Waste management, waste resources facilities and waste conversion process. *Energy Conversion and Management*, *52*(2), 1280–1287. doi:10.1016/j.enconman.2010.09.025

E&P Forum/UNEP. (1997). *Environmental management in oil and gas exploration and production-An over view of issues and management approaches*. Joint E&P Forum/UNEP Technical Report 37.

EPA, Environmental Protection Agency. (1997). *Test Methods for Evaluating Solid Waste. n-Hexane Extractable Material (HEM) for Sludge, Sediment and Solid Samples, Method 9071B, Revision 2. Environmental Protection Agency, OSW 486*. Office of Solid Waste.

ESCAP. (1997). *Guidelines and manual on land use planning and practices in watershed management and disaster reduction. Economic and social commission for Asia and Pacific*. Bangkok: United Nations.

ESCAP. (1997). *Road transport and the environment: Areas of concern for the Asia and Pacific region* (ST/ESCAP/1840). ESCAP.

Gidde, M. R., Todkar, V. V., & Kokate, K. K. (2008). Municipal solid waste management in emerging mega cities: a case study of Pune city. *Proceedings of Indo Italian Conference on Green and Clean Environment.*

Gupta, Yadav, Kumar, Gupta, Yadav, & Kumar. (2015). A review on current status of municipal solid waste management in India. *Journal of Environmental Sciences, 37*, 206–217. Retrieved from www.journals.elsevier.com/journal-of-environmental-sciences

Gupta, R., Mutiyar, P. K., Rawat, N. K., Saini, M. S., & Garg, V. K. (2007). development of a water hyacinth based vermireactor using an epigeic earthworm Eiseniafoetida. *Bioresource Technology, 98*(13), 2605–2610. doi:10.1016/j.biortech.2006.09.007 PMID:17071077

Gupta, S., Mohan, K., Prasad, R., Gupta, S., & Kansal, A. (1998). Solid waste management in India: Options and opportunities. *Resources, Conservation and Recycling, 24*(2), 137–154. doi:10.1016/S0921-3449(98)00033-0

Higham, I. (1998). Advanced thermal conversion technologies for energy from solid waste. IEA Caddett Centre for Renewable Energy, ETCU.

Hunt, C. (1996). Child Waste Pickers in India; The Occupation and its Health Risks. *Environment and Urbanization, 8*(2), 111–118. doi:10.1177/095624789600800209

Imam, A., Mohammed, B., Wilson, D. C., & Cheeseman, C. R. (2008). Solid waste management in Abuja, Nigeria. *Waste Management (New York, N.Y.), 28*(2), 468–472. doi:10.1016/j.wasman.2007.01.006 PMID:17379496

Indris, A., Inane, B., & Hassan, M. N. (2004). Overview of waste disposal and landfills/dumps in Asian countries. *Journal of Material Cycles and Waste Management, 6*, 104–110.

Jafari, A., Godini, H., & Mirhousaini, S. H. J. (2010). Municipal solid waste management in Khoram Abad city and experiences. *World Academy of Science, Engineering and Technology, 4*, 163–168.

Joshi, R., & Ahmed, S. (2016). Status and challenges of municipal solid waste management in India: A review. *Cogent Environmental Science, 2*(1), 1139434. doi:10.1080/23311843.2016.1139434

Kansal, A. (2002). Solid Waste management Strategies for India. *Indian Journal of Environment Protection, 22*(4), 444–448.

Khan, R. R. (1994). Environmental management of municipal solid wastes. *Indian J. Environ. Prot., 14*(1), 26–30.

Kiser, D.M.R.E., Johnson, N.E., Russell, J..T.G., & MacDonald. (1998). *Low level class A/C waste, Near Surface Land Disposal Level Feasibility Design Description.* INEEL/EXT-98-00051, Lock head Martin Idaho Technologies Company.

Koe, L. C. C., & Aziz, M. I. A. (1995). *Suggested approach to control marine population from land-based sources.* Paper presented at Asia Pacific Regional conference on Prevention of Marine Population From Land Sources and Sea Sources, Singapore.

Manyuchi, M. M., & Phiri, A. (2013a). Vermi-composting in solid waste management: A review. *Int. J. Sci. Eng. Technol.*, *2*(12), 1234–1242.

Manyuchi, M. M., Phiri, A., & Muredzi, P. (2013). Effect of Vermi-compost, Vermi-wash and Application Time on Soil Micronutrients Composition. *International Journal of Engineering and Advanced Technology.*, *2*(5), 215–218.

Maudgal, S. (1995). Waste management in India. *J. India Assoc. Environ. Manag.*, *22*(3), 203–208.

Mohanty, C. R., Mishra, U., & Beuria, P. R. (2014). Municipal solid waste management in Bhubaneswar, India— a review. *Int. J. Latest Trends Eng. Technol.*, *3*(3), 303–312.

Moore, S. A. (2012). Garbage matters: Concepts in new geographies of waste. *Progress in Human Geography*, *36*(6), 780–799. doi:10.1177/0309132512437077

Noorjahan, A., Dhanakumar, S., Mohanraj, R., & Ravichandran, M. (2012). Status of heavy metals distribution in municipal solid waste in Tiruchirappalli City, India. *International Journal of Applied Biology and Pharmaceutical Technology*, *3*(3), 252–258.

Rajput,, R., & Prasad,, G., & Chorea. (2009). Scenario of solid waste management in present Indian context. *Caspian Journal of Environment Science*, *7*(1), 45-53.

Rathi, S. (2006). Alternative approaches for better municipal solid waste management in Mumbai, India. *J. Waste Manag.*, *26*(10), 1192–1200. doi:10.1016/j.wasman.2005.09.006 PMID:16288861

Sahu, Nair, & Sharma. (2014). Review on Solid Waste Management Practice in India: A State of Art. *International Journal of Innovative Research and Development*, *3*(3).

Saxena, S., Srivastava, R. K., & Samaddar, A. B. (2010). Sustainable waste management issues in India. *IUP J. Soil Water Sci.*, *3*(1), 72–90.

Sharholy, M. (2005). Analysis of municipal solid waste management systems in Delhi—a review. *Proceedings for the second International Congress of Chemistry and Environment*, 773–777.

Sharholy, M., Ahmad, K., Mahmood, G., & Trivedi, R. C. (2008). Municipal solid waste management in Indian cities – A review. *Waste Management (New York, N.Y.)*, *28*(2), 459–467. doi:10.1016/j.wasman.2007.02.008 PMID:17433664

Sharholy, M., Ahmad, K., Vaishya, R. C., & Gupta, R. D. (2007). Municipal solid waste characteristics and management in Allahabad, India. *J. Waste Manag.*, *27*(4), 490–496. doi:10.1016/j.wasman.2006.03.001 PMID:16766176

Shazwin, T. M., & Nakagoshi, N. (2010). Sustainable waste management through international cooperation: Review of comprehensive waste management technique and training course. *J. Int. Dev. Coop.*, *16*(1), 23–33.

Siddiqui, T. Z., Siddiqui, F. Z., & Khan, E. (2006). Sustainable development through integrated municipal solid waste management (MSWM) approach— a case study of Aligarh District. *Proceedings of National Conference of Advances in Mechanical Engineering*.

Sridevi, P., Modi, M., Lakshmi, M. V. V. C., & Kesavarao, L. (2012). A review on integrated solid waste management. *Int. J. Eng. Sci. Adv. Technol.*, *2*, 1491–1499.

Supreme Court Committee Report. (1999). *Report of the committee on Solid Waste Management in Class I cities in India*.

Vij, D. (2012). Urbanization and solid waste management in India: Present practices and future challenges. *Procedia: Social and Behavioral Sciences*, *37*, 437–447. doi:10.1016/j.sbspro.2012.03.309

World Bank. (1995). Lebanese republic solid waste/environmental management project, Draft Report. Washington, DC: Author.

World Bank. (1998). Sectoral Analysis in Bangladesh, Draft Report. Washington, DC: Author.

Zaman, A. U. (2009). Life Cycle Environmental Assessment of Municipal Solid Waste to Energy Technologies. Global Journal of Environmental Research, 3(3), 155-163.

Chapter 2
Impacts of Hazardous Waste on Soil Health:
Sources of Hazardous Wastes

Jatinder Kaur Katnoria
Guru Nanak Dev University, India

Priyanka Sharma
Guru Nanak Dev University, India

ABSTRACT

Soil, a natural medium for plant growth, provides the anchorage to the plants and supplies various nutrients required for the plant growth. It is an important ecosystem that not only provides the shelter to various organisms but also participates directly or indirectly in various biogeochemical cycles. However, in recent years, the earth's soil has been stripped away, rendered sterile, and contaminated with toxic chemicals due to various anthropogenic activities. This increasing wide spread pollution has caused vast areas of land to become non-arable and hazardous for both wildlife and human populations. Unlike many other organic pollutants, which are degraded in the soils, some of the hazardous compounds like heavy metals, pesticides, polycyclic aromatic compounds, and phthalates are of special concern as most of these are conserved. The presence of such compounds in soil ecosystem causes severe toxicity and imbalances in the soil ecosystem as a whole. The chapter focuses on various sources of soil pollutants and the effects of hazardous compounds on soil health.

INTRODUCTION

Soil is an essential component for all living beings and is designated as basic living factor because it provides the essential nutrients for growth of plants. Intensifying global industrialization in both public and private sectors along with population explosion and green revolution has resulted in varying degrees of environmental pollution (Watanabe & Hirayama, 2001; Singh et al., 2018). The progress of urbanization as well as industrialization led to the development of new technologies and products, whose manufacturing involves utilization of various raw materials as well as synthetic compounds, resulting

DOI: 10.4018/978-1-7998-0031-6.ch002

in release of huge quantities of hazardous wastes (Przybylinska & Wyszkowski, 2016). Vehicular emissions and open dumping of solid waste have also consequenced in deterioration of soil quality as well as decrease in vegetation abundance (Ali et. al., 2014). Migration of various contaminants into non-polluted areas as dust or leachates through soil and spreading of pollutants through sewage sludge are the factors contributing towards contamination of soil ecosystem (Tangahu et al., 2011). The ultimate threat lies in the accumulation of toxic/hazardous compounds, released from various sources, in the food chains via crops grown on the contaminated lands because most of these synthetic compounds are either non-biodegradable or the slow degradable. Environmental contaminants adversely poses the negative influences on the soil properties which further result in toxic effects to the native flora and fauna along with other forms of life (Ziolkowska & Wyszkowski, 2010; Wyszkowska & Sivitskaya, 2012). Unlike many other organic pollutants which are anthropogenic and often degraded in the soils, potent carcinogens like heavy metals and pesticides are of special concern because are conserved (Wade *et al.*, 1993). Moreover, these contaminants cannot be destroyed biologically (Knasmuller *et al.*, 1998). Global concern over the magnitude of soil pollution calls for immediate remedial actions (Garbisu, 2001). The present study pertains to report the sources of soil contaminants and their impacts on soil health.

SOURCES

There are numbers of factors responsible for release of hazardous chemicals to air, water and soil. The present part of the chapter deals with various sources of hazardous chemicals leading to soil pollution.

Dumping of Solid Waste

The major source of the soil pollution is the waste generated from the households which is dumped in the outskirts of the city or town. Municipal solid waste (MSW) is an inevitable byproduct of human activities and is presently, causing the menace of environmental pollution in most of the developing countries. This waste is generated from domestic and commercial activities, human settlements, the small-scale industries along with wastes from additional sources like hospitals and clinics, especially due to non-availability of specific techniques to handle these wastes (Pattnaik & Reddy, 2009). As MSW is a mixture of both organic and inorganic wastes, its dumping not only alters the physico-chemical characteristics of soil but also results in inducing toxicity including damage to the gene pool of exposed population.

Impacts of open dumping of municipal solid waste on soil and vegetation diversity were studied by Ali et al. (2014). Surface soil samples (n=12) from the depth of 9 inches were collected from the waste disposal site, Islamabad, Pakistan. Wide variability existed in all samples for various parameters with respect to the control soil samples. Various elements (μg/g) were observed as Ca (14196.32), Mg (24704.85), Na (282.09), K (19229.5), Pb (133.23), Ni (101.92) and Zn (63249) and the values were found to be higher than those of the control soil. The study revealed the accumulation of heavy metals in soil through the dumping of solid waste. The authors stated that MSW services were recognized as the third priority in municipal commitments in their state after water supply and sanitation. The main environmental problem associated with the open disposal sites was the potential health risks posed to the soil ecosystem. Some earlier studies revealed the elevated levels of average trace elements in soil samples directly due to deposition of municipal waste (McCalla et al., 1977). Direct toxicity of the leachates from the dumping sites has also been documented. Feng et al. (2006) reported the genotoxic effects of leachates from municipal solid waste employing micronucleus assay in *Vicia faba* root tip cells.

Geschwind et al. (1992) investigated the incidence of birth defects in relation to residence near hazardous waste sites and reported several types of birth defects in the residents of areas near to hazardous waste dumping yards. Cabrera et al. (1999) also reported that samples of compost from municipal garbage showed high genotoxicity. Kudyakov et al. (2004) conducted a survey on rates of hospitalization for respiratory diseases in relation to residences in zip codes with hazardous waste sites, as well as socio-economic status. They observed that chronic bronchitis was elevated in people who lived in zip codes containing persistent pollutants (POPs, PCBs and pesticides) as compared to clean zip codes without hazardous waste sites.

Sewage Amended Soils

Many cities in the third world are not having complete sewerage system and hence the sewage is directly discharged into large lakes, rivers, canals along with agricultural fields. A number of studies have been conducted to evaluate the effects of sewage amendment on the soils. Some authors reported the benefits of direct discharges to the agricultural soil while some others reported the harmful consequences. Epstein (1974) found that physical properties of soil were improved due to high organic matter in the sewage effluents while Feigin and Kipnis (1980) suggested that sewage irrigated soils resulted in long term sustenance of soil fertility due to the presence of N, P and K in significant quantities in sewage effluents. Hence, application of sewage water to agricultural land was demonstrated to be a practicable alternative for the disposal of sludge. Besides, sludge amendment was found to improve soil properties like organic content, soil aeration, water holding capacity and aggregation (Epstein *et al.*, 1976). Apart from these, it has also been reported that sewage sludge acted as a good fertilizer and/or soil conditioner due to high nutrient and organic matter contents present in sewage (Jones, 1981). One major constraint in land application of sewage sludge for soil amendment was supposed to be the potential metal accumulation in crops, which in turn would transfer up the food chain posing potential health hazard to human beings (Sims & Kline, 1991). The concentrations of trace metals like Cd, Mn, Pb, Zn, Cu and Fe in surface soils have been found to increase, when irrigated with concentrated and contaminated sewage and industrial effluents (Sakal et al., 1992; Adhikari et al., 1993, Wong et al., 2001; Katnoria et al., 2011).

Discharges From Industries

Besides the sewage effluents being prime source of contamination of soil, implications of industrial wastewater in agriculture have caught attention of researchers in the recent past regarding their usage and disposal. The phenomenal growth of chemical industry has led the whole world today to produce chemicals faster than their management. There is no doubt that this rapid progress of the industrial sector has brought immense economic and social benefits but at the same time, it has accentuated the environmental, social and health problems (Moldoveanu et al., 2014). Klein *et al.* (1974) documented heavy metal contamination of soils collected from within the industrial districts in U. S. and also compared it with urban non-industrial or nearby agricultural areas. They found that the soil samples contained high amounts of heavy metals including silver, cadmium, cobalt, chromium, copper, iron, nickel, lead, zinc and mercury.

To study the effects of heavy metals which are major pollutants of industrial effluents, some authors have also estimated the effects of heavy metals by treating the soils with different types of heavy metals. For example, Sinha (1999) carried out a study to estimate accumulation and toxicity of Cu, Cd, Cr, Mn and Pb in *Bacopa monnieri* grown on artificially contaminated soils. The growing shoots were planted in garden subsoil containing 3, 16, 32, 64 and 160 μM each of above metals. During phytotoxicity study, the author found that metal accumulation was considerably high in shoots and was in the order Mn>Cr>Cu>Cd>Pb. The chlorophyll content was also reduced by 62% after 12 weeks and further high concentrations of combined metals were found lethal to plants at 16 weeks.

In some other studies, scientists have estimated heavy metal content in different polluted soils and water bodies due to direct industrial effluent discharge. Tatsumi et al. (1983) reported increased levels of some heavy metals *viz.* zinc, copper, cadmium and lead in the surface soils of Sakai city, Japan due to industrial waste disposal. Tripathi et al. (1990) observed that lead, copper and zinc were higher in surface soils of industrial areas while it was minimum in that of residential areas. Ostrander and Clark (1991), Alloway (1995) has found an increase in heavy metal concentrations in soils of urban areas when compared to those of rural areas which could probably be due to presence of industries in urban areas. Heavy metals were found in the vicinity of industries discharging effluents bearing heavy metals (Prusty *et al.*, 1994). In another study, Rao and Sarma (1998) estimated trace metals (Zn, Pb, Cu, Ni, Fe and Mn) in soils near an industrial belt in Visakhapatnam and found Zn, Pb and Cu to be predominant. They also observed decreasing trend of all the metals with depth of soil. Increase in concentrations of soluble salts and other chemical constituents in agricultural soil due to effluent contamination have been reported by Kumar (1998). Although most of the reports showed the harmful effects of industrial effluents, Rao and Rao (2001) reported that the concentrations of metals in various industrial wastewaters vary considerably both in quality and quantity depending upon types of waste. The concentrations of heavy metals like Cd, Cu, Mn, Ni, Pb and Zn in soil samples and plants were found to be very high in the vicinity of industries in Bangladesh by Kashem and Singh (1999).

Abrol et al. (2003) conducted a field survey to determine the nutrient status of different sewage and industrial effluents, their irrigation potential and influence on soil properties. Majority of sewage and effluents were neutral with a few showing acidic character. Trace metals Zn, Mn, Fe and B in irrigational sources were below permissible limits with mean concentration of 0.02, 0.05, 0.10 and 0.11 mg/kg, respectively while Cd and Ni exceeded permissible limits. The metals were found to be in increasing order of Cd< Zn< Cu< Mn< Fe. Assessment of heavy metal pollution and health risks in the soil-plant-human system in Yangtze River delta, China was done by Hu et al. (2017). The author reported that dynamics of heavy metals in soil and their absorption by plants were strongly affected by soil properties. The influence of metal pollution on soil microbial diversity was observed in some studies (Khan et al., 2001; Garcia et al., 2001).

Emissions From Vehicles

Apart from the industrial effluents, there are some reports of vehicular pollution as one of the causes of soil pollution. Most of the pollutants released from vehicular emissions either directly fall on the surface soil or get settled on ground after being remained suspended in the ambient air for long period of time. The concentrations of heavy metals in soils and plants, especially Pd, depended on distance from road (Onyari et al. 1991; Munch 1993; Fatoki, 1996; Chronopolous et al. 1997; Olajire & Ayodele, 1997; Othman *et al.*, 1997; Sanchez-Martin *et al.,* 2000). Sadiq et al. (1989), Chronopolous et al. (1997), and

Olajire and Ayodele (1997) reported that traffic-related metals originated mainly from combustion of leaded fuel, lubricating motor oil, vehicle tire wear and brake pads (*eg*. Cd, Mn, Zn, Cu). Many studies and surveys have been conducted to assess heavy metal contamination of roadside soils and plants (Ho & Tai 1988; Fatoki, 1996; Sanchez-Martin *et al.,* 2000). Cuny et al. (2001) monitored the trace elements in air and soil compartments along the major motorway in France. The study was conducted in 12 areas located along a 55 km stretch of motorway. Cd, Pb and Zn concentrations in the topsoil horizon (0-10 cm) were found to exceed concentrations at the control regions. Sutherland and Tack (2000), while studying metal phase associations in an urban watershed, Honolulu, Hawaii, reported that roadside soils received considerable amounts of traffic-generated pollutants. Swaileh et al. (2004) reported high amounts of heavy metal content in the roadside surface soils and vegetations from the West Bank, Palestine. Contamination of soil samples of Madina Town of Faisalabad, Metropolitan with heavy metals like cadmium, nickel and lead were reported by Farid et al. (2015).

Disposal of Hospital Waste

Hospital waste disposal is one of the burning environmental issues due to presence of infectious materials, high plastic content and trace metals (Alappat & Dikshit, 1999). Though the incineration of hospital waste solves the problem of disinfection and reduction of waste but the problem of heavy metal contamination still remains due to their non degradable nature. In combustible fraction of U. S. municipal incinerators, chromium metal has been reported to be in the range of 22-100 ppm, with a medium value of 55ppm (Jacko & Neuendrof, 1977). The improper hospital waste disposal has been suggested to be the main reason for soil pollution (Cresser et al., 1993). Banerjee et al. (1999) reported soil pollution due to disposal of hospital wastes to municipality dustbins, as very few incinerators were present in the area. Rampal et al. (2002) too suggested proper collection and solid waste disposal to be very important to keep check on the pollution. Kulkarni and Agarwal (2003) studied physico-chemical characteristics like: texture, colour, pH, conductance, salinity, chloride, sulphates, nitrates, nitrites, and salinity of topsoil of a hospital to determine the deterioration due to ongoing hospital activities. It was observed that levels of chlorides, nitrates and calcium were more in soil. It was suggested that proper management of hospital solid waste was essential, lest it would cause irreversible changes in the hospital soil chemistry.

Mining Units

Pollution by toxic elements is also associated with various other aspects of industries like metal mining and processing. Mining produces ore as a product, along with a considerable amount of metal contaminated wastes. The ore is taken to the mill where it is crushed and separated from mineral fractions, producing a metal rich concentrate plus a large quantity of waste tailing. Pollution around metal mine sites is caused by dumping of contaminated ore burden, excavation wastes etc. Because of toxicity from the large metal concentrates, development of vegetation on the waste generally limits to an early succession grass land herb community only. In some cases, the toxicity can be so severe that even after long periods of time, only a few pioneer species of plants could manage to establish like the case with mine wastes from Roman lead working in Britain (Bradshaw & Chadwick, 1981).

The mining products like coal are the major causes of pollution when burnt for various domestic or industrial purposes as the ash contains secondary pollutants like heavy metals. Klein and Russel (1973) reported increase in chromium content in soils around coal-burning power plant. Doubleday (1974) opined that excess of toxic metals in soil is a major hurdle for the vegetative stabilization of coal mine waste. However, numerous evidences are also available to show the evolution of metal tolerance in plants growing on soils contaminated by mining and smelting activity (Antonovics et al. 1971; Wu & Bradshaw, 1972; Wu *et al.,* 1975). Hashiguchi *et al* (1976) estimated heavy metal contents of vegetables grown in Sukeye area near a mine and found that lead content was highest in bell pepper (2.065ppm) and other vegetables such as white radish, potato, eggplant, sweet potato and squash (0.0-0.72ppm). In a study by Bradshaw and Chadwick (1980) on mine waste, most of the plants grew well on mine waste while some species showed abnormal growth due to presence of excess heavy metals along with nutrient deficiency. Deo (2004) studied heavy metal accumulation by plant species from a coal mining area in Orissa. They also observed destruction of natural vegetation suggesting that coal mining resulted in reduced productivity, low water holding capacity, acidity and lack of nutrients in soil.

Loureiro et al. (2006) conducted a test to evaluate the bioavailability of heavy metals in two soils coded as JNC (soil containing low concentrations of heavy metals) and JC (soil containing high concentrations of heavy metals) from Jales mine by using bioassays with plants *Brassica rapa* and *Avena sativa*. They also measured biochemical parameters (protein and malondialdehyde content, catalase and peroxidase activities). They reported that *Brassica* showed toxic effects with JC, having higher concentrations of heavy metal contents, while *Avena sativa* showed no toxic effects with any of the two soil samples studied.

Agricultural Practices

Pesticides have been existed as an indispensable part of agriculture and have played a crucial role in protecting crops (Sun et al., 2018). The enviable use of pesticides and fertilizers has also resulted in soil pollution. All these pesticides and inorganic fertilizers pose the adverse health effects not only on the human population but also on the local micro flora and fauna. The effects of lambda-cyhalothrin, a non-systemic pyrethroid insecticide, on the earthworm *Eisenia fetida* were studied by Garcia et. al. (2011). The EC_{50} values in two temperate soils were reported to be varying from 54.1 – 60.2 mg a.i./kg DW. The authors stated that the effects of insecticides depended on many factors such as characteristics and fate of the compound, frequency of its application, soil properties like pH, soil texture, organic matter along with climatic factors and soil microorganism community.

Khan and Singh (1999) studied the influence of certain insecticides such as endosulfan, dichlorvos and monocrotophos on the mobility of some heavy metals like Cr, Fe, Cu and Ni in soils saturated with different cations such as Na^+, K^+, Mg^{++} and Ca^{++}. The mobility order was Cr > Fe > Cu > Ni. The mobility was found to be influenced by the nature and quantity of insecticides application. Yokel and Delistraty (2003) investigated arsenic and lead concentrations in orchid soil contaminated with lead arsenate pesticides at the Hansford site in Washington State (USA) and found that As and Pb concentrations were greater than background levels. Singh and Kalamdhad (2011) studied the effects of metals on soil, plants and human health and reported that in aquatic ecosystems, metals stimulated the production of reactive oxygen species (ROS) that could damage fish and other aquatic organisms.

E-Waste

It is estimated that approximately 2.7 million tonnes of e-waste is produced every year in India which designates India to be the 5th largest producer of e-waste. The disposal of e-waste has become a major problem in most of the developing countries (Wath et al., 2011). Burning of e-waste in open yards to obtain precious metals like gold causes severe toxic effect to the soil ecosystem. The presence of metals like lead and mercury is of serious concern as most of the e-waste is dumped unsafely in municipal yards (Dharini et al., 2017). Growth reduction as a result of changes in physiological and biochemical processes in plants exposed to such metal polluted areas has also been well documented (Chibuike & Obiora, 2014).

Some Hazardous Compounds of Miscellenous Sources

Polycyclic Aromatic Hydrocarbons

Marked increase in polycyclic aromatic hydrocarbons (PAH) contents in various environmental samples has been witnessed over past decades (Wilcke, 2007). PAHs, a group of chemical compounds with two or more fused aromatic rings, have been categorized as carcinogenic and mutagenic (Ma et al. 2014). The major inputs of PAHs are fossil oil discharges and incomplete combustion and pyrolysis of biomass, fossil fuels and municipal wastes. Both natural (volcanic eruptions, forest fires) and anthropogenic sources (vehicular emissions, oil, wood, petrol and coal burning, industrial combustions, petroleum catalytic cracking etc.) add to discharge of PAHs to environment (Abdel-Shafy & Mansour, 2016). Once entered to soil ecosystem, PAHs being hydrophobic in nature settle persistently in the particular environmental compartment. However, PAHs can be easily adsorbed onto organic matter of soil, thereby forming the persistent micropollutants in environment (Gan et. al., 2009). Soil contaminated with PAHs exhibit potential risks to the ecological health. Li et al. (2010) estimated the PAH contents in water, soil sediment and plant samples of Aojiang River wateray in Wenzhou, China and reported that total PAHs ranged from 910 - 1520ng/l in water samples. The content of PAHs in soil was dominant with 4-5 ring PAHs. Ma et al. (2014) estimated polycyclic aromatic hydrocarbons of 55 soil samples from wetland in Liao River Delta, Northeast China. High contents of alkyl PAHs were observed in the soil of LRD wetlands.

Phthalates

Phthalates are the esters of Phthalic acid which are added to polyvinyl chloride to improve the flexibility and hardness of plastics. It is widely exploited in almost every industry and can be found in varnish, latex paints, textiles, cosmetics, food packaging, printing inks as well as pharmaceuticals. The use of these plasticizers is also witnessed in medical products such as intravenous cannulas (Przybylinska & Wyszkowski, 2016). Due to their wide usage and wastes produced during daily use, this group of chemicals has turned out to be a potent environmental pollutant and their presence is detected in air water and soil. Di-n-butyl phthalate (DBP) has been reported to be responsible for the steady decrease in number of reptilian species around the world (Przybylinska & Wyszkowski, 2016) and has been reported to impair the development of internal reproductive organs in animals leading to underdevelopment of vas deferns, vacuolization of sertoli cell cytoplasm and gonadal dysgenesis (Belowska-Blen & Zdrojewicz, 2013). Some classes of phthalates have been reported to be more toxic in young children. In a study on rodents, phthalates caused kidney, thyroid and liver damage and contributed to liver and spleen cancer (Koniecki et al., 2011). However, some of the soil microorganisms have the tendency to degrade complex organics like phthalates. For example, *Micrococcus kristinae*, a bacterium isolated from soil resulted in 100% degradation of DEP in 48 h (Jackson et al., 1996) while *Rhodococcus ruber* G17 isolated from soil was found to degrade DBP (100%) in 2 - 3 days (Chao et al., 2006). Distribution of phthalate esters in urban soils of subtropical city, Guangzhou, China was reported in a study (Zeng et al., 2009). Pattern, sources and toxic potential of PAH in agricultural soil of Delhi, India was given by Agarwal et al. (2009).

Dioxins

Kaisarevic et al. (2007) detected dioxin-like contaminants in soil from area of oil refeneries in Vojvodina region of Serbia. Nording et al. (2006) evaluated the dioxin (polychlorinated dibenzo-*p*-dixin) in contaminated soil samples following immunochemical detection method. Polychlorinated dibenzo-p-dioxins (PCDDs) comprises a group of ubiquitous ecological contaminants, dioxin, which poses a hazard to health of human beings and other organisms because of their non-degradation both biologically and chemically and their potential to bioaccumulate in the ecosystem. This class of compounds shows toxic properties in terms of causing the reproductive disruption and cancer in wildlife (Nording et al., 2006). Dioxins are produced during chlorine bleaching of pulp, combustion processes, the synthesis of chlorine gas and various organochlorine chemicals. Consequently, soil especially near the industrial sites and the dumping sites is often heavily contaminated with dioxins. The major causes of soil contamination with dioxin and dioxin-like compounds are from accidental spillages in the production, transport, storage and use of various chlorinated compounds and throwing away of these compounds to unregulated sanitary landfills (Nording et al., 2006).

INTERCATIONS OF PHYSICOCHEMICAL CHARACTERISTICS AND AVAILABILTY OF HAZARDOUS CHEMICALS IN SOIL

The bioavailability of hazardous pollutants depends on their solubility, size of particles, sedimentation rates and retention as well as the absorption by the plants. In addition to these factors, the availability of pollutants strongly depends on various physicochemical characteristics like soil texture, pH, bulk density, water holding capacity, moisture content, alkalinity etc. Soil physico-chemical parameters which were most important in influencing the solubility of pollutants included the soil composition (inorganic or organic), pH, type/texture and density of charge on soil colloids, and reactive surface area. However, the solubility phenomena would also be dependent upon soil properties, including pollutant concentrations and form, particle size distribution, quantity and reactivity of hydrous oxides, mineralogy, degree of aeration and microbial activity (Jenne & Louma, 1977; Keeney & Wildung, 1972; Wildung *et al.*, 1976). It has also been reported that many pollutants including heavy metal uptake by plants was found to be influenced by various physicochemical parameters like pH, cation exchange capacity, redox potential and the chemical compositions of the soil solutions (Chaney, 1973; Yassoglou *et al.*, 1987; Chambers & Siddle, 1991).

French *et al.* (1999) studied the influence of physical and chemical characteristics on bioavailability of mercury in aquatic sediments. The study revealed that site specific factors such as acidity and dissolved organic carbon affected the mobilization of mercury and methylation process. Distribution of organic matter and toxic metals in sediments of Ramganga River at Moradabad were estimated by Pande and Sharma (1999). They reported that due to alkaline nature of river water in this stretch most of the heavy metals have precipitated and settled to riverbed as carbonates, oxides and hydroxides. Saison et al. (2006) reported that compost amendments resulted in alterations in the physicochemical characteristics of soil ultimately leading to alterations and resilience of soil microbial community. The contents of heavy metals and their bioavailability to the soil-plant system are considered as the main limitation to the application of sewage sludge compost in soil. A study on evolution of heavy metal speciation during composting process of sewage sludge was conducted by Liu et al. (2006). They observed direct correlation between heavy metal fractions and changes of parameters like pH, composting temperature and organic matter.

As soil is a dynamic and complex system, any change in physico-chemical characteristics alters or affects the availability of pollutants to the plant and microbial systems. Reduction-oxidation potential and pH were recognized as critical parameters controlling the fate of pollutants in the environment in many studies.

EFFECTS ON SOIL HEALTH

The accumulation of hazardous chemicals in the soil complex matrix results in ecological imbalances due to toxicities induced to microflora and fauna. Concentrated levels of hazardous compounds in soil result in alteration of nutrient contents such as nitrates, phosphates, potassium and other essential elements thereby affecting the natural vegetation. Apart from this, accumulation of these toxic compounds in soil proceeds their entry to the food chain and thus leading to biomagnifications, another matter of serious concern.. Increased level or high concentration of pollutants in lithosphere also result in deleterious/lethal effects on exposed populations. As soil also acts as a sink for pollutants, many a times, these pollutants leach into the ground water or make their entry to surface water bodies via surface run offs that in turn, causes total ecological imbalance.

The prevalence of hazardous compounds in soil ecosystem from various sounces has been shown in Figure 1.

Figure 1. Prevalence of hazardous compounds upon entry to soil ecosystem through various sources

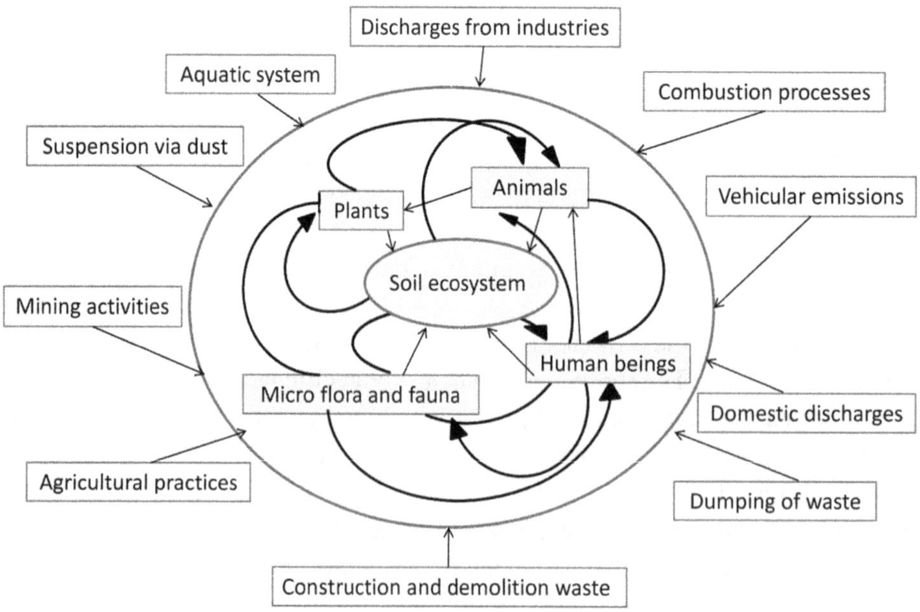

Toxicity of hazardous pollutants depends upon their concentration experienced by living beings in the environment. At higher concentrations, these pollutants interfere with the normal functioning of the body, which due to the stress, may finally result in loss in production in case of plant systems or death in case of organisms. Many studies of crops growing in contaminated soils show that levels of pollutants like heavy metals are generally higher in roots than in edible grains or leaves, but still there are reports on toxicity caused by heavy metals in human beings through food chains.

Food crops represent important pathways for movement of potentially toxic elements from soil to human beings. Pollutants in soils apparently move by forming organic complexes (Schinitzer & Hanson, 1970; Lund *et al.*, 1968). Crop plants could absorb and accumulate toxic amounts of hazardous pollutants resulting in severe health problems for human beings and animals (Cook, 1977; Shrivastava *et al.*, 1989). Agarwal *et al.* (1977) observed that heavy metals inhibit the synthesis of chlorophyll as well as alter the activities of catalase and peroxidase. Genotoxic agents in soil may gain entry into human beings through pathways such as inhalation of dust, ingestion of crops growing on contaminated soil, drinking water contaminated by leaching of compounds from soil to ground water or surface water. Exposed populations are expected to exhibit adverse health effects like arthrosclerosis, premature aging, sterility, cancer and even death.

CONCLUSION

Soil is one of the important ecological media for survival of all living beings. Increasing soil pollution in the past years has posed various types of adverse effects on health of plant and animal beings. Any hazardous compound which is non-biodegradable remains in environment over long periods of time and pose adverse effects. At present, it is the need of an hour to take necessary and strict actions to control the environmental pollution. Espousing of remedial techniques to combat the pollution should be the priority in every regulation concerning environmental protection.

REFERENCES

Abdel-Shafy, I. H., & Monsour, M. S. M. (2016). A review on polycyclic aromatic hydrocarbons:Source, environmental impact, effect on human health and remediation. *Egyptian Journal of Petroleum*, *25*(1), 107–123. doi:10.1016/j.ejpe.2015.03.011

Abrol, V., Wali, P., Pareek, N., Mondal, A. K., & Jalali, V. K. (2003). Edaphic impact of sewage and industrial effluents on soil resources. *Indian Journal of Environmental Protection*, *23*, 741–748.

Adhikari, S., Gupta, S. K., & Banerjee, S. K. (1993). Trace metal content of city sewage and sludge. *Journal of the Indian Society of Soil Science*, *41*, 170–172.

Agarwal, S. C., Bisht, S. S., & Sharma, C. P. (1977). Relative effectiveness of certain heavy metals in producing toxicity and symptoms of iron deficiency in barley. *Canadian Journal of Biotechnology*, *55*, 1299–1307.

Agarwal, T., Khillare, P. S., Shridhar, V., & Ray, S. (2009). Pattern, sources and toxic potential of PAHs in the agricultural soils of Delhi, India. *Journal of Hazardous Materials*, *163*(2-3), 1033–1039. doi:10.1016/j.jhazmat.2008.07.058 PMID:18757133

Alappat, B. J., & Dikshit, A. K. (1999). Management of plastic wastes. *Indian Journal of Environmental Protection*, *19*, 892–897.

Ali, S. M., Pervaiz, A., Afzal, B., Hamid, N., & Yasmin, A. (2014). Open dumping of municipal solid waste and its hazardous impacts on soil and vegetation diversity at waste dumping sites of Islamabad city. *Journal of King Saud University Science*, *26*(1), 59–65. doi:10.1016/j.jksus.2013.08.003

Alloway, B. J. (1995). Soil Processes and the behaviour of heavy metals. In B. J. Alloway (Ed.), *Heavy Metals in Soils* (Vol. 354, pp. 11–37). London, UK: Blackie Academic and Professional. doi:10.1007/978-94-011-1344-1_2

Antonovics, J., Bradshaw, A. D., & Turner, R. Q. (1971). Heavy metal tolerance in plants. In J. B. Cragg (Ed.), *Advances in ecological research* (Vol. 7, pp. 1–85). New York: Academic Press.

Belowska-Bien, K., & Zdrojewicz, Z. (2006). Phthalates – structure, properties, clinical significance. *Advances in Clinical and Experimental Medicine*, *15*, 677–681.

Benerjee, S. K., & Bagchi, K. (1999). Hospital solid waste and its management approach- A case study of a hospital in Calcutta. *Indian Journal of Environmental Protection*, *19*, 932–938.

Bradshaw, A. D., & Chadwich, M. J. (1981). *The restoration of land*. Oxford, UK: Blackwell.

Bradshaw, A. D., & Chadwick, M. J. (1980). *The construction of land*. Blackwell.

Caberara, G. L., Rodriguez, D. M. G., & Maruri, A. B. (1999). Genotoxicity of the extracts from the compost of organic and the total municipal garbage using three plant bioassays. *Mutation Research*, *426*(2), 201–206. doi:10.1016/S0027-5107(99)00068-8 PMID:10350598

Chambers, J. C., & Siddle, R. C. (1991). Fate of heavy metals in abandoned lead zinc tailing ponds: Vegetation. *Journal of Environmental Quality*, *20*(4), 745. doi:10.2134/jeq1991.00472425002000040005x

Chaney, R. L. (1973). Crop and food chain effects of toxic elements in sludge and effluents: Recycling municipal sludges and effects on land. Academic Press.

Chao, W. L., Lin, C. M., Shiung, I. I., & Kuo, Y. L. (2006). Degradation of di-butyl-phthalate by soil bacteria. *Chemosphere*, *63*(8), 1377–1383. doi:10.1016/j.chemosphere.2005.09.021 PMID:16289698

Chibuike, G. U., & Obiora, S. C. (2014). Heavy Metal Polluted Soils:Effect on Plants and Bioremediation Methods. *Applied and Environmental Soil Science*, *2014*, 1–12. doi:10.1155/2014/752708

Chronopoulos, J., Haidouti, C., Chronopoulousereli, A., & Massas, I. (1997). Variations in plant and soil lead and cadmium content in urban parks in Athens. Greece. *The Science of the Total Environment*, *196*(1), 91–98. doi:10.1016/S0048-9697(96)05415-0

Cook, J. (1977). Environmental pollution by heavy metals. *International Journal of Environmental Qulaity*, *9*, 253–266.

Cresser, M., Killham, K., & Edwards, T. (1993). *Soil chemistry and its applications*. Cambridge, UK: Cambridge University Press. doi:10.1017/CBO9780511622939

Cuny, D., Haluwyn, C. V., & Pesch, R. (2001). Biomonitoring of trace elements in air and soil compartments along the major motorway in France. *Water, Air, and Soil Pollution*, *125*(1/4), 273–289. doi:10.1023/A:1005278900969

Deo, B. (2004). Heavy metal accumulation by plant species from a coal mining area in Orissa. *Journal of Environmental Biology*, *25*, 163–166. PMID:15529873

Dharini, K., Cynthia, J. B., Kamalambikai, B., Celestina, J. P. A. S., & Muthu, D. (2017). Hazardous E-waste and its impact on soil structure. *Earth and Environmental Science*, *80*, 1–5.

Doubleday, G. P. (1974). The reclamation of land after coal mining. *Outlook on Agriculture*, *8*(3), 156–162. doi:10.1177/003072707400800307

Epstein, E. (1974). Effect of sewage sludge on some soil physical properties. *Journal of Environmental Quality*, *4*(1), 139–142. doi:10.2134/jeq1975.00472425000400010033x

Epstein, E., Taylor, J. M., & Chaney, R. L. (1976). Effects of Sewage Sludge and Sludge Compost Applied to Soil on some Soil Physical and Chemical Properties1. *Journal of Environmental Quality*, *5*(4), 422. doi:10.2134/jeq1976.00472425000500040021x

Farid, G., & Sarwar, N. (2015). Heavy metals (Cd, Ni & Pb) contamination of soils, plants and waters in Madina town of Faisalabad Metropolitan and preparation of GIS based maps. *Advances in Crop Science and Technology*, *4*(01), 1–7. doi:10.4172/2329-8863.1000199

Fatoki, O. S. (1996). Trace zinc and copper concentrations in roadside surface soils and vegetation measurement of local atmospheric pollution in Alice, South Africa. *Environment International*, *22*(6), 759–762. doi:10.1016/S0160-4120(96)00068-2

Feigin, A., & Kipnis, T. (1980). Agrochemicals in soils. Pergamon Press.

Feng, S., Wang, X., Wei, G., Peng, P., Yang, Y., & Cao, Z. (2006). Leachates of municipal solid waste incineration bottom ash from Macao: Heavy metal concentrations and genotoxicity. *Chemosphere*, *67*(6), 1133–1137. doi:10.1016/j.chemosphere.2006.11.030 PMID:17217988

French, K. J., Siruton, D. A., & Anderson, M. R. (1999). Influence of physical and chemical characteristics on mercury in aquatic sediments. *Water, Air, and Soil Pollution*, *110*(3/4), 347–362. doi:10.1023/A:1005007402287

Gan, S., Lau, E. V., & Ng, H. K. (2009). Remediation of soils contaminated with polycyclic aromatic hydrocarbons (PAHs). *Journal of Hazardous Materials*, *172*(2-3), 532–549. doi:10.1016/j.jhazmat.2009.07.118 PMID:19700241

Garbisu, C., & Alkorta, I. (2001). Phytoextraction: A cost-effective plant based technology for the removal of metals from the environment. *Bioresource Technology*, *77*(3), 229–236. doi:10.1016/S0960-8524(00)00108-5 PMID:11272009

Garcia, M., Scheffczyk, A., Garcia, T., & Römbke, J. (2011). The effects of the insecticide lambda-Cyhalothrin on the earthworm *Eisenia fetida* under experimental conditions of tropical and temperate regions. *Environmental Pollution, 159*(2), 398–400. doi:10.1016/j.envpol.2010.10.038 PMID:21106283

Garcia, M. J. M., Moreno-Grau, S., Garcia, J. J. M., Moreno, J., Bayo, J., Perez, J. J. G., & Moreno-clavel, J. (2001). Distribution of the metals lead, cadmium, copper and zinc in the top soil of cartagena, spain. *Water, Air, and Soil Pollution, 131*(1/4), 329–347. doi:10.1023/A:1011916218755

Geschwind, S. A., Stolwijk, J. A. J., Bracken, M., Fitzgeralg, E., Stark, A., Oslen, C., & Melius, J. (1992). Risk of congenital malformations associated with proximity to hazardous waste sites. *American Journal of Epidemiology, 135*(11), 1197–1207. doi:10.1093/oxfordjournals.aje.a116226 PMID:1626538

Hashiguchi, T., Toya, M., Uemura, K., Fuchigami, T., & Ito, K. (1976). Detection of heavy metals in vegetables grown in the Sukeye area. *Kageshima Ken Kegai Kenkyushe Ho, 12*, 55–56.

Ho, Y. B., & Tai, K. M. (1988). Elevated levels of level and other metals in roadside soil and grass and their use to monitor aerial metal deposition in Hong Kong. *Environ. Pollut., 49*(1), 37–51. doi:10.1016/0269-7491(88)90012-7 PMID:15092672

Hu, B., Jia, X., Hu, J., Xu, D., Xia, F., & Li, Y. (2017). Assessment of heavy metal pollution and Health risks in the Soil-Plant-Human System in the Yangtze River Delta, China. *International Journal of Environmental Research and Public Health, 14*(9), 1–18. doi:10.3390/ijerph14091042 PMID:28891954

Jacko, R. B., & Neurendrof, D. W. (1977). Trace metal particulate emission test results from a number of industrial and municipal point sources. *Journal of the Air Pollution Control Association, 27*(10), 989–994. doi:10.1080/00022470.1977.10470522

Jenne, E. A., & Louma, S. N. (1997). Forms of trace elements in soils and associated waters: an overview of their determination and biological availability. In Biological implications of metals in the environment. Conf. 750929.

Jones, G. E. (1981). Sludge Characteristics. In G. E. Jones & R. T. Sprague (Eds.), *Sludge and Its Ultimate Disposal*. Ann Arbor Science.

Kaisarevic, S., Andric, N., Bobic, S., Trickovic, J., Teodorovic, I., Miloradov, M. V., & Kovacevic, R. Z. (2007). Detection of Dioxin like Contaminants in Soil from the Area of oil Refineries in Vojvodina Region of Serbia. Springer Science Business Media, 79, 422-426.

Kashem, M. A., & Singh, B. (1999). Heavy metal contamination of soil and vegetation in the vicinity of industries in Bangladesh. *Water, Air, and Soil Pollution, 115*(1/4), 347–361. doi:10.1023/A:1005193207319

Katnoria, J. K., Bhardwaj, R., & Nagpal, A. (2011). Evaluation of genotoxic potential of industrial waste contaminated soil extracts of Amritsar, India. *Journal of Environmental Biology, 32*, 363–367. PMID:22167950

Keeny, D. R., & Wildung, R. E. (1972). Chemical properties of soils. In Soils for management of organic wastes and waste waters. American Soc. Agron.

Khan, A. G. (2001). Relationships between chromium biomagnification ratio, accumulation factor and mycorrhizae in plants growing on tannery effluent-polluted soil. *Environment International*, *26*(5-6), 417–423. doi:10.1016/S0160-4120(01)00022-8 PMID:11392761

Khan, S., & Singh, J. (1999). Mobility of heavy metals as affected by some insecticides through soils, saturated with different cations. *Pollution Research*, *18*, 133–136.

Klein, D. H., & Russel, P. (1973). Heavy metals: Fall out around a power plant. *Environmental Science & Technology*, *7*(4), 357–358. doi:10.1021/es60076a004

Klein, L. A., Aang, M., Nash, N., & Krishchner, S. L. (1974). Sources of metals in New York city waste water. *Journal - Water Pollution Control Federation*, *46*, 26–53.

Knasmuller, S., Gottmann, E., Steinkellner, H., Fomin, A., Pickl, C., Paschki, A., ... Kundi, M. (1998). Detection of genotoxic effects of heavy metal contaminated soils with plant bioassays. *Mutation Research*, *420*(1-3), 37–48. doi:10.1016/S1383-5718(98)00145-4 PMID:9838036

Koniecki, D., Wang, R., Moody, R. P., & Zhu, J. (2011). Phthalates in cosmetics and personal care products:concentrations and possible dermal exposure. *Environmental Research*, *111*(3), 329–336. doi:10.1016/j.envres.2011.01.013 PMID:21315328

Kudyakov, R., Baibergenova, A., Zdeb, M., & Carpenter, D. O. (2004). Respiratory disease in relation to patient residence near to hazardous waste sites. *Environmental Toxicology and Pharmacology*, *18*(3), 249–257. doi:10.1016/j.etap.2004.06.005 PMID:21782756

Kulkarni, N. P., & Agrawal, Y. K. (2003). Physico-chemical characteristics of the topsoil of a hospital. *Indian Journal of Environmental Protection*, *23*, 503–507.

Kumar, A. (1998). Effect of mixed industrial effluents on properties of ground water and irrigated soils. *Journal of Indian Soiety for Soil Sciences*, *46*, 427–429.

Li, J., Shang, X., Zhao, Z., Tanguay, R. L., Dong, Q., & Huang, C. (2010). Polycyclic aromatic hydrocarbons in water, sediment, soil, and plants of the Aojiang River waterway in Wenzhou, China. *Journal of Hazardous Materials*, *173*(1-3), 75–81. doi:10.1016/j.jhazmat.2009.08.050 PMID:19726127

Liu, Y., Ma, L., Li, Y., & Zheng, L. (2006). Evolution of heavy metal speciation during the aerobic composting process of sewage sludge. *Chemosphere*, *67*(5), 1025–1032. doi:10.1016/j.chemosphere.2006.10.056 PMID:17184823

Loureiro, S., Santos, C., Pinto, G., Costa, A., Monteiro, M., Nogueira, A. J., & Soares, A. M. (2006). Toxicity assessment of two soils from Jales mine (Portugal) using plants: Growth and biochemical parameters. *Archives of Environmental Contamination and Toxicology*, *50*(2), 182–190. doi:10.100700244-004-0261-3 PMID:16307215

Lund, L. J., Page, A. L., & Nelon, C. O. (1968). Movement of heavy metals below sewage disposal pond. *Journal of Environmental Quality*, *5*(3), 330–334. doi:10.2134/jeq1976.00472425000500030023x

Ma, C., Ye, S., Lin, T., Ding, X., Yuan, H., & Guo, Z. (2014). Sorce apportionment of polycyclic aromatic hydrocarbons in soils of wetlands in the Liao River Delta, Northeast China. *Marine Pollution Bulletin*, *80*(1-2), 160–167. doi:10.1016/j.marpolbul.2014.01.019 PMID:24461691

McCalla, T. M., Paterson, J. R., & Lue-Hing, C. (1977). *Properties of agricultural and municipal wastes. In soils for management of organic waste and wastewater* (R. J. Stevenson, Ed.). Madison, WI: Amer. Soc. Agron.

Moldoveanu, A. M. (2014). Assessment of Soil Pollution with Heavy Metals in Romania: Environmental Risk Assessment of Soil Contamination. INTECH.

Munch, D. (1993). Concentration profiles of arsenic, cadmium, chromium, copper, lead, mercury, nickel, zinc, vanadium and polynucleated aromatic hydrocarbons in forest soil beside an urban road. *The Science of the Total Environment*, *138*(1-3), 47–55. doi:10.1016/0048-9697(93)90404-T

Nording, M., Nichkova, M., Spinnel, E., Persson, Y., Gee, S. J., Hammock, B. D., & Haglund, P. (2006). Rapid screening of dioxin-contaminated soil by accelerated solvent extraction/purification followed by immunochemical detection. *Analytical and Bioanalytical Chemistry*, *385*(2), 357–366. doi:10.100700216-006-0402-9 PMID:16598456

Olajire, A. A., & Ayodele, E. T. (1997). Contamination of roadside soil and grass with heavy metals. *Environment International*, *23*(1), 91–101. doi:10.1016/S0160-4120(96)00080-3

Onyari, J. M., Wandiga, S. O., & Njentede, J. O. (1991). Lead contamination in street soils of Nairoubi city, Mombassa Island, Kenya. *Bulletin of Environmental Contamination and Toxicology*, *46*(5), 782–789. doi:10.1007/BF01689968 PMID:2039875

Ostrander, R., & Clark, S. C. (1991). Soil heavy metal concentrations and erosion damage in upland Grasslands in the Pennines, England. *Water, Air, and Soil Pollution*, *59*(1-2), 81. doi:10.1007/BF00283173

Othman, I., Al-Qudat, M., & Al-Masri, M. S. (1997). Lead levels in roadside soils and vegetation of Damascus city. *The Science of the Total Environment*, *207*(1), 43–48. doi:10.1016/S0048-9697(97)00243-X PMID:9397598

Pande, K. S., & Sharma, S. O. (1999). Distribution of organic matter and toxic metals in the sediments of Ramganga river at Moradabad. *Pollution Research*, *18*, 43–47.

Pattnaik, S., & Reddy, M. V. (2009). Assesment of municipal solid waste management in Puducherry (Pondicherry), India. *Resources, Conservation and Recycling*, *54*(8), 512–520. doi:10.1016/j.resconrec.2009.10.008

Prusty, B. G., Sahu, K. C., & Godgul, G. (1994). Metal contamination due to mixing and milling activities at Zamar zinc mine, Rajasthan, India (Part I). *Chemical Geology*, *112*(3-4), 275–291. doi:10.1016/0009-2541(94)90029-9

Przybylinska, P. A., & Wyszkowski, M. (2016). Environmental contamination with phthalates and its impacts on living organisms. *Ecological Chemistry and Engineering. S*, *23*(2), 347–356. doi:10.1515/eces-2016-0024

Rampal, R. K., Kour, J., & Jamwal, R. (2002). Soild waste generation in government hospitals of Jammu city, India. *Pollution Research*, *21*, 39–43.

Rao, A. S., & Rao, P. R. (2001). Heavy metals concentrations in the sediments from Kolleru lake, India. *Indian Journal of Environmental Health*, *43*, 148–153. PMID:12395518

Rao, S. L. N., & Sarma, D. R. R. (1998). Trace metals in soils near an industrial belt in Visakhapatnam, India. *Indian Journal of Environmental Protection*, *17*, 377–380.

Sadiq, M., Alam, I., El-Mubarek, A., & Al-Mohdhar, H. M. (1989). Preliminary evaluation of metal pollution from wear auto tires. *Bulletin of Environmental Contamination and Toxicology*, *42*(5), 743–749. doi:10.1007/BF01700397 PMID:2743004

Saison, C., Degrange, V., Oliver, R., Millard, P., Commeaux, C., Montange, D., & Le Roux, X. (2006). Alteration and resilience of the soil microbial community following compost amendment: Effects of compost level and compost-borne microbial community. *Environmental Microbiology*, *8*(2), 247–257. doi:10.1111/j.1462-2920.2005.00892.x PMID:16423013

Sakal, R., Prasad, A. K., & Singh, A. P. (1992). Depth wise distribution of trace metals in soils receiving sewage and sludge. *Journal of the Indian Society of Soil Science*, *40*, 732–737.

Sanchez-Martin, M. J., Sanchez-Camazano, M., & Lorenzo, L. F. (2000). Chromium and lead contents in suburban and urban soils of medium-sized cities of spain: Influence of traffic. *Environmental Contamination and Toxicology*, *64*(2), 250–257. doi:10.1007001289910037 PMID:10656892

Schnitzer, M., & Hanson, E. H. (1970). Organo-metallic interactions in soils and evaluation of the methods for the determination of stability constants of metal-fulvic acid complexes. *Soil Science*, *109*(6), 333–339. doi:10.1097/00010694-197006000-00001

Shrivastva, V. S., Rai, A. K., & Mehrotra, R. C. (1989). Rapid acid dissolution of wheat plant tissue for iron, copper and cadmium determination. *Indian Journal of Environmental Protection*, *10*, 198–202.

Sims, J. T., & Kline, J. S. (1991). Chemical fraction and plant uptake of heavy metals in soils amended with co-composed sewage sludge. *Journal of Environmental Quality*, *20*(2), 387–395. doi:10.2134/jeq1991.00472425002000020009x

Singh, J., & Kalamdhad, A. S. (2011). Effect of heavy metals on soil, plants, human health and aquatic life. *International Journal of Research in Chemistry and Environment*, *1*, 15–21.

Singh, R., Ahirwar, N.K., Tiwari, J., & Pathak, J. (2018). Review on sources and effect of heavy metal in soil: its bioremedition. *Impact Journals,* 1-22.

Sinha, S. (1999). Accumulation of Cu, Cd, Cr, mn and Pb from artificially contaminated soil by *Bacopa monnieri*. *Environmental Monitoring and Assessment*, *57*(3), 253–264. doi:10.1023/A:1006091215529

Sun, S., Sidhu, V., Rong, Y., & Zheng, Y. (2018). Pesticide Pollution in Agricultural soils and Sustainable Remediation Methods: A Review. Springer International Publishing AG.

Sutherland, R. A., & Tack, F. M. (2000). Metal phase association in an urban watershed. Honolulu. Hawaii. *The Science of the Total Environment*, *4*(2-3), 103–113. doi:10.1016/S0048-9697(00)00458-7 PMID:10902838

Swaileh, K. M., Hussein, R. M., & Elhaj, S. A. (2004). Assessment of heavy metal contamination in roadside surface soil and vegetation from the west bank. *Archives of Environmental Contamination and Toxicology*, *47*(1), 23–30. doi:10.100700244-003-3045-2 PMID:15346775

Tangahu, B. V., Abdullah, S. R. S., Basri, H., Idris, M., Anuar, N., & Mukhlisin, M. (2011). A review on heavy metals (As, Pb, & Hg) uptake by plants through Phytoremediation. *International Journal of Chemical Engineering*, *2011*, 1–31. doi:10.1155/2011/939161

Tatasumi, Y., Ikeda, A., & Yoda, K. (1983). Soil pollution by heavy metals in Sakai city, Japan. *Journal of Ecology*, *32*, 241–249.

Tripathi, R. M., Khandekar, R. N., & Mishra, U. C. (1990). Toxic trace metals in the atmosphere of Mordabad (India). *Indian Journal of Environmental Health*, *32*, 140–147.

Wade, M. J., Davis, B. K., Carlisle, J. S., Klein, A. K., & Valoppi, L. M. (1993). Environmental transformation of toxic metals. *Occupational Medicine*, *8*, 575–601. PMID:8272980

Watanabe, T., & Hirayama, T. (2001). Genotoxicity of soil. *Journal of Health Science*, *47*(5), 433–438. doi:10.1248/jhs.47.433

Wath, S. B., Dutt, P. S., & Chakrabarti, T. (2011). E-Waste scenario in India, its management and implications. *Environmental Monitoring and Assessment*, *172*(1-4), 249–262. doi:10.100710661-010-1331-9 PMID:20151189

Wilcke, W. (2007). Global patterns of polycyclic aromatic hydrocarbons (PAHs) in soil. *Science Direct*, *141*, 157–166.

Wildung, R. E., Garland, T. R., & Drucker, H. (1976). *Transformation of trace metals by soil microorganisms*. Agronomy abstracts, No. 28.

Wu, L., & Bradshaw, A. D. (1972). Actual pollution and the rapid evolution of copper tolerance. *Nature*, *238*(5360), 167–169. doi:10.1038/238167a0

Wu, L., Bradshaw, A. D., & Thurraan, D. A. (1975). The potential for evolution of heavy metal tolerance in plants. III. The rapid evolution of copper tolerance in *Agrostis stolonifera*. *Hereditas*, *34*(2), 165–187. doi:10.1038/hdy.1975.21

Wyszkowski, M., & Sivitskaya, V. (2012). Changes in the content of organic carbon and available forms of macronutrients in soil under the influence of soil contamination with fuel oil and application of different substances. *Journal of Elementology*, *17*, 139–148.

Yassoglou, N., Kosmas, C., Asimakopoulos, J., & Kallinou, C. (1987). Heavy metal contamination of roadside soils in the greater Athens area. *Environmental Pollution*, *47*(4), 293–304. doi:10.1016/0269-7491(87)90149-7 PMID:15092703

Yokel, J., & Delistraty, D. A. (2003). *Arsenic, lead and other trace elements in soils contaminated with pesticide residues at the Hansford site*. Wiley Periodicals.

Zeng, F., Cui, K., Xie, Z., Wu, L., Luo, D., Chen, L., ... Sun, G. (2009). Distribution of phthalate esters in urban soils of subtropical city, Guangzhou, China. *Journal of Hazardous Materials*, *164*(2-3), 1171–1178. doi:10.1016/j.jhazmat.2008.09.029 PMID:18963455

Ziolkowska, A., & Wyszkowski, M. (2010). Toxicity of petroleum substances to microorganisms and plants. *Ecological Chemistry and Engineering. S*, *17*, 73–82.

KEY TERMS AND DEFINITIONS

Dioxins: Dioxins are ubiquitous ecological contaminants that are chemically related to polychlorinated dibenzo-p-dioxins (PCDDs) and are persistent environmental pollutants (POPs). This class of compounds exhibit various toxic effects such as hormonal problems, infertility, cancer, and possibly diabetes. High levels of exposure can even lead to chloracne, a severe skin disease with acne-like lesions mainly on the face and upper body. The most toxic compound of this group is 2,3,7,8-tetrachlorodibenzo-p-*dioxin* or TCDD. Dioxins are produced during chlorine bleaching of pulp, combustion processes, the synthesis of chlorine gas and various organochlorine chemicals.

Heavy Metals: Heavy metals are generally defined as metals with density of more than 5 g/cm^3. These metals have relatively high densities, atomic weights, as well as atomic numbers. In the field of metallurgy, a heavy metal is defined on the basis of density, whereas in physics, the distinguishing criterion is their atomic number and in chemistry, their distinguishing chemical behavior. Some of the examples of heavy metals are copper, chromium, lead, nickel, mercury, and zinc.

Pesticides: Pesticides are substances which are used to control pests and weeds in order to increase the crop yield. The term pesticide includes various forms like herbicide, insecticides, Molluscicdes, Ovicides, weedicides, fungicides nematicides, repellants and defoliage agents.

Phthalates: Phthalate esters are the esters of phthalic acid and are mainly used as plasticizers (i.e., substances added to plastics to increase their flexibility, transparency, durability, and longevity). They are used primarily to soften polyvinyl chloride (PVC). Studies have shown their potential to induce severe reproductive disorders.

Polycyclic Aromatic Hydrocarbons (PAHs): PAHs also Known as polyaromatic hydrocarbons or polynuclear aromatic hydrocarbons are the compounds containing two or more aromatic rings. They only contain carbon and hydrogen and are composed of multiple organic rings in which the electrons are delocalized. The simplest form of PAH is naphthalene which contains only two aromatic rings followed by anthracene and phenanthrene which contain three-ring compounds. Other examples are Phenalene, Tetracene, Chrysene, Triphenylene and pyrene etc. Most of the PAHs have been documented to be potent mutagens and carcinogens.

Chapter 3
Biodegradable Waste a Nutrient Asset for Upholding Soil Health and Crop Production

Ifra Ashraf
SKUAST-K, India

Shazia Ramzan
University of Kashmir, India

Nowsheeba Rashid
Amity University, India

Ikhlaq A. Mir
SKUAST-K, India

Asima Jillani
SKUAST-K, India

ABSTRACT

Management of solid wastes is a grave concern because of its associated significant negative impacts on quality of the environs. Accretion and putrefaction of solid wastes have potent hazardous effects on biotic and abiotic factors of the environment including human beings. Unmanaged solid wastes especially organic in nature add efficient quantity of greenhouse gases in the atmosphere. For dealing with wastes purely organic in nature, there is a need of an adequate waste management technology to reduce the quantity of organic waste being disposed of traditionally. Composting is an environmentally sound and sustainable approach to manage biodegradable fractions of solid waste. It has received considerable attention in the last few decades because of its potential of redressing the environmental pollution concerns associated with other waste disposal methods. This chapter is aimed to review supremacy of composting over other waste disposal methods.

DOI: 10.4018/978-1-7998-0031-6.ch003

INTRODUCTION

Globally, steady growth in the population, urbanization and industrialization has triggered generation of mammoth quantity of wastes. These waste spurts have steered a number of socio-economic and environmental challenges principally in the developing countries (Awasthi et al., 2014; Sukholthaman & Sharp, 2016). Annually, about 2 billion tonnes of municipal solid waste (MSW) is produced at global level (Wilson et al., 2015). India being densely populated generates massive amount of wastes of the order of 48 million tonnes of solid waste which had exceeded to 52 million tonnes in 2016 (Kumar et al., 2017).

Solid waste management has been a matter of grave concern since years, particularly for developing countries like India. It is considered a predicament of every civilization (Wolny-Koładka & Malinowski, 2015), wastes being mostly disposed off by landfilling and incineration globally (Tweib et al., 2011). Landfilling has been widely harnessed because of many advantages associated with it, be its simple approach or low investment and operating cost or greater handling capacity (Li et al., 2017; Wang et al., 2018), non-obligation of skilled labor (Ali et al., 2014). A landfill is a storehouse for waste which is deposited in a runs of tamped layers in specifically fabricated compartments either superficially on the land surface or beneath the surface in pits (Crowley et al., 2003). Although the land filling is most prevalent method of waste disposal, yet there are many environmental hazards associated with it particularly production of highly contaminating leachate and gases including methane gas, carbon dioxide and some other toxic gases at low concentrations (Crowley et al., 2003; Damgaard et al., 2011; Danthurebandara et al., 2012; Vaverková et al., 2012). Many environmental problems like global warming, eutrophication of water bodies, depletion of ozone layer, acidification, degradation of human health and ecosystem are caused due to the induction of these by-products of landfilled wastes in environment (Emeryet al., 2007; Damgaard et al., 2011). The leachate compounds pollute the groundwater (Szymański et al., 2018) and surface water bodies if nearby (Giang et al., 2018). Groundwater contamination is more precarious than surfacewater body contamination because recuperation of the aquifers entail mammoth period (Crowley et al., 2003). The 40% of the waste generated at global level finds its way in dumpsites; 38 out of 50 largest dumpsites directly imperiling the marine and coastal areas. Due to faulty management of wastes at dumping sites death of more than 750 people occurred in the first half of 2016 (Somani et al., 2018).

Another widely used approach of getting rid of wastes is incineration which involves the thermal oxidation of wastes at high temperatures in excess of 850 °C. Combustion of wastes yields a variety of volatile and gasiform emanations, which negatively affect the environmental quality and human health (Crowley et al., 2003). An instance of negative implication of burning of wastes reported by Gadde et al., 2009, is 0.05% contribution to greenhouse gases (GHG) emission due to burning of 23% of rice straw (RS) in the farms in India. But incineration in India is unwarranted because of high organic, inert or moisture content in Indian MSW of the order of about 30% to 60% each and low calorific value (800–1,100 kcal/kg) in MSW (Joardar, 2000; Kansal, 2002; Sharholy et al., 2008). Burning of organic matter (OM) generate huge amount of GHG emissions and the extra fuel need to be expended for burning the wastes having high moisture content or low calorific value (Joshi & Ahmed, 2016).

Due to rapid industrialization and urbanization, not only the quantity of wastes has increased but their characteristics have also been altered. Consequently, the upgradation of solid waste management system (SWMS) necessitates to be done to harmonize with the quality, quantity and composition of wastes (Manaf et al., 2009). A conventional SWMS in developing countries depicts an assemblage of problems, encompassing inefficient waste collection and pooling, crude open dumping, multiplication of flies and mosquitoes, burning in absence of air, control of environmental degradation, and the regulation and control of unauthorized waste picking (Ogawa, 2000). In particular in urban areas, the problem of management of solid wastes has been constrained with the growing urbanization. In order to surmount these problems; an effective management system needs to be adopted to safely dispose off the wastes with minimal negative consequences (Moh & Manaf, 2017). Additionally, the wastes, predominantly the biodegradable wastes which can be degraded by the biochemical action of the microbial organisms, should be utilized to their maximum potential and then followed by their disposal. This forms the fundamental principle of the sustainable SWMS (Wagner & Raymond, 2015). The pledge towards the cost-effective and environmentally acceptable approach for treatment of biodegradable wastes is fulfilled by the method of composting (Qian et al., 2014; Gonawala & Jardosh, 2018), which involves different techniques for treating the organic wastes in consonance with the sustainable agriculture (Adbrecht et al., 2011).

Composting is "sensustricto, a self-heating, aerobic solid phase biodegradative process of organic materials under controlled conditions, which distinguishes it from natural rotting or putrefaction" (Ryckeboer et al., 2003). Insam And de Bertoldi (2007) also defined the composting process as "a biodegradation process of a mixture of substrates carried out by a microbial community composed of various populations in aerobic condition in the solid state". Composting transmutes the organic constituents into biologically stable and humic-like constituents, which can be easily handled. It is a heterogeneous process, biochemical in nature encompassing the mineralization of OM to CO_2, H_2O, NH_3, and partial humification ensuring a stabilized product-compost with lessened toxicity and pathogenicity (Das et al., 2011).

DIFFERENT COMPOSTING TECHNOLOGIES

Composting has been put into practice in many metropolises owing to its economic and functioning practicability (Soobhany, 2018). The systematic disposal of wastes gets constrained due to absence of proper segregation facilities at source, causing scanty amount of wastes to be processed via composting (Lou & Nair, 2009). In India, a marginal proportion of waste (12.45%) is handled scientifically while rest finds its way in open dumps. The high organic content (biodegradable) in wastes produced in India i.e., 50% in contrast to the developed nations i.e., 30% (Joshi & Ahmed, 2016), makes it suitable for treatment with composting (Guidoni et al., 2018). There are numerous types of composting methods for generating superior quality compost out of which four customary methods espoused are aerated static pile, windrow method, in-vessel composting and vermicomposting (Cooperband, 2002). Some of the practices of composting adopted across India are described in this section.

Aerobic Composting

In aerobic composting, the specific microbial organisms decompose the OM into compost in presence of air under the controlled environmental conditions, i.e., specific temperature and pH. Aeration is the core factor which influences the rate of composting (Raut et al., 2008), which when controlled at optimum rate can normalize the problem of odor engendered during the composting sequel (Li et al., 2015). The aeration rate of 0.25L/min is commended per kg of volatile solids in order to get the least amount of odor (Li et al., 2008). Aeration rate not only controls the odor generation but the production of GHG i.e., CH_4 and CO_2 is also regulated by lowering the rate of aeration (Bernal et al., 2009).

Windrow System

In a windrow system, the waste material is piled up and arranged parallelly in long rows for microbial degradation. These rows are flipped around to aid the air and moisture conditioning, porosity and heat redistribution (Sánchez et al., 2017). The triangular or trapezoidal shaped windrows can be used depending upon the equipments used to mix the heaps. No mechanical equipment is used for turning the windrows, instead it is done manually e.g., by pitchfork (Diaz et al., 2007). Shammas and Wang (2009) recommended the turning of piles at least once or twice per week to provide the required aeration to the microbes. The schematic diagram for windrow composting is depicted in Figure1.

Aerated Static Pile Composting

This system was developed in 1975 by research team at Beltsville (Epstein et al., 1976). In this system, the waste is piled and the air circulation is maintained by the distribution system employing the perforated pipe/tubes at the bottom of the mounds (Figure 2). These systems can be windrows, covered, open or in closed vessels. The pivotal to this composting system is the aeration factor similar to that of aerobic composting. The temperature gradient is created vertically across the top and bottom of the pile as governed by the intensity of aeration (Sánchez et al., 2017; Kumar et al., 2018). This system has a benefit that it allows the lesser space exploitation for composting as the waste need not to be moved to another place for aeration purposes (Shammas & Wang, 2009).

In-Vessel Composting

In-vessel composting method is one of the effectual and fast composting systems (Manyapu et al., 2018), treating the biodegradable waste in closed containers with the temperature, moisture and oxygen regulation (Mohee & Mudhoo, 2005; Shammas & Wang, 2009). Different bioreactors/vessels employed in in-vessel composting include containers, tunnels, rotating drums, agitated bags, and enclosed halls (Kumar et al., 2018). For large bioreactors, the temperature, moisture and oxygen regulation gets constrained due to hydrodynamic restraints stemming due to bigger size (De Guardia et al., 2012), which limits the use of this method to either laboratory or small scale experiments (Sánchez et al., 2017). The additives have been exploited to accelerate the process of composting, among which the jaggery has revealed the

Figure 1. Schematic diagram of windrow composting process (Ambade et al., 2011)

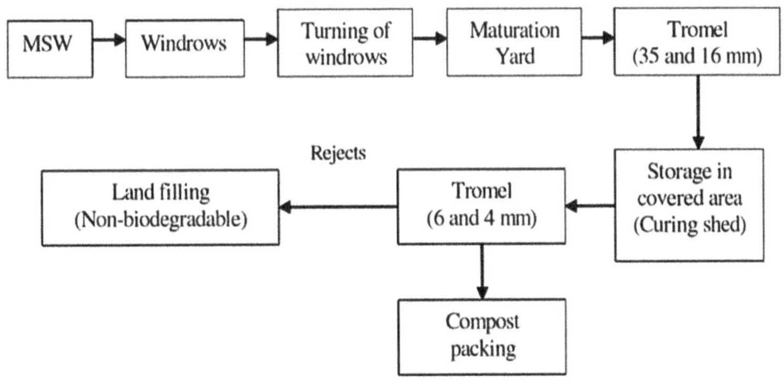

Figure 2. Illustration of aerated static pile composting (Diaz et al., 2007)

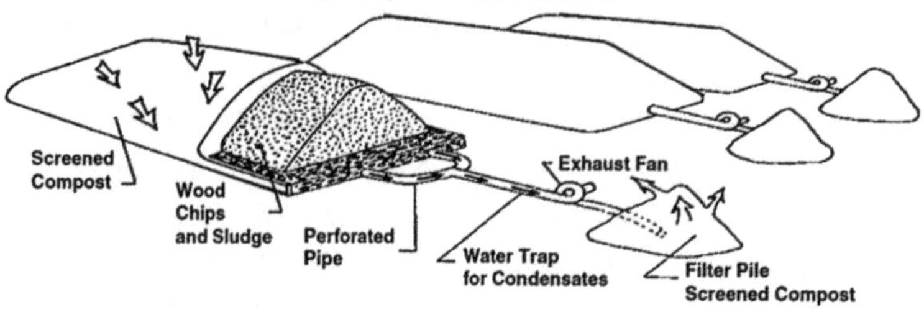

best outcomes (Gabhane et al., 2012). Manyapu et al., 2018 recommended the combination of 20% fly ash, 50% biomass, 30% kitchen waste and 5% jiggery for reducing the bio-availability of heavy metals, thereby enhancing the soil productivity.

Vermicomposting

Vermicomposting is a biotechnological technique of composting (Adhikary, 2012) in which the earthworms in conjunction with the aerobic microbes stabilize the organic waste. Primarily, the biodegradable waste is acted upon by the microbial organisms via extra cellular enzymatic action, termed as primary decomposition, and this partially festered organic waste is further broken down by earthworms (Bhat et al., 2018). Earthworms on daily basis consume about five times of their body weight (Kumar, 2011). Commonly, the epigeic species of *Eudrilus eugeniae*, *Eisenia foetida* and *Perionyx excavates* are used for composting the organic waste. The microorganisms present in the gut of earthworms are potent enough to digest the organic substances and polysaccharides (Aira et al. 2007). Although, the vermicomposting is reported to produce greater CO_2 emissions in comparison with an aerobic composting (Lim et al., 2016), however, the inclusion of mineral bulking agents like red mud, a product attained in the progress of alumina abstraction from bauxite, in the waste showed the reduction in the CO_2 emanations (Barthod et al., 2018). This reduction of CO_2 releases due to addition of bulking agents can be attributed to a reason that worms simultaneously ingest the organic waste along with the mineral thereby boosting the carbon stability (Barthod et al., 2016). The biggest plant for vermicomposting in India is located in Bengaluru having the capacity of 100 million tonnes on daily basis (Kumar et al., 2018), while there are smaller plants in Bangalore, Mumbai, Hyderabad, and Faridabad (Joshi & Ahmed, 2016).

Co-Composting

Co-composting is an approach of enhancing the activity of composting by adding two or more substances. Co-composting with a particular bulking agent can endow organic waste with the appropriate particle structure, density and porosity needed to enhanceactivity of aerobic bacteria, thereby increasing the microbial decomposition rate (Awasthi et al., 2018). Few instances of co-composting are presented in Table 1.

Table 1. Co-composting of some solid wastes

Solid Waste	Bulking Agent	Benefit	Reference
Rice straw (RS)	Cattle manure	Optimization of the C/N ratio which forms one of the keyparameters for the digestion.	Li et al., 2008; Qian et al., 2014; Das et al., 2011; Kong et al., 2018; Negi et al., 2018
Lettuces, apple flesh, residual maize leaves, ground spent coffee and pieces of cardboard	Red mud and fly ash	CO_2 Emissionswere lessened due to red mud addition, whereas fly ash had no influence	Barthod et al., 2018
Sewage sludge	Biochar	Biocharhas exceptional capability to adsorbNH_3, and GHG	Malińska et al., 2014; Jiang et al., 2016
Sewage sludge	Biochar	Biochar intensifies the humification of organic waste	Dias et al., 2010
Sewage sludge	Zeolite	Decrease the heavy metals concentration	Zorpas &Loizidou,2008
Sewage sludge	Saw dust	As the saw dust concentration was increased, the concentration of humic substances in compost also increased	Zorpas & Loizidou, 2008
Food waste	Fly ash and lime	Optimization of pH and C/N ratio	Wong et al., 2009.

ROLE OF COMPOSTING FOR UPGRADING SOIL HEALTH

The biodegradable fraction of MSW is a potent repository of plant nutrients, and apposite practices of composting can transform it to the highly nutritious compost (Sanasam and Talukdar, 2017). In the present epoch, from agricultural sustainability point of view, compost is opined to be the best source available for restoration and enrichment of the soil potential so as to maximize the soil and crop throughput (Bera et al., 2013).The composting recycles the organic matter back into the soil and the compost produced thereof helps to improve the physical, chemical and biological properties of the soil, hence, fertility and acts as a soil conditioning agent (Tweib et al., 2011; Kumar et al., 2018).

The adverse repercussions of industrialization on the soil properties are reported to have been alleviated by composting and compost appliance, which mediates the improvisation of the environmental amenities of green infrastructure in cities encompassing the carbon pooling, nutrient cycling, nutrient use proficiency, soil water holding capacity and even the bio-control of the disease-causing microbes (Milinković et al., 2019). During the process of composting, the humic substances are produced which help in promoting the soil fertility and the organic content of the soil is augmented (Bertoncini et al., 2008). The humic fraction of composts not only heightens the plant growth and soil health, but also wields the biocontrol on diverse soil-borne phytopathogens for instance fungi (Traversa et al., 2010; Dionne et al., 2012; Sanasam & Talukdar, 2017; Milinković et al., 2019). Even though the mechanism of suppression of plant disease by composts is not yet clear, but it is assumed to have chemical, physical and biological characteristics (Boulter et al., 2002). No doubt the MSW do contain the pathogenic organisms, which if applied to the soil may proliferate and cause plant diseases. This nuisance of pathogenic organisms in waste is eliminated during the process of composting itself due to the temperature build-up which can even raise upto 70 °C in the course of composting (Neklyudov et al., 2006; Pandey et al., 2016). Composting has also been attributed for causing the reduction in the weed germination and annihilation of stinking compounds (Bernal et al., 2009). Vermicompost clenches the hormones and enzymes that increases plant growth and reduces plant diseases (Abbasi & Ramasamy, 1999; Hussain et al., 2017). Many researchers have reported plant growth hormones (auxin, cytokinins, humic substances) produced by microbes from vermicompost (Muscolo et al., 1999; Atiyeh et al., 2002)

The upshot of composting, i.e., compost has also been reported to have many values. In accord with Hermann et al. (2011), composts provide with a benefit of acting as a source for the substitution of the soil conditioners as espoused by the humus formation, which can't be attained by artificial means. Composts have also been reported to enhance the water-holding capacity of the soil and provide with better tilts (Mohammad et al., 2012). Compost on integration with fly ash supplements soil with essential nutrients like alumina, magnesium oxide, potassium oxide, silica, iron oxide, calcium oxide, sulphate and sodium oxide (Singh et al., 2011). The nutrient values embedded in fly ash get accrued with the compost without distressing the biological activity. On contrary, fly ash if present in huge amount can hamper the respiration of microbes (Wong & Wong, 1986).

COMPOST: A REPLACEMENT TO THE IN-ORGANIC FERTILIZERS

Researchers have irrefutably corroborated that organic manure application on long term basis contends competently in terms of productivity with direct chemical fertilizer therapy (Briggs & Courtney, 1985). The generation of organic wastes at the rate of 100-300 kg per person on annual basis at global level (Jones and Healey, 2010) is in line with the swelling population of more than 7.5 billion (Population Reference Bureau, 2017). So, with the same land available for agricultural production, the rate of production is fast-tracked to meet the increasing demands for food for such burgeoning population. This is accomplished by the over-dosage of chemical fertilizers and pesticides which deteriorate the soil health by contamination with heavy metals (Manyapu et al., 2018), causing soil erosion (Sánchez et al., 2017) and imbalancing the physico-chemical and micro-biological equilibria (Altomare & Tringovska, 2011). The soil degradation is occurring at an alarming rate which needs to be addressed by enriching the soil with essential nutrients employing eco-friendly approaches.

Annually, more than 620 million tonnes of agrarian waste is produced in India (Singh and Sidhu, 2014). More or less 39-54% of MSW produced in India is compostable (Kumar et al., 2009). Such a massive amount of organic waste can be potentially tapped to fortify the soils with nutrients supplemented eco-friendly compost (Sánchez et al., 2017). Composts acquired from organic wastes have been substantiated to outweigh the commercially used inorganic fertilizers in quality (Chowdhury et al., 2015). Chander et al., 2018 postulated that compost not only enhance the crop yield but also saved the chemical fertilizers up to 25%. The apt blend of organic and inorganic nutrients is indispensable for sustainable food production that not only guarantees high quality food production, but also fortifies the soil with essential nutrients (Nambiar, 1990; Ali et al., 2009).

Vermiwash, a vermicomposting derivative, has been exhausted as liquid fertilizers by many researchers with good results, some of which are presented in Table 2.

Table 2. Outline of vermiwash as liquid fertilizer in agricultural utilization from different types of substrates

Substrate Used for Vermiwash	Plant	Findings	References
Leaf litter and Cow dung	Rice	Increased root and leaf length, plant height and number of leaves were chronicled from pot treated with vermicompost and vermiwash	Tharmaraj et al., 2011
Kitchen waste and animal waste	-	Results showed potential as a bio-fertilizer for particular nutrient deficient soil by using combination of substrate.	Nath et al., 2009
Coconut leaf litter and cow dung	Maize, Cowpea and Okra	High concentration of P, K, Ca and Mg were found in vermiwash. Vermiwash from pure cow dung showed highest in N, P, Ca, Mg and indoleacetic acid (IAA) concentration. Seedling vigour index were highest in 1:10 dilution for cowpea.	Gopal et al., 2010
Cow manure	Indian Borage (*Plectranthusambionicus*)	High concentration of potassium (K) concentration was determined. Diluted vermiwash (10%) shows highest in germination percentage, root and shoot length, total chlorophyll and carotenoids content.	Quaik et al., 2012

Source: Quaik & Ibrahim, 2013

CONCLUSION

With the growing population, the escalation in the waste generation is blatant which demands safe disposal without any negative impact on environment. Undoubtedly, prevalent conventional practices of landfilling and incineration relieve the burden of these wastes, but simultaneously negatively impact the environment especially due to the organic carbon content present in the MSW. The organic fraction of MSW which forms the major portion of waste can be managed by a productive system of composting which not only rid the environment of these wastes but also utilize the wastes to their best possible potential. Composting and its product compost have positive impact on soil biological, physical and chemical properties, add the nutrients to the soil, which in turn boost the productivity potential of soil and crops. Compost can serve as a potent replacement for the inorganic fertilizers which have negative impacts on environs. Although, the replacement can't be 100%, but integrated use of both organic and inorganic forms of nutrients can serve towards the sustainable management of agriculture. Thus, biodegradable forms of waste can serve as a nutrient asset and these reserves must be tapped to their maximum potential.

REFERENCES

Adhikary, S. (2012). Vermicompost, the story of organic gold: A review. *Agricultural Sciences*, *3*(7), 905–917. doi:10.4236/as.2012.37110

Aira, M., Monroy, F., & Domínguez, J. (2007). Earthworms strongly modify microbial biomass and activity triggering enzymatic activities during vermicomposting independently of the application rates of pig slurry. *The Science of the Total Environment*, *385*(1-3), 252–261. doi:10.1016/j.scitotenv.2007.06.031 PMID:17628641

Albrecht, R., Le Petit, J., Terrom, G., & Périssol, C. (2011). Comparison between UV spectroscopy and nirs to assess humification process during sewage sludge and green wastes co-composting. *Bioresource Technology, 102*(6), 4495–4500. doi:10.1016/j.biortech.2010.12.053 PMID:21239169

Ali, M. E., Islam, M. R., & Jahiruddin, M. (2009). Effect of integrated use of organic manures with chemical fertilizers in the rice-rice cropping system and its impact on soil health. *Bangladesh Journal of Agricultural Research, 34*(1), 81–90. doi:10.3329/bjar.v34i1.5757

Ali, S. M., Pervaiz, A., Afzal, B., Hamid, N., & Yasmin, A. (2014). Open dumping of municipal solid waste and its hazardous impacts on soil and vegetation diversity at waste dumping sites of Islamabad city. *Journal of King Saud University-Science, 26*(1), 59–65. doi:10.1016/j.jksus.2013.08.003

Altomare, C., & Tringovska, I. (2011). Beneficial soil microorganisms, an ecological alternative for soil fertility management. In *Genetics, biofuels and local farming systems* (pp. 161–214). Dordrecht: Springer. doi:10.1007/978-94-007-1521-9_6

Ambade, B., Sharma, S., Sharma, Y., & Sharma, Y. (2011). Assessment of windrow composting plant's performance at Keru, Jodhpur, India. *Asian Journal of Environmental Science, 6*(2), 107–111.

Awasthi, M. K., Pandey, A. K., Khan, J., Bundela, P. S., Wong, J. W., & Selvam, A. (2014). Evaluation of thermophilic fungal consortium for organic municipal solid waste composting. *Bioresource Technology, 168*, 214–221. doi:10.1016/j.biortech.2014.01.048 PMID:24507579

Awasthi, M. K., Wang, Q., Awasthi, S. K., Li, R., Zhao, J., Ren, X., ... Zhang, Z. (2018). Feasibility of medical stone amendment for sewage sludge co-composting and production of nutrient-rich compost. *Journal of Environmental Management, 216*, 49–61. doi:10.1016/j.jenvman.2018.01.032 PMID:29357991

Barthod, J., Rumpel, C., Calabi-Floody, M., Mora, M. L., Bolan, N. S., & Dignac, M. F. (2018). Adding worms during composting of organic waste with red mud and fly ash reduces CO 2 emissions and increases plant available nutrient contents. *Journal of Environmental Management, 222*, 207–215. doi:10.1016/j.jenvman.2018.05.079 PMID:29857313

Barthod, J., Rumpel, C., Paradelo, R., & Dignac, M. F. (2016). The effects of worms, clay and biochar on CO 2 emissions during production and soil application of co-composts. *Soil (Göttingen), 2*(4), 673–683. doi:10.5194oil-2-673-2016

Bera, R., Datta, A., Bose, S., Dolui, A.K., Chatterjee, A.K., Dey, G.C., … Seal, A. (2013). Comparative Evaluation of Compost Quality, Process Convenience and Cost under Different Composting Methods to assess their Large Scale Adoptability Potential as also Complemented by Compost Quality Index. *International Journal of Scientific and Research Publications, 3*(6).

Bernal, M. P., Alburquerque, J. A., & Moral, R. (2009). Composting of animal manures and chemical criteria for compost maturity assessment. A review. *Bioresource Technology, 100*(22), 5444–5453. doi:10.1016/j.biortech.2008.11.027 PMID:19119002

Bhat, R. A., Dar, S. A., Dar, D. A., & Dar, G. H. (2018). Municipal Solid Waste Generation and current Scenario of its Management in India. *International Journal of Advance Research in Science and Engineering, 7*(2), 419-431.

Boulter, J. I., Trevors, J. T., & Boland, G. J. (2002). Microbial studies of compost: Bacterial identification, and their potential for turfgrass pathogen suppression. *World Journal of Microbiology & Biotechnology, 18*(7), 661–671. doi:10.1023/A:1016827929432

Briggs, D. J., & Courtney, F. M. (1985). *Agriculture and environment*. Longman Inc.

Chander, G., Wani, S. P., Gopalakrishnan, S., Mahapatra, A., Chaudhury, S., Pawar, C. S., ... Rao, A. K. (2018). Microbial consortium culture and vermi-composting technologies for recycling on-farm wastes and food production. *International Journal of Recycling of Organic Waste in Agriculture, 7*(2), 99–108. doi:10.100740093-018-0195-9

Chowdhury, A. K. M. M. B., Konstantinou, F., Damati, A., Akratos, C. S., Vlastos, D., Tekerlekopoulou, A. G., & Vayenas, D. V. (2015). Is physicochemical evaluation enough to characterize olive mill waste compost as soil amendment? The case of genotoxicity and cytotoxicity evaluation. *Journal of Cleaner Production, 93*, 94–102. doi:10.1016/j.jclepro.2015.01.029

Crowley, D., Staines, A., Collins, C., Bracken, J., Bruen, M., Fry, J., … Thunhurst, C. (2003). *Health and environmental effects of landfilling and incineration of waste-A literature review*. Academic Press.

Damgaard, A., Manfredi, S., Merrild, H., Stensøe, S., & Christensen, T. H. (2011). LCA and economic evaluation of landfill leachate and gas technologies. *Waste Management (New York, N.Y.), 31*(7), 1532–1541. doi:10.1016/j.wasman.2011.02.027 PMID:21435856

Danthurebandara, M., Van Passel, S., Nelen, D., Tielemans, Y., & Van Acker, K. (2012). Environmental and socio-economic impacts of landfills. *Linnaeus Eco-Tech, 2012*, 40–52.

Das, M., Uppal, H. S., Singh, R., Beri, S., Mohan, K. S., Gupta, V. C., & Adholeya, A. (2011). Co-composting of physic nut (Jatrophacurcas) deoiled cake with rice straw and different animal dung. *Bioresource Technology, 102*(11), 6541–6546. doi:10.1016/j.biortech.2011.03.058 PMID:21489784

De Guardia, A., Petiot, C., Benoist, J. C., & Druilhe, C. (2012). Characterization and modelling of the heat transfers in a pilot-scale reactor during composting under forced aeration. *Waste Management (New York, N.Y.), 32*(6), 1091–1105. doi:10.1016/j.wasman.2011.12.028 PMID:22301461

Dias, B. O., Silva, C. A., Higashikawa, F. S., Roig, A., & Sánchez-Monedero, M. A. (2010). Use of biochar as bulking agent for the composting of poultry manure: Effect on organic matter degradation and humification. *Bioresource Technology, 101*(4), 1239–1246. doi:10.1016/j.biortech.2009.09.024 PMID:19796932

Diaz, L. F., Savage, G. M., Eggerth, L. L., & Chiumenti, A. (2007). Systems used in composting. In Compost science and technology (pp. 67-87). Elsevier.

Dionne, A., Tweddell, R. J., Antoun, H., & Avis, T. J. (2012). Effect of non-aerated compost teas on damping-off pathogens of tomato. *Canadian Journal of Plant Pathology, 34*(1), 51–57. doi:10.1080/07060661.2012.660195

Emery, A., Davies, A., Griffiths, A., & Williams, K. (2007). Environmental and economic modelling: A case study of municipal solid waste management scenarios in Wales. *Resources, Conservation and Recycling, 49*(3), 244–263. doi:10.1016/j.resconrec.2006.03.016

Epstein, E., Willson, G. B., Burge, W. D., Mullen, D. C., & Enkiri, N. K. (1976). A forced aeration system for composting wastewater sludge. *Journal - Water Pollution Control Federation*, *48*, 688–694.

Gabhane, J., William, S. P., Bidyadhar, R., Bhilawe, P., Anand, D., Vaidya, A. N., & Wate, S. R. (2012). Additives aided composting of green waste: Effects on organic matter degradation, compost maturity, and quality of the finished compost. *Bioresource Technology*, *114*, 382–388. doi:10.1016/j.biortech.2012.02.040 PMID:22444633

Gadde, B., Menke, C., & Wassmann, R. (2009). Rice straw as a renewable energy source in India, Thailand, and the Philippines: Overall potential and limitations for energy contribution and greenhouse gas mitigation. *Biomass and Bioenergy*, *33*(11), 1532–1546. doi:10.1016/j.biombioe.2009.07.018

Giang, N. V., Kochanek, K., Vu, N. T., & Duan, N. B. (2018). Landfill leachate assessment by hydrological and geophysical data: Case study NamSon, Hanoi, Vietnam. *Journal of Material Cycles and Waste Management*, *20*(3), 1648–1662. doi:10.100710163-018-0732-7

Gonawala, S. S., & Jardosh, H. (2018). Organic Waste in Composting: A brief review. *International Journal of Current Engineering and Technology*, *8*(1), 36–38. doi:10.14741/ijcet.v8i01.10884

Gopal, M., Gupta, A., Palaniswami, C., Dhanapal, R., & Thomas, G. V. (2010). Coconut leaf vermiwash: A bio-liquid from coconut leaf vermicompost for improving the crop production capacities of soil. *Current Science*, *98*(9), 1202–1210.

Guidoni, L. L. C., Marques, R. V., Moncks, R. B., Botelho, F. T., da Paz, M. F., Corrêa, L. B., & Corrêa, É. K. (2018). Home composting using different ratios of bulking agent to food waste. *Journal of Environmental Management*, *207*, 141–150. doi:10.1016/j.jenvman.2017.11.031 PMID:29156437

Hermann, B. G., Debeer, L., De Wilde, B., Blok, K., & Patel, M. K. (2011). To compost or not to compost: Carbon and energy footprints of biodegradable materials' waste treatment. *Polymer Degradation & Stability*, *96*(6), 1159–1171. doi:10.1016/j.polymdegradstab.2010.12.026

Insam, H., & de Bertoldi, M. (2007). Microbiology of the composting process. In L. F. Diaz, M. de Bertoldi, W. Bidlingmaier, & E. Stentiford (Eds.), *Compost science and technology. Waste management series 8*. Amsterdam: Elsevier.

Jiang, T., Ma, X., Tang, Q., Yang, J., Li, G., & Schuchardt, F. (2016). Combined use of nitrification inhibitor and struvite crystallization to reduce the NH_3 and N_2O emissions during composting. *Bioresource Technology*, *217*, 210–218. doi:10.1016/j.biortech.2016.01.089 PMID:26865057

Joardar, S. D. (2000). Urban residential solid waste management in India: Issues related to institutional arrangements. *Public Works Management & Policy*, *4*(4), 319–330. doi:10.1177/1087724X0044006

Jones, D. L., & Healey, J. R. (2010). Organic amendments for remediation: Putting waste to good use. *Elements*, *6*(6), 369–374. doi:10.2113/gselements.6.6.369

Joshi, R., & Ahmed, S. (2016). Status and challenges of municipal solid waste management in India: A review. *Cogent Environmental Science*, *2*(1), 1139434. doi:10.1080/23311843.2016.1139434

Joshi, R., & Ahmed, S. (2016). Status and challenges of municipal solid waste management in India: A review. *Cogent Environmental Science*, *2*(1), 1139434. doi:10.1080/23311843.2016.1139434

Kansal, A. (2002). Solid waste management strategies for India. *Indian Journal of Environmental Protection, 22*(4), 444–448.

Kong, Z., Wang, X., Liu, Q., Li, T., Chen, X., Chai, L., ... Shen, Q. (2018). Evolution of various fractions during the windrow composting of chicken manure with rice chaff. *Journal of Environmental Management, 207*, 366–377. doi:10.1016/j.jenvman.2017.11.023 PMID:29182981

Kumar, S. (2011). Composting of municipal solid waste. *Critical Reviews in Biotechnology, 31*(2), 112–136. doi:10.3109/07388551.2010.492207 PMID:20854128

Kumar, S., Bhattacharyya, J. K., Vaidya, A. N., Chakrabarti, T., Devotta, S., & Akolkar, A. B. (2009). Assessment of the status of municipal solid waste management in metro cities, state capitals, class I cities, and class II towns in India: An insight. *Waste Management (New York, N.Y.), 29*(2), 883–895. doi:10.1016/j.wasman.2008.04.011 PMID:18595684

Kumar, S., Negi, S., Mandpe, A., Singh, R. V., & Hussain, A. (2018). Rapid composting techniques in Indian context and utilization of black soldier fly for enhanced decomposition of biodegradable wastes- A comprehensive review. *Journal of Environmental Management, 227*, 189–199. doi:10.1016/j.jenvman.2018.08.096 PMID:30193208

Kumar, S., Smith, S. R., Fowler, G., Velis, C., Kumar, S. J., Arya, S., ... Cheeseman, C. (2017). Challenges and opportunities associated with waste management in India. *Royal Society Open Science, 4*(3), 160764. doi:10.1098/rsos.160764 PMID:28405362

Li, G., Yun, Y., Li, H., & Sang, N. (2008). Effect of landfill leachate on cell cycle, micronucleus, and sister chromatid exchange in Triticumaestivum. *Journal of Hazardous Materials, 155*(1–2), 10–16. doi:10.1016/j.jhazmat.2007.10.106 PMID:18068897

Li, J., Wang, C., Du, L., Lv, Z., Li, X., Hu, X., ... Zhang, Y. (2017). Did municipal solid waste landfill have obvious influence on polychlorinated dibenzo-p-dioxins and polychlorinated dibenzofurans (PCDD/Fs) in ambient air: A case study in East China. *Waste Management (New York, N.Y.), 62*, 169–176. doi:10.1016/j.wasman.2017.02.014 PMID:28223077

Li, Z., Huang, G., Yu, H., Zhou, Y., & Huang, W. (2015). Critical factors and their effects on product maturity in food waste composting. *Environmental Monitoring and Assessment, 187*(4), 217. doi:10.100710661-015-4430-9 PMID:25822329

Lim, P. N., Wu, T. Y., Shyang Sim, E. Y., & Lim, S. L. (2011). The potential reuse of soybean husk as feedstock of Eudriluseugeniae in vermicomposting. *Journal of the Science of Food and Agriculture, 91*(14), 2637–2642. doi:10.1002/jsfa.4504 PMID:21725978

Malińska, K., Zabochnicka-Świątek, M., & Dach, J. (2014). Effects of biochar amendment on ammonia emission during composting of sewage sludge. *Ecological Engineering, 71*, 474–478. doi:10.1016/j.ecoleng.2014.07.012

Manaf, L.A., Samah, M. A. A., & Zukki, N. I. M. (2009). Municipal solid waste management in Malaysia: Practices and challenges. *Waste Management, 29*, 2902–2906.

Manyapu, V., Mandpe, A., & Kumar, S. (2018). Synergistic effect of fly ash in in-vessel composting of biomass and kitchen waste. *Bioresource Technology*, *251*, 114–120. doi:10.1016/j.biortech.2017.12.039 PMID:29274517

Milinković, M., Lalević, B., Jovičić-Petrović, J., Golubović-Ćurguz, V., Kljujev, I., & Raičević, V. (2019). Biopotential of compost and compost products derived from horticultural waste—Effect on plant growth and plant pathogens' suppression. *Process Safety and Environmental Protection*, *121*, 299–306. doi:10.1016/j.psep.2018.09.024

Moh, Y. C., & Manaf, L. A. (2017). Solid waste management transformation and future challenges of source separation and recycling practice in Malaysia. *Resources, Conservation and Recycling*, *116*, 1–14. doi:10.1016/j.resconrec.2016.09.012

Mohammad, N., Alam, M. Z., Kabbashi, N. A., & Ahsan, A. (2012). Effective composting of oil palm industrial waste by filamentous fungi: A review. *Resources, Conservation and Recycling*, *58*, 69–78. doi:10.1016/j.resconrec.2011.10.009

Mohee, R., & Mudhoo, A. (2005). Analysis of the physical properties of an in-vessel composting matrix. *Powder Technology*, *155*(1), 92–99. doi:10.1016/j.powtec.2005.05.051

Nambiar, K. K. M. (1990). Long-term fertility effects on wheat productivity. In . *International Conference on Wheat for the Nontraditional Warm Areas*. CIMMYT.

Nath, G., Singh, K., & Singh, D. K. (2009). Chemical analysis of vermicomposts/vermiwash of different combinations of animal, agro and kitchen wastes. *Australian Journal of Basic and Applied Sciences*, *3*(4), 3671–3676.

Negi, S., Dhar, H., Hussain, A., & Kumar, S. (2018). Biomethanation potential for co-digestion of municipal solid waste and rice straw: A batch study. *Bioresource Technology*, *254*, 139–144. doi:10.1016/j.biortech.2018.01.070 PMID:29413914

Neklyudov, A. D., Fedotov, G. N., & Ivankin, A. N. (2006). Aerobic processing of organic waste into composts. *Applied Biochemistry and Microbiology*, *42*(4), 341–353. doi:10.1134/S0003683806040016

Ogawa, H. (2000). Sustainable solid waste management in developing countries. In *7th ISWA International Congress and Exhibition*. World health Organization.

Pandey, P. K., Cao, W., Biswas, S., & Vaddella, V. (2016). A new closed loop heating system for composting of green and food wastes. *Journal of Cleaner Production*, *133*, 1252–1259. doi:10.1016/j.jclepro.2016.05.114

Population Reference Bureau. (2017). *World Population Data Sheet.* PRB Publ. 1–22.

Qian, X., Shen, G., Wang, Z., Guo, C., Liu, Y., Lei, Z., & Zhang, Z. (2014). Co-composting of livestock manure with rice straw: Characterization and establishment of maturity evaluation system. *Waste Management (New York, N.Y.)*, *34*(2), 530–535. doi:10.1016/j.wasman.2013.10.007 PMID:24188923

Quaik, S., Embrandiri, A., Rupani, P. F., Singh, R. P., & Ibrahim, M. H. (2012). Effect of vermiwash and vermicomposting leachate in hydroponics culture of Indian Borage (Plectranthus ambionicus) plantlets. In *UMT 11th International Annual Symposium on Sustainability Science and Management* (pp. 210-214). Academic Press.

Quaik, S., & Ibrahim, M. H. (2013). A review on potential of vermicomposting derived liquids in agricultural use. *International Journal of Scientific and Research Publications*, *3*(3), 1–6.

Raut, M. P., William, S. P., Bhattacharyya, J. K., Chakrabarti, T., & Devotta, S. (2008). Microbial dynamics and enzyme activities during rapid composting of municipal solid waste– a compost maturity analysis perspective. *Bioresource Technology*, *99*(14), 6512–6519. doi:10.1016/j.biortech.2007.11.030 PMID:18155903

Ryckeboer, J., Mergaert, J., Vaes, K., Klammer, S., De Clercq, D., Coosemans, J., ... Swings, J. (2003). A survey of bacteria and fungi occurring during composting and self-heating processes. *Annals of Microbiology*, *53*(4), 349–410.

Sanasam, S. D., & Talukdar, N. C. (2017). Quality compost production from municipality biowaste in mix with rice straw, cow dung, and earthworm EiseniaFetida. *Compost Science & Utilization*, *25*(3), 141–151. doi:10.1080/1065657X.2016.1249312

Sánchez, Ó. J., Ospina, D. A., & Montoya, S. (2017). Compost supplementation with nutrients and microorganisms in composting process. *Waste Management (New York, N.Y.)*, *69*, 136–153. doi:10.1016/j.wasman.2017.08.012 PMID:28823698

Shammas, N. K., & Wang, L. K. (2009). Biosolids composting. In L. K. Wang, N. C. Pereira, Y. T. Hung, & N. K. Shammas (Eds.), *Biological Treatment Processes. Handbook of Environmental Engineering* (Vol. 8, pp. 669–713). Totowa, NJ: Humana Press. doi:10.1007/978-1-60327-156-1_16

Sharholy, M., Ahmad, K., Mahmood, G., & Trivedi, R. C. (2008). Municipal solid waste management in Indian cities–A review. *Waste Management (New York, N.Y.)*, *28*(2), 459–467. doi:10.1016/j.wasman.2007.02.008 PMID:17433664

Singh, S., Ram, L. C., Masto, R. E., & Verma, S. K. (2011). A comparative evaluation of minerals and trace elements in the ashes from lignite, coal refuse, and biomass fired power plants. *International Journal of Coal Geology*, *87*(2), 112–120. doi:10.1016/j.coal.2011.05.006

Somani, M., Datta, M., Ramana, G. V., & Sreekrishnan, T. R. (2018). Investigations on fine fraction of aged municipal solid waste recovered through landfill mining: Case study of three dumpsites from India. *Waste Management & Research*, *36*(8), 744–755. doi:10.1177/0734242X18782393 PMID:29939108

Soobhany, N. (2018). Preliminary evaluation of pathogenic bacteria loading on organic Municipal Solid Waste compost and vermicompost. *Journal of Environmental Management*, *206*, 763–767. doi:10.1016/j.jenvman.2017.11.029 PMID:29161678

Sukholthaman, P., & Sharp, A. (2016). A system dynamics model to evaluate effects of source separation of municipal solid waste management: A case of Bangkok, Thailand. *Waste Management (New York, N.Y.)*, *52*, 50–61. doi:10.1016/j.wasman.2016.03.026 PMID:27026497

Tharmaraj, K., Ganesh, P., Kolanjinathan, K., Suresh Kumar, R., & Anandan, A. (2011). Influence of vermicompost and vermiwash on physico chemical properties of rice cultivated soil. *Current Biotica, 2*, 18–21.

Traversa, A., Loffredo, E., Gattullo, C. E., & Senesi, N. (2010). Water-extractable organic matter of different composts: A comparative study of properties and allelochemical effects on horticultural plants. *Geoderma, 156*(3-4), 287–292. doi:10.1016/j.geoderma.2010.02.028

Tweib, S. A., Rahman, R. A., & Kalil, M. S. (2011). A Literature Review on the Composting. *Proceedings of International Conference on Environment and Industrial Innovation (ICEII 2011).*

Vaverková, M., Toman, F., Adamcová, D., Kotovicová, J., Vaverková, M., Stejskal, B., ... Yapsakli, K. (2013). Verifying research of waste landfill environmental impact using bioindicators. *Journal of Environmental Studies, 4*(7), 8.

Wagner, T. P., & Raymond, T. (2015). Landfill mining: Case study of a successful metals recovery project. *Waste Management (New York, N.Y.), 45*, 448–457. doi:10.1016/j.wasman.2015.06.034 PMID:26152366

Wang, K., Li, L., Tan, F., & Wu, D. (2018). Treatment of Landfill Leachate Using Activated Sludge Technology: A Review. *Archaea (Vancouver, B.C.).* PMID:30254508

Wilson, D. C., Rodic, L., Modak, P., Soos, R., Carpintero, A., Velis, K., ... Simonett, O. (2015). *Global waste management outlook.* UNEP.

Wolny-Koładka, K., & Malinowski, M. (2015). Assessment of the microbiological contamination of air in a municipal solid waste treatment company. *Ecological Chemistry and Engineering A, 22*(2), 175–183.

Wong, J. W. C., Fung, S. O., & Selvam, A. (2009). Coal fly ash and lime addition enhances the rate and efficiency of decomposition of food waste during composting. *Bioresource Technology, 100*(13), 3324–3331. doi:10.1016/j.biortech.2009.01.063 PMID:19268581

Wong, M. H., & Wong, J. W. C. (1986). Effects of fly ash on soil microbial activity. *Environmental Pollution. Series A. Ecological and Biological, 40*(2), 127–144. doi:10.1016/0143-1471(86)90080-2

Zorpas, A. A., & Loizidou, M. (2008). Sawdust and natural zeolite as a bulking agent for improving quality of a composting product from anaerobically stabilized sewage sludge. *Bioresource Technology, 99*(16), 7545–7552. doi:10.1016/j.biortech.2008.02.014 PMID:18439819

KEY TERMS AND DEFINITIONS

Aeration: Aeration (also called aerification or aeriation) is the introduction of air into a material.

Leachate: Leachate is a by-product derived from municipal solid wastes due to their physical, chemical, and biological changes and will be formed in landfills, incineration plants, composting plants, and transfer stations, with high strength and toxicity. Leachate usually contains various toxic organic pollutants, heavy metals, ammonia nitrogen compounds, and other components, which are much more complex than domestic sewage.

Phytopathogen: A microorganism which produces disease in plants.

Putrefaction: The decomposition of organic matter due to the action of microorganisms with the formation of foul-smelling incompletely oxidized products.

Vermiwash: Vermiwash is a liquid that is collected after the passage of water through a column of worm action and is very useful as a foliar spray. It is a collection of excretory products and mucus secretion of earthworms along with micronutrients from the soil organic molecules.

Chapter 4
White Pollution:
A Hazard to Environment and Sustainable Approach to Its Management

Mehvish Hameed
National Institute of Technology, India

Rouf Ahmad Bhat
Sher-e-Kashmir University of Agricultural Sciences and Technology, India

Dig Vijay Singh
Sher-e-Kashmir University of Agricultural Sciences and Technology, India

Mohammad Aneesul Mehmood
Cluster University Srinagar, India

ABSTRACT

Plastic derived from the petrochemical industry with a high molecular weight constitutes about 9-13% of total solid waste. Since the industrial revolution, the use of plastic has increased manifold without improving its adequate management as a waste. Most of the plastic waste produced in the world is mainly from packaging industry followed by building and construction. Plastic is a non-degradable deadly pollutant to degrade environmental quality and are known to remain in water and soil for years without making any change in their structure. Due to enormous generation, open burning of plastic is also preferred due to the lack of resource in the developing countries thus releasing toxic gases thereby causing air pollution. Plastic disturbs the balance of the environment by acting as physical barrier leading to the drainage of the drains, degrading soil properties, and are often ingested by the organisms ultimately leading to their death. Thus, it becomes more important to manage the plastic pollution keeping in view its detrimental impacts on the environment.

DOI: 10.4018/978-1-7998-0031-6.ch004

INTRODUCTION

Population explosion as well as urbanization leads to the enormous quantity of waste generation (Camill, 2010; Bhat et al., 2014; Bhat et al., 2018) and this waste generated is known as municipal solid waste (Kumar et al., 2016). The availability of the different alternative choices of the product has resulted in the short life span of the single product. Apart from that, the new trend is the use of disposable products which are produced, used and disposed easily (Ahmed and Ali, 2004). The generation of waste is directly related to the economic background (Bhat et al., 2012) as the economically sound families produce more quantity of the waste compared to poor families (Ambulkar & Shekdar, 2004). The waste is generated in such a large quantity if it is allowed to be disposed of in landfill than whole earth will soon get covered with waste (Buenrostro & Bocco, 2003) and if the waste is treated by incineration, toxic pollutant are released which will ultimately destroy the climate of the earth (Kumar et al., 2009). Plastic waste constitutes the major proportion of the solid waste (Sangawar & Deshmukh, 2012) thus it became important to look for good management techniques like recycling which help to minimize the quantity of the waste (Zmak & Hartmann, 2017).

The word 'plastic' comes from the Greek word 'plasticos', which means on heating, changes the shape and form a totally different product. Plastic invented a century ago (Baekeland, 1909) is one of the widely used human made product and can be found in every nook and corner of the country. Plastic has found its use in every field of the life and each person on an average produce 52 kg of plastic waste annually (Jambeck et al., 2015). Geologists are now considering a plastic horizon in the world's soils and sediments as one of the key indicators marking the current geological epoch, the anthropocene (Waterset al., 2016). 'Polymers' the generic term for all kinds of plastic materials, produced mostly from the petro industry is a giant molecule of the carbon-based organic compounds (Ismail & Hashmi, 2008). The use of polymers in alloys and blends has increased its uses for the production of different kinds of products as blending of polymers increase performance as well as life of the product (Gupta et al., 1998). Plastics waste is mainly produced from the packaging of the material, daily use as carry bags and discarded items (Staniskis, 2005). Plastic are having extensive industrial applications (Heskett et al., 2012) because of their cheapness, strength, durability and many other properties which make them one of the most commonly used products on the earth (Andrady, 2011). The single largest consumer is the packaging industry which uses more than 1/3rd of the plastic (Schmidt et al., 2017). The good quantity (0.2- 0.3%) of the plastic waste ultimately reach to the oceans thereby leads to the degradation of the ocean ecosystem (Andrady & Neal, 2009). The products produced from plastic are the major part of human life and due to their increased use, the production on global level has reached more than 150 million tonnes annually (Rillig, 2012). The comparison of per capita plastic consumption with the rest of the word is presented in Table 1.

Further, as per the data from World Economic Forum, the quantity of waste and green house gases generation increases with the production and consumption of the plastic (Wohlleben & Neubauer, 2016)

Table 1. Plastic waste consumption

S.N.	Country/Continent	Per Capita/Year Consumption (Kg)
1	India	6.0
2	East Europe	10.0
3	South East Asia	10.0
4	China	24.0
5	West Europe	65.0
6	North America	90.0
7	World Average	25.0

Figure 1. "Time trends in total plastics production worldwide. (a) Global production in million metric tons per year (Mt year^{-1}). (b) Usage patterns of plastic in 2016 are estimated from available sources (Plastics Europe, 2016). Three phases are seen: Phase 1 signifies slow development and invention of most plastics commonly used today (innovation phase), Phase 2 is marked by rapid global expansion and exponential growth (growth phase), and Phase 3 shows more linear dynamics that more closely mirrors global economic growth (consolidation phase)"(Plastics Europe, 2016; Plastics Europe, 2012; ACS, 1993)

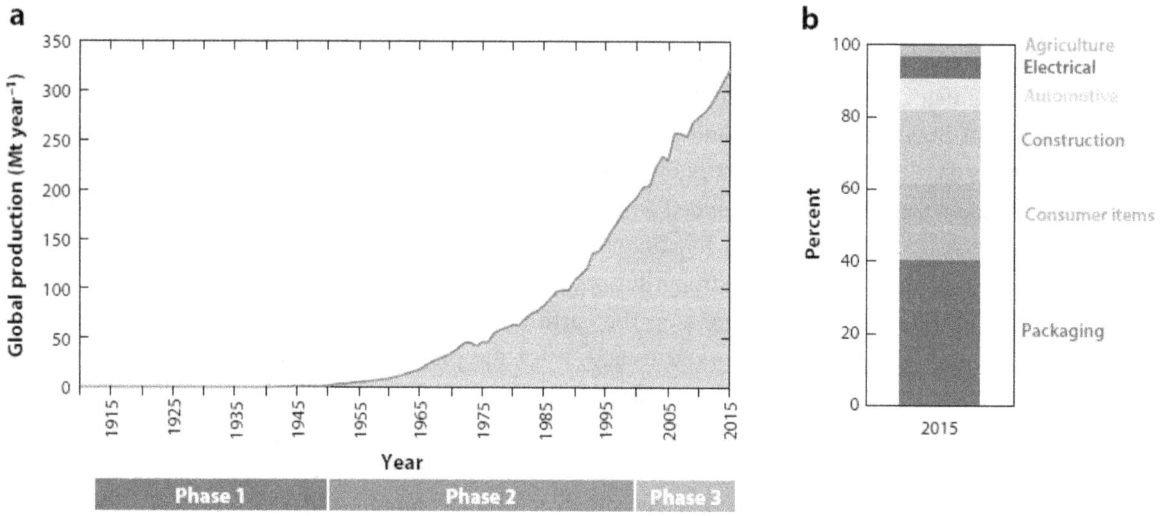

PLASTIC AND HEALTH IMPLICATIONS

The use, recycling or keeping of plastic untouched in the environment results in the breakdown (Barnes et al., 2009) and release of heavy metals (cadmium and lead) and other harmful chemicals like benzene, dioxins which are the serious environmental pollutants (Wagner & Oehlmann, 2009). The chemicals use to enhance the properties of the plastic are released back into the environment making it more harmful to the health of living organisms (Kakuta, 2008). The large quantity of the synthetic chemicals are (>82,000) are available for use but most of them are untested and from the tested one, few chemical

Figure 2. "Cumulative plastic waste generation and disposal (in million metric tons)". Solid lines show historical data from 1950 to 2015; dashed lines show projections of historical trends to 2050" (Geyer et al., 2017)

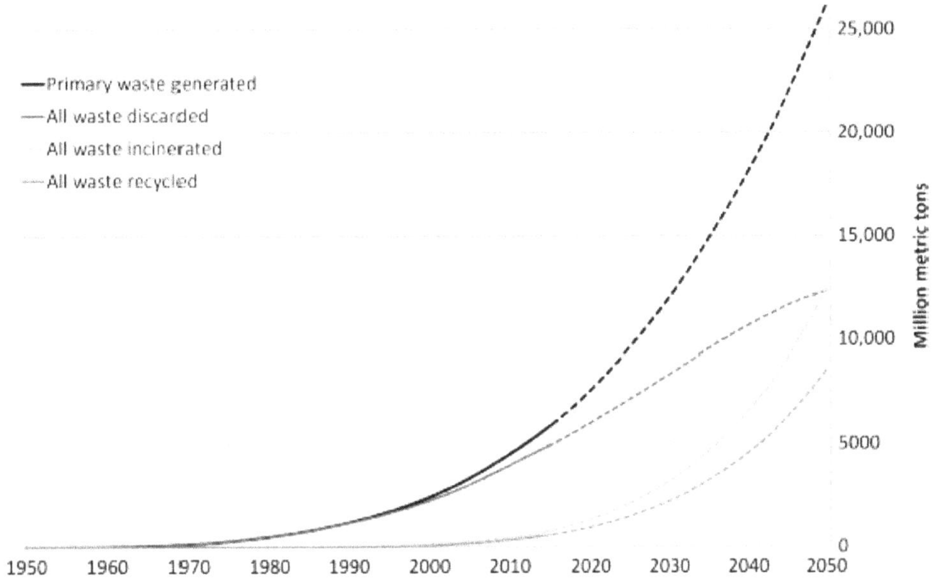

are totally banned for manufacturing (Landrigan et al., 2002). BPA and phthalates having different uses can easily leach to the food chain and can directly affect human health (Adibi, et al., 2003; Anastas & Beach, 2007). The leaching of these chemicals increases with the age of the plastic and also due to contact with fatty food on heating (Mato et al., 2001). The burning of plastics releases various harmful gases like carbon monoxide gas (CO) and carbon dioxide (CO_2) which have serious impact on the human health (Moore, 2008). During the incineration of PVC, colorless gas and strong poisonous gas called 'Phosgene' is produced having serious effects on human health (Waters et al., 2016). The combustion of the plastic waste also results in the release of ashes and soot which ultimately settle on the plants or on the surface of water bodies or on the ground surface thereby effecting the almost all the component of the environment (Waggoner, 2006). The functioning of the aquatic ecosystem is disturbed as some of the pollutant released react directly with the water and results in the change in pH of the system (Mato et al., 2001). The compounds generated by the incineration of PVC and its harmful health effects are presented in the following Table 2

PLASTIC IN OCEANS

The improper management of the waste has resulted in the accumulation of the plastic waste in the oceans. The major quantity of the waste accumulated in the ocean is the plastic followed by metals, glass and paper (Galgani et al., 2010). The accumulation of the plastic waste in ocean is considered global problem by United Nations Environment Assembly (Gesamp, 2016; Werner et al., 2016). The largest sector consuming plastic in Europe currently (Plastics Europe, 2016) is single-use packaging (40%), followed by consumer goods (22%) construction materials (20%) and automotive (9%), electrical (6%),

Table 2. Generated compounds and their harmful effects during the incineration of PVC (Pal, 1990)

Name of Compound	Health Effects
'Acetaldehyde"	"It damages the nervous system, causing lesions"
"Acetone"	"Irritates the eyes, the respiratory tract"
"Benzaldehyde"	"Irritates the eyes, skin, respiratory system, limits brain function"
"Benzole"	"Carcinogenic, adversely affects the bone marrow, the liver, the immune system"
"Formaldehyde"	"Serious eye damage, carcinogenic, may cause pulmonary oedema"
"Phosgene"	"The gas used in the WWI. Corrosive to the eyes, skin and respiratory organs"
"Polychlorinated dibenzo-dioxin"	"Carcinogenicirritate the skin, eyes and respiratory system. It damages the circulatory, digestive and nervous system, liver, bone marrow"
"Polychlorinated Dibenzofuran"	"Irritates the eyes and the respiratory system, causes asthma"
"Hydrochloric acid"	"Corrosive to the eyes, the skin and the respiratory tract"
"Salicyl-aldehyde"	"Irritates the eyes, the skin, and the respiratory tract. It can also affect the central nervous system"
"Toluene"	"Irritates the eyes and the respiratory tract, can cause depression"
"Xylene"	"Irritates the eyes. It can also affect the central nervous system, reduces the level of consciousness and impairs learning ability"
"Propylene"	"Damages the central nervous system by lowering of consciousness"
"Vinyl chloride"	"Carcinogenic, irritating to eyes, skin and respiratory system. Effect on the central nervous system, liver, spleen, blood-forming organs"

and agriculture applications (3%). The continuous rise in the use of plastic is a serious issue as use of plastic is expected to grow 4% annually (Plastics Europe, 2016). The annual production of the plastic in the world is comparable to the average weight (45 Kg) of the human population on the earth. Almost half (49%) of the plastic production was done in Asia with China being the world's largest producer (28%), followed by Europe and North America, each contributing 19% (Figure 3). The non-degradable nature and improper management of the plastic has resulted in the entering into the ocean from land-based sources, often via rivers, wastewater outflows, and transport by wind or tides (Jambeck et al., 2015). The constant increase in plastic production followed by a continuous supply of waste in oceans is expected to continue the long-term accumulation of plastic in marine ecosystems. In 2010, the amount of plastic waste (4.8-12.7 Mt) (Jambeck et al., 2015), released into the marine environment is estimated to be equal to the dumping of garbage truck of plastic every minute (Newscentre, 2017). The difference in the accumulation of plastic in different countries is dependent on several factors like coastal population density, quantity of plastic produced annually and management practices. The major proportion of the waste in the ocean comes from the coastline of China and United States (Jambeck et al., 2015). The enormous quantity of waste generated is not managed properly as in individual countries about 2-90% of the waste is mismanaged and 2-25% of mismanagement of waste is contributed by plastic (Jambeck et al., 2015). In India and US, coastal areas of the both countries is covered by almost same population (188 and 113 million people) and generation of waste per person is more in US compared to India but

Figure 3. "Spatial patterns of plastic production and pollution. Shown are the percentage contributions of different regions to global plastic production and the estimated mass of mismanaged plastic waste in million tons (Mt) generated in 2010 by populations living within 50 km of the coast" (Jambeck et al., 2015)

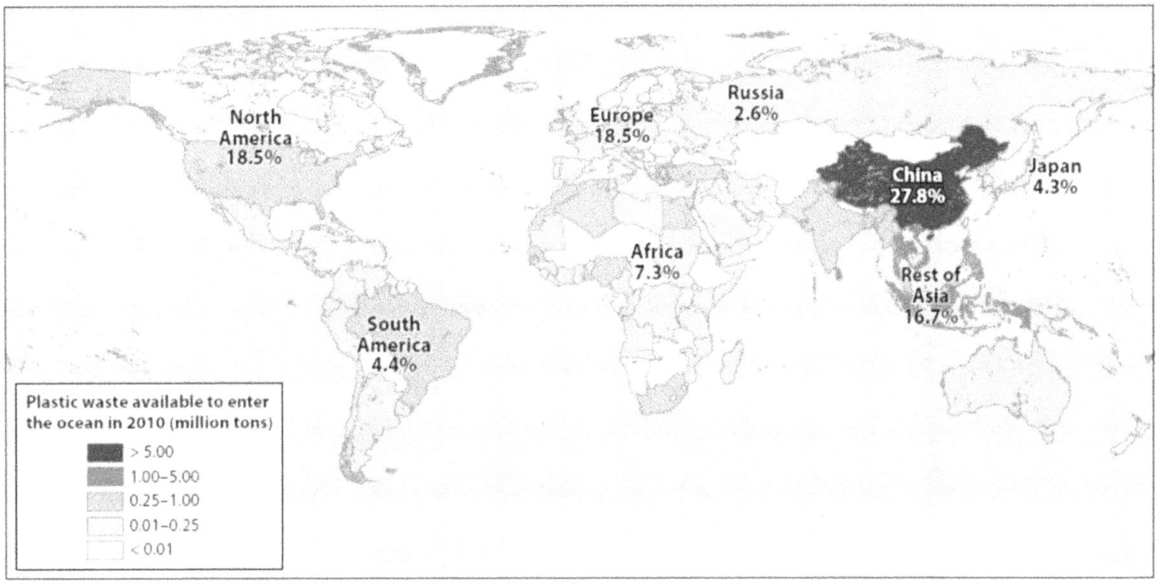

mismanagement of the waste is more in India (88%) compared to US (2%). Due to the improper management of the waste, India contributes more in the marine plastic pollution compared to US (Jambeck et al., 2015). The majority of the marine plastic pollution (44%) is contributed by three countries namely China, Indonesia and the Philippines as coastal areas are mostly inhibited by peoples, rapidly increasing consumption of plastics and following poor waste management practices. Jambeck et al., (2015) predicted a three-fold increase in the number of plastics in the ocean between 2015 and 2025 (Figure 4).

EFFECTS OF MICROPLASTICS AND NANO-PLASTIC ON MARINE ORGANISMS

Microplastic has large surface area and has capability to cross cellular boundaries thereby effecting the various functioning within the body of the living organisms (Gesamp, 2016). The entering of the microplastic in the body effects the growth and development as it directly effects feeding habits and reproduction ability of the organism. Scientists have predicted that microplastic may alter the functioning by effecting the growth and development of the organisms (Jambeck et al., 2015). Some animals have capacity to carry out carry carbon to the different ocean depths but increasing plastic accumulation can have impact on such capability of the organism thus effects the balance of the marine ecosystem (Thompson et al., 2009).

"Nanoplastics" are the fragment of the plastic having dimension smaller than 100 nm and can be present in marine water as single particles or aggregates of particles. The impact of nanoplastic on the health of different organism is very scarce and need more extensive research to know about the impact

Figure 4. "The estimated mass of plastic waste that enters the ocean because of inadequate waste management from populations living within 50 km of coast in 192 countries, plotted as a cumulative sum from 2010 to 2025. Estimates reflect assumed conversion rates of mismanaged plastic waste to marine debris (high, 40 percent; mid, 25 percent; low, 15 percent)" (Jambeck et al., 2015)

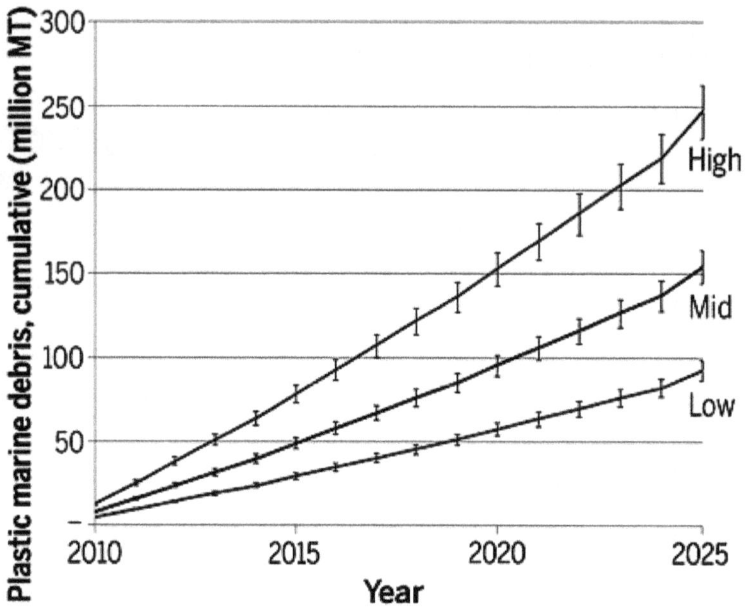

Figure 5. "Potential sources of microplastics to the marine environment" (Gesamp, 2016)

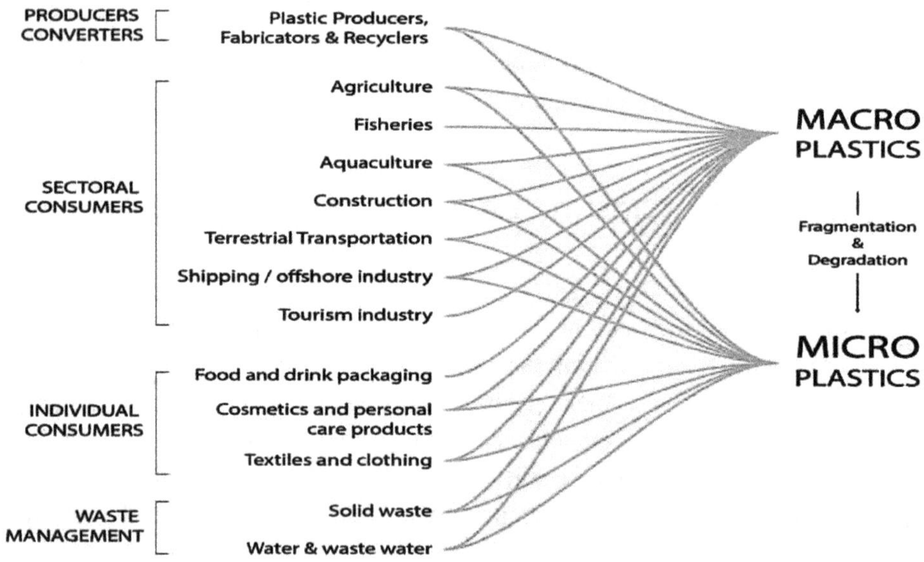

on the various bodies functioning of the organisms (Canesi et al., 2015). As per laboratory experiment, nanoparticles caused behavioral disorder as nanoplastic were able to pass from algae to zooplankton and then to fish in the food web (Mattson et al., 2017). During whole lifetime, animals are exposed to nanoplastic although in very small amount and are known to show synergistic effect (Canesi et al., 2015) with different toxins like chromium toxicity in fish increased in the presence of microplastic (Luıs et al., 2015).

PLASTICS AND THE ENVIRONMENT

Impact of plastic pollution on the environment is now recognized global issue (Meeker et al., 2009; Oehlmann et al., 2009). The problem of plastic pollution is rising as almost $4/5^{th}$ of the waste which accumulates is the plastic waste (Barnes et al., 2009). The different properties and uses like in paints, cable covering (Andrady & Neal, 2009; Koch & Calafat, 2009; Meeker et al., 2009; Oehlmann et al., 2009; Talsness et al., 2009; Wagner & Oehlmann 2009) has made plastic as one of the important part of human life but pose serious threats after use to living organisms as well as to their environment (Thompson et al., 2009). Plastic being cheap and easily available has numerous uses like in food industry can be used to pack different kinds of products which has significantly increased the quantity of the plastic waste production (ECOSS, 2014). The increased use of the plastic bags has made them one of the toxic pollutants in recent decades as plastic are made up of toxic chemicals which results in the pollution of air, water and soil (ECOSS, 2014). The chemicals used for the production has serious impact as some of them are carcinogenic ("benzene and vinyl chloride") while the gases ("ethylene oxide, benzene, and xylenes") (Swan et al., 2005; Swan 2008) released on the combustion of the plastic are responsible for the degradation of the environment (Rudel et al., 2001,2003; Oehlmann et al., 2009).The multiple uses of the plastic has resulted in the enormous production of the plastic waste which often get mixed with the general waste and cause problems in the management of the waste (UNEP, 2009; Kaewlue et al., 2012). Plastic known for its stability and durability (ECOSS, 2014) does not undergo degradation and disposing to landfill sites means allowing toxic pollutants to accumulate forever (UNEP, 2009; Kaewlue et al., 2012) in the water and soil of the surrounding area.

The rate of degradation of the plastic is very slow and took years to degrade in favorable conditions (Massardier-Nageotte et al., 2006) like high humidity and high oxygen supply (Allen et al., 1988). The plastic bottles can persist in the wet soil between 35-180 years as plastic bottle loses viscosity with time and finally disintegrate into micro-plastic (Allen et al., 1994; Edge et al., 1991). In marine environment, the degradation of the plastic starts in few weeks (Kedzierski et al., 2018) but took few decades or centuries or more for total degradation of the plastic waste due to presence of low temperature, low oxygen concentration and absence of sunlight (Barnes et al., 2009; Ioakeimidis et al., 2016). The materials like wood or pollen have high degradation rate but can be preserved for numerous years under such conditions (Blanchette et al., 1991; Boswijk et al., 2006; Haberle & Maslin, 1999).

ENVIRONMENTAL IMPACTS

Marine environment contaminated with plastic affects the flora and fauna of the marine ecosystem (Werner et al., 2016). Plastic accumulation in the marine ecosystem results in physical damage to the various organisms as direct ingestion of the plastic by various organism like turtle by mistake as food source (Derraik, 2002). The fragile marine ecosystem is also disturbed by the continuous accumulation of the plastic waste at the bottom and removal of debris certainly disturbs the balance of the marine ecosystem (Goldberg, 1994; Kanehiro et al., 1995). Human are also indirectly affected as ingestion of plastic by marine organism will get accumulated in the human body feeding on marine organisms for food (Mato et al., 2001). The persistent nature of the plastic demands strong steps for management of the plastic waste otherwise the frequency of the impact will increase manifolds. Thus, it is important not to disturb the balance of marine ecosystem so that the various species can be conserved, and the only available solution is to shun the production as well as use of plastic (Kanehiro et al., 1995).

Impacts on Marine Wildlife

The increase in the development in science has resulted in the production of the product which has many benefits for the human population but the impacts on the environment are not taken into consideration (Holmes et al., 2012). With the increase of the population, the use of the plastic is also increasing but absence of the management practice has led to the accumulation of the plastic in marine ecosystem (Browne et al., 2015). The accumulation of the plastic by the birds has increased with the quantity of the waste accumulation in the marine ecosystem. The marine plastic debris is affecting the number of species (267 species) throughout the world including 86% of all sea turtle species, 44% of all seabird species and 43% of all marine mammal species (Derraik, 2002). Seabed debris consists of plastic (80–85%) that are buoyant and buoyant debris can inhibit the gas exchange process between bottom and top surface of the marine ecosystem thus results in the creation of the hypoxia or anoxia condition (Goldberg, 1994; Kanehiro et al., 1995).The impact varies with the size and type of the debris and also varies with the different levels of organization in different habitat (Browne et al., 2015; UNEP, 2016; Werner et al., 2016). The ingestion of the plastic by various species of the marine environment increases if the size of the particle is in the range of < 5mm. As per studies conducted in laboratory, microplastic can easily transfer from prey to the predator (Watts et al., 2015) and can also facilitate in the transfer of toxic chemicals (Holmes et al., 2012; Rochman & Browne, 2013; Rochman et al., 2013).

Impacts on Sea Turtles

Almost all the species of sea turtle are prone to plastic pollution in the marine environment as turtle are harmed by feeding on plastic. Sea turtle feed mainly on jelly fish but floating plastic bags seem them like jellyfish and feeding on plastic bags results in the accumulation of plastic in their body (Mrosovsky et al., 2009). Sublethal impacts of plastics on sea turtles can be substantial, yet mortality resulting from interactions with plastic is much more difficult to quantify.

Ingestion of Plastic

The different organisms like seabirds, sea turtles, fish and mammals living in the marine environment ingest plastic by mistake for food (Thompson et al., 2009). For example, whales and sea turtles habitually mistake plastic bags for squid and birds usually mistake plastic pellets for fish eggs (Lee et al., 2001; Moore, 2002). A study conducted on 38 green turtles found that more than half (61%) of the turtles swallowed the accumulated debris like plastic bags, cloth and rope (Bugoni et al., 2001). Another study conducted in USA observed that 568 out of 1033 species were found to have plastic material in their body (Moser & Lee, 1992). The marine plastic debris contain some chemical (Moore, 2002) which can cause reproductive failure and even result in the death of the organism. The plastic material in the body of seabird is directly due to their scavenging strategies and diet (Azzarello &Van-Vleet, 1987; Ryan, 1987a; Moser & Lee, 1992; Laist, 1997). Plastic debris ingested by the marine organisms can be the way for entrance of PCBs into the food chain (Carpenter & Smith, 1972; Carpenter et al., 1972; Rothstein, 1973; Zitko & Hanlon, 1991; Mato et al., 2001) and even in low concentration can poses a serious risk to the life of different organisms by changing the level of hormones in the body or directly attacking reproductive organs (Ryan et al., 1988; Lee et al., 2001).

INDIRECT ENVIRONMENTAL IMPACTS

The impact is not restricted to animal only, but plants and other immobile organism are also affected to great extent by plastic debris (Zitko & Hanlon, 1991; Mato et al., 2001). Coral reefs are typically spoiled by dilapidated fishing equipment that breaks or suffocates coral. Plants can be smothered by plastic bags and fishing nets. The ocean floor ecosystems can be damaged and altered by the movement of an abandoned vessel or other marine debris (Lee et al., 2001).

Ecosystem Alteration

The removing of the plastic debris from the bottom of the marine ecosystem by using tractor can disturb the balance of marine flora and fauna. This technique of removal is usually unsafe for aquatic flora, birds, sea turtles, and other marine organisms. Erosion of shore-line and disruption of natural vegetation is a serious consequence of beach raking when conducted too close to a dune (Laist, 1997; Moser & Lee, 1992)

Invasive Species

Plastic debris from the marine environment can result in the transfer of the invasive species. The floating plastic debris carries dangerous species from one location to the other and result in the spreading of dangerous species (Barnes, 2002). In 2002, a study conducted by British Antarctic Survey estimated that due to the increased accumulation of the plastic waste in the marine environment, the number of the species doubled in the subtropics (Barnes, 2002).

Economic Impacts

The main important component of our economy is directly affected by the plastic pollution in the marine environment. The main components are tourism, fishing and navigation.

Tourism

The impacts of plastic pollution in marine environment are loss of key species, decrease the quality of the water, fish population decreases, foul smell from the water and create problems in navigation (Figure 7) thereby results in the less tourist influx in the area (Browne et al., 2015; UNEP, 2016; Werner et al., 2016). However, evaluating individual socio-economic costs of litter is challenging because of limited understanding about impacts and as a consequence of the variety of approaches that can be used to place economic values on the natural environment. The cost of removing debris from the marine ecosystem is laborious, time consuming, costly and also poses risk to the survival of the marine organisms (UNEP, 2016; Werner et al., 2016). The enormous accumulation of the plastic waste results in foul smell and even sometime cause beach closure which is ultimately affecting the number of tourist coming in the area (Barnes, 2002).

Fishing

The most commonly affected due to plastic debris is the fisheries (Rothstein, 1973; Zitko & Hanlon, 1991; Mato et al., 2001). The enormous quantity of the waste accumulated prevents the movement of the fish as fishes may get trapped in the polythene bags (Mato et al., 2001). The chemical released in the marine environment by the plastic waste affect the reproductive potential and also cause feminizing in the fish. The vessels and fishing gear get damaged by the marine debris which leads to extra cost on the repairing and wastage of the days of fisherman without any output in the business (Werner et al., 2016)

Navigation

Floating marine debris is a navigational hazard that entangles propellers and clogs cooling water intake valves (Reinhard et al., 2012). A survey by Scottish trawlers obtained the data on most common material that get trapped in the fishing gear are ropes and plastic (Figures 5 and 6). The floating debris also prevents the smooth movement of the water boats thus consume time as well as fuel thereby adding extra financial burden on the boatmen (Browne et al., 2015; UNEP, 2016; Werner et al., 2016). Sometimes boats are also damaged by the marine debris and repairing of boats is expensive as well as time consuming. Repairing boats damaged by marine debris is both time-consuming and expensive. The impacts related to marine debris is going to increase with time as the management of the plastic waste is lacking the effective techniques necessary to deal the enormous quantity of plastic waste generation (Reinhard et al., 2012).

Figure 6. Impacts of the beach and marine litter on socio-economic activities (Reinhard et al., 2012)

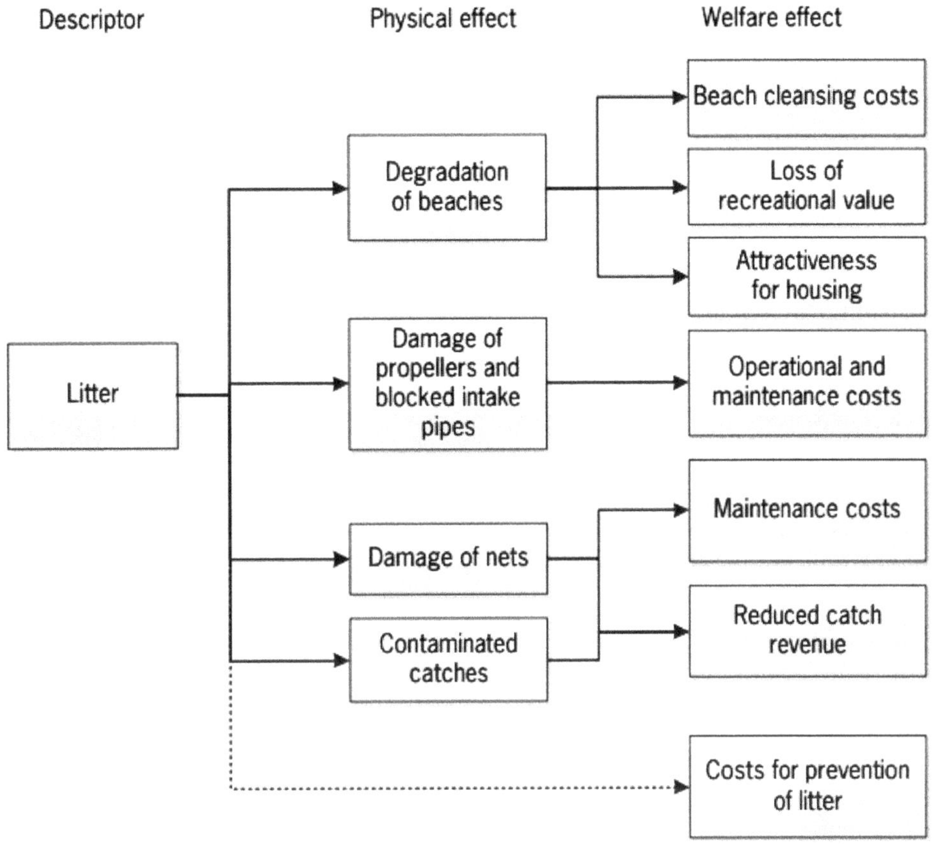

Figure 7. Potential impacts of marine litter on fisheries (Mouat et al., 2010)

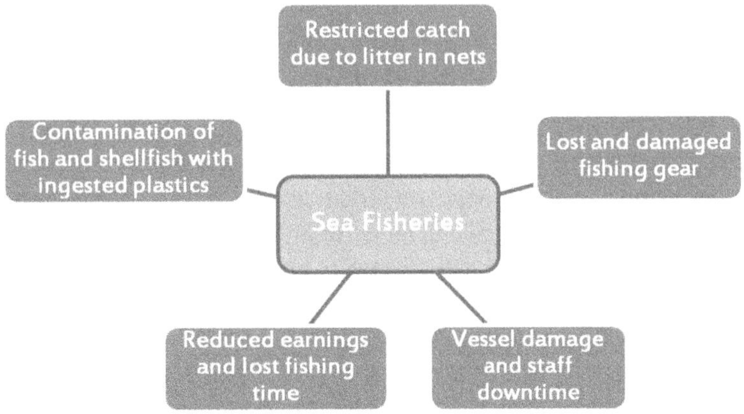

DISPOSAL AND WASTE MANAGEMENT SOLUTIONS

The consequences related to the plastic debris in marine environment are numerous but avoidable. The accumulation of plastic waste in marine environment can be prevented by improving the management techniques (WRAP, 2006, 2008; Defra, 2007). Disposal of plastic waste in improper manner has also increased with the use of disposal material. The plastic waste can also be managed by using the concept of 3 R's (Reduce, Reuse, Recycle) in order to reduce the enormous quantity of the waste generated (Hopewell et al., 2009). Energy recovery and molecular redesign are the emerging strategy for the management of the plastic waste hence instead of 3R's, 5R's (Reduce, Reuse, Recycle, Recover and Redesign) strategy can play efficient role in the management of the increasing plastic waste.

The life cycle analysis calculated that using recycled PET can help to reduce the emission of carbon dioxide by 27% compared to the use of virgin PET (WRAP, 2006, 2008; Defra, 2007). Recycling is easy for the products made up of single polymer compared to products made up of composite items. Public awareness and support is necessary to achieve good recycling targets as in some countries like in America and UK, public support for recycling is 80% and 57% respectively (Hopewell et al., 2009). The awareness and support for the recycling of the plastic waste is lacking in developing countries which results in the accumulation of plastic waste in almost every ecosystem.

Degradable Polymers

The hazard associated with the use of plastic has forced to look for the alternatives in place of oil-based plastic. The best alternative is the use of degradable polymers and their production has increased in the recent decades. Materials with functionality comparable to conventional plastics can now be produced on an industrial scale; they are more expensive than conventional polymers and account for less than 1 percent of plastics production (Song et al., 2009). Biopolymer is natural polymer as their source is biomass which is renewable compared to oil based polymer. The use of biopolymer is beneficial only if they are disposed though proper system by using their biodegradable features. Other recovery processes are following:

Applications of Recycled Thermoplastic Polymers

The waste generation is increasing with the each passing day but management practices are lacking far behind in the proper management of the waste (Bhattacharya et al., 2018). Different kinds of waste are produced but among them plastic waste can be recycled and various products can be obtained. The various applications of the plastic are presented in Table 3.

Recycled Polymers for Food Industry

The use of recycled polymer increases the chances of contamination which may occur during the process of production or during packaging. The contamination of these materials may be of a chemical or microbiological nature (Kolek et al., 2001). The issues related to the use of recycled plastic material raised by Food and Drug Administration (FDA) is:

Table 3. Applications of recycled plastics (Bhattacharya et al., 2018)

Plastics	Applications	References
"Polyethylene terephthalate" (PET)	"Water and soft drink bottles, food jar"	Siddique et al., 2008
Polyvinyl chloride (PVC)	"Cables, plumbing pipes"	Zare, 2017
"High-density polyethylene" (HDPE)	"Shampoo bottles, packaging"	Rahimi et al., 2017; Siddique et al., 2008
"Low-density polyethylene" (LDPE)	"Grocery bags, packaging"	Rahimiet al. 2017;Achilias et al., 2009
"Polypropylene" (PP)	"Bottle caps, medicine bottles, chips packs"	Hopewell et al., 2009; Siddique et al., 2008
"Polystyrene" (PS)	"Disposal cups, cutlery, packaging foam"	Gallop et al., 2009
"Polycarbonate" (PC)	"Food packaging, electronic goods, and defense Gadgets"	www.natureworksllc.com/~/media
"Nylon"	"Fishing nets, clothing, ropes"	www.natureworksllc.com/~/media

- Contaminants, pollutants and chemicals attached to the recycled material can get easily transferred to the food material thus can ultimately affects the health of the human beings (Siddique et al., 2008)

The following methods are recommended to avoid migration of potential contamination:

- Washing of plastic material
- Depolymerization of plastic (converting polymers into monomers)
- The migration of the contaminants can be prevented by using material having few layers in which the first layer if of recycled material and second one of pure material. The process of removing contaminants from the bottle is costly thus PET is kept on outer side of the bottle and pure plastic on inner side which acts as the barrier for the transfer of the contaminants (Kolek et al., 2001).

Recycled Polymers for Indoor Applications

The increase in the use of electrical appliances has increased the risk to the environment (Aizawa et al., 2008). So, to avoid these risks many countries have introduced recycling systems. Electronic waste includes used air conditioners, refrigerators, computers and cellular phones can be recycled and the rate of recycling of the electronic waste is increasing (> 70%) gradually (Aizawa et al., 2008; Bianchini et al., 2012).

Pyrolysis

Pyrolysis is the process which occurs in the absence of oxygen in order to break down complex polymer into simple one. It is thermal decomposition of material at temperature ranging between 300–400 °C in the presence of a catalyst (such as aluminum oxides, fly ash, red mud, and calcium hydroxide) (Miandad et al., 2016). The pyrolysis of the plastic waste results in the production of oil (45%–50%), gases (35%–40%) and tar (10%–20%) (Wong et al., 2015). The oil produced by the process of pyrolysis has similarity with that of conventional diesel. Under restricted condition, the yield of oil can be also increased approximately 80% (UNEP, 2009) Therefore, this may be an effective way to recycle plastic waste into fuels. Hence, the need is to develop the necessary facility so that huge quantity of the plastic waste generated can be converted into fuel at industrial as well as domestic level (Bhattacharya et al., 2018)

Gasification

Gasification is the thermo chemical decomposition of the plastic at high temperature. Gasification is very simple process and has gained attention as it involves air as gasification agent instead of oxygen (Vermeulen et al., 2001). In this process under controlled condition, the material is oxidized to produce syngas, which is a mixture of carbon monoxide and hydrogen with minor quantities of hydrocarbons (Scheirs, 1998; Vermeulen et al., 2001). The syngas produced can be used as an alternative in place of natural gas thus can be used for heating purpose, as a source of light and power generation (Bhattacharya et al., 2018)

Plastic Waste in Road Laying

The new concept of utilizing of the plastic for improving the quality of the roads is helpful in reducing the quantity of the plastic waste (Manju et al., 2017). Cost and the performance of the modifier determine the type of modifier to be used for the roads (Manju et al., 2017). The two procedure of modification namely dry and wet process are used for improving the quality of the roads. In dry process, plastic is firstly mixed with aggregates and then added to bitumen while plastic and bitumen is blended at the same time in wet process (Bhattacharya et al., 2018). The utilization of the plastic waste for the modification of the roads involves segregation and then shredding of the plastic waste in to the size of 2-4 mm. The waste after shredding is added to aggregate and the bitumen is heated to 160 °C in order to get good biding. Jambulingam Street in Chennai was one of India's first plastic roads built in 2002 (Manju et al., 2017).. Thus, utilization of the plastic for the construction of roads is helpful in reducing the quantity of the waste going for the treatment and is also helpful in improving the quality of the road.

Co-Processing of Plastic

Co-processing refers to the use of waste materials as an alternative fuel or raw material in industrial processes such as cement plants (Manju et al., 2017). In cement industry, the source of energy for the manufacturing of cement is provided by fossil fuels but using waste as a source of energy depends upon number of factors like composition of the waste, moisture content of the waste and economic viability (Scheirs, 1998; Vermeulen et al., 2001). The waste material received cannot be used due to its mixed nature and needs processing in order to maintain the necessary specification (calorific value, moisture content and heavy metal concentration) for the material to be used as an alternative source of energy (Bhattacharya et al., 2018)

Number of Times to be Recycled

Reusing and recycling the material again and again is helpful in increasing the life of the plastic material but can be done for limited times. Plastic cannot be recycled after crossing the optimum limit (7-9 recycling) and in some polymers, only 1-2 time recycling can be done before converting into lesser value product. The items such as clothing, fleece or even lumber cannot be recycled and may lastly end up in a landfill (Bhattacharya et al., 2018).

Bacteria and Worms That Can Degrade Plastic

Micro-organisms (bacteria, algae or fungi) may use plastic as carbon source by breaking the polymer chains. Microbial communities break the polymer chain by forming biofilms on the surface of plastic debris (Zettler et al., 2013). Some species secrete enzymes which are helpful in the cleavage of the polymer chains thus by breaking chains, the process of plastic degradation starts (Shah et al., 2008). On breaking of the plastic into smaller fragments by the process of mechanical and photo-degradation processes (Albertsson et al., 1980), the smaller fragment are easily attacked as compared to the larger one (Kawai et al., 2004). For example, the degradation of the polyethylene films increased on inoculating *Streptomyces* sp. as compared to uninoculated one (Lee et al., 1991). Similarly, the plastic waste on inoculation with some fungal increases the degradation as well as biomass accumulation in the soil (Orhan et al., 2000). Some macrofauna like wax moth (*Galleria mellonella*) can breakdown and also digest the polyethylene plastic material. may possess the ability to break up and digest certain plastic materials. The larvae of *Galleria mellonella*, can eat polyethylene, which along polypropylene is the main type of plastic found in waste (Bombelli et al., 2017). In 2016, the team of scientist from Japan identified wild bacteria (*Ideonellasa kaiensis*) which can feed on PET (polyethylene terephthalate) mostly used in the manufacturing of bottles. Therefore, more research is needed to search such organism which have the capacity to feed on the plastic thus can be helpful in the management of plastic waste (O'Brine et al., 2010).

CONCLUSION

Plastics offer numerous benefits to the society but its continuous production with improper management is adding problems to the present as well as future generations. Plastic has now entered into our every system and has also disturbed the balance of these fragile ecosystems. The need is to look for alternative or to keep check on the production as well as the consumption of the plastic in the world. The hazards related to the plastic waste can be minimized by the combined efforts of citizens, civil societies and local government. The awareness camps should be organized at mass level so the maximum number of citizens can be reached in minimum time. The problem of the generation of the plastic waste is growing day by day and is becoming a global issue therefore, it become necessary to take timely measure in order to keep this growing problem under check.

REFERENCES

Achilias, D., Giannoulis, A., & Papageorgiou, G. (2009). Recycling of polymers from plastic packaging materials using the dissolution–reprecipitation technique. *Polymer Bulletin*, *63*(3), 449–465. doi:10.100700289-009-0104-5

Adekomaya, O., & Ojo, K. (2016). Adaptation of plastic waste to energy development in lagos: An overview assessment. *Nigerian Journal of Technology*, *35*(4), 778–784. doi:10.4314/njt.v35i4.12

Adibi, J. J., Perera, F. P., Jedrychowski, W., Camann, D. E., Barr, D., Jacek, R., & Whyatt, R. M. (2003). Prenatal exposures to phthalates among women in New York City and Krakow, Poland. *Environmental Health Perspectives*, *111*(14), 1719–1722. doi:10.1289/ehp.6235 PMID:14594621

Ahmed, S. A., & Ali, M. (2004). Partnerships for solid waste management in developing countries: Linking theories to realities. *Habitat International*, *28*(3), 467–479. doi:10.1016/S0197-3975(03)00044-4

Aizawa, H., Yoshida, H., & Sakai, S. I. (2008). Current results and future perspectives for Japanese recycling of home electrical appliances. *Resources, Conservation and Recycling*, *52*(12), 1399–1410. doi:10.1016/j.resconrec.2008.07.013

Albertsson, A. C. (1980). The shape of the biodegradation curve for low and high density polyethenes in prolonged series of experiments. *European Polymer Journal*, *16*(7), 623–630. doi:10.1016/0014-3057(80)90100-7

Albrecht, M. A., Evans, C. W., & Raston, C. L. (2006). Green chemistry and the health implications of nanoparticles. *Green Chemistry*, *8*(5), 417–432. doi:10.1039/b517131h

Allen, N. S., Edge, M., Appleyardt, J. R., Jewitt, T. S., & Rorie, C. V. (1988). Degradation of Historic Cellulose Triacetate Cinematographic Film: Influence of Various Film Parameters and Prediction of Archival Life. *The Journal of Photographic Science*, *36*(6), 194–198. doi:10.1080/00223638.1988.11736999

Allen, N. S., Edge, M., Mohammadian, M., & Jones, K. (1994). Physicochemical aspects of the environmental degradation of polyethylene terephthalate. *Polymer Degradation & Stability*, *43*(2), 229–237. doi:10.1016/0141-3910(94)90074-4

Ambulkar, A. R., & Shekdar, A. V. (2004). Prospects of biomethanation technology in the Indian context: A pragmatic approach. *Resources, Conservation and Recycling*, *40*(2), 111–128. doi:10.1016/S0921-3449(03)00037-5

American Chemistry Society. (1993). *The Bakelizer*. Washington, DC: ACS.

Amin, S., & Amin, M. (2011). Thermoplastic elastomeric (TPE) materials and their use in outdoor electrical insulation. *Reviews on Advanced Materials Science*, *29*, 15–30.

Anastas, P. T., & Beach, E. S. (2007). Green chemistry: The emergence of a transformative framework. *Green Chemistry Letters and Reviews*, *1*(1), 9–24. doi:10.1080/17518250701882441

Anastas, P. T., Bickart, P. H., & Kirchhoff, M. M. (2000). *Designing safer polymers*. New York, NY: John Wiley and Sons, Wiley-Interscience.

Anastas, P. T., & Crabtree, R. H. (Eds.). (2009). Handbook of green chemistry green catalysis.: Vol. I. *Homogenous catalysis.* John Wiley & Sons.

Anastas, P. T., & Warner, J. C. (1998). *Green chemistry: theory and practice.* Oxford, UK: Oxford University Press.

Andrady, A. L. (2011). Microplastics in the marine environment. *Marine Pollution Bulletin, 62*(8), 1596–1605. doi:10.1016/j.marpolbul.2011.05.030 PMID:21742351

Andrady, A. L. (2011). Microplastics in the marine environment. *Marine Pollution Bulletin, 62*(8), 1596–1605. doi:10.1016/j.marpolbul.2011.05.030 PMID:21742351

Andrady, A. L., & Neal, M. A. (2009). Applications and societal benefits of plastics Philos. *Trans. R. Soc. Lond. B Biol. Sci., 364*, 1977–1984.

Andrady, A. L., & Neal, M. A. (2009). Application and societal benefits of plastics. *Philosophical Transactions of the Royal Society of London. Series B, Biological Sciences, 364*(1526), 1977–1984. doi:10.1098/rstb.2008.0304 PMID:19528050

Baekeland, L. H. (1909). The synthesis, constitution, and uses of Bakelite. *Industrial & Engineering Chemistry, 1*(3), 149–161. doi:10.1021/ie50003a004

Barnes, D. K. A., Galgani, F., Thompson, R. C., & Barlaz, M. (2009). Accumulation and fragmentation of plastic debris in global environments. *Philosophical Transactions of the Royal Society of London. Series B, Biological Sciences, 364*(1526), 1985–1998. doi:10.1098/rstb.2008.0205 PMID:19528051

Bello, M., Lawan, M. K., Aluwong, T., & Sanusi, M. (2015). Management of slaughter houses in northern Nigeria and the safety of meat produced for human consumption. *Food Control, 49*, 34–39. doi:10.1016/j.foodcont.2013.09.007

Benachour, N., & Aris, A. (2009). Toxic Effects of Low Doses of Bisphenol A on Human Placental Cells. *Toxicology and Applied Pharmacology, 241*(3), 322–328. doi:10.1016/j.taap.2009.09.005 PMID:19769995

Bhat, R.A., Dar, G.A., Jehangir, A., Bhat, B.A., & Yousuf, A. R. (2012). Municipal solid waste generation and present scenario of waste management during Yatra season in Pahalgam: a tourist health resort of Kashmir valley. *International Journal of Current Research 4*, 4-9.

Bhat, R.A., Dar, S.A., Dar, D.A., & Dar, G.H. (2018). Municipal Solid Waste generation and current scenario of its management in India, *International Journal of Advance Research in science and Engineering, 7*(2), 419-431.

Bhat, R. A., Kamili, A. N., & Bandh, S. A. (2013). Characterisation and composition of municipal solid waste (MSW) generated in Yusmarg: A health resort of Kashmir valley. A Glance at the World. *Waste Management (New York, N.Y.), 33*, 774–777.

Bhat, R. A., Nazir, R., Ashraf, S., Ali, M., Bandh, S. A., & Kamili, A. N. (2014). Municipal solid waste generation rates and its management at Yusmarg forest ecosystem, a tourist resort in Kashmir. *Waste Management & Research, 32*(2), 165–169. doi:10.1177/0734242X13518089 PMID:24519231

Bianchini, F., & Hewage, K. (2012). How "green" are the green roofs? Lifecycle analysis of green roof materials. *Building and Environment*, *48*, 57–65. doi:10.1016/j.buildenv.2011.08.019

Blanchette, R. A., Cease, K. R., Abad, A., Koestler, R. J., Simpson, E., & Sams, G. K. (1991). An evaluation of different forms of deterioration found in archaeological wood. *International Biodeterioration*, *28*(1-4), 3–22. doi:10.1016/0265-3036(91)90030-U

Bombelli, P., Howe, C. J., & Bertocchini, F. (2017). Polyethylene bio-degradation by caterpillars of the wax moth *Galleria mellonella*. *Current Biology*, *27*(8), 292–293. doi:10.1016/j.cub.2017.02.060 PMID:28441558

Boswijk, G., Fowler, A., Lorrey, A., Palmer, J., & Ogden, J. (2006). Extension of the New Zealand kauri (*Agathis australis*) chronology to 1724 BC. *The Holocene*, *16*(2), 188–199. doi:10.1191/0959683606hl919rp

Browne, M. A., Dissanayake, A., Galloway, T. S., Lowe, D. M., & Thompson, R. C. (2008). Ingested Microscopic Plastic Translocates to the Circulatory System of the Mussel, *Mytilus edulis* (L.). *Environmental Science & Technology*, *42*(13), 5026–5031. doi:10.1021/es800249a PMID:18678044

Buenrostro, O., & Bocco, G. (2003). Solid waste management in municipalities in Mexico: Goals and perspectives. *Resources, Conservation and Recycling*, *39*(3), 251–263. doi:10.1016/S0921-3449(03)00031-4

Camill, P. (2010). Global Change. *Nature Education Knowledge*, *3*(10), 49.

Canesi, L., Ciacci, C., & Balbi, T. (2015). Interactive effects of nanoparticles with other contaminants in aquatic organisms: Friend or foe? *Marine Environmental Research*, *111*, 128–134. doi:10.1016/j.marenvres.2015.03.010 PMID:25842999

Carpenter, E. J., & Smith, K. L. Jr. (1972). Plastic on the Sargasso Sea surface. *Science*, *175*(4027), 1240–1241. doi:10.1126cience.175.4027.1240 PMID:5061243

Cheung, W. M., & Pachisia, V. (2015). Facilitating waste paper recycling and repurposing via cost modelling of machine failure, labour availability and waste quantity. *Resources, Conservation and Recycling*, *101*, 34–41. doi:10.1016/j.resconrec.2015.05.011

Cole, M., & Galloway, T. S. (2015). Ingestion of nanoplastics andmicroplastics by Pacific oyster larvae. *Environmental Science & Technology*, *49*(24), 14625–14632. doi:10.1021/acs.est.5b04099 PMID:26580574

Cole, M., Lindeque, P., Fileman, E., Halsband, C., & Galloway, T. S. (2015). The impact of polystyrene microplastics on feeding, function and fecundity in the marine copepod *Calanushel golandicus*. *Environmental Science & Technology*, *49*(2), 1130–1137. doi:10.1021/es504525u PMID:25563688

Cozar, A., Echevarría, F., Gonzalez-Gordillo, J. I., Irigoien, X., Úbeda, B., Hernandez-Leon, S., ... Duarte, C. M. (2014). Plastic debris in the open ocean. *Proceedings of the National Academy of Sciences of the United States of America*, *111*(28), 10239–10244. doi:10.1073/pnas.1314705111 PMID:24982135

da-Costa, J. P., Santos, P. S. M., Duarte, A. C., & Rocha-Santos, T. (2016). (Nanoplastics in the environment sources, fates and effects. *The Science of the Total Environment*, *566–567*, 15–26. doi:10.1016/j.scitotenv.2016.05.041 PMID:27213666

Deanin, R. D. (1975). Additives in plastics. *Environmental Health Perspectives*, *11*, 35–39. doi:10.1289/ehp.751135 PMID:1175566

Defra, W. S., & Hannan, M. (2006). Review of England's waste strategy. In Environmental report under the 'SEA' directive, (p. 96). London, UK: DEFRA.

Defra. (2007). *Waste strategy for England*. Norwich, UK: Department of Environment food and Rural Affairs, HMSO.

Drumright, R. E., Gruber, P. R., & Henton, D. E. (2000). Polylactic acid technology. *Advanced Materials*, *12*(23), 1841–1846. doi:10.1002/1521-4095(200012)12:23<1841::AID-ADMA1841>3.0.CO;2-E

Edge, M., Hayes, M., Mohammadian, M., Allen, N. S., Jewitt, T. S., Brems, K., & Jones, K. (1991). Aspects of polyethylene terephthalate degradation for archival life and environmental degradation. *Polymer Degradation & Stability*, *32*(2), 131–153. doi:10.1016/0141-3910(91)90047-U

EnCams. (2006). Litter segmentation 2006. Wigan, UK: Environmental Campaigns Limited (ENCAMS).

Eriksen, M., Lebreton, L. C., Carson, H. S., Thiel, M., Moore, C. J., Borerro, J. C., ... Reisser, J. (2014). Plastic pollution in the world's oceans:more than 5 trillion plastic pieces weighing over 250,000 tons afloat at sea. *PLoS One*, *9*(12), e111913. doi:10.1371/journal.pone.0111913 PMID:25494041

Europe, P. (2012). *Plastics—the Facts 2012. An Analysis of European Plastics Production, Demand and Waste Data for 2011*. Brussels: Plastic Eur.

Europe, P. (2015). *Plastics the Facts 2015: An Analysis of European Plastics Production, Demand and Waste Data, 30*. Brussels: Plastics Europe.

Plastics Europe. (2016). *Plastics - the Facts 2016. An analysis of European plastics production, demand and waste data*. Brussels, Belg.: Plast. Eur.

European Association of Plastics Recycling and Recovery. (2017). Available online: http://www.epro-plasticsrecycling.org/pages/75/epro_statistics

Franz, R., & Welle, F. (2003). Recycling packaging materials A2—Ahvenainen, Raija. In *Novel Food Packaging Techniques* (pp. 497–518). Cambridge, UK: Woodhead Publishing. doi:10.1533/9781855737020.4.497

Galgani, F., Fleet, D., van Franeker, J., Katsanevakis, S., Maes, T., Mouat, J., ... Janssen, C. (2010). Marine Strategy Framework Directive, Task Group 10 Report: Marine Litter. In N. Zampoukas (Ed.), *JRC Scientific and Technical Reports*. Ispra: European Comission Joint Research Centre.

Gall, S. C., & Thompson, R. C. (2015). The impact of debris on marine life. *Marine Pollution Bulletin*, *92*(1-2), 170–179. doi:10.1016/j.marpolbul.2014.12.041 PMID:25680883

Gallop, W. A., Evans, M.G., & Mithal, A.K. (2009). *Knife Having Superior Functionality and Appeal*. Academic Press.

Galloway, T.S., Lewis, C.N. (2016). *Marine microplastics spell big problems for future generations*. Academic Press.

Gesamp. (2016). Sources, Fate and Effects of Microplastics in the Marine Environment – Part Two of a Global Assessment. In *IMO/FAO/UNESCO-IOC/UNIDO/WO/IAEA/UN/UNEP/UNDP Joint Group of Experts on the Scientific Aspects of Marine Environmental Protection*. Academic Press.

Geyer, R., Jambeck, J. R., & Law, K. L. (2017). Production, use, and fate of all plastics ever made. *Science Advances, 3*(7), e1700782. doi:10.1126ciadv.1700782 PMID:28776036

Giboz, J., Copponnex, T., & Mélé, P. (2007). Microinjection molding of thermoplastic polymers: A review. *Journal of Micromechanics and Microengineering, 17*(6), R96–R109. doi:10.1088/0960-1317/17/6/R02

Green, D. S. (2016). Effects of Microplastics on European Flat Oysters, Ostrea Edulis and their Associated Benthic Communities. *Environmental Pollution, 216*, 95–103. doi:10.1016/j.envpol.2016.05.043 PMID:27239693

Green, D. S., Boots, B., Blockley, D. J., Rocha, C., & Thompson, R. (2015). Impacts of Discarded Plastic Bags on Marine Assemblages and Ecosystem Functioning. *Environmental Science & Technology, 49*(9), 5380–5389. doi:10.1021/acs.est.5b00277 PMID:25822754

Green, D. S., Boots, B., O'Connor, N., & Thompson, R. C. (2017). Microplastics Affect the Ecological Functioning of an Important Biogenic Habitat. *Environmental Science & Technology, 51*(1), 68–77. doi:10.1021/acs.est.6b04496 PMID:27936642

Gregory, M. R. (2009). Environmental Implications of Plastic Debris in Marine Settings – Entanglement, Ingestion, Smothering, Hangers-on, Hitch-hiking, and Alien Invasions. *Philosophical Transactions of the Royal Society of London. Series B, Biological Sciences, 364*(1526), 2013–2026. doi:10.1098/rstb.2008.0265 PMID:19528053

Grósz, Z. (2003). Az ABV Védelemalapjai. Budapest: Zrínyi Miklós Nemeztvédelmi Egyetem, (Tankönyv).

Gupta, S., Mohan, K., Prasad, R., Gupta, S., & Kansal, A. (1998). Solid waste management in India: Options and opportunities. *Resources, Conservation and Recycling, 24*(2), 137–154. doi:10.1016/S0921-3449(98)00033-0

Haberle, S. G., & Maslin, M. A. (1999). Late Quaternary Vegetation and Climate Change in the Amazon Basin Based on a 50,000 Year Pollen Record from the Amazon Fan, ODP Site 932. *Quaternary Research, 51*(1), 27–38. doi:10.1006/qres.1998.2020

Halász, L., & Nagy, K. (2001). Mérgezőanyagokkémiája. Budapest: Zrínyi Miklós Nemeztvédelmi Egyetem, (Egyetemijegyzet)

Hartley, B. L., Pahl, S., & Thompson, R. C. (2013). Baseline Evaluation of Stakeholder Perceptions and Attitudes Towards Issues Surrounding Marine Litter, Deliverable D2.1 Report hazardous. *Nature, 494*, 169–171. PMID:23407523

Heap, B. (2009). Preface. *Philosophical Transactions of the Royal Society of London. Series B, Biological Sciences, 364*(1526), 1971–1971. doi:10.1098/rstb.2009.0030 PMID:19528048

Heskett, M., Takada, H., Yamashita, R., Yuyama, M., Ito, M., Geok, Y. B., ... Mermoz, J. (2012). Measurement of persistent organic pollutants (POPs) in plastic resin pellets from remote islands: Toward establishment of background concentrations for International Pellet Watch. *Marine Pollution Bulletin*, *64*(2), 445–448. doi:10.1016/j.marpolbul.2011.11.004 PMID:22137935

Hopewell, J., Dvorak, R., & Kosior, E. (2009). Plastics recycling: Challenges and opportunities. *Philosophical Transactions of the Royal Society of London. Series B, Biological Sciences*, *364*(1526), 2115–2126. doi:10.1098/rstb.2008.0311 PMID:19528059

Ioakeimidis, C., Fotopoulou, K. N., Karapanagioti, H. K., Geraga, M., Zeri, C., Papathanassiou, E., ... Papatheodorou, G. (2016). The degradation potential of PET bottles in the marine environment: An ATR-FTIR based approach. *Scientific Reports*, *6*(1), 1–8. doi:10.1038rep23501 PMID:27000994

Ioakeimidis, C., Zeri, C., Kaberi, H., Galatchi, M., Antoniadis, K., Streftaris, N., ... Papatheodorou, G. (2014). A comparative study of marine litter on the seafloor of coastal areas in the Eastern Mediterranean and Black Seas. *Marine Pollution Bulletin*, *89*(1-2), 296–304. doi:10.1016/j.marpolbul.2014.09.044 PMID:25440189

Isacsson, U., & Lu, X. (1995). Testing and Appraisal of Polymer Modified Road Bitumen. *Materials and Structures*, *28*(3), 139–159. doi:10.1007/BF02473221

Ismail, Z. Z., & Hashmi, E. A. A. L. (2008). Use of waste plastic in concrete mixture as aggregate replacement. *Waste Management (New York, N.Y.)*, *28*(11), 2041–2047. doi:10.1016/j.wasman.2007.08.023 PMID:17931848

Jambeck, J. R., Geyer, R., Wilcox, C., Siegler, T. R., Perryman, M., Andrady, A., ... Law, K. L. (2015). Plastic Waste Inputs from Land into the Ocean. *Science*, *347*(6223), 768–771. doi:10.1126cience.1260352 PMID:25678662

Jobling, S., Reynolds, T., White, R., Parker, M. G., & Sumpter, J. P. (1995). A Variety of Environmentally Persistent Chemicals, Including Some Phthalate Plasticizers, are Weakly Estrogenic. *Environmental Health Perspectives*, *10*(6), 582–587. doi:10.1289/ehp.95103582 PMID:7556011

Joint Group of Experts on the Scientific Aspects of Marine Environmental Protection (GESAMP). (Ed.). (2015). *Sources, Fate and Effects of Microplastics in the Marine Environment: A Global Assessment*. London: Int. Mar. Org.

Joint Group of Experts on the Scientific Aspects of Marine Environmental Protection (GESAMP). (Ed.). (2016). *Sources, Fate and Effects of Microplastics in the Marine Environment: Part Two of a Global Assessment*. London: Int. Mar. Org.

Kaewlue, W. (2012). *Solid Waste Management of Tha Khon Yang Municipality, Kantharawichai District, Maha Sarakham Province* (Master's Thesis). Mahasarakham University, Talat, Thailand.

Kakuta, Y., Hirano, K., Sugano, M., & Mashimo, K. (2008). Study on chlorine removal from mixture of waste plastics. *Waste Management (New York, N.Y.)*, *28*(3), 615–621. doi:10.1016/j.wasman.2006.12.023 PMID:17482803

Kawai, F., Watanabe, M., Shibata, M., Yokoyama, S., Sudate, Y., & Hayashi, S. (2004). Comparative study on biodegradability of polyethylene wax by bacteria and fungi. *Polymer Degradation & Stability, 86*(1), 105–114. doi:10.1016/j.polymdegradstab.2004.03.015

Kedzierski, M., D'Almeida, M., Magueresse, A., Le Grand, A., Duval, H., César, G., ... Le Tilly, V. (2018). Threat of plastic ageing in marine environment. Adsorption/desorption of micro pollutants. *Marine Pollution Bulletin, 127*, 684–694. doi:10.1016/j.marpolbul.2017.12.059 PMID:29475712

King, G. N., & King, H. W. (1986). Polymer Modified Asphalts: An Overview. American Society of Civil Engineers.

Koch, H. M., & Calafat, A. M. (2009). Human body burdens of chemicals used in plastic manufacture. *Philosophical Transactions of the Royal Society of London. Series B, Biological Sciences, 364*(1526), 2063–2078. doi:10.1098/rstb.2008.0208 PMID:19528056

Kolek, Z. (2001). Recycled polymers from food packaging in relation to environmental protection. *Polish Journal of Environmental Studies, 10*, 73–76.

Kuhn, S. B., Rebolledo, E.L., & van Franeker, J.A. (2015). Deleterious Effects of Litter on Marine Life. In Marine Anthropogenic Litter (pp. 75–116). Heidelberg, Germany: Springer.

Kumar, S., Bhattacharyya, J. K., Vaidya, A. N., Chakrabarti, T., Devotta, S., & Akolkar, A. B. (2009). Assessment of the status of municipal solid waste management in metro cities, state capitals, class I cities, and class II towns in India: An insight. *Waste Management (New York, N.Y.), 29*(2), 883–895. doi:10.1016/j.wasman.2008.04.011 PMID:18595684

Kumar, S., Dhar, H., Nair, V. V., Bhattacharyya, J. K., Vaidya, A. N., & Akolkar, A. B. (2016). Characterization of municipal solid waste in high-altitude sub-tropical regions. *Environmental Technology, 37*(20), 2627–2637. doi:10.1080/09593330.2016.1158322 PMID:26915419

Laist, D. W. (1997). Impacts of marine debris: entanglement of marine life in marine debris including a comprehensive list of species with entanglement and ingestion records. In J. M. Coe & D. B. Rogers (Eds.), *Marine Debris: Sources, Impacts, and Solutions* (pp. 99–140). New York: Springer-Verlag. doi:10.1007/978-1-4613-8486-1_10

Landrigan, P. J., Schechter, C. B., Lipton, J. M., Fahs, M. C., & Schwartz, J. (2002). Environmental pollutants and disease in American children: Estimates of morbidity, mortality, and costs for lead poisoning, asthma, cancer, and developmental disabilities. *Environmental Health Perspectives, 110*(7), 721–728. doi:10.1289/ehp.02110721 PMID:12117650

Lang, I. A., Galloway, T. S., Scarlett, A., Henley, W. E., Depledge, M., Wallace, R. B., & Melzer, D. (2008). Association of urinary bisphenol A concentration with medical disorders and laboratory abnormalities in adults. *Journal of the American Medical Association, 300*(11), 1303–1310. doi:10.1001/jama.300.11.1303 PMID:18799442

Law, K. L., Morét-Ferguson, S., Maximenko, N. A., Proskurowski, G., Peacock, E. E., Hafner, J., & Reddy, C. M. (2010). Plastic Accumulation in the North Atlantic Subtropical Gyre. *Science, 329*(5996), 1185–1188. doi:10.1126cience.1192321 PMID:20724586

Law, K. L., Morét-Ferguson, S. E., Goodwin, D. S., Zettler, E. R., DeForce, E., Kukulka, T., & Proskurowski, G. (2014). Distribution of Surface Plastic Debris in the Eastern Pacific Ocean from an 11-Year Data Set. *Environmental Science & Technology*, *48*(9), 4732–4738. doi:10.1021/es4053076 PMID:24708264

Lebreton, L. C. M., van der Zwet, J., Damsteeg, J. W., Slat, B., Andrady, A., & Reisser, J. (2017). River plastic emissions to the world's oceans. *Nature Communications*, *8*(1), 15611. doi:10.1038/ncomms15611 PMID:28589961

Lee, B., Pometto, A. L., Fratzke, A., & Bailey, T. B. (1991). Biodegradation of degradable plastic polyethylene by Phanerochaete and *Streptomyces* species. *Applied and Environmental Microbiology*, *57*, 678–685. PMID:16348434

Likens, G. E., & Bormann, F. H. (2004). Acid Rain: A Serious Regional Environmental Problem. *Science*, *184*(4142), 1176–1179. doi:10.1126cience.184.4142.1176 PMID:17756304

Lithner, D., Larsson, A., & Dave, G. (2011). Environmental and health hazard ranking and assessment of plastic polymers based on chemical composition. *The Science of the Total Environment*, *409*(18), 3309–3324. doi:10.1016/j.scitotenv.2011.04.038 PMID:21663944

Lobelle, D., & Cunliffe, M. (2011). Early microbial biofilm formation on marine plastic debris. *Marine Pollution Bulletin*, *62*(1), 197–200. doi:10.1016/j.marpolbul.2010.10.013 PMID:21093883

Lonnstedt, O. M., & Eklov, P. (2016). Environmentally relevant concentrations of microplastic particles influence larval fish ecology. *Science*, *352*(6290), 1213–1216. doi:10.1126cience.aad8828 PMID:27257256

Luıs, L. G., Ferreira, P., Fonte, E., Oliveira, M., & Guilhermino, L. (2015). Does the presence of microplastics influence the acute toxicity of chromium (VI) to early juveniles of the common goby (*Pomato schistusmicrops*)? A study with juveniles from two wild estuarine populations. *Aquatic Toxicology (Amsterdam, Netherlands)*, *164*, 163–174. doi:10.1016/j.aquatox.2015.04.018 PMID:26004740

Macfadyen, G., Huntington, T., & Cappell, R. (2009). Abandoned, Lost or Otherwise Discarded Fishing Gear. In *UNEP Regional Seas Reports and Studies, 115*. Rome: UNEP/FAO.

Madalina, E. G. R. (2017). Methods of Recycling. *Properties and Applications of Recycled Thermoplastic Polymers Recycling*, *2*, 24. doi:10.3390/recycling2040024

Manju, R., Sathya, S., & Sheema, K. (2017). Use of Plastic Waste in Bituminous Pavement. *International Journal of Chemtech Research*, *8*, 804–811.

Mato, Y., Isobe, T., Takada, H., Kanehiro, H., Ohtake, C., & Kaminuma, T. (2001). Plastic resin pellets as a transport medium for toxic chemicals in the marine environment. *Environmental Science & Technology*, *35*(2), 318–324. doi:10.1021/es0010498 PMID:11347604

Mattsson, K., Elyse, V., Malmendal, J. A., Linse, S., Hansson, L., & Cedervall, T. (2017). Brain damage and behavioural disorders in fish induced by plastic nanoparticles delivered through the food chain. *Scientific Reports*, *7*(1), 11452. doi:10.103841598-017-10813-0 PMID:28904346

McDermid, K. J., & McMullen, T. L. (2004). Quantitative analysis of small-plastic debris on beaches in the Hawaiian archipelago. *Marine Pollution Bulletin*, *48*(7-8), 790–794. doi:10.1016/j.marpolbul.2003.10.017 PMID:15041436

Meeker, J. D., Sathyanarayana, S., & Swan, S. H. (2009). Phthalates and other additives in plastics: Human exposure and associated health outcomes. *Philosophical Transactions of the Royal Society of London. Series B, Biological Sciences, 364*(1526), 2097–2113. doi:10.1098/rstb.2008.0268 PMID:19528058

Miandad, R., Rehan, M., Nizami, A., El-FetouhBarakat, M. A., & Ismail, I. M. (2016). The Energy and Value-Added Products from Pyrolysis of Waste Plastics. In Recycling of Solid Waste for Biofuels and Bio-chemicals. Environmental Footprints and Eco-design of Products and Processes, (pp. 333–355). Academic Press.

Michler, G. H., & Balta-Calleja, F. J. (2016). *Mechanical Properties of Polymers Based on Nanostructure and Morphology* (Vol. 71). Roca Raton, FL: CRC Press. doi:10.1201/9781420027136

Moore, C. J. (2008). Synthetic polymers in the marine environment: A rapidly increasing, long-term threat. *Environmental Research, 108*(2), 131–139. doi:10.1016/j.envres.2008.07.025 PMID:18949831

Moore, C. J., Moore, S. L., Leecaster, M. K., & Weisberg, S. B. (2001). A comparison of plastic and plankton in the North Pacific central gyre. *Marine Pollution Bulletin, 42*(12), 1297–1300. doi:10.1016/S0025-326X(01)00114-X PMID:11827116

Mouat, T., Lopez-Lozano, R., & Bateson, H. (2010). *Economic Impacts of Marine Litter*. KIMO.

Murphy, J. D., & McKeogh, E. (2004). Technical, economic and environmental analysis of energy production from municipal solid waste. *Renewable Energy, 29*(7), 1043–1057. doi:10.1016/j.renene.2003.12.002

Mutha, N. H., Patel, M., & Premnath, V. (2006). Plastics materials flow analysis for India. *Resources, Conservation and Recycling, 47*(3), 222–244. doi:10.1016/j.resconrec.2005.09.003

Nagy, A., & Kuti, R. (2016).. . *The Environmental Impact of Plastic Waste Incineration, 15*(3), 231–237.

Nelms, S. E., Coombes, C., Foster, L. C., Galloway, T. S., Godley, B. J., Lindeque, P. K., & Witt, M. J. (2017). Marine Anthropogenic Litter on British beaches: A 10-year Nationwide Assessment using Citizen Science Data. *The Science of the Total Environment, 579*, 1399–1409. doi:10.1016/j.scitotenv.2016.11.137 PMID:27913017

Nicholson, J. (2017). *The Chemistry of Polymers*. London, UK: Royal Society of Chemistry.

Nicolau, L., Marcalo, A., Ferreira, M., Sa, S., Vingada, J., & Eira, C. (2016). Ingestion of marine litter by loggerhead sea turtles, Carettacaretta, in Portuguese continental waters. *Marine Pollution Bulletin, 103*(1-2), 179–185. doi:10.1016/j.marpolbul.2015.12.021 PMID:26763321

O'Brine, T., & Thompson, R. C. (2010). Degradation of plastic carrier bags in the marine environment. *Marine Pollution Bulletin, 60*(12), 2279–2283. doi:10.1016/j.marpolbul.2010.08.005 PMID:20961585

Oehlmann, J., Schulte-Oehlmann, U., Kloas, W., Jagnytsch, O., Lutz, I., Kusk, K. O., ... Tyler, C. R. (2009). A critical analysis of the biological impacts of plasticizers on wildlife. *Philosophical Transactions of the Royal Society of London. Series B, Biological Sciences, 364*(1526), 2047–2062. doi:10.1098/rstb.2008.0242 PMID:19528055

Pavani & Rajeswari. (2014). Impact of Plastics On Environmental Pollution. *Journal of Chemical and Pharmaceutical Sciences*.

Plastic Coding Guidelines in the United States. (2017). Available online: http://www.natureworksllc.com/~/media/The_Ingeo_Journey/EndofLife_Options/plastic_codes/2008_04_10_plastic_code_guidelines_pdf.pdf

Porta, D., Milani, S., Lazzarino, A. I., Perucci, C. A., & Forastiere, F. (2009). Systematic review of epidemiological studies on health effects associated with management of solid waste. *Environmental Health*, *8*(1), 60. doi:10.1186/1476-069X-8-60 PMID:20030820

Porter, M. E., & Kramer, M. R. (2006). Strategy and society: The link between competitive advantage and corporate social responsibility. *Harvard Business Review*, *84*(12), 78–92. PMID:17183795

Rahimi, A., & García, J.M. (2017). Chemical recycling of waste plastics for new materials production. *Nat. Rev. Chem.,* 1.

Rillig, M. C. (2012). Microplastic in terrestrial ecosystems and the soil. *Environmental Science & Technology*, *46*(12), 6453–6454. doi:10.1021/es302011r PMID:22676039

Rudel, R. A., Brody, J. G., Spengler, J. C., Vallarino, J., Geno, P. W., Sun, G., & Yau, A. (2001). Identification of selected hormonally active agents and animal mammary carcinogens in commercial and residential air and dust samples. *Journal of the Air & Waste Management Association*, *51*(4), 499–513. doi:10.1080/10473289.2001.10464292 PMID:11321907

Rudel, R. A., Camann, D. E., Spengler, J. D., Korn, L. R., & Brody, J. G. (2003). Phthalates, alkylphenols, pesticides, polybrominated diphenyl ethers, and other endocrine-disrupting compounds in indoor air and dust. *Environmental Science & Technology*, *37*(20), 4543–4553. doi:10.1021/es0264596 PMID:14594359

Sangawar & Deshmukh. (2012). a short overview on development of the plastic waste management: environmental issues and challenges. *Sci. Revs. Chem. Commun., 2*(3), 349-354.

Scarlat, N., Motola, V., Dallemand, J. F., Monforti-Ferrario, F., & Mofor, L. (2015). Evaluation of energy potential of Municipal Solid Waste from African urban areas. *Renewable & Sustainable Energy Reviews*, *50*, 1269–1286. doi:10.1016/j.rser.2015.05.067

Scheirs, J. (1998). *Polymer Recycling*. New York: Wiley.

Schlosser, E., Nass, B., & Wanzke, W. (2003). *Flame Retardant Combination for Thermoplastic Polymers L.* U.S. Patent 6,547,992.

Schmidt, C., Krauth, T., & Wagner, S. (2017). Export of plastic debris by rivers into the sea. *Environmental Science & Technology*, *51*(21), 12246–12253. doi:10.1021/acs.est.7b02368 PMID:29019247

Shah, A. A., Hasan, F., Hameed, A., & Ahmed, S. (2008). Biological degradation of plastics: A comprehensive review. *Biotechnology Advances*, *26*(3), 246–265. doi:10.1016/j.biotechadv.2007.12.005 PMID:18337047

Shaw, D. G., & Day, R. H. (1994). Colour- and form-dependent loss of plastic micro-debris from the North Pacific Ocean. *Marine Pollution Bulletin*, *28*(1), 39–43. doi:10.1016/0025-326X(94)90184-8

Siddique, R., Khatib, J., & Kaur, I. (2008). Use of recycled plastic in concrete: A review. *Waste Management (New York, N.Y.)*, *28*(10), 1835–1852. doi:10.1016/j.wasman.2007.09.011 PMID:17981022

Song, J.H., Murphy, R.J., Narayan, R., & Davies, G.B.H. (2009). Biodegradable and compostable alternatives to conventional plastics. *Phil. Trans. R. Soc. B, 364*, 2127–.2139.

Staniskis, J. (2005). Integrated waste management: Concept and implementation. *Environmental Research, Engineering and Management, 33*(3), 40–46.

Sussarellu, R., Suquet, M., Thomas, Y., Lambert, C., Fabioux, C., Pernet, M. E. J., ... Huvet, A. (2016). Oyster Reproduction is Affected by Exposure to Polystyrene Microplastics. *Proceedings of the National Academy of Sciences of the United States of America, 113*(9), 2430–2435. doi:10.1073/pnas.1519019113 PMID:26831072

Swan, S. H. (2008). Environmental phthalate exposure in relation to reproductive outcomes and other health endpoints in humans. *Environmental Research, 108*(2), 177–184. doi:10.1016/j.envres.2008.08.007 PMID:18949837

Swan, S. H., Main, K. M., Liu, F., Stewart, S. L., Kruse, R. L., Calafat, A. M., ... Teague, J. L. (2005). Decrease in anogenital distance among male infants with prenatal phthalate exposure. *Environmental Health Perspectives, 113*(8), 1056–1061. doi:10.1289/ehp.8100 PMID:16079079

Talsness, C. E., Andrade, A. J. M., Kuriyama, S. N., Taylor, J. A., & vom Saal, F. S. (2009). Components of plastic: Experimental studies in animals and relevance for human health. *Philosophical Transactions of the Royal Society of London. Series B, Biological Sciences, 364*(1526), 2079–2096. doi:10.1098/rstb.2008.0281 PMID:19528057

Ter Halle, A., Jeannau, L., Martignac, M., Jarde, E., Pedrono, B., Brach, L., & Gigault, J. (2017). Nanoplastic in the North atlantic subtropical gyre. *Environmental Science & Technology, 51*(23), 13689–13697. doi:10.1021/acs.est.7b03667 PMID:29161030

Ter Halle, A., Ladirat, L., Gendre, X., Goudouneche, D., Pusineri, C., Routaboul, C., ... Perez, E. (2016). Understanding the fragmentation pattern of marine plastic debris. *Environmental Science & Technology, 50*(11), 5668–5675. doi:10.1021/acs.est.6b00594 PMID:27138466

Teuten, E., Rowland, S., Galloway, T., & Thompson, R. (2007). Potential for plastics to transport hydrophobic contaminants. *Environmental Science & Technology, 41*(22), 7759–7764. doi:10.1021/es071737s PMID:18075085

Teuten, E. L., Saquing, J. M., Knappe, D. R. U., Barlaz, M. A., Jonsson, S., Björn, A., ... Takada, H. (2009). Transport and release of chemicals from plastics to the environment and to wildlife. *Philosophical Transactions of the Royal Society of London. Series B, Biological Sciences, 364*(1526), 2027–2045. doi:10.1098/rstb.2008.0284 PMID:19528054

The Ecotourism and Conservation Society of Sikkim (ECOSS). (2014). *Plastics and the Environment Assessing the Impact of the Complete Ban on Plastic Carry Bag*. Retrieved from http://www.toxicslink.org

Thompson, R. C., Moore, C. J., vomSaal, S. F., & Swan, S. H. (2009). Plastics, the environment and human health: Current consensus and future trends. *Philosophical Transactions of the Royal Society of London. Series B, Biological Sciences, 364*(1526), 2153–2166. doi:10.1098/rstb.2009.0053 PMID:19528062

Thompson, R. C., Olsen, Y., Mitchell, R. P., Davis, A., Rowland, S. J., John, A. W. G., ... Russell, A. E. (2004). Lost at sea: Where is all the plastic? *Science*, *304*(5672), 838–838. doi:10.1126cience.1094559 PMID:15131299

Thompson, R. C., Swan, S. H., Moore, C. J., & vom Saal, F. S. (2009). Our plastic age. *Philosophical Transactions of the Royal Society of London. Series B, Biological Sciences*, *364*(1526), 1973–1976. doi:10.1098/rstb.2009.0054 PMID:19528049

Thompson, R. C., Swan, S. H., Moore, C. J., & Vomsaal, F. S. (2009b). Our plastic age. *Philosophical Transactions of the Royal Society of London. Series B, Biological Sciences*, *364*(1526), 1973–1976. doi:10.1098/rstb.2009.0054 PMID:19528049

Topçu, E. N., Tonay, A. M., Dede, A., Öztürk, A. A., & Öztürk, B. (2013). Origin and abundance of marine litter along sandy beaches of the Turkish Western Black Sea Coast. *Marine Environmental Research*, *85*, 21–28. doi:10.1016/j.marenvres.2012.12.006 PMID:23290790

Uddin, A., Koizumi, K., Murata, K., & Sakata, Y. (1997). Thermal and catalytic degradation of structurally different types of polyethylene into fuel oil. *Polymer Degradation & Stability*, *56*(1), 37–44. doi:10.1016/S0141-3910(96)00191-7

UNEP. (2016). *Marine Plastic Debris and Microplastics – Global Lessons and Research to Inspire Action and Guide Policy Change, 192*. Nairobi: United Nations Environment Programme.

United Nations Newscentre. (2017). *UN declares war on ocean plastic*. Retrieved from http://web.unep.org/newscentre/un-declares-war-ocean-plastic

United Nations Environment Programme (UNEP). (2009). *Converting Waste Plastics into Fuel: Report on Situation Analysis of Existing Solid Waste Management System for Bangkok Metropolitan Administration* (p. 40). Bangkok, Thailand: International Environmental Technology Centre.

United Nations Environmental Program. Converting Waste Plastic into a Resource. (2009). Retrieved from http://www.unep.or.jp/

Vergara, S. E., & Tchobanoglous, G. (2012). Municipal Solid Waste and the Environment: A Global Perspective. *Annual Review of Environment and Resources*, *37*(1), 277–309. doi:10.1146/annurev-environ-050511-122532

Vermeulen, I., Van Caneghem, J., Block, C., Baeyens, J., & Vandecasteele, C. (2011). 'Automotive Shredder Residue (ASR): Reviewing Its Productions from End-of-Life Vehicles (ELVs) and Its Recycling, Energy and Chemicals Valorization. *Journal of Hazardous Materials*, *190*(1-3), 8–27. doi:10.1016/j.jhazmat.2011.02.088 PMID:21440364

Villarrubia-Gómez, P., Cornell, S. E., & Fabres, J. (2018). Marine plastic pollution as a planetary boundary threat–The drifting piece in the sustainability. *Marine Policy*, *96*, 213–220. doi:10.1016/j.marpol.2017.11.035

von Moos, N., Burkhardt-Holm, P., & Köhler, A. (2012). Uptake and effects of microplastics on cells and tissue of the blue mussel *Mytilusedulis* L. after an experimental exposure. *Environmental Science & Technology*, *46*(20), 11327–11335. doi:10.1021/es302332w PMID:22963286

Waggoner, Miller, De Roo, & Henry. (2006). Plastic mulching: Principles and benefits. *Bulletin of the Connecticut Agricultural Experiment Station, 634*, 1–44.

Wagner, M., & Oehlmann, J. (2009). Endocrine disruptors in bottled mineral water: Total estrogenic burden and migration from plastic bottles. *Environmental Science and Pollution Research International, 16*(3), 278–286. doi:10.100711356-009-0107-7 PMID:19274472

Waters, C. N., Zalasiewicz, J., Summerhayes, C., Barnosky, A. D., Poirier, C., Ga uszka, A., ... Wolfe, A. P. (2016). The Anthropocene is functionally and stratigraphically distinct from the Holocene. *Science, 351*(6269), 137. doi:10.1126cience.aad2622 PMID:26744408

Watts, A. (2015). *Are we really "choking the ocean with plastic"? Tracing the creation of an eco-myth.* Available at: https://wattsupwiththat.com/2015/12/24/are-we-really-choking-theocean- with-plastic-tracing-the-creation-of-an-eco-myth/

Webb, H., Arnott, J., Crawford, R., & Ivanova, E. (2013). Plastic Degradation and Its Environmental Implications with Special Reference to Polyethylene terephthalate. *Polymers, 5*(1), 1–18. doi:10.3390/polym5010001

Werner, S., Budziak, A., van Franeker, J., Galgani, F., Hanke, G., Maes, T., ... Vlachogianni, T. (2016). *Harm Caused by Marine Litter. MSFD GES TG Marine Litter–Thematic Report. JRC Technical Report.* Luxembourg: European Union.

Wilcox, C., Hardesty, B. D., Sharples, R., Griffin, D. A., Lawson, T. J., & Gunn, R. (2013). Ghost net Impacts on Globally Threatened Turtles, a Spatial Risk Analysis for Northern Australia. *Conservation Letters, 6*(4), 247–254. doi:10.1111/conl.12001

Wohlleben, W., & Neubauer, N. (2016). Quantitative rates of release from weathered nanocomposites are determined across 5 orders of magnitude by the matrix, modulated by the embedded nanomaterial. *Nano Impact, 1*, 39–45.

Wong, S. L., Ngadi, N., Abdullah, T. A. T., & Inuw, I. M. (2015). Current State and Future Prospects of Plastic Waste as Source of Fuel: A Review'. *Renewable & Sustainable Energy Reviews, 50*, 1167–1180. doi:10.1016/j.rser.2015.04.063

Woodall, L. C., Sanchez-Vidal, A., Canals, M., Paterson, G. L., Coppock, R., Sleight, V., ... Thompson, R. C. (2014). The deep sea is a major sink for microplastic debris. *Royal Society Open Science, 1*(4), 140317. doi:10.1098/rsos.140317 PMID:26064573

Wrap. (2006). *Environmental benefits of recycling: an international review of life cycle comparisons for key materials in the UK recycling sector*. Banbury, UK: WRAP.

Wrap. (2008). *The carbon impact of bottling Australian wine in the UK: PET and glass bottles*. Banbury, UK: WRAP.

Wright, S. L., Rowe, D., Reid, M. J., Thomas, K. V., & Galloway, T. S. (2015). Bioaccumulation and biological effects of cigarette litter in marine worms. *Scientific Reports, 5*(1), 14119. doi:10.1038rep14119 PMID:26369692

Wright, S. L., Rowe, D., Thompson, R. C., & Galloway, T. S. (2013). Microplastic Ingestion Decreases Energy Reserves in Marine Worms. *Current Biology*, *23*(23), 1031–1033. doi:10.1016/j.cub.2013.10.068 PMID:24309274

Yamashita, R., & Tanimura, A. (2007). Floating plastic in the Kuroshio Current area, western North Pacific Ocean. *Marine Pollution Bulletin*, *54*(4), 485–488. doi:10.1016/j.marpolbul.2006.11.012 PMID:17275038

Yang, Y., Boom, R., Irion, B., van Heerden, D. J., Kuiper, P., & de Wit, H. (2012). Recycling of composite materials. Chem. Eng. Process. *Process Intensif.*, *51*, 53–68. doi:10.1016/j.cep.2011.09.007

Yarsley, V. E., & Couzens, E. G. (1945). *Plastics*. Middlesex, UK: Penguin Books Limited.

Yoshida, S., Hiraga, K., Takehana, T., Taniguchi, I., Yamaji, H., Maeda, Y., ... Oda, K. (2017). A bacterium that degrades and assimilates polyethylene terephthalate. *Science*, *351*(6278), 1196–1199. doi:10.1126cience.aad6359 PMID:26965627

Zare, Y. (2017). *3Recycled Polymers: Properties and Applications*. Available online: https://www.smithersrapra.com/ SmithersRapra/media/Sample-Chapters/Recycled-Polymers-Properties-and-Applications,-Volume-2.pdf

Zettler, E. R., Mincer, T. J., & Amaral-Zettler, L. A. (2013). Life in the "plastisphere": Microbial communities on plastic marine debris. *Environmental Science & Technology*, *47*(13), 7137–7146. doi:10.1021/es401288x PMID:23745679

Žmak & Hartmann. (2017). Current State Of The Plastic Waste Recycling System In The European Union And In Germany. *Technical Journal*, *11*(3), 138–142.

Zorrob, S. E., & Suparama, L. B. (2004). Laboratory Design and Investigation of Proportion of Bituminous Composite Containing Waste Recycled Plastics Aggregate Replacement (Plastiphalt). *Cement and Concrete Composites*, *22*(4), 233–242. doi:10.1016/S0958-9465(00)00026-3

Zubris, K. A. V., & Richards, B. K. (2005). Synthetic fibers as an indicator of land application of sludge. *Environ. Pollut.*, *138*(2), 201–211. doi:10.1016/j.envpol.2005.04.013 PMID:15967553

KEY TERMS AND DEFINITIONS

Pyrolysis: It is the thermal decomposition of materials at elevated temperatures in an inert atmosphere and involves a change of chemical composition and is irreversible.

Solid Waste: Solid or semisolid, non-soluble material (including gases and liquids in containers) such as agricultural refuse, demolition waste, industrial waste, mining residues, municipal garbage, and sewage sludge.

White Pollution: The plastic things viz., plastic bottles, plastic bags, plastic silverware and so on kills the animals and degrades the quality of the environment on the earth.

Chapter 5
Electronic Waste:
Implications on Environs and Management Strategies

Moonisa A. Dervash
Sher-e-Kashmir University of Agricultural Sciences and Technology, India

Syed Maqbool Geelani
Sher-e-Kashmir University of Agricultural Sciences and Technology, India

Rouf Ahmad Bhat
Sher-e-Kashmir University of Agricultural Sciences and Technology, India

Dig Vijay Singh
Sher-e-Kashmir University of Agricultural Sciences and Technology, India

Akhlaq Amin Wani
Sher-e-Kashmir University of Agricultural Sciences and Technology, India

ABSTRACT

Electronic waste (e-waste) is one of the swift waste streams and comprises of end of life electronic products. The Western countries as alternative destinations for disposal ship the wastes to underdeveloped and developing countries where labor cost is reasonably meager and environmental laws are feebly implemented. When not recycled, the e-waste is either incinerated or landfilled. These methods involve not only wasting valuable metals but also creating potential risk for the environment. These substances are detrimental to nervous system, kidneys, bones, reproductive system, and endocrine system, and some of them are even carcinogenic and neurotoxic. Thus, extensive research is needed to evolve sophisticated technology which may help to curb environmental pollution and contribute towards sustainable development in terms of recycling of precious metals.

DOI: 10.4018/978-1-7998-0031-6.ch005

INTRODUCTION

"Waste of electrical and electronic equipment; includes all the machinery, subassemblies and consumables which are part of the product at the time of discarding" (ED, 2012). E-waste encompasses multiple arrays of electronic appliances ranging from large household electronic gadgets (like refrigerators, AC's, cell phones, stereo systems and consumable electronic items) to computers discarded by their respective users (Kumar & Singh, 2019). The 'Organization for Economic Co-operation and Development' (OECD) defines E-waste as "any appliance using an electric power supply that has reached its end-of-life" (DTIE, 2007). In general, "domestic machinery represents the largest proportion of E-waste (50%), followed by information and communications technology equipment (30%) and consumer electronics (10%)". The E-waste characterization is reasonably diverse which varies throughout the scheme of products. Overall, it contains more than 1000 diverse substances which plunge into "hazardous" and "non-hazardous" groups. Considerably, the toxicity of majority of the chemicals in E-waste is indefinite (Kumar and Singh, 2019). Generally, "electronic products consist of ferrous and non-ferrous metals, plastics, glass, wood and plywood, printed circuit boards, concrete and ceramics, rubber and other items, whereas, iron and steel constitutes about 50% of E-waste followed by plastics (21%), nonferrous metals (13%) and other constituents" (DTIE, 2007). E-waste constitutes end of life electronic and electrical gadgets (Sankhla et al., 2016) which contains toxic elements and has to be treated in an environmentally friendly approach (Sankhla et al., 2016). E-waste comprises of "electrical gadgets such as fridges, air conditioners, washing machines, microwave ovens and florescent light bulbs; and electronic products such as computers and accessories, mobile phones, television sets and stereo equipment" (Radha, 2002). Data security concerns are also associated with E-waste handling and management in order to safeguard extraction of commerce, economics and legal data by unscrupulous recyclers; therefore, it should be handed over to recyclers/operators who have the mandate for E-waste management (Radha, 2002). Besides, E-waste is less energy intensive and cheaper source for base and precious metals like silver, gold, palladium and platinum (Lucier & Gareau, 2006) thus it is often referred to as urban mine. Moreover, as the demand for metals is growing (Lucier & Gareau, 2006) recycling would play a major part in ensuring sustainable development (Schluep *et al.*, 2009).

In the last 20 years, the worldwide development in electrical and electronic equipment production and consumption has been exponential (Schluep *et al.*, 2009). This is chiefly because of escalating marketplace dissemination of products in developing countries (DTIE, 2007); progression of a alternate marketplace in developed countries which is usually high product obsolescence rate (DTIE, 2007), coupled with a price drop and the drift towards internet access. Currently, E-waste is the fastest growing waste stream (4% growth a year) (Radha, 2002; DIT, 2003; Sankhla et al., 2016). Approximately 40×10^6 tonnes of E-waste is created each year (Schluep *et al.*, 2009). Environmentally accountable waste management strategies are hi-tech that entail high monetary speculation.

There is currently a high level of trans-boundary, often illegitimate, movement of E-waste into developing countries for affordable recycling (Strategic Approach to International Chemicals Management, SAICM, 2009). Trans-boundary movement of E-waste is principally profit oriented. Waste recyclers are taking due advantage of lower recycling expenditure in developing countries and at the same time avoiding disposal at home. It is estimated that up to 80% of all E-waste sent for recycling in developed countries ends up in informal E-waste recycling sites in developing countries, primarily in Africa and Asia. In receiving countries, crude and hazardous methods of recycling are used, jeopardizing people's health and the environment (Smith *et al.*, 2006). This raises the coherent issues pertaining to a dispropor-

Table 1. Valuable metals in electronic waste (Umicore Precious Metals Refining, 2007)

"Electronic"	"Copper (%)"	"Silver (%)"	"Gold (%)"	"Palladium (%)"
"TV Board"	"50"	"7"	"22"	"7"
"PC Board"	"18"	"5"	"61"	"15"
"Mobile Phone"	"9"	"13"	"64"	"14"
"DVD-player"	"42"	"5"	"32"	"5"

tionate burden of a global problem in developing countries, without having the technology to deal with it. Globalization of E-waste has adverse environmental and health implications as developing countries face economic challenges and lack the infrastructure for sound hazardous waste management, including recycling, or effective regulatory frameworks for hazardous waste management (SAICM, 2009).

Electronic products often contain several persistent, bio-accumulative and toxic substances including heavy metals such as lead, nickel, chromium and mercury, and persistent organic pollutants (POPs) such as polychlorinated biphenyls (PCBs) and brominated flame retardants (Bertram et al., 2002). The exigency of the problem is evident worldwide; as from 1994-2003, about 500 million personal computers containing approximately 718,000 tonnes of lead, 1,363 tonnes of cadmium and 287 tonnes of mercury, reached their end-of-life (Smith *et al.*, 2006). There is generally low public awareness of the hazardous nature of E-waste and the crude waste management techniques used in developing countries (Brigden et al., 2008). The focus on scientific temper, managerial services and associated software has sidetracked the public, who distinguish those employed chunk as "white collars workers" in the consumer electronics industry (Bertram et al., 2002). A supplementary but veiled aspect of E-waste stream is that the loss of rare metals present in E-waste has to be compensated by intensified mining activities; and it is well identified that the rapid augmentation in demand for raw materials used in electronic products has turned into mounting stimuli to conflicts over resources worldwide (Bertram et al., 2002). The E-waste recycling industry could play an imperative role in reducing such conflict risks by lowering the pressure on primary mining sites, such as those of gold, palladium and tantalum (Brigden et al., 2008).

"According to the EPA, one metric ton of circuit boards can contain 40 to 800 times the amount of gold, and 30 to 40 times the amount of copper mined from one metric ton of ore in the U.S".

THE INTERNATIONAL FLOW OF E-WASTE

- "E-waste recycling operations have been identified in numerous locations in China and India (Ni & Zeng, 2009)".
- "Less examined locations are in the Philippines, Nigeria (in the city of Lagos), Pakistan (Karachi) and Ghana (Accra) (Ni & Zeng, 2009)".
- "In general, small-scale exports are shipped to West Africa, whereas, the larger and sometimes more structurally organized transports are drifted to South East Asia".

Table 2. Recovery of precious metals from electronic waste (Umicore Precious Metals Refining, 2007)

"Potential Metal"	"Primary Mining (g/ton in ore)"	"Urban Mining (g/ton)"
"Au"	"5"	"200-250 in PC PWBs " "300-350 in cell phones"
"Cu"	"4,500-9,100"	"112,500-131,250 in cell phones"

- "It is estimated that China receives the highest proportion of all E-waste -about 70% and rising. However, there are no confirmed figures available on how substantial these trans-boundary E-waste streams are (Ni & Zeng, 2009)".
- "In addition, due to the recent impetus of regulations in Asia, it is estimated that more E-waste will flow into West African countries in the future".
- "Common Asian destinations for E-waste include China, India, Pakistan, Malaysia, the Philippines, Singapore, Sri Lanka, Thailand and Vietnam. Exporters of E-waste to China can avoid detection by routing container ships through Hong Kong, Taipei or the Philippines, and then transshipping them to smaller ports in mainland China, where customs officials are willing to look the other way in exchange for a share of profits. Dubai and Singapore often serve as transit points for E-waste from developed countries (Kalra, 2004)".

GENERATION OF E-WASTE

- "Generation of E-waste 4.1 million tonnes and expected to reach 8 Mt/yr by 2025 (Toxics Link, 2015)".
- "It is growing at the rate of 10% per annum and it constitutes 3-8% of municipal solid waste (Toxics Link, 2015)".
- "The global quantity of E-waste in 2014 was comprised of 1.0 Mt lamps, 3.0 Mt of Small IT, 6.3 Mt of screens and monitors, 7.0 Mt of temperature exchange equipment (cooling and freezing equipment), 11.8 Mt large equipment, and 12.8 Mt of small equipment (Toxics Link, 2015)".
- "The amount of E-waste is expected to grow to 49.8 Mt in 2018, with an annual growth rate of 4 to 5% (Toxics Link, 2015)".
- "The per inhabitant amount generated in 2015 was nearly as high as 15.2 kg/inh in Europe and lowest in Africa (1.7 kg/inh) (Kalra, 2004)".

Tables 3 to 6 show generation of E-waste in elaborative manner.

ENVIRONMENTAL IMPLICATIONS OF E-WASTE

Electronic products often contain several persistent, bio-accumulative and toxic substances including heavy metals such as lead, nickel, chromium and mercury, and persistent organic pollutants (POPs) such as polychlorinated biphenyls (PCBs) and brominated flame retardants (StEP WP Standard, 2014). The urgency of the problem is evident: worldwide, in the decade between 1994 and 2003, about 500 million

Table 3. Global quantity of e-waste generated

"Year"	"E-Waste Generated (MT/Year)"	"Population (Billion)"	"E-Waste Generated (Kg/Capita)"
"2010"	"33.8"	"6.8"	"5"
"2011"	"35.8"	"6.9"	"5.2"
"2012"	"37.8"	"6.9"	"5.4"
"2013"	"39.8"	"7.0"	"5.7"
"2014"	"41.8"	"7.1"	"5.9"
"2015"	"43.8"	"7.2"	"6.1"
"2016"	"45.7"	"7.3"	"6.3"
"2017"	"45.7"	"7.4"	"6.5"
"2018"	"49.8"	"7.4"	"6.7"

Source: http://www.toxicslink.org (2015)
Note: Figures 2015 onwards are forecasts

Table 4. E-waste generation in different continents (Toxiclink.org, 2015)

S.NO.	"E-Waste Generator"	"Quantity of E-Waste Generated"
1	"Asia"	"16 Mt"
2	"Europe"	"11.6 Mt"
3	"North America"	"7.9 Mt"
4	"South America"	"2.7 Mt"
5	"Africa"	"1.9 Mt"
6	"Central America"	"1.1 Mt"
7	"Oceania"	"0.6 Mt"

Table 5. Countrywise e-waste generation (toxiclink.org, 2015)

S. No.	"Country"	Kilotons/Year
1	"China"	"6033"
2	"Japan"	"2200"
3	"India"	"1641"
4	"United Kingdom"	"1511"
5	"Italy"	"1077"
6	"Spain"	"817"
7	"Norway"	"146"
8	"Finland"	"118"
9	"Kenya"	"44"
10	"Nepal"	"15"
11	"Switzerland"	"4"

Table 6. E-waste generation in India (Toxiclink.org, 2015)

S.No.	State	E-Waste Generated (Million tonnes per annum)	Metropolitan Cities and Others	E-Waste Generated (Million tonnes per annum)
1	"Maharashtra"	"20270.59"	"Mumbai"	"11017.1"
2	"Tamil Nadu"	"13486.24"	"Delhi"	"9729.15"
3	"Andhra Pradesh"	"12780.33"	"Bangaluru"	"4648.4"
4	"Uttar Pradesh"	"10381.11"	"Chennai"	"132.2"
5	"West Bengal"	"10059.36"	"Kolkata"	"4025.3
6	"Delhi"	"9729.15"	"Ahmadabad"	"3287.5"
7	"Karnataka"	"9118.74"	"Hyderabad"	"2833.5"
8	"Gujarat"	"8994.33"	"Pune"	"2584.2"
9	"Madhya Pradesh"	"7800.62"	"Surat"	"1836.5"
10	"Punjab"	"6958.46"	"Nagpur"	"1768.9"

Source: http://www.toxicslink.org (2015)

personal computers containing approximately 718,000 tonnes of lead, 1,363 tonnes of cadmium and 287 tonnes of mercury, reached their end-of-life (Sonnenfeld & Naguib, 2006). There is generally low public awareness of the hazardous nature of E-waste and the crude waste management techniques used in developing countries (UNEP, 1989).

Impact of E-Wastes on Atmosphere

Open burning of E wastes pollutes the air quality by releasing various quantitative as well as qualitative pollutants into the air (Kumar and Singh, 2019):

- "**Sulphur**: Found in lead-acid batteries, increases the problem of acid rain (Liu et al. 2008)".
- "**Chlorofluorocarbons** (CFC): used in E wastes in their Cooling units, insulation foam depletes Ozone layer (Liu et al. 2008)".
- "Dioxins and furans are released into the atmosphere from treatment facilities (incinerators) (ATSDR, 1995) which are potent environmental estrogens".

Impact of E-Wastes on Aquatic Systems

Contamination of surface water systems by surface run off is the primary problem of E-waste disposal followed by the contamination of ground water by leaching of toxic metals near E-waste landfill sites. The contamination of aquatic systems by E-waste disposal causes the change in ecological, physical, chemical and aesthetic characteristics of water (Kumar & Singh, 2019). Ecological bio-magnification of recalcitrant elements into aquatic food chain is also one of the catastrophic effects (Kumar and Singh, 2019).

Table 7.

"Chemical Complexes"	"Effects"
"Poly brominated diphenyl ethers, Poly brominated biphenyls"	"Under thermal treatment possible formation of dioxines and furanes- environmental estrogens"
"Polychlorinated biphenyls"	"Cancer, effects on the immune system, nervous system"
"Chlorofluorocarbons"	"Increased incidence of skin cancer / genetic damage"
"Americium"	"Radioactive element, mutagenic"
"Antimony"	"Carcinogenic potential"
"Arsenic"	"Skin diseases, decrease nerve conduction velocity, lung cancer"
"Barium"	"Brain swelling damage to the heart, liver and spleen"
"Cadmium"	"Lung cancer and kidney damage"
"Chromium VI"	"Irritation of eyes, skin and mucous membranes, DNA damage"
"Lead"	"Brain damage, Vomiting, diarrhea, convulsions, coma or even death, appetite loss and abdominal pain"
"Mercury"	"Brain and liver damage"

Table 8.

S.No	"Heavy Metals"	"Workshop Dust (mg/kg)"	"Roadside Dust (mg/kg)"
1	"Pb"	"110000"	"22600"
2	"Cu"	"8360"	"6170"
3	"Zn"	"4420"	"2370"
4	"Ni"	"1500"	"304"

Source: http://www.toxicslink.org (2014)

Table 9.

Heavy Metals Site	"Pb"	"Cd"	"Ni"	"Hg"	"Cr(VI)"	"Zn"
Ppm						
"LONI"	"710.07"	"2.88"	"24.48"	"0.85"	"6.18"	"212.49"
"MANDOLI"	"872.44"	"5.94"	"76.6"	"4.9"	"32.97"	"2291.66"

Source: http://www.toxicslink.org (2014)

Impact of E-Wastes on Soil

E-waste if not properly recycled can leach into the soil, harms microorganisms and disrupt the soil ecosystem (Wong et al. 2007). It makes soil sterile as directly impacts the physico-chemical properties (Yu et al. 2007). It makes soil toxic cocktail of heavy metals. Bioaccumulation of persistent materials into the various trophic levels is again a problem (Jinhui et al. 2011).

Impact of E-Wastes on Health

The detrimental effects of E-wastes disposal (popularly known as Toxic mine) on health of human beings (ATSDR, 2003) can be summed up into a tabular form (Fu et al. 2008; Sankhla, et al. 2016; Kumar & Singh, 2019) as under:

HEAVY METAL CONCENTRATION FROM E-WASTE SOUTHEAST CHINA: A CASE STUDY

- "The recycling of printed circuit boards in Guiyu, China, a village intensely involved in E-waste processing presents a significant environmental and possible human health risks (Toxics link, 2014)".
- "The concentration of various heavy metals is given in table exceed the permissible limits which pose severe threat to society in Guiyu, China (Toxics link, 2014)".

IMPACT OF E-WASTE RECYCLING ON SOIL IN DELHI: A CASE STUDY

- "In India, 95% E-waste is handled in the informal sector and recycled without any environmental safeguards (Toxics link, 2014)".
- "The Loni and Mandoli sites show the concentration of heavy metals well above the permissible limits and Mandoli exhibits high risk factor than Loni (Toxics link, 2014)".

GLOBAL MANAGEMENT STRATEGIES

Official Take-Back Systems

- "In this scenario, usually under the requirement of national E-waste legislation, E-waste is collected by designated organizations, producers and/or by the government. This happens via retailers, municipal collection points and/or pick-up services".
- "The final destination of E-waste is state-of-the-art treatment facilities, which recover the valuable materials in an environmentally-sound way and reduce the negative impacts (Kumar and Singh, 2019)".
- "In the European Union, roughly 40% of the annually generated E waste is officially reportedly treated in this manner; in the United States and Canada, the level is around 12%; for China and Japan, it is around 24 to 30% and in Australia, is around 1% (Toxics link, 2014). That said, the scope of collected products differs among the countries, depending on the priority setting at the national level. Usually, product categories with significant potential for resource recovery or con-

taining a significant amount of toxic elements are collected, such as temperature exchange equipment (cooling and freezing equipment), screens and monitors, lamps, large equipment and small IT and telecommunication equipment".

- "This disposal scenario exists in both developed and developing countries (Kumar and Singh, 2019)".

Disposal of E-Waste in Mixed Residual Waste

- "In this scenario, consumers directly dispose of E-waste through the normal dustbins together with other types of household waste (SAICM, 2009)".
- "As a consequence, the disposed of E-waste is then treated with the regular mixed waste from households. Depending on the region, it can either be sent to landfill or municipal solid waste incineration with a low chance of separation prior to these final destinations (Kumar and Singh, 2019)".
- "Neither of these two destinations is regarded as an appropriate technique to treat e waste, because it leads to resource loss and has the potential to negatively impact the environment (SAICM, 2009; Kumar and Singh, 2019)".
- "The E waste in landfill can lead to toxin leaching, and if E-waste is incinerated, emissions into air occur (Toxics Link, 2014; Jinhui et al. 2011; Wang et al. 2012). This disposal scenario exists in both developed and developing countries".
- "Products commonly thrown away in dustbins include small equipment, small IT equipment and lamps. In most developing countries, valuable E-waste is hardly seen in dustbins, but invaluable E-waste like lamps and small products can be easily disposed of in dustbins and then sent to landfill or incinerator".
- "There are no official statistics in countries about the quantity of E-waste that is disposed with mixed waste in dustbins. For all data that was found, about 1 to 2 kg per inhabitant was disposed in the waste bin in Europe. This represents roughly 8% of the total European E-waste generation (Toxics Link, 2015)".

Collection of E-Waste Outside Official Take Back Systems in Developed Countries

- "In developed countries, E-waste is also collected by individual waste dealers or companies and then traded through various channels. Possible destinations for E-waste in this scenario include metal recycling, plastic recycling, and specialized E-waste recycling and also export".
- "Usually, E-waste handled in this scenario is not reported as part of official treatment amount by the established take-back systems (Scenario 1)".
- "E-waste categories that are typically handled by the informal collection are temperature exchange equipment, large equipment, screens and IT products".

- "The main feature of this scenario is that E-waste is traded freely, and usually, its quantity is not systematically documented or reported to authorities, due to lack of specific reporting framework or requirements (Kumar and Singh, 2019)".
- "In this scenario, E-waste is often not treated in the state-of-the art facilities, and there is a potential that E-waste is shipped off to developing countries. There is a substantial amount of E-waste being collected in developed countries and then traded to developing countries for further treatment".
- "The demand for inexpensive second-hand equipment and raw materials in less-developed regions is the biggest driver for the interregional and global trade of E-waste. Trading of second hand equipment is legal only if it is allowed by both sending and receiving countries. However, the dumping of waste occurs exists in practice, is illegal".
- "If the exporting has ratified the Basel convention, exports of hazardous waste must comply with the Basel Convention. The Basel Convention is meant to prevent developed countries from illegally dumping waste in developing countries, where recycling infrastructure is typically absent".

Informal Collection and Recycling in Developing Countries

- "In most developing countries, there are an enormous number of self-employed people engaged in the collection and recycling of E-waste. They usually work on a door to door basis to buy E-waste from consumers at home, and then they sell it to refurbishers and recyclers".
- "These types of informal collection activities provide the basic means necessary for many unskilled workers to pay for their living".
- "Apart from domestic collection, the demand for inexpensive second-hand goods and secondary materials is an incentive of to import E-waste from developed countries (as explained in Scenario 3)".
- "After informal collection, when electronic products do not have any reuse value, they are mostly recycled by through "backyard recycling" or substandard methods, which can cause severe damage to the environment and human health (Kumar and Singh, 2019)".
- "Such substandard treatment techniques include open burning to extract metals, acid leaching for precious metals, unprotected melting of plastics and direct dumping of hazardous residuals (Ceballos et al. 2015)".
- "Lacking legislation, treatment standards, environmental protection measures and recycling infrastructure, are the main reasons that E-waste is recycled in a crude manner (Ceballos et al. 2015)".
- "Typical E-waste categories handled by the informal collection include temperature exchange equipment, large equipment, screens and IT products (Jain et al. 2009)".

MODERN RECYCLING METHODS

Since E-waste contains valuable metals which can be recovered by using various innovative recycling technologies. Some of the recycling methods used are:

1. **Hydrometallurgy:** It is also called as chemical leaching method. The method involves using a suitable chemical reagent for the dissolution of metals from E-waste.
2. **Pyro-Metallurgy:** The process involves heating the electronic waste in the absence of oxygen in a specially designed reactor, resulting in zero emission and produces an energy rich liquid of good calorific value.
3. **Bio-Metallurgy:** The method uses preparing a bio-adsorbent like activated carbon from biological organic matter and using it for adsorbing the valuable metals present in E-waste.
4. **Electro-Metallurgy:** It uses electro refining method. The method involves preparing an electrolyte and using suitable electrodes for the deposition of valuable metals on the surface of the electrode from E-waste.

E-WASTE RECYCLING IN INDIA

India's Informal E-Waste Recycling Sector

- "About 80% of imported E-waste is imported from the US and the remaining share comes mainly from the EU. Women and children also constitute a significant proportion of the workforce (Mohite, 2013)".
- "Although deemed illegal by Indian authorities, the informal sector is believed to survive through local political patronage systems (Ngoc et al. 2009)".
- "Work carries on for 12-14 hours a day with workers sitting on the ground amongst piles of electronics parts. Most work is carried out with bare hands, without the use of masks, cleaning, crushing or heating the parts (Kalra, 2004)".
- "Many workers also sit cramped in unventilated rooms with inadequate lighting and no clean drinking water or toilets. Many of these workers complain of eye irritation, breathing problems and constant headaches (Kumar and Singh, 2019)".
- "Most people involved in informal recycling are the urban poor with low literacy levels, and hence have very little awareness regarding the hazards of E-waste and the recycling processes. India's informal E-waste recycling sector; 95% of all recycling whereas only 5% of all recycling follow MoEF Guidelines, 2008 but these 5% also merge with informal recycled sector in a closed loop mechanism".

CONVENTIONAL DISPOSAL METHODS

- "Land filling is frequently used in the disposal of E-waste; this involves excavating soil from trenches and burying the waste, usually covered with a liner of plastic (Jinhui et al. 2011)".
- "However, it is virtually impossible to prevent some leaching of hazardous substances which contaminate soil and water resources, even from the most technologically-advanced facilities (Jinhui et al. 2011; Wang et al. 2012)".
- "Another commonly used method of E-waste disposal is incineration; a controlled combustion process in which waste is burned at high temperature in specially designed incinerators. This

method enables volumes of waste to be reduced and some energy to be recovered from the combustion process (Kumar and Singh, 2019)".

- "However, incineration also involves the release of hazardous substances into the atmosphere (dioxins and furans) (ATSDR, 1995) and leaves hazardous waste in the form of slag for disposal (ATSDR, 1995; Liu et al. 2008; Kumar and Singh, 2019)".

REGULATIONS FOR MANAGEMENT OF E-WASTE IN INDIA

- E-waste (Management and Handling) Rules, 2011.
- Draft of E-waste (Management) Rules, 2015.

Objectives of E-Waste (Management and Handling) Rules, 2011

1. "Minimize illegal recycling / recovery operations."
2. "Environmentally Safe & Sound Recycling by channelizing E-waste to registered E-waste recyclers."
3. "Extended Responsibilities to producers to manage a system of E-waste collection/take back and channelizing to a registered dismantler/recycler."
4. "Responsibilities to Urban Local Bodies for orphan products and for waste found mixed with MSW."
5. "To Create an E-waste collection channelization system."
6. "Reduce Hazardous substances in Electrical and Electronic components."

Responsibility of State Pollution Control Boards/ Pollution Control Committees

1. "Preparation of inventory of E-waste."
2. "Granting authorization."
3. "Granting registration."
4. "Monitoring of compliance of authorization and registration conditions."
5. "Maintaining information on the conditions imposed for authorization."
6. "Taking action against violations of these rules."

Ensure that collection centre should not store E-waste for a period exceeding one hundred and eighty days.

Responsibility of Central Pollution Control Boards

- "Coordination with State Pollution Control Boards/Committees of Union territories."
- "The preparation of guidelines for environmentally sound management of E-waste."
- "To conduct assessment of E-waste generation and processing. Recommend standards and specifications for processing and recycling E-waste."
- "Documentation, compilation of data on E-waste and uploading on websites of Central Pollution Control Board. Conducting training & awareness programmes."
- "Submit Annual Report to the Ministry of Environment & Forests."

- "Enforcement of provisions regarding reduction in use of hazardous substances in manufacture of electrical and electronic equipment."
- "Initiatives for IT industry for reducing hazardous substances. And to set targets for compliance to the reduction in use of hazardous substance in manufacture of electrical and electronic equipment."

Responsibility of Municipal Authority

- "The Municipal Authorities are responsible for channelization of E-waste after segregation to either authorized collection centre or dismantler or recycler if found to be mixed with MSW".
- "Collection and channelization of E-waste generated from non-branded or assembled electrical and electronic equipment to either authorized collection centre or dismantler or recycler".

CONCLUSION

E-waste is much more hazardous than many other municipal wastes because electronic gadgets contain thousands of components made of deadly chemicals and metals like lead, cadmium, etc. Long term exposure to these substances damages the nervous systems, kidney, bones, and the reproductive and endocrine systems, and some of them are carcinogenic and neurotoxin. Primitive recycling or disposal of E-waste to landfills and incinerators causes irreversible environmental damage by polluting water and soil, and contaminating air. The E-waste recycling is one of the better steps towards sustainable development only if treated in ecofriendly and formalized manner. The need of the hour is to enforce the laws and regulations pertaining to E-wastes and to adopt advanced and efficient technologies for the ecofriendly disposal of E-waste. And extensive research is needed to study effects of E-waste disposal on health of humans and other living creatures.

REFERENCES

ATSDR, Agency for Toxic Substances and Disease Registry. (1995). *Toxicological profile for polycyclic aromatic hydrocarbons (PAHs)*. Atlanta, GA: US Department of Health and Human Services, Public Health Service.

ATSDR, Agency for Toxic Substances and Disease Registry. (2003). *Toxicological profiles on CD-ROM. Agency for Toxic Substances and Disease Registry*. Public Health Service.

Bertram, M., Graedel, T. E., Rechberger, H., & Spatari, S. (2002). The contemporary European copper cycle: Waste management subsystem. *Ecological Economics*, *42*(1-2), 43–57. doi:10.1016/S0921-8009(02)00100-3

Brigden, K., Labunska, I., Santillo, D., & Johnston, P. (2008). *Chemical contamination at e-waste recycling and disposal sites in Accra and Korforidua, Ghana*. Greenpeace Research Laboratories Technical Note. Greenpeace International. Retrieved from http://www.greenpeace.org/international/press/reports/chemicalcontamination-at-e-wa

Ceballos, D., Gong, W., & Page, E. (2015). A pilot assessment of occupational health hazards in the U.S. electronic scrap recycling industry. *Journal of Occupational and Environmental Hygiene*, *12*(7), 482–488. doi:10.1080/15459624.2015.1018516 PMID:25738822

Chatterjee, S. (2012). Sustainable electronic waste management and recycling process. *American Journal of Environmental Engineering*, *2*(1), 23–33. doi:10.5923/j.ajee.20120201.05

DIT. (2003). *Environmental management for Information Technology industry in India*. Department of Information Technology, Government of India.

Divya, K., Singh, T., & Kumar, A. (2012). E-waste management: An overview. *International Journal of Research in IT & Management*, *2*, 763–770.

DTIE, Division of Technology, Industry and Economics. International Environmental Technology Centre. (2007). E-*waste: Volume 1: Inventory assessment manual (Osaka/Shiga)*. Available at: http://www.unep.or.jp/ietc/Publications/spc/EWasteManual_Vol1.pdf

European Union. (2012). Directive 2012/19/EU of the European Parliament and of the Council of 4 July 2012 on Waste Electrical and Electronic Equipment (WEEE) recast. *Official Journal of the European Union*, *55*, 197.

Ficeriova, J., Balaz, P., Dutkova, E., & Gock, E. (2008). Leaching of gold and silver from crushed Au-Ag wastes. *The Open Chemical Engineering Journal*, *2*(1), 6–9. doi:10.2174/1874123100802010006

Fu, J., Zhou, Q., Liu, J., Liu, W., Wang, T., Zhang, Q., & Jiang, G. (2008). High levels of heavy metals in rice (*Oryza sativa* L.) from a typical E-waste recycling area in southeast China and its potential risk to human health. *Chemosphere*, *71*(7), 1269–1275. doi:10.1016/j.chemosphere.2007.11.065 PMID:18289635

Gupta, Sangita, & Kaur. (2009). Electronic waste: A case study. *Research Journal of Chemical Sciences*, *1*, 49–56.

Jain, P., Rani, B. M., & Bhawsar, J. (2009). Study recycling of plastic and CRTs from electronic waste. *Oriental Journal of Computer Science & Technology*, *2*, 51–56.

Jinhui, L., Huabo, D., & Pixing, S. (2011). Heavy metal contamination of surface soil in electronic waste dismantling area: Site investigation and source-apportionment analysis. *Waste Management & Research*, *29*(7), 727–738. doi:10.1177/0734242X10397580 PMID:21406506

Kalra, V. (2004). E-*waste in India: System failure imminent: Take action now! (Delhi, Toxics Link)*. Available at: http://www.toxicslink.org/pub-view.php?pubnum=40.pdf

Kumar, S., & Singh, V. (2019). E-Waste: Generation, Environmental and Health Impacts, Recycling and Status of E-Waste Legislation. *Journal of Emerging Technologies and Innovative Research*, *6*(1), 592–600.

Liu, G., Niu, Z., Van Niekerk, D., Xue, J., & Zheng, L. (2008). Polycyclic aromatic hydrocarbons (PAHs) from coal combustion: Emissions, analysis, and toxicology. *Reviews of Environmental Contamination and Toxicology*, *192*, 1–28. doi:10.1007/978-0-387-71724-1_1 PMID:18020302

Lucier, C. A., & Gareau, B. J. (2009). *Solving the E-Waste Problem. Recycling - From E-Waste to Resources: Sustainable Innovation and Technology Transfer Industrial Sector Studies* Available from: http:// www.unep.fr/shared/publications/pdf/ DTIx1192xPA-Recycling%20from%20 ewaste%20to%20 Resources.pdf

Mohite, B. J. (2013). Issues and strategies in managing E-waste in India. *Indian Journal of Research in Management. Business and Social Sciences, 1*, 46–51.

Ngoc, H. N., Agusa, T., Ramu, K., Phuc, C. T. N., Murata, S., & Bulbule, K. A. (2009). Contamination by trace elements at E-waste recycling sites in Bangalore, India. *Chemosphere, 76*(1), 9–15. doi:10.1016/j.chemosphere.2009.02.056 PMID:19345395

Ni, H. G., & Zeng, E. Y. (2009). Law enforcement and global collaboration are the keys to containing e-waste tsunami in China. *Environmental Science & Technology, 43*(11), 3991–3994. doi:10.1021/es802725m PMID:19569320

Radha, G. (2002). *A Study of the Performance of the Indian IT Sector*. Available at: www.nautilus.org

SAICM, Strategic Approach to International Chemicals Management. (2009). *Background information in relation to the emerging policy issue of electronic waste*. Paper presented at the International Conference on Chemicals Management, Geneva.

Sankhla, M. S., Kumari, M., Nandan, M., Mohril, S., Singh, G., Chaturvedi, B. & Kumar, R. (2016). Effect of Electronic waste on Environmental & Human health-A Review. *IOSR Journal of Environmental Science, Toxicology and Food Technology, 10*, 2319-2399. doi:10.9790/2402-10090198104

Schluep, M., Hagelueken, C., Kuehr, R., Magalini, F., Maurer, C., Meskers, C., . . . Wang, F. (2009). *Recycling: From E-waste to resources, Sustainable Innovation and Technology Transfer Industrial Sector Studies (Nairobi and Bonn, UNEP and STeP)*. Available at: http://www.unep.org/PDF/PressReleases/EWaste_publication_screen_FINALVERSION sml.pdf

Smith, T., Sonnenfeld, D. A., & Naguib, P. D. (2006). *Challenging the chip: Labor rights and environmental justice in the global electronics industry*. Philadelphia, PA: Temple University Press.

Standard, StE. P. W. P. (2014). *Solving the E-Waste Problem. Recommendations for Standards Development for the Collection, Storage, Transport and Treatment of E-Waste*. Available from: http://www.step-initiative.org/files/_documents/whitepapers/StEP_WP_Standard_20140602.pdf

Terazono, A., Murakami, S., Abe, N., Inanc, B., Moriguchi, Y., Sakai, S., ... Williams, E. (2006). Current status and research on E-waste issues in Asia. *Journal of Material Cycles and Waste Management, 8*(1), 1–12. doi:10.100710163-005-0147-0

Toxic Link Organisation (T.L.O). (2014). Available at: http://www.toxicslink.org

Toxic Link Organisation (T.L.O). (2015). Available at: http://www.toxicslink.org

Umicore Precious Metals Refining. (2007). *Metals Recovery from e-scrap in a global environment.* Available at: http://archive.basel.int/industry/sideevent030907/umicore.pdf

UNEP, United Nations Environment Programme. (1989). *Basel convention on the control of transboundary movements of hazardous waste and their disposal*. http://www.basel.int/Portals/4/Basel%20Convention/docs/text/BaselConventionText-e. pdf

Wang, Y., Tian, Z., Zhu, H., Cheng, Z., Kang, M., Luo, C., ... Zhang, G. (2012). Polycyclic aromatic hydrocarbons (PAHs) in soils and vegetation near an e-waste recycling site in South China: Concentration, distribution, source, and risk assessment. *The Science of the Total Environment*, *439*, 187–193. doi:10.1016/j.scitotenv.2012.08.018 PMID:23063924

Wong, C. S., Duzgoren-Aydin, N. S., Aydin, A., & Wong, M. H. (2007). Evidence of excessive releases of metals from primitive e-waste processing in Guiyu, China. *Environ Pollut*, *148*(1), 62–72. doi:10.1016/j.envpol.2006.11.006 PMID:17240013

Yu, X. Z., Gao, Y., Wu, S. C., Zhang, H. B., Cheung, K. C., & Wong, M. H. (2006). Distribution of polycyclic aromatic hydrocarbons in soils at Guiyu area of China, affected by recycling of electronic waste using primitive technologies. *Chemosphere*, *65*(9), 1500–1509. doi:10.1016/j.chemosphere.2006.04.006 PMID:16713614

KEY TERMS AND DEFINITIONS

Acid Rain: Rain or any form of precipitation that have pH less than 5.6. The acidic pH is due to burning of coal and other fossil fuels. Acid rain can be detrimental to aquatic as well as terrestrial vegetation, animals, microorganisms, and even alters physicochemical composition of water and soil environs.

Ecological Bio-Magnification: It is the increase in the quantum of a xenobiotic concentration (synthetic substance) with each successive trophic level of the food chain.

Environmental Estrogens: These are xenobiotics entities (e.g., dioxins and furans) which impersonates the estrogen levels (female sex hormones) in humans eventually leads to alteration of hormonal activity and result in estrogen dominance conditions such as uterine fibroid tumors, fibrocystic breasts (adenosis), glandular dysfunction, hair loss, weight gain, depression, etc.

Recalcitrant Compounds: These are the synthetic compounds that are non-biodegradable in nature.

Chapter 6
Generation and Management of Biomedical Waste

Kalpana
Dr. Harisingh Gour Vishwavidyalaya, India

Rifat Azam
Babasaheb Bhimrao Ambedkar, India

Nazia Parveen
Babasaheb Bhimrao Ambedkar, India

Dig Vijay Singh
Babasaheb Bhimrao Ambedkar, India

Zaffar Azam
Dr. Harisingh Gour Vishwavidyalaya, India

ABSTRACT

Biomedical wastes (BMWs) are generated by the treatment of living organism as well as during research activities. With the increase of population in developing countries, generation of BMWs is increasing, but with inappropriate disposal management technologies. BMWs pose threats to the humans and also degrade the quality of environment. Handling of the BMWs by inexperienced and untrained staff increases the risk of accumulation of waste at the source as well as increases the chances of the spread of the dreadful diseases. The implementation of BMWs regulations is inadequate and unsatisfactory in developing countries: a major concern. Safe and reliable methods for management of BMWs are of utmost importance. The proper management of BMWs can be helpful in maintaining the overall quality of the environment. In order to get good results, the need is to increase the involvement of the ecofriendly techniques as well as increase awareness among the citizens, staff, and the workers directly involved in the management of the BMWs.

DOI: 10.4018/978-1-7998-0031-6.ch006

INTRODUCTION

Biomedical waste is produced from various activities (Chatraborty et al., 2014) of human beings such as treatment or immunization of patients (Rutala & Mayhall, 1992). The waste can also be generated from diagnosis of human beings, animals or testing of biological activities (Liu et al., 2016). The health care facilities like dialysis availed at household level are also the source of biomedical waste (Szycher, 2017). Medical waste includes all the waste produced from hospitals, clinics and laboratories that is discarded and is of no use (Anurag et al., 2013). The waste produced in the hospital is mostly nonhazardous (85%) while as infectious (10%) and noninfectious waste (5%) contribute in small amount to overall biomedical waste (Mathur et al., 2012). The biomedical waste contains small quantity of infectious waste but has high potential to transmit diseases (Ghasemi & Yusuff, 2016). Infectious waste is hazardous in nature and if mixes up with the general waste, entire waste will be treated as infectious waste (Chowdhary, 2018). The waste poses great risk to the living organisms and particularly to the workers involved in the management of the hazardous waste (Li & Jenq, 1993). Peoples having poor nutrition, inability to afford healthcare facility and lack of timely vaccination are more susceptible to infection from improper management of biomedical waste (Patwary et al., 2011b). Therefore, the hospital waste in addition to the risk for patients and workers associated with the management of wastes poses a serious threat to public health and almost all components of the environment (Mathur et al., 2012). Activities undertaken to achieve the aims of reducing health related problems leads to the production of the biomedical waste which is hazardous as well as infectious (Patil & Pokhrel, 2005). Different types of the waste are managed with different treatment techniques but are lacking the effective new techniques in the management of biomedical waste. Biomedical waste was not considered a major problem (Hegde et al., 2007) but the exposure to viruses from waste has raised concern about the risk associated with the medical waste (Gershon et al., 1995). The raising concerns related to the biomedical waste is putting pressure on the existing government policy, citizens and environmentalist to think for safe disposal of the biomedical waste (Misra & Pandey, 2005). Developed countries are well aware about the concerns associated with the biomedical waste thus are enforcing strict guidelines related to waste segregation, storage and transportation from time to time (Marinkovic et al., 2008; Tudor et al., 2005) but developing countries are lacking the timely measures needed for the proper management of the medical waste (Pittet et al., 2006). There are strict guidelines for the segregation of biomedical waste by using different types of dustbins (Table 1.) within the hospitals so that the disposal of the waste can be done in an economical way (Da Silva et al., 2005). Developing countries are not only lacking the new and cheap techniques but are also lacking skills necessary for the proper waste management (Porter and Van der Linde, 1995). The timely measure necessary for efficient management of the waste (waste minimization and segregation) can be helpful in reducing the biomedical waste footprint (Jain et al., 2017). Thus, hospital waste management is of prime importance as the impact of the improper management is not only restricted to patients and hospital staff but all the components of environment as well as human being are affected (Hegde et al., 2007).

Table 1. Color coding for the disposal of biomedical waste (Rajan et al., 2018)

Color coding	Type of waste	Treatment method
Yellow	All those waste that can be burned	Plasma pyrolysis, hydro calving, deep burial, micro-waving,
Red	Recyclable waste	Autoclaving or micro-waving
Blue	Glass disposed in cardboard boxes	Disinfection, autoclaving, microwaving, hydroclaving
White	Sharps	Autoclaving or Dry Heat Sterilization

SOURCES OF BIOMEDICAL WASTE

The biomedical waste is generated from different sources (Figure 1) like hospitals, laboratories and pharmaceutical industries (Hall et al., 2006). Different hospital rooms as well as department like ICU, maternity room, emergency room, pathological section, anatomy section and laboratory section are dealing with different kinds of patients and are the massive source of biomedical waste. The waste produced in these section consist of needles, syringes, flacon, catheters, disposable gloves, masks, bottles (Pasupathi et al., 2011) cloth with blood or body fluids (Kim et al., 1998), dressings former, cotton, blood transfusion hose, former bandage and alcohol swab, pot urine or faeces, reagents, chemicals, glass slides, serum, cotton, filter paper, petri dish, autoclave bags, specimen containers, expired drugs, metals and toxic chemicals (Coronel et al., 2002).

Figure 1. Sources of biomedical waste

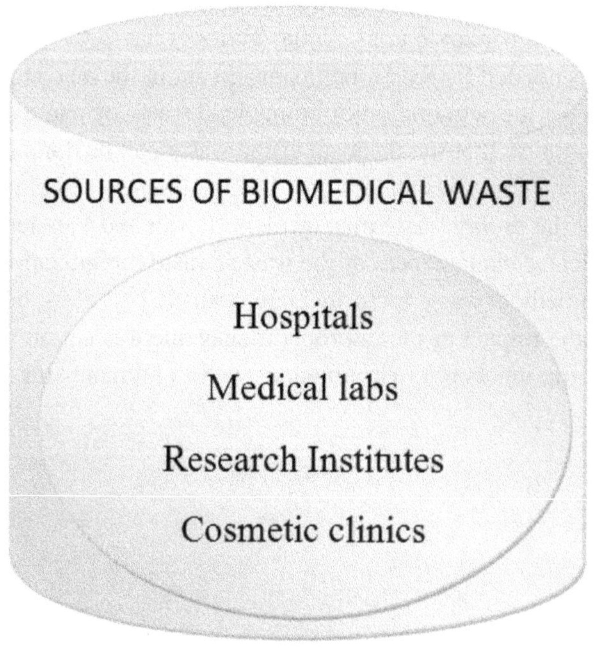

COMPOSITION

Health-care wastes are categorized as infectious and non-infectious (Miyazaki & Une, 2005). Infectious waste is the major source disease causing organisms that ultimately leads to the transfer of various diseases (Rutala & Mayhall, 1992). Biomedical waste that is very infected and contaminated include body parts, organs, tissues, blood and body fluids cotton, bandage, syringes, needles (Tiwari & Kadu, 2013) and other materials containing blood (Chartier, 2014). Hospitals are the big source of noninfectious waste and the quantity of the produced waste is much more than that of infectious wastes (Patil & Shekdar, 2001). The large quantity of the waste like needles, syringes (Ali et al., 2017) produced in the hospitals is recyclable but before sending the waste for recycling (Patil & Shekdar, 2001), it should be separately stored and sterilized as it may be contaminated with infectious pathogens (Jang et al., 2006). Every year, an estimated 16000 million injections are administered worldwide but not all of the needles and syringes are properly disposed of afterward (Oli et al., 2016). The recycling of disposable waste without sterilization is responsible for spreading of the diseases like hepatitis, HIV (Mathur et al., 2012) thus it is mandatory to sterilize the waste before going for recycling. It is very essential to properly collect, segregate, store, transport, treat and dispose the biomedical waste in scientific manner (Tiwari & Kadu, 2013). Mixed items produced on handling of patient should be burned so that the source of infection can be completely eliminated (Gupta, 2014). However, sharp items infected with cytotoxic, radioactive items or prion should be dealt with relevant treatment mechanisms (Askarian et al., 2004). The chemical waste contain harmful chemical which are utilized in the day to day procedure of the hospital are released directly whisg results in the pollution of the environment (Diaz et al., 2005). Thus the waste water released from the hospitals should be properly filtered before disposing into public sewers (Thomas et al., 2007).

PROBLEMS RELATING TO BIOMEDICAL WASTE

The quantity of the generation of the different kinds of the waste is increasing day by day (Oweis et al., 2005) but the treatment of waste generated is not done in proper scientific way (Dutta et al., 2006). The improper management of the waste not only increases the quantity of the waste but also become the cause of other related issues like unpleasant smell, environmental pollution, rapid growth and multiplication of pathogens (CEET, 2008). The dumping of hospital waste in and around the hospitals can be the source of various communicable diseases (plague and rabies) as the dumped waste become the living place for insects, flies, mosquitoes, rodents, cats and dogs (Sheavly & Register, 2007). Improper handling of waste adversely affects the hospital (Oli et al., 2016) as well as human environment (Nwachukwu et al., 2013) as various airborne pathogen increases with the improper management of the waste. Biomedical waste also presents a high risk to doctors, nurses, technicians, sweepers, hospital visitors and patients due to improper management of the waste (Kumar et al., 2010). Thus, it is the primary responsibility of health administrators to manage hospital waste in eco-friendly manner and also safeguard the life of the people associated with management of the waste (Wittmann, 2010).

Enough rules and regulation are for the management of the waste but implementation part is lacking leading to the increasing accumulation of the waste in the developing countries (Tiwari & Kadu, 2013). Due to improper management of the biomedical waste, infectious and hazardous waste gets mixed with solid waste (Coker et al., 2009) and ground water during rainy season (Tiwari & Kadu, 2013) resulting in more spreading of various diseases. The mixing of the hospital waste with the general waste not only increases the incidence of infectious diseases but also increase the management cost of the waste (Patwary et al., 2011a). Developed countries used to took timely measure for the various processes (segregation, storage and transportation) involved in the management of the biomedical waste (Marinkovic et al., 2008; Tudor et al., 2005). Developed countries face challenges with the sheer volume of waste from the use of disposable items (Clapp, 2002) but developing countries are facing challenges of sorting and disposing of all types of medical waste in a sanitary manner (Diaz et al., 2005). Developing countries are also lacking the resource and recent techniques necessary for the effective management of the biomedical waste (Hossain et al., 2011).

Handling Methods

In India, biomedical waste without any disinfection is collected in the open containers. Different types of waste produced should not be mixed (Nema et al., 2011) and workers handling the waste should be trained so that risk associated with biomedical waste can be minimized (Kumar et al., 2015). Appropriate equipment should be used by the workers dealing collection and disposal of the waste (Mostafa et al., 2009). The quantity of the generated waste is helpful in determining the frequency of waste collection (Gellynck et al., 2011) and collection of waste should be at least done once in a day (Angelidaki et al., 2009). The proper label on the waste collection bags must at least contain the information like date, area, floor, weight and types of waste. The bags should be tied tightly so that no leakage occur during the transportation of the waste (Bordner et al., 1978) and if there is cut on the polythene, the double bagging should be done in order to avoid leakage (Poon et al., 2004). The disposal of the sharp waste after disinfection and mutilation may be helpful in minimizing the spreading of infectious diseases after their reuse (Radha et al., 2009). The collection of the waste in India is mostly done in mixed form without taking into care the nature of the waste and its effects after their mixed disposal (Shekdar, 2009). The collected waste is segregated by the workers and the recyclable material separated is then dispatched for recycling (Sharholy et al., 2008).

Waste Minimization

The huge amount of waste is generated from hospitals all over the world (Mastorakis et al., 2011). The amount of the waste generated varies with the country as well as also from region to region and the amount of waste accumulated depends mostly on the management techniques (Alhumoud & Alhumoud, 2007). The generation of the waste can be minimized by increasing the use of reusable items (Song et al., 2015), using steam sterilization instead of chemicals (Patil & Shekdar, 2001) and having efficient recycling policies for the plastic waste (Wilson et al., 2006). It becomes responsibility of the government, citizens and hospital authority to devise new techniques which are ecofriendly and economical in nature for the treatment of biomedical waste (Roseland, 2012). Effective biomedical waste management (BMWM) is mandatory for healthy humans and cleaner environment (Tiwari & Kadu, 2013). The basic steps in the effective management of the waste are the source segregation into different components (reusable and non-reusable, hazardous and non-hazardous) of the waste and good transportation facility from waste generation point to disposal sites. Thus, proper management of the waste is helpful in minimizing the risk associated with the biomedical waste (Hegde et al., 2007).

Waste minimization can be done by adopting and purchasing environmentally friendly products and by applying the 4R'sprinciple *i.e.* reduce, reuse, recycle and recover (Dennis, 2017). The best BMW management (BMWM) methods aim at avoiding generation of waste or recovering as much as waste as possible rather than disposing (Tiwari & Kadu, 2013). Therefore, the various methods of BMW disposal, according to their desirability are prevent, reduce, reuse, recycle, recover, treat, and lastly dispose (Patil & Shekdar, 2001). Hence, the waste should be tackled at source rather than "end of pipe approach (Larsen et al., 2004). 4 R's help in safe and sound management of medical waste by reducing and managing the waste and waste streams effectively (Pruss-Ustun, 1999).

1. **Reduce**: Reducing the waste is a basic step in the minimization of the waste generation and should utilize the product whose impact on the environment is less and the product should be recyclable (Ngoc and Schnitzer, 2009).
2. **Reuse**: Re-use is using the product either for the same purpose or for the other use (Woolridge et al., 2006). The reuse principle of waste management focuses on the use of the material again and again rather than selection of such materials that is disposed after single use (Gupta, 2017). This principle can be applied whenever possible and standards for disinfection and sterilization have to be strictly followed if equipment's and materials have to be reused (Rutala & Weber, 2007). The equipment to be reused should be sterilized (autoclaving, disinfection) in the proper way so that chances of the transmission of the infection source can be minimized (Beilenhoff et al., 2008). Some items like syringes and hypodermic needles in the hospital waste are prohibited for the reuse as there are high chances of cross infection (Heffner, 2004).
3. **Recovery and Recycling of Waste**: Resource recovery can be achieved by selective extraction of disposed materials from waste stream for a specific use by diverting it to resource recovery stream (Perrin & Barton, 2001). Recycling is reprocessing of a used item into a new raw material (King et al., 2006). Reuse is the utilization of the product for the same purpose or for different purpose while recycling is the changing the product from its original form to new form and using it for the new purpose (De Brito & Dekker, 2004).

The different kinds of the waste are generated by the various health care facilities provided by the hospitals (Patil & Shekdar, 2001). Different waste management practices like private contractor (Yong et al., 2009) manage waste by incineration or landfilling (Singh et al., 2011) is used by different hospitals or in some cases municipalities are responsible for the management of the waste (Turan et al., 2009).

MANAGEMENT OF BIOMEDICAL WASTE

Biomedical waste generated by different activities of the hospitals need to managed in the proper way so that adverse effects of waste on the environment can be minimized (Capoor *et al.*, 2017). Biomedical waste management being universal problem and the quantity of the waste is increasing due to increase in the number of health care institutes (Cheng et al., 2009). Today, with increasing number of health care institutions and careless attitude of health care personnel, the disposal of waste has become even more challenging (Berry & Parasuraman, 2004). The bio-medical waste shall be segregated into containers or bags at the point of generation in accordance with Schedule I prior to its storage, transportation, treatment and disposal (Radha et al., 2009). Segregation of the waste at the source of generation is known as primary step and the "HEART" of biomedical waste management (Jayathilakan et al., 2012). The waste segregated at source should remain in the same category of the waste till the final disposal of the waste (Gupta et al., 1998). Nearly 75-90% of waste generated at health care facilities is non-hazardous or general waste in nature hence most of the waste produced is recyclable (Vasistha *et al.*, 2018). Segregation of the waste not only help to separate the hazardous and recyclable waste but also helps in minimizing the load of the landfill and the waste going to incinerators for final disposal (Chandler et al., 1997). The small amount (10-15%) of the hazardous waste generated should be handled, stored and disposed of in the manner that has very less impact on the environment (Sharholy et al., 2008). Non-hazardous pharmaceuticals can be drained in general sewers in limited quantities in each month or should be returned to manufacturers for proper treatment and disposal of the nonhazardous waste (Dias-Ferreira et al., 2016; Smith, 2002).

PROCESS FOR THE MANAGEMENT OF BIOMEDICAL WASTE

The BMW management process includes the following stages:

Handling

It is very important to handle and dispose biomedical waste properly because health care workers, sanitation workers and the general public are at a risk of exposure to dangerous diseases (Babanyara et al., 2013). Bright color bags or boxes should be used for the collection and disposal of the waste and should be clearly marked with the symbols (Hirani et al., 2014).

Segregation

The generated biomedical waste from different health care facilities (Datta et al., 2018) need to be segregated at the generation point but not in the later stage (Ananth et al., 2011) as segregation of waste at later stages makes the process complicated and increase the cost of waste management. "Point of Generation" is the initial place of the waste generation and is under the supervision of the doctor as well as nursing staff providing health care facility to the patient (Jang *et al.*, 2006, Marinkovic *et al.*, 2008). The colour coded bins and bags should be made available in good number (Nemathaga *et al.*, 2008) as per Plastic Waste Management Rules, 2016. Hence lack of source segregation, color coding, record keeping and carelessness of the staff are some of the main issues leading to poor segregation practices across hospitals in the developing countries (Ali et al., 2017). Segregation of waste would result in a clean solid waste stream which could be easily, safely and cost effectively managed through recycling, composting and landfilling (Ngwuluka et al., 2009). The lack of segregation of waste result in the mixing of the different kinds of waste which may result in accidental injuries (Wilburn & Eijkemans, 2013) and leads to the outbreak of various diseases (Pruss-Ustun *et al.*, 2005) but proper segregation of the waste at the source can help to reduce the risk of these diseases (Mbongwe et al., 2008). Segregation of the waste at the source of generation not only reduce the quantity of waste to be incinerated or landfilled but also help to reduce the financial burden which is one of the main hurdle in the management of the waste (Alvim-Ferraz & Afonso, 2005). Thus, segregation of waste helps to increase the life of the system responsible for the final disposal of the waste (Tabata et al., 2011).

Mutilation

The reusing of the biomedical waste without proper treatment is also the source of infection (Hassan et al., 2008) but the waste to be recycled is first autoclaved and then mutilated into pieces in order to avoid the spreading of infection (Diaz et al., 2005). Shredders are used to destroy plastic and paper waste to prevent their reuse (Hirani et al., 2014). Shredders also help in reducing the volume of the waste thus make transportation and disposal of the waste easy (Zurbrugg et al., 2012). The shredder consists of the set of blades that help to cut the waste (Miller, 1997) and then passed though mesh in which large particle are retained and again circulated through the set of moving blades (Saad, 2013). The main problem of the shredder is the blades which need replacement from time to time due to continuous wear and tear during the cutting of the different types of the waste. Biomedical waste to be shredded is syringe, saline bottles, IV tube, catheters and plastic tubing (Zhang *et al.*, 2013).

Disinfection

Disinfection of the waste is done by using chlorine compounds and has been used from years for the disinfection process (Rutala & Weber, 2008). Chlorine compounds are also used in drinking water treatment and swimming pools to reduce the risk of disease transmission by killing the micro-organisms present in the water (Jaffery, 2013). Ethylene oxide is also used in the disinfection of equipment (Herwaldt & Rutala, 1996) and is sometimes also used in treatment of medical waste (Tsakona et al., 2007). Environmental protection agency identifies chemical disinfection as the appropriate method for medical wastewater treatment (Sreekanth et al., 2014) and after treatment wastewater from hospitals can be discharged into sewer system (Saad, 2013).

Storage

Biomedical waste should be stored in the well-ventilated separate room (El-Salam, 2010) with access to waste and sewerage (Abdulla et al., 2008). It is important to store medical waste in secure locations keeping in view the health of the people and safety of the environment (Sankoh et al., 2013). The necessary safety equipment needs to be kept nearby so that impact of sudden accident can be minimized (Askarian *et al.*, 2004).

Transportation

Different mechanisms are followed in the developing countries for the transportation of the waste from the source of generation to the site of final disposal (Yong et al., 2009). In developing countries, the transportation of waste can be carried out by hospital staff (Askarian *et al.*, 2004), private contractors (Da Silva *et al.*, 2005) or local municipilities (El-Salam et al., 2010). The various problems in the transportation of the waste are lack of protective equipment (Abd El-Salam, 2010; Al-Khatib *et al.*, 2009; Askarian *et al.*, 2004), trolleys (Bdour *et al.*, 2007) and suitable vehicles (Tsakona et al., 2007)

TRANSPORTATION WITHIN THE HEALTHCARE FACILITY

General Requirements

The fixed route, time (Friedman et al., 2001), elevators (Marinkovic *et al.*, 2008) for the transportation of waste should be used and the workers involved in the transportation of the waste should be provided with best possible facilities as they are always at the risk of infection (Yong et al., 2009).

Transport Systems

The trolleys for collection of general waste should be black painted, labeled properly and only used for the collection of general or nonhazardous waste. Each trolley can have separate compartment based on type of the waste designated by the colour (yellow, red and blue) of the compartment (Mandal & Dutta, 2009).

Transport Trolleys

The trolley used for the transportation of the waste should have wheels so that transportation can be done in easy way. The trolley used for waste collection should not be used for any other purpose (Rao et al., 2004). Trolleys should not be having sharp edges as sharp edges could damage the containers (Al-Khatib & Sato, 2009) and should not be overloaded by waste (WHO, 2015). Hazardous waste should not be transported by hands and trolleys should be used for this purpose in order to avoid risk of accidents (Tsakona et al., 2007). The spare trolleys need to be kept for transportation in case of sudden accidents (Jang *et al.*, 2006) and all trolleys need to be cleaned and disinfected daily (Rao, et al., 2004).

Routing

Separate routes are to be followed for the different types of the waste and collection of the waste should start from high risk areas like intensive care units, dialysis units and operation theaters (Abd El-Salam, 2010).

Offsite Transportation

The operator of the vehicle should transport the bio-medical waste only in vehicles allowed to be used for transportation (Mandal & Dutta, 2009). The drivers handling the medical waste must have training certificate on handling of the healthcare waste (Omar et al., 2012).

Vehicle Requirements

The vehicle should have separate compartment, one for the driver's cabin and the other for waster (Abdullah et al., 2011). The waste should be secured and should not fall out from the vehicle (Blowers, 2013). The vehicle should contain special items like protective clothing, plastic bag, disinfectant and kits for liquid spillage in order to deal any troublesome situation (Harbec, 2009). The vehicle should contain the name and address of the waste carrier (Pruss-Ustun, 1999). The vehicle must be covered from all the sides (Ali *et al.*, 2017) and the symbol used for hazardous waste should be clearly visible (Verma et al., 2008).

Cleaning of Container and Vehicle

The vehicle should be cleaned and disinfected daily after use. Soaps and detergents should be used for cleaning and should be serviced regularly (WHO, 2015).

DIFFERENT PROCESSES OF DISINFECTION

Microwave

Microwave radiation as a heat source is used to treat medical waste (Ahn et al., 2009) and the treatment can be done by onsite installation or by mobile vans for the treatment of medical waste at different sites (Heberlein & Murphy, 2008). The processing usually includes front-end shredding of the waste, both to increase the efficacy of the microwave treatment and to reduce the volume of the end waste for disposal (Worrell & Vesilind, 2011). Microwave radiation works efficiently on the wet waste compared to dry waste and if the waste is dry, water is sprinkled to make the waste wet as microwave radiation disinfection works only on water but not on dry solid waste (Appleton et al., 2005). Thus, the microwave chamber is provided with humidifier in order to maintain the moisture content of the medical waste (Sankoh et. al., 2013).

Irradiation

Irradiation process disinfects the waste as gamma rays are fatal to micro-organisms thus help to completely eliminate the microbes from the waste (Daly, 2009). In irradiation process, radioactive isotope of cobalt is used which is also used in the treatment of cancer (Swinwood et al., 1994). Irradiation process only remove microbes from the waste and does not change the shape and size of waste thus mechanical gridding and shredding are also provided in order to increase the efficiency of the irradiation process (Gulyurt, 2012).

Nitrification

Nitrification means production of glass by solid waste mixing and is rarely utilized for the treatment of biomedical waste (Ternes et al., 2004). The high temperatures result in the burning of combustible material as well as killing of the pathogens in the medical waste (Hossain et al., 2011). The material that remains after treatment is encapsulated in glass of low diffusivity (WHO, 2015) as leaching of material from the glass is very low.

Electric Arc Plasma

Plasma treatment has been developed and proposed as an alternative to incineration but has not been used widely for the treatment of the medical waste (Pruss et al., 1998).

FINAL DISPOSAL

The conventional techniques and equipment available (Kovalova et al., 2013) for disposal waste are quite expensive and developing countries are lacking the resource necessary for the implementation for these techniques (Ali et al., 2017). Different methods for the disposal of the waste are followed in different countries of the world (Coronel, 2002). Incineration, autoclave, hydroclave are the technologies to reduce the harm of bio medical waste (Gupta et al., 2009). Among them incineration is a common approach that can be used on site or off site to both treat and dispose waste at the same time (Zhang et al, 2013; Kumar et al., 2015) but the waste can also be decontaminated by using thermal processing, irradiation, chemical or biological treatments (Mathur et al., 2012).

Chemical Treatment

In chemical treatment, the waste is treated by 1% hypochlorite solution for a minimum time of about 30 minutes (Mathur et al., 2012). Chemical treatment is commonly used in USA and the waste after shredding is treated with hypochlorite solution followed by shredding, drying and then landfilling of the waste (Devine et al., 2006).

Secured Landfills

Landfills are for the final disposal of the waste and all the waste after different treatments is disposed to landfills (Cherubini et al., 2009). Different layers of different material are used and the waste is kept in between these layers in order to prevent the leaching of the leachate released after the decomposition of the waste (Kjeldsen et al., 2002). The lining materials like absorbent material, plastic, geotextile are often used to prevent the percolation of waste to soil and water (Devine et al., 2006).

Incineration

Incineration is a process that converts the organic combustible waste into inorganic, incombustible materials by using high temperature (Gomez et al., 2009). This process works under controlled conditions at high temperature and converts the waste into inert material (Kosky et al., 1995). The other main aim of doing incineration of the waste is to destroy the pathogens present in the waste (Lee et al., 1991). Incineration is usually selected to treat wastes that cannot be recycled, reused or disposed off in a land fill site (Giusti, 2009). About 59–60% of medical waste is treated through incineration, 37–20% by steam sterilization (autoclaving) and 4–5% by other treatment methods (landfilling, microwaving, plasma pyrolysis) (Ghasemi & Yusuff, 2016).

Autoclaving

The second most preferred technique for the treatment of the waste is autoclaving (Messerle et al., 2018). It is a very effective technique as it works on the application of heat and pressure for a recommended period of time (Taylor et al., 1996). The metal-based vessel works on the principle of pressure cooker (Mathur et al., 2012) and its designing is such that it can easily sustain high temperature and pressure simultaneously (Messerle et al., 2018). Autoclaving is one of the best sterilization methods which under high temperature and pressure releases steam to inactivate the micro-organisms present in the waste material (Hossain et al., 2011). Autoclave treatment has been recommended for microbiology and biotechnology waste, waste sharps, soiled and solid wastes (Mathur et al., 2012).

Microwave Irradiation

The microwave generates high frequency waves (Mathur et al., 2012) which vibrate the particles and leads to the generation of the heat within the waste (Leonelli & Mason, 2010). This heat generated kills all pathogens present in the waste. Microwave irradiation of hospital waste is alternative technology to that of incineration (Ghasemi & Yusuff, 2016). Microwave irradiation generates steam and electromagnetic waves to destroy the pathogens in the waste (Held et al., 1996). The types of waste generally treated in microwave systems are the same as those treated in autoclaves (Diaz et al., 2005). In microwave irradiation, the wet waste is heated by conduction and all the harmful pathogens in this process are completely destroyed (Zhou et al., 2010).

Hydroclaving

Hydroclave is the process of inactivating most of the microorganisms from the infected waste by exposing them to high temperature and high-pressure steam (Maamari et al., 2016). The operating cost and investment is low and it is more environments friendly. Hydroclaving is required for the shredded infectious waste (Rao et al., 2004).

Plasma Gasification

Plasma gasification is efficient technology for management of the biomedical waste (Huang & Tang, 2007; Ruj & Ghosh, 2014). Plasma is electrically conductive gas in which an important fraction of the atoms is ionized, so that the electrons and ions are separated free (Lieberman & Lichtenberg, 2005). Plasma may be created in the laboratory by a variety of ways including passing a gas which serves as a dielectric between objects with large electrical potential differences or by exposing gases to high temperatures as in the case of arc welding or graphite electrode torches (Anyaegbunam, 2013). The potential difference and subsequent electric field in plasma heating causes ionization of the gas and electrons are pulled toward the anode while the nucleus pulled towards cathode (Lal & Singh, 2012). The plasma gasification is best suited for processing of all kinds of waste like municipal solid waste (MSW), industrial waste and chemical waste (Maczulak, 2010). The energy required to break the chemical bond of the waste material is provided by plasma torch (Materazzi et al., 2014; Camacho, 1995) as extremely intense energy produced in this process is powerful enough to disintegrate the hazardous waste (Gomez et al., 2009). In the limited supply of oxygen, the syngas is produced while the glass like substance at the bottom can be collected and used for different purposes (Nielsen et al., 2015). Medical waste on coming in contact with high temperature plasma arc is converted into CO, H_2 and hydrocarbons (Mathur et al., 2012). Incineration of the waste results in emissions of dioxins, furans and benzo[a]pyrene while plasms gasification helps in the reduction in emission of these compounds into the atmosphere thus making it good option for the treatment of medical waste compared to incineration (Messerle et al., 2018). At this high temperature, all material even the cytotoxic and cytostatic drugs are destroyed (Tsakona, et al., 2007). Due to the strong heat generated by the plasma, it can dispose all types of waste in a safe and reliable manner (Nema & Ganeshprasad, 2002). However, the produced syngas can be used as a feedstock in the Fischer-Tropsch process for liquid fuel production, a gas turbine or a fuel cell for electricity production or chemical products such as ammonia, methanol and hydrogen (Messerle et al., 2018). Thermal plasma process is modern (Nema & Ganeshprasad, 2002) and efficient technology for the treatment of the waste in anoxic condition and during this process intense heat is produced that causes the total destruction of the waste (Anyaegbunam, 2013; Evans Steve, 2009).

CONCLUSION

With the rise of population, the biomedical waste generation has increased as awareness about the health-related issues has increased but the most of the countries are failing in the proper management of the generated biomedical waste. The technology available for the treatment of biomedical waste in most of the countries are not environment friendly and are lacking the innovative techniques which are environment friendly as well as economical for the developing countries. It is crucial to segregate the waste at the source level so that the final disposal of waste does not take time and ultimately helps in curbing the spread of infection. Both infectious and noninfectious waste generated as biomedical waste and if not properly disposed can increase the spread of various infectious diseases and also results in mixing with the general municipal waste. Infectious waste should be burned so that source of infection can be eliminated and noninfectious waste should be properly managed so that effect on local environment can be minimized. There is also need to spread awareness about the biomedical waste and its hazards to the local public, staff as well as workers dealing with such waste. The problem of biomedical waste in the coming years may become one of the main issues and thus it become important to look for the methods which not only decrease the burden of the increasing waste but also be environment friendly.

REFERENCES

Abd El-Salam, M. M. (2010). Hospital waste management in El-Beheira Governorate, Egypt. *Journal of Environmental Management, 91*(3), 618–629. doi:10.1016/j.jenvman.2009.08.012 PMID:19931245

Abdulla, F., Qdais, H. A., & Rabi, A. (2008). Site investigation on medical waste management practices in northern Jordan. *Waste Management, 28*(2), 450-458.

Abdullah, M. O., Tan, I. A. W., & Lim, L. S. (2011). Automobile adsorption air-conditioning system using oil palm biomass-based activated carbon: A review. *Renewable & Sustainable Energy Reviews, 15*(4), 2061–2072. doi:10.1016/j.rser.2011.01.012

Ahn, J. H., Shin, S. G., & Hwang, S. (2009). Effect of microwave irradiation on the disintegration and acidogenesis of municipal secondary sludge. *Chemical Engineering Journal, 153*(1-3), 145–150. doi:10.1016/j.cej.2009.06.032

Al-Khatib, I. A., Al-Qaroot, Y. S., & Ali-Shtayeh, M. S. (2009). Management of healthcare waste in circumstances of limited resources: A case study in the hospitals of Nablus city, Palestine. *Waste Management & Research, 27*(4), 305–312. doi:10.1177/0734242X08094124 PMID:19423582

Al-Khatib, I. A., & Sato, C. (2009). Solid health care waste management status at health care centers in the West Bank–Palestinian Territory. *Waste Management (New York, N.Y.)*, *29*(8), 2398–2403. doi:10.1016/j.wasman.2009.03.014 PMID:19398317

Alhumoud, J. M., & Alhumoud, H. M. (2007). An analysis of trends related to hospital solid wastes management in Kuwait. *Management of Environmental Quality*, *18*(5), 502–513. doi:10.1108/14777830710778274

Ali, M., Wang, W., & Chaudhry, N. (2016). Management of wastes from hospitals: A case study in Pakistan. *Waste Management & Research*, *34*(1), 87–90. doi:10.1177/0734242X15616474 PMID:26628050

Ali, M., Wang, W., Chaudhry, N., & Geng, Y. (2017). Hospital waste management in developing countries: A mini review. *Waste Management & Research*, *35*(6), 581–592. doi:10.1177/0734242X17691344 PMID:28566033

Alvim-Ferraz, M. C. M., & Afonso, S. A. V. (2005). Incineration of healthcare wastes: Management of atmospheric emissions through waste segregation. *Waste Management (New York, N.Y.)*, *25*(6), 638–648. doi:10.1016/j.wasman.2004.07.017 PMID:15993348

Ananth, A. P., Prashanthini, V., & Visvanathan, C. (2010). Healthcare waste management in Asia. *Waste Management (New York, N.Y.)*, *30*(1), 154–161. doi:10.1016/j.wasman.2009.07.018 PMID:19726174

Angelidaki, I., Alves, M., Bolzonella, D., Borzacconi, L., Campos, J. L., Guwy, A. J., ... Van Lier, J. B. (2009). Defining the biomethane potential (BMP) of solid organic wastes and energy crops: A proposed protocol for batch assays. *Water Science and Technology*, *59*(5), 927–934. doi:10.2166/wst.2009.040 PMID:19273891

Anyaegbunam Felix, N. C. (2013). Plasma Gasification for waste management and sustainable renewable clean energy generation. Proceedings of Nigerian Academy of Science, 6(1), 33-50.

Appleton, T. J., Colder, R. I., Kingman, S. W., Lowndes, I. S., & Read, A. G. (2005). Microwave technology for energy-efficient processing of waste. *Applied Energy*, *81*(1), 85–113. doi:10.1016/j.apenergy.2004.07.002

Askarian, M., Vakili, M., & Kabir, G. (2004). Results of a hospital waste survey in private hospitals in Fars province, Iran. *Waste Management, 24*(4), 347-352.

Babanyara, Y. Y., Ibrahim, D. B., Garba, T., Bogoro, A. G., & Abubakar, M. Y. (2013). Poor Medical Waste Management (MWM) practices and its risks to human health and the environment: A literature review. *International Journal of Environmental Health and Engineering*, *11*(7), 1–8.

Bdour, A., Altrabsheh, B., Hadadin, N., & Al-Shareif, M. (2007). Assessment of medical wastes management practice: A case study of the northern part of Jordan. *Waste Management (New York, N.Y.)*, *27*(6), 746–759. doi:10.1016/j.wasman.2006.03.004 PMID:16713238

Beilenhoff, U., Neumann, C. S., Rey, J. F., Biering, H., Blum, R., Cimbro, M., ... Schmidt, V. (2008). ESGE–ESGENA guideline: Cleaning and disinfection in gastrointestinal endoscopy. *Endoscopy*, *40*(11), 939–957. doi:10.1055-2008-1077722 PMID:19009486

Berry, L. L., & Parasuraman, A. (2004). *Marketing services: Competing through quality*. Simon and Schuster.

Blowers, A. (2013). Pollution and waste-a sustainable burden? In *Planning for a Sustainable Environment* (pp. 81–104). Routledge. doi:10.4324/9781315070315

Bordner, R., Winter, J. A., & Scarpino, P. (Eds.). (1978). *Microbiological methods for monitoring the environment: water and wastes*. Environmental Protection Agency, Office of Research and Development, Environmental Monitoring and Support Laboratory.

Camacho, S. L. (1995). The Plasma Arc Torch, Its Electrical and Thermal Characteristics. In *Proceedings of the International Symposium on Environmental Technologies: Plasma Systems and Applications*. Georgia Institute of Technology Research Corporation.

Capoor, M. R., & Bhowmik, K. T. (2017). Current perspectives on biomedical waste management: Rules, conventions and treatment technologies. *Indian Journal of Medical Microbiology*, *35*(2), 157. PMID:28681801

CEET. (2008). *Biomedical Waste Management- Burgeoning issue*. CEET.

Chakraborty, S., Veeregowda, B., Gowda, L., Sannegowda, S. N., Tiwari, R., Dhama, K., & Singh, S. V. (2013). Biomedical waste management. *Interaction*.

Chandler, A. J., Eighmy, T. T., Hjelmar, O., Kosson, D. S., Sawell, S. E., Vehlow, J., ... Hartlén, J. (1997). *Municipal solid waste incinerator residues* (Vol. 67). Elsevier.

Chartier, Y. (Ed.). (2014). *Safe management of wastes from health-care activities*. World Health Organization.

Cheng, Y. W., Sung, F. C., Yang, Y., Lo, Y. H., Chung, Y. T., & Li, K. C. (2009). Medical waste production at hospitals and associated factors. *Waste Management (New York, N.Y.)*, *29*(1), 440–444. doi:10.1016/j.wasman.2008.01.014 PMID:18359619

Cherubini, F., Bargigli, S., & Ulgiati, S. (2009). Life cycle assessment (LCA) of waste management strategies: Landfilling, sorting plant and incineration. *Energy*, *34*(12), 2116–2123. doi:10.1016/j.energy.2008.08.023

Chowdhary, A. (2018). Study of knowledge, behaviour and practice of biomedical waste among health personnel. *International Journal of Community Medicine and Public Health*, *5*(8), 3330–3334. doi:10.18203/2394-6040.ijcmph20183056

Clapp, J. (2002). The distancing of waste: Overconsumption in a global economy. *Confronting Consumption*, 155-176.

Coker, A., Sangodoyin, A., Sridhar, M., Booth, C., Olomolaiye, P., & Hammond, F. (2009). Medical waste management in Ibadan, Nigeria: Obstacles and prospects. *Waste Management (New York, N.Y.)*, *29*(2), 804–811. doi:10.1016/j.wasman.2008.06.040 PMID:18835151

Coronel, B., Duroselle, P., Behr, H., Moskovtchenko, J. F., & Freney, J. (2002). In situ decontamination of medical wastes using oxidative agents: A 16-month study in a polyvalent intensive care unit. *The Journal of Hospital Infection*, *50*(3), 207–212. doi:10.1053/jhin.2002.1188 PMID:11886197

Da Silva, C. E., Hoppe, A. E., Ravanello, M. M., & Mello, N. (2005). Medical wastes management in the south of Brazil. *Waste Management (New York, N.Y.)*, *25*(6), 600–605. doi:10.1016/j.wasman.2004.03.002 PMID:15993344

Daly, M. J. (2009). A new perspective on radiation resistance based on Deinococcus radiodurans. *Nature Reviews. Microbiology*, *7*(3), 237–245. doi:10.1038/nrmicro2073 PMID:19172147

Datta, P., Mohi, G. K., & Chander, J. (2018). Biomedical waste management in India: Critical appraisal. *Journal of Laboratory Physicians*, *10*(1), 6. doi:10.4103/JLP.JLP_89_17 PMID:29403196

De Brito, M. P., & Dekker, R. (2004). A framework for reverse logistics. In *Reverse logistics* (pp. 3–27). Berlin: Springer. doi:10.1007/978-3-540-24803-3_1

Dennis, P. (2017). *The Roles of Strategic Environmental Assessment and Learning in Planning for Successful Community-Based Solid Waste Management in Kenya*. Academic Press.

Devine, T. J., & Clark, B. T. (2006). *Mobile apparatus and process for treating infectious waste*. U.S. Patent 7,144,550.

Dias-Ferreira, C., Valente, S., & Vaz, J. (2016). Practices of pharmaceutical waste generation and discarding in households across Portugal. *Waste Management & Research*, *34*(10), 1006–1013. doi:10.1177/0734242X16639388 PMID:27026517

Diaz, L. F., Savage, G. M., & Eggerth, L. L. (2005). Alternatives for the treatment and disposal of health-care wastes in developing countries. *Waste Management (New York, N.Y.)*, *25*(6), 626–637. doi:10.1016/j.wasman.2005.01.005 PMID:15993347

Dutta, S. K., Upadhyay, V. P., & Sridharan, U. (2006). Environmental management of industrial hazardous wastes in India. *Journal of Environmental Science & Engineering*, *48*(2), 143. PMID:17913193

El-Salam, M. M. A. (2010). Hospital waste management in El-Beheira Governorate, Egypt. *Journal of Environmental Management*, *91*(3), 618–629. doi:10.1016/j.jenvman.2009.08.012 PMID:19931245

Evans Steve, D. (2009). *Plasma Gasification Plant Benefits for Municipal Waste Management*. Available at http://www.articlesbase.com/literaturearticles/plasma-gasificationplant-benefits-for-municipal-waste-management-850915

Friedman, M. S., Powell, K. E., Hutwagner, L., Graham, L. M., & Teague, W. G. (2001). Impact of changes in transportation and commuting behaviors during the 1996 Summer Olympic Games in Atlanta on air quality and childhood asthma. *Journal of the American Medical Association*, *285*(7), 897–905. doi:10.1001/jama.285.7.897 PMID:11180733

Gellynck, X., Jacobsen, R., & Verhelst, P. (2011). Identifying the key factors in increasing recycling and reducing residual household waste: A case study of the Flemish region of Belgium. *Journal of Environmental Management*, *92*(10), 2683–2690. doi:10.1016/j.jenvman.2011.06.006 PMID:21704444

Gershon, R. R., Vlahov, D., Felknor, S. A., Vesley, D., Johnson, P. C., Delcios, G. L., & Murphy, L. R. (1995). Compliance with universal precautions among health care workers at three regional hospitals. *American Journal of Infection Control*, *23*(4), 225–236. doi:10.1016/0196-6553(95)90067-5 PMID:7503434

Ghasemi, M. K., & Yusuff, R. B. (2016). Advantages and Disadvantages of Healthcare Waste Treatment and Disposal Alternatives: Malaysian Scenario. *Polish Journal of Environmental Studies*, *25*(1).

Giusti, L. (2009). A review of waste management practices and their impact on human health. *Waste Management (New York, N.Y.)*, *29*(8), 2227–2239. doi:10.1016/j.wasman.2009.03.028 PMID:19401266

Gomez, E., Rani, D. A., Cheeseman, C. R., Deegan, D., Wise, M., & Boccaccini, A. R. (2009). Thermal plasma technology for the treatment of wastes: A critical review. *Journal of Hazardous Materials*, *161*(2-3), 614–626. doi:10.1016/j.jhazmat.2008.04.017 PMID:18499345

Gulyurt, M. (2012). Biomedical instrument application: Medical waste treatment technologies. In *A Roadmap of Biomedical Engineers and Milestones*. IntechOpen. doi:10.5772/39276

Gupta, A. (2014). Human hair "waste" and its utilization: gaps and possibilities. *Journal of Waste Management*.

Gupta, S. (2017). Biomedical waste in India a review. *Journal of Advanced Medical and Dental Sciences Research*, *5*(4), 99.

Gupta, S., Boojh, R., Mishra, A., & Chandra, H. (2009). Rules and management of biomedical waste at Vivekananda Polyclinic: A case study. *Waste Management (New York, N.Y.)*, *29*(2), 812–819. doi:10.1016/j.wasman.2008.06.009 PMID:18684608

Gupta, S., Mohan, K., Prasad, R., Gupta, S., & Kansal, A. (1998). Solid waste management in India: Options and opportunities. *Resources, Conservation and Recycling*, *24*(2), 137–154. doi:10.1016/S0921-3449(98)00033-0

Hall, R., Belson, D., Murali, P., & Dessouky, M. (2006). Modeling patient flows through the healthcare system. In Patient flow: Reducing delay in healthcare delivery (pp. 1-44). Academic Press. doi:10.1007/978-0-387-33636-7_1

Harbec, S. (2009). *The Uptake of Sustainable Wastes Management: the Case of Electronic Medical Equipment* (Doctoral dissertation). École Polytechnique de Montréal.

Hassan, A., Tudor, T., & Vaccari, M. (2018). Healthcare waste management: A case study from Sudan. *Environments*, *5*(8), 89. doi:10.3390/environments5080089

Heberlein, J., & Murphy, A. B. (2008). Thermal plasma waste treatment. *Journal of Physics. D, Applied Physics*, *41*(5), 053001. doi:10.1088/0022-3727/41/5/053001

Heffner, H. A. (2004). *Analysis of post-use hypodermic needle medical waste disposal* (Doctoral dissertation). Georgia Institute of Technology.

Hegde, V., Kulkarni, R. D., & Ajantha, G. S. (2007). Biomedical waste management. *Journal of Oral and Maxillofacial Pathology*, *11*(1), 5. doi:10.4103/0973-029X.33955

Held, J. S., & Sharp, J. W. (1996). *Apparatus and method for processing medical waste*. U.S. Patent 5, 508, 004.

Herwaldt, L. A., & Rutala, W. A. (1996). Disinfection and sterilization of patient-care items. *Infection Control and Hospital Epidemiology*, *17*(6), 377–384. doi:10.2307/30141141 PMID:8805073

Hirani, D. P., Villaitramani, K. R., & Kumbhar, S. J. (2014). Biomedical waste: An introduction to its management. *International Journal of Innovative Research in Advanced Engineering*, *1*(8), 82–87.

Hossain, M. S., Santhanam, A., Norulaini, N. N., & Omar, A. M. (2011). Clinical solid waste management practices and its impact on human health and environment–A review. *Waste Management (New York, N.Y.)*, *31*(4), 754–766. doi:10.1016/j.wasman.2010.11.008 PMID:21186116

Huang, H., & Tang, L. (2007). Treatment of organic waste using thermal plasma pyrolysis technology. *Energy Conversion and Management*, *48*(4), 1331–1337. doi:10.1016/j.enconman.2006.08.013

Jaffery, S. (2013). *Medical waste illegally sold off from Pakistan hospital*. Academic Press.

Jain, S., Agarwal, A., Jani, V., Singhal, S., Sharma, P., & Jalan, R. (2017). Assessment of carbon neutrality and sustainability in educational campuses (CaNSEC): A general framework. *Ecological Indicators*, *76*, 131–143. doi:10.1016/j.ecolind.2017.01.012

Jang, Y. C., Lee, C., Yoon, O. S., & Kim, H. (2006). Medical waste management in Korea. *Journal of Environmental Management*, *80*(2), 107–115. doi:10.1016/j.jenvman.2005.08.018 PMID:16338054

Jayathilakan, K., Sultana, K., Radhakrishna, K., & Bawa, A. S. (2012). Utilization of byproducts and waste materials from meat, poultry and fish processing industries: A review. *Journal of Food Science and Technology*, *49*(3), 278–293. doi:10.100713197-011-0290-7 PMID:23729848

Kim, K. T., Graves, P. B., Safadi, G. S., Alhadeff, G., & Metcalfe, J. (1998). Implementation recommendations for making health care facilities latex safe. *AORN Journal*, *67*(3), 615–632. doi:10.1016/S0001-2092(06)62830-5 PMID:9541704

King, A. M., Burgess, S. C., Ijomah, W., & McMahon, C. A. (2006). Reducing waste: Repair, recondition, remanufacture or recycle. *Sustainable Development*, *14*(4), 257–267. doi:10.1002d.271

Kjeldsen, P., Barlaz, M. A., Rooker, A. P., Baun, A., Ledin, A., & Christensen, T. H. (2002). Present and long-term composition of MSW landfill leachate: A review. *Critical Reviews in Environmental Science and Technology*, *32*(4), 297–336. doi:10.1080/10643380290813462

Kosky, J. P., Jones, J. F., & Corp, M. E. I. (1995). *System and method for the treatment of hazardous waste material*. U.S. Patent 5, 471,937.

Kumar, R., Khan, E. A., Ahmed, J., Khan, Z., Magan, M., Nousheen, N., & Mughal, M. I. (2010). Healthcare waste management (HCWM) in Pakistan: Current situation and training options. *Journal of Ayub Medical College, Abbottabad*, *22*(4), 101–106. PMID:22455273

Kumar, R., Shaikh, B. T., Somrongthong, R., & Chapman, R. S. (2015). Practices and challenges of infectious waste management: A qualitative descriptive study from tertiary care hospitals in Pakistan. *Pakistan Journal of Medical Sciences*, *31*(4), 795. doi:10.12669/pjms.314.7988 PMID:26430405

Kumar, R., Somrongthong, R., & Shaikh, B. T. (2015). Effectiveness of intensive healthcare waste management training model among health professionals at teaching hospitals of Pakistan: A quasi-experimental study. *BMC Health Services Research*, *15*(1), 81. doi:10.118612913-015-0758-7 PMID:25889451

Lal, P. M., & Singh, C. J. (2012). Plasma Gasification: A Sustainable Solution for the Municipal Solid Waste Management in the State of Madhya Pradesh, India. *International Journal of Environmental Sciences, 3*(1), 306.

Larsen, T. A., Lienert, J., Joss, A., & Siegrist, H. (2004). How to avoid pharmaceuticals in the aquatic environment. *Journal of Biotechnology, 113*(1-3), 295–304. doi:10.1016/j.jbiotec.2004.03.033 PMID:15380662

Lee, C. C., Huffman, G., & Nalesnik, R. (1991). Medical waste management. *Environmental Science & Technology, 25*(3), 360–363. doi:10.1021/es00015a607

Leonelli, C., & Mason, T. J. (2010). Microwave and ultrasonic processing: now a realistic option for industry. *Chemical Engineering and Processing: Process Intensification, 49*(9), 885-900.

Li, C. S., & Jenq, F. T. (1993). Physical and chemical composition of hospital waste. *Infection Control and Hospital Epidemiology, 14*(3), 145–150. doi:10.1086/646700 PMID:8478527

Lieberman, M. A., & Lichtenberg, A. J. (2005). *Principles of plasma discharges and materials processing* (Vol. 2). New York: Wiley-interscience. doi:10.1002/0471724254

Liu, Y. Y., Wang, Y., Walsh, T. R., Yi, L. X., Zhang, R., Spencer, J., ... Yu, L. F. (2016). Emergence of plasmid-mediated colistin resistance mechanism MCR-1 in animals and human beings in China: A microbiological and molecular biological study. *The Lancet Infectious Diseases, 16*(2), 161–168. doi:10.1016/S1473-3099(15)00424-7 PMID:26603172

Maamari, O., Mouaffak, L., Kamel, R., Brandam, C., Lteif, R., & Salameh, D. (2016). Comparison of steam sterilization conditions efficiency in the treatment of Infectious Health Care Waste. *Waste Management (New York, N.Y.), 49*, 462–468. doi:10.1016/j.wasman.2016.01.014 PMID:26803472

Maczulak, A. E. (2010). *Waste treatment: reducing global waste* (Vol. 2). Infobase Publishing.

Mandal, S. K., & Dutta, J. (2009). Integrated bio-medical waste management plan for Patna city. Institute of Town Planners. *India Journal, 6*(2), 1–25.

Marinkovic, N., Vitale, K., Holcer, N. J., Džakula, A., & Pavic, T. (2008). Management of hazardous medical waste in Croatia. *Waste Management, 28*(6), 1049-1056.

Mastorakis, N. E., Bulucea, C. A., Oprea, T. A., Bulucea, C. A., & Dondon, P. (2011). Holistic approach of biomedical waste management system with regard to health and environmental risks. Development, Energy, Environment. *Economics, 5*(3), 287–295.

Materazzi, M., Lettieri, P., Mazzei, L., Taylor, R., & Chapman, C. (2014). Tar evolution in a two stage fluid bed–plasma gasification process for waste valorization. *Fuel Processing Technology, 128*, 146–157. doi:10.1016/j.fuproc.2014.06.028

Mathur, P., Patan, S., & Shobhawat, S. (2012). Need of biomedical waste management system in hospitals-An emerging issue-a review. *Current World Environment, 7*(1), 117–124. doi:10.12944/CWE.7.1.18

Mbongwe, B., Mmereki, B. T., & Magashula, A. (2008). Healthcare waste management: Current practices in selected healthcare facilities, Botswana. *Waste Management (New York, N.Y.)*, *28*(1), 226–233. doi:10.1016/j.wasman.2006.12.019 PMID:17350817

Messerle, V. E., Mosse, A. L., & Ustimenko, A. B. (2018). Processing of biomedical waste in plasma gasifier. *Waste Management (New York, N.Y.)*, *79*, 791–799. doi:10.1016/j.wasman.2018.08.048 PMID:30343813

Misra, V., & Pandey, S. D. (2005). Hazardous waste, impact on health and environment for development of better waste management strategies in future in India. *Environment International*, *31*(3), 417–431. doi:10.1016/j.envint.2004.08.005 PMID:15734194

Miyazaki, M., & Une, H. (2005). Infectious waste management in Japan: A revised regulation and a management process in medical institutions. *Waste Management, 25*(6), 616-621.

Mostafa, G. M., Shazly, M. M., & Sherief, W. I. (2009). Development of a waste management protocol based on assessment of knowledge and practice of healthcare personnel in surgical departments. *Waste Management, 29*(1), 430-439.

Nema, A., Pathak, A., Bajaj, P., Singh, H., & Kumar, S. (2011). A case study: Biomedical waste management practices at city hospital in Himachal Pradesh. *Waste Management & Research*, *29*(6), 669–673. doi:10.1177/0734242X10396753 PMID:21382874

Nema, S. K., & Ganeshprasad, K. S. (2002). Plasma pyrolysis of medical waste. *Current Science*, 271–278.

Nemathaga, F., Maringa, S., & Chimuka, L. (2008). Hospital solid waste management practices in Limpopo Province, South Africa: A case study of two hospitals. *Waste Management (New York, N.Y.)*, *28*(7), 1236–1245. doi:10.1016/j.wasman.2007.03.033 PMID:18060760

Ngoc, U. N., & Schnitzer, H. (2009). Sustainable solutions for solid waste management in Southeast Asian countries. *Waste Management (New York, N.Y.)*, *29*(6), 1982–1995. doi:10.1016/j.wasman.2008.08.031 PMID:19285384

Ngwuluka, N., Ochekpe, N., Odumosu, P., & John, S. A. (2009). Waste management in healthcare establishments within Jos Metropolis, Nigeria. *African Journal of Environmental Science and Technology*, *3*(12).

Nielsen, M. C., DePuy, R. A., Kumar, A., Lyons, J. P. F., Lissianski, V., Shi, R., . . . Zamansky, V. (2015). *Plasma-assisted waste gasification system*. U.S. Patent 9,074,152.

Nwachukwu, N. C., Orji, F. A., & Ugbogu, O. C. (2013). Health care waste management–public health benefits, and the need for effective environmental regulatory surveillance in federal Republic of Nigeria. In Current topics in public health. InTechOpen.

Oli, A. N., Ekejindu, C. C., Adje, D. U., Ezeobi, I., Ejiofor, O. S., Ibeh, C. C., & Ubajaka, C. F. (2016). Healthcare waste management in selected government and private hospitals in Southeast Nigeria. *Asian Pacific Journal of Tropical Biomedicine*, *6*(1), 84–89. doi:10.1016/j.apjtb.2015.09.019

Omar, D., Nazli, S. N., Subramaniam, A., & Karuppannan, L. (2012). Clinical waste management in district hospitals of Tumpat, Batu Pahat and Taiping. *Procedia: Social and Behavioral Sciences*, *68*, 134–145. doi:10.1016/j.sbspro.2012.12.213

Oweis, R., Al-Widyan, M., & Ohood Al-Limoon, O. (2005). *Medical waste management in Jordan: a study at the King Hussein*. Academic Press.

Pasupathi, P., Sindhu, S., & Ponnusha, B. S. (2011). Biomedical waste management for health care industry: A review. *International Journal of Biological & Medical Research*, *2*(2), 472–486.

Patil, A. D., & Shekdar, A. V. (2001). Health-care waste management in India. *Journal of Environmental Management*, *63*(2), 211–220. doi:10.1006/jema.2001.0453 PMID:11721600

Patil, G. V., & Pokhrel, K. (2005). Biomedical solid waste management in an Indian hospital: A case study. *Waste Management (New York, N.Y.)*, *25*(6), 592–599. doi:10.1016/j.wasman.2004.07.011 PMID:15993343

Patwary, M. A., O'Hare, W. T., & Sarker, M. H. (2011). Assessment of occupational and environmental safety associated with medical waste disposal in developing countries: A qualitative approach. *Safety Science*, *49*(8-9), 1200–1207. doi:10.1016/j.ssci.2011.04.001

Perrin, D., & Barton, J. (2001). Issues associated with transforming household attitudes and opinions into materials recovery: A review of two kerbside recycling schemes. *Resources, Conservation and Recycling*, *33*(1), 61–74. doi:10.1016/S0921-3449(01)00075-1

Pittet, D., Allegranzi, B., Storr, J., & Donaldson, L. (2006). 'Clean care is safer care': The global patient safety challenge 2005–2006. *International Journal of Infectious Diseases*, *10*(6), 419–424. doi:10.1016/j.ijid.2006.06.001 PMID:16914344

Poon, C. S., Yu, A. T., & Jaillon, L. (2004). Reducing building waste at construction sites in Hong Kong. *Construction Management and Economics*, *22*(5), 461–470. doi:10.1080/0144619042000202816

Porter, M. E., & Van der Linde, C. (1995). Toward a new conception of the environment-competitiveness relationship. *The Journal of Economic Perspectives*, *9*(4), 97–118. doi:10.1257/jep.9.4.97

Pruss, A., & Townend, W. K. (1998). *Teacher's guide: management of wastes from health-care activities* (No. WHO/EOS/98.6). Geneva: World Health Organization.

Pruss-Ustün, A., Rapiti, E., & Hutin, Y. (2005). Estimation of the global burden of disease attributable to contaminated sharps injuries among health-care workers. *American Journal of Industrial Medicine*, *48*(6), 482–490. doi:10.1002/ajim.20230 PMID:16299710

Radha, K. V., Kalaivani, K., & Lavanya, R. (2009). A case study of biomedical waste management in hospitals. *Global Journal of Health Science*, *1*(1), 82–88.

Rajan, R., Robin, D. T., & Vandanarani, M. (2018). Biomedical waste management in Ayurveda hospitals e current practices & future prospective. *Journal of Ayurveda and Integrative Medicine*, 1–8. PMID:29555257

Rao, R. K., & Sharm, V. R. (2004). Biomedical Waste Management: An Infrastructural Survey of Hospitals. *MJAFI*, *60*, 379-382.

Rao, S. K. M., Ranyal, R. K., Bhatia, S. S., & Sharma, V. R. (2004). Biomedical waste management: An infrastructural survey of hospitals. *Medical Journal, Armed Forces India*, *60*(4), 379–382. doi:10.1016/S0377-1237(04)80016-9 PMID:27407678

Roseland, M. (2012). *Toward sustainable communities: Solutions for citizens and their governments*. New Society Publishers.

Ruj, B., & Ghosh, S. (2014). Technological aspects for thermal plasma treatment of municipal solid waste—A review. *Fuel Processing Technology*, *126*, 298–308. doi:10.1016/j.fuproc.2014.05.011

Rutala, W. A., & Mayhall, C. G. (1992). Medical waste. *Infection Control and Hospital Epidemiology*, *13*(1), 38–48. doi:10.2307/30146966 PMID:1545111

Rutala, W. A., & Weber, D. J. (2007). How to assess risk of disease transmission to patients when there is a failure to follow recommended disinfection and sterilization guidelines. *Infection Control and Hospital Epidemiology*, *28*(2), 146–155. doi:10.1086/511700 PMID:17265395

Rutala, W. A., & Weber, D. J. (2010). Guideline for disinfection and sterilization of prion-contaminated medical instruments. *Infection Control and Hospital Epidemiology*, *31*(2), 107–117. doi:10.1086/650197 PMID:20055640

Saad, S. A. G. (2013). Management of hospitals solid waste in Khartoum State. *Environmental Monitoring and Assessment*, *185*(10), 8567–8582. doi:10.100710661-013-3196-1 PMID:23644667

Sankoh, F. P., Yan, X., & Tran, Q. (2013). Environmental and health impact of solid waste disposal in developing cities: A case study of granville brook dumpsite, Freetown, Sierra Leone. *Journal of Environmental Protection*, *4*(07), 665–670. doi:10.4236/jep.2013.47076

Sharholy, M., Ahmad, K., Mahmood, G., & Trivedi, R. C. (2008). Municipal solid waste management in Indian cities–A review. *Waste Management (New York, N.Y.)*, *28*(2), 459–467. doi:10.1016/j.wasman.2007.02.008 PMID:17433664

Sheavly, S. B., & Register, K. M. (2007). Marine debris & plastics: Environmental concerns, sources, impacts and solutions. *Journal of Polymers and the Environment*, *15*(4), 301–305. doi:10.100710924-007-0074-3

Shekdar, A. V. (2009). Sustainable solid waste management: An integrated approach for Asian countries. *Waste Management (New York, N.Y.)*, *29*(4), 1438–1448. doi:10.1016/j.wasman.2008.08.025 PMID:19081236

Singh, R. P., Singh, P., Araujo, A. S., Ibrahim, M. H., & Sulaiman, O. (2011). Management of urban solid waste: Vermicomposting a sustainable option. *Resources, Conservation and Recycling*, *55*(7), 719–729. doi:10.1016/j.resconrec.2011.02.005

Smith, C. A. (2002). Managing pharmaceutical waste. *The Journal of the Pharmacy Society of Wisconsin: JPSW*, *5*, 17–22.

Song, Q., Li, J., & Zeng, X. (2015). Minimizing the increasing solid waste through zero waste strategy. *Journal of Cleaner Production*, *104*, 199–210. doi:10.1016/j.jclepro.2014.08.027

Sreekanth, K., Vishal, G. N., Raghunandan, H. V., & Nitin, K. U. (2014). A review on managing of pharmaceutical waste in industry. *International Journal of Pharm Tech Research*, *6*(3), 899–907.

Swinwood, J. F., Waite, T. D., Kruger, P., & Rao, S. M. (1994). Radiation technologies for waste treatment: A global perspective. *International Atomic Energy Agency Bulletin*, *36*(1), 11–15.

Szycher, M. (2017). Medical pharmaceutical markets for medical plastics. In High Performance Biomaterials (pp. 3-30). Academic Press. doi:10.1201/9780203752029-1

Tabata, T., Hishinuma, T., Ihara, T., & Genchi, Y. (2011). Life cycle assessment of integrated municipal solid waste management systems, taking account of climate change and landfill shortage trade-off problems. *Waste Management & Research*, *29*(4), 423–432. doi:10.1177/0734242X10379493 PMID:20699290

Ternes, T. A., Joss, A., & Siegrist, H. (2004). *Peer reviewed: scrutinizing pharmaceuticals and personal care products in wastewater treatment*. Academic Press.

Thomas, K. V., Dye, C., Schlabach, M., & Langford, K. H. (2007). Source to sink tracking of selected human pharmaceuticals from two Oslo city hospitals and a wastewater treatment works. *Journal of Environmental Monitoring*, *9*(12), 1410–1418. doi:10.1039/b709745j PMID:18049781

Tiwari, A. V., & Kadu, P. A. (2013). Biomedical Waste Management Practices in India-A Review. *International Journal of Current Engineering and Technology*, *3*(5), 2030–2033.

Tsakona, M., Anagnostopoulou, E., & Gidarakos, E. (2007). Hospital waste management and toxicity evaluation: A case study. *Waste Management (New York, N.Y.)*, *27*(7), 912–920. doi:10.1016/j.wasman.2006.04.019 PMID:16872822

Tudor, T. L., Noonan, C. L., & Jenkin, L. E. (2005). Healthcare waste management: A case study from the National Health Service in Cornwall, United Kingdom. *Waste Management (New York, N.Y.)*, *25*, 606–615. doi:10.1016/j.wasman.2004.10.004 PMID:15993345

Turan, N. G., Çoruh, S., Akdemir, A., & Ergun, O. N. (2009). Municipal solid waste management strategies in Turkey. *Waste Management (New York, N.Y.)*, *29*(1), 465–469. doi:10.1016/j.wasman.2008.06.004 PMID:18644708

Vasistha, P., Ganguly, R., & Gupta, A. K. (2018). Biomedical Waste Generation and Management in Public Sector Hospital in Shimla City. In *Environmental Pollution* (pp. 225–232). Singapore: Springer. doi:10.1007/978-981-10-5792-2_19

Verma, L. K., Mani, S., Sinha, N., & Rana, S. (2008). Biomedical waste management in nursing homes and smaller hospitals in Delhi. *Waste Management (New York, N.Y.)*, *28*(12), 2723–2734. doi:10.1016/j.wasman.2007.12.013 PMID:18358710

Wilburn, S. Q., & Eijkemans, G. (2004). Preventing needle stick injuries among healthcare workers: A WHO-ICN collaboration. *International Journal of Occupational and Environmental Health*, *10*(4), 451–456. doi:10.1179/oeh.2004.10.4.451 PMID:15702761

Wilson, D. C., Velis, C., & Cheeseman, C. (2006). Role of informal sector recycling in waste management in developing countries. *Habitat International*, *30*(4), 797–808. doi:10.1016/j.habitatint.2005.09.005

Wittmann, M. (2010). *Sustainable healthcare design*. In *Evidence-based Design for Healthcare Facilities*. Sigma Theta Tau International.

Woolridge, A. C., Ward, G. D., Phillips, P. S., Collins, M., & Gandy, S. (2006). Life cycle assessment for reuse/recycling of donated waste textiles compared to use of virgin material: An UK energy saving perspective. *Resources, Conservation and Recycling*, *46*(1), 94–103. doi:10.1016/j.resconrec.2005.06.006

World Health Organization. (2015). *Guidance on regulations for the transport of infectious substances 2015–2016: Applicable as of 1 January 2015*. Author.

Worrell, W. A., & Vesilind, P. A. (2011). *Solid waste engineering*. Cengage Learning.

Yong, Z., Gang, X., Guanxing, W., Tao, Z. & Dawei, J. (2009). Medical waste management in China: a case study of Nanjing. *Waste Management, 29*(4), 1376-1382.

Zhang, H. J., Zhang, Y. H., Wang, Y., Yang, Y. H., Zhang, J., Wang, Y. L., & Wang, J. L. (2013). Investigation of medical waste management in Gansu Province, China. *Waste Management & Research*, *31*(6), 655–659. doi:10.1177/0734242X13482161 PMID:23567844

Zhou, B. W., Shin, S. G., Hwang, K., Ahn, J. H., & Hwang, S. (2010). Effect of microwave irradiation on cellular disintegration of Gram positive and negative cells. *Applied Microbiology and Biotechnology*, *87*(2), 765–770. doi:10.100700253-010-2574-7 PMID:20393700

Zurbrugg, C., Gfrerer, M., Ashadi, H., Brenner, W., & Küper, D. (2012). Determinants of sustainability in solid waste management–The Gianyar Waste Recovery Project in Indonesia. *Waste Management (New York, N.Y.)*, *32*(11), 2126–2133. doi:10.1016/j.wasman.2012.01.011 PMID:22330265

Chapter 7
Bioconversion of Seri Waste to Value-Added Products:
Innovations in Sericulture Industry

Khursheed Ahmad Wani
Government Degree College Bijbehara, India

Azad Gull
P4 Basic Seed Farm, India

Ashaq Ahmad Dar
Pondicherry University, India

Shubeena Nazir
Government of Jammu and Kashmir, India

ABSTRACT

The conversion of waste into vermicompost is seen as an opportunity by different workers that may be advantageous in terms of improving the structure and physical conditions of the soil, increase the water retention capacity and availability of phosphorous to the growing plants, prevention of leaching loss of soluble inorganic nitrogen, and addition of nutrients/essential micro nutrients. It is observed that the vermicompost obtained from seri waste has better nutrient value as compared to conventional fertilizers. The success of sericulture industry in future will depend on sustainability of organic fertilizers both in terms of nutrient management and environmental protection. Similarly, the waste can be used for biogas production, as activated carbon, feed for the livestock, etc. are the opportunities of converting waste into value-added products. This chapter will review the different types of waste that is produced from sericulture and can be used to form value-added services.

DOI: 10.4018/978-1-7998-0031-6.ch007

INTRODUCTION

Waste is 'any substance or object which the holder discards or intends to discard' (Waste Framework Directive 75/442/EEC 1975).The compelling need to harness the potentials of the numerous agro-industrial by-products and the so-called "Wastes" as part to increase the economy of farmers. Recycling is the retrieval of ingredients to style novel outputs and condense the magnitude of virgin untreated constituents desirable to encounter customer demands (REI Report, 2016). It involves Recycling includes the three steps, collection and processing, manufacturing, and purchasing new products made from recycled materials, which create a continuous loop. Recycling has been reported from ancient times as early as when broken American rock equipment were reused for different purposes (Blakeman, 2004). It is considered as an indispensable chunk of Sustainable Material Management, a process that accentuates the fruitful and sustainable material use throughout their life span, while curtailing their ambient effects at the same time. Moreover, as per the report, recycling and recycled goods frolic a portentous social and economic duty. It was estimated that there are 1.57, $76030 and $14101 jobs, wages and tax revenues, respectively for 907 tons of recyclable waste for collection and recycling.

The domestication of plants and animals with modern technology has caused a conspicuous impact on atmosphere, hydrosphere and lithosphere (Larson & Fuller, 2014). It characterizes on a particular sort of evolutionary process motivated by human selection, hence performs a critical part in Darwin's theory of evolution (Jensen, 2014). The major domesticated insect, which has been commercially exploited is the mulberry silkworm (Bombyx mori L.), accounting for the major share in total silk production (89%) (Patil et al., 2009). Sericulture, is a cost effective technique practiced in about 25 countries all over the world. The main sericulture product is the natural silk fiber. Over the years man have utilized silk to produce textiles of great value and beauty. Silk Industry provides employment to 30 million families in countries such as China, India, Brazil, Bulgaria, Vietnam, Korea and Thailand (Kim et al., 2010).

Sericulture is an agro-based employment-intensive occupation for rural and semi urban areas, and mostly practiced by average economy people. Mulberry is an essential crop for silkworm production. The main use of mulberry globally is to produce silk but in some countries, it is also appreciated for its fruit delicious vegetable, medicinal properties in infusions, landscaping and as animal feed etc. sericulture had been such an important industry that mulberry was not used for any other purposes in Japan. However, with the recent decline of sericulture industry, mulberry has been re-evaluated for other purposes, such as medicinal, fruit and animal production (Singhal et al. 2005a, 2005b, 2005a). Recently, importance of natural products is being invigorated to lessen various health discrepancies. The association between health and diet is well documented and the consumers' trend reflects conscience towards their dietary habits. Probing these links has led to the emergence of functional, nutraceuticals and pharma foods, now taking hold over global nutrition market. The different phytochemicals in diet could provide protection against several threats like free radical formation, degenerative disorders and lifestyle related diseases but still role of active ingredients should be unveiled. It has a exciting nutritional profile containing proteins, phenolics, flavonoids and anthocyanins that increases its significance as promising nature's functional tonic.

The breeding of insects is associated to the problem of waste, such as excreta and leaf waste. A major issue of the silk industry is the need to make the most efficient use of natural fibers, but also to utilize the waste products. Besides the main product of cocoons, a series of secondary products have unique economic value, if properly done. It involves a large scale of inter-reliant technologies from which results different by products and wastes. These ones may however be turned into new commercial products with a high useful value. Huge quantity of papal waste and silkworm litter are produced per year as a by-product of sericulture industry. It has been observed that three kinds of waste generates from sericulture. Bed waste from silkworm rearing, mulberry twigs, surplus leaves and fruits and waste from cocoon reeling such as pupae. Out of these major portions of sericulture waste comes from mulberry farm. 1 Kg of raw silk 8.014 Kg wet and 2 Kg of dry pupa is formed. It has been argued that the silkworm digests only 55% of the ingested leaf and the rest comes out as litter (Patil et al., 2013). It is estimated that about 2794 kg of litter per acre per year could be obtained by rearing 300 layings (one laying = 400eggs) per acre. Marginal farmers may produce 250–300 kg of silkworm waste, which is equal to 2500 kg farm manure and may be used to fertilize 0.067 ha farmland (Wenhua, 2001).In many of the sericulture producing countries after the silk reeling, the pupae are discarded without any utilization though they are rich in proteins (80%), fats, carbohydrates and vitamins. A little work has been done on the utilization of secondary wastes of silk industry, which will definitely enhance the profit of silk industry (Majumder, 1997). Research from China, India and Japan, has collected the information on existing status of secondary waste products utilization and processing. There is a dire need to describe an approach for proper utilization of secondary waste products of silk industry i.e., sericin, pupae, moths, silkworm rearing wastes and silk fiber waste from all varieties of silk that may be useful for the future as well. The different by-products that are discarded as waste can be put to better use for monetary gains and generation of value-based products. This chapter will review the different types of waste that is produced from sericulture and can be used form value added services.

SILKWORM FAECES BASED ACTIVATED CARBON

Active carbons are considered the resourceful adsorbents that are used comprehensively for the removal of undesirable odor, color, taste, and other organic and inorganic impurities from domestic and industrial wastewater (Meidl, 1997). Because of the application of active carbon for the treatment of waste water, the consumption of activated carbon is the uppermost in the U.S. and Japan, which together consume two to four times more active carbons than European and other Asian countries. The per capita consumption of activated carbon per year is 0.5 kg in Japan, 0.4 kg in the U.S., 0.2 kg in Europe, and 0.03 kg in the rest of the world (Bansal & Goyal, 2005). Other uses of activated carbon include solvent recovery, removal of color from various syrups and decontamination of many pharmaceutical and chemical products.

The absorbent structure of an activated carbon is a purpose of the chemical nature of the precursor used in its preparation, activation method, and the extent of activation which is dependent on the surface area and pore volume of activated carbon which is different in all cases (Haimour & Emeish, 2006; Bansode et al., 2003). The substances with a high content of lignin build up activated carbon with macro porous structure, while the raw materials with a higher content of cellulose yield activated carbon with a predominantly microporous structure (Ioannidou & Zabaniotou, 2007). Cellulose and hemicelluloses are the carbonaceous materials that contain significant proportion of fibrous structure which are highly useful in this process (Daud & Ali, 2004). A critical role is played by the less robust structure and more labile nature of cellulose and hemicelluloses as compared to lignin play the major role in this context (Toles et al., 1997).

The silk industry is generates waste in tons due to the development of advanced silk-based industrialization around the world, huge amount of silkworm excrements (feces) are expected, causing severe problems in the community. Silkworms' feces contain organic matters (83.77-90.44%) and ash (16.23-

9.56%) according to the age of silkworms (Bergmann, 1934; Neelagund, 2007; Hirakiet al., 1996). Huge amount of related wastes are thus expected as a result of the wide scope of significant applications and industries based on silkworms and its excrements.

Silkworm feces were used as predecessor for production of activated carbon via chemical activation method on the one hand, and using new activating agents on the other hand. Chemical activation of silkworm feces use different activating agents results in solids possessing different surface functional groups of basic and acidic nature. In addition to that, obtained solids with microporous texture of high internal surface area showed high adsorption activity of methylene blue from aqueous solution in the pH range 4-10, and cadmium at pH =8. However, efficiency levels recorded in these adsorption processes in addition to the cheap precursor used in obtaining the activated carbons are encouraging to conduct further experiments (ElShafei et al., 2013).

BIOGAS PRODUCTION

The demand for environmental safety and atmosphere due to adverse effects by the use of non-renewable sources of energy has generated the awareness towards the use of renewable sources of energy including the biomass. The threat of increasing energy demand and global warming has resistant to develop alternative and renewable energy resources (Medipally et al., 2015; Ellabban, 2014). Use of renewable energy includes the use of eco-friendly raw materials derived from the ecosystem and their waste (Hastik et al., 2015). Among different sources, biogas has been identified as one of the most gifted and dependable sources of energy (Scarlat et al., 2018). The benefits of using biogas include economic viability having the lowest monetary input of output energy, eco-friendly in nature and unlimited in potential (Lyytimäki et al., 2018). This is now a well-established fact that the total organic waste is utilized without any negative impact (Paolini et al., 2018). The use of biogas has resolved the issue of energy resource in a 'zero-waste' manner *via* generation of energy and bio-manure.

The world is moving towards organic food and organic agriculture is prefaced now a day's all over the world. The demand of organic agriculture, sericulture focuses not only on the cocoon production, but also on other ways that can benefit the farm's economy.

It is very important to find new sources of income for small-scale farmers not only through cocoon selling, but also by the multiple uses of by-products. Insect farming technology provides a cheap source of biomass, which may be a good material in biogas production. The advantages from transforming organic waste matter into biogas are numerous. The biogas production is not only a financial gain, but also the manure is much more homogenous. Defatted pupae of the mulberry silkworm were already tested as a feedstock of the biogas production (Viswanath & Nand, 1994). Insect technology prepares waste mainly as food residue mixed with excreta. It is a high quality substrate to produce biogas through an anaerobic fermentation process, since it ensures a favorable environment for the development and optimal metabolic activity of bacteria involved in the process. This substrate contains only biodegradable organic matter, its C/N ratio isaround the optimal 15–35 (Mao et al., 2015); additionally, it does not contain inhibitory compounds (Dobre et al., 2014).However, obtaining an effective hydraulic retention time and biogas production depend not only on the substratecomposition, but also on the organic loading rate, appropriate process temperature (25 – 35.5° C), and pH (6.7 – 9.4). The process efficiency increases with an increase in organic loading rate; however, an extremely high organic loading rate inhibits bacterial activity (Mao et al., 2015). Approximately, 80% of accumulated methane production was achieved in the first 10 days. However, breeding waste was characterized by a higher energy potential than silkworm excrement. Thanks to the relatively short time of methane production andthe hydraulic retention time, the high content of dry matter, high methane content, which significantly increases the calorific value of biogas production (Łochyńska & Frankowski, 2018).

VERMICOMPOST

Vermicomposting is a process of converting waste into vermicompost with the help of earthworms (Lim et al., 2016; Lee et al., 2018). Furthermore, there's growing focus and inclination of farmers closer to organic farming with very little use of chemical fertilizers. The mulberry cultivation can be made sustainable and low-cost via recycling all the natural residues of sericultural beginning as compost and vermicompost. although, the reaction of vegetation to chemical fertilizers is immediately, the function of natural manure in increasing the yield can't be under envisioned. Vermicompost is an eco-friendly era and has a top notch scope within the recycling of sericultural residue.

Vermicompost technology includes earthworms and composting of any organic waste that's completed through 50-60 days after introducing earthworms in vermicompost pits, while anaerobic composting in pits takes 4-5 months. The material can be without difficulty dealt with. It consists of beneficial micro-organisms and rich plant nutrients consisting of micro-nutrients.

it has been estimated that from one hectare of mulberry farm, about 15 MT of sericultural waste is generated yearly in the form of silkworm rearing waste and other farm wastes which is equivalent to 280-three hundred Kg of nitrogen, 90-one hundred Kg of phosphorus and 150-2 hundred Kg of potash(Das et al.,1997). proper usage of these raw materials as organic manure can substantially bring-down the expenditure on chemical fertilizers. Vermicompost technology become, consequently, opted for rapid conversion of those wastes into nutrient rich compost for making use of to mulberry garden to bring down the cost of cultivation as well as to improve the soil fitness and mulberry leaf yield and nice.

The protocol of vermicomposting is organic residue comprising silkworm clutter, bed refuse and farm weeds etc., They are first loaded in a open pit/trench and charged with sufficient amount of water mixed with cowdung or biogas plant slurry and left for 25-30 days for semi decomposition. This serves as starter feed for earthworms. During this process, the temperature of the semi-decomposing fabric will rise to50-60 °C. To reduce the temperature, once or twice the material should be turned up and down, and then the temperature will come to ordinary country. Later, the chamber needs to be crammed up with two hundred-300 Kg semi-decomposed Seri-cultural wastes having 30-forty% moisture. The vermiculture devices in recommended model are then loaded with semi decomposed feed as consistent with the size and potential of the fashions as vermin mattress. Combined subculture of earthworms in juvenile stage @ 1.five kg populace in keeping with MT of residue is then released in each case. Water is sprinkled as soon as in each 2-3 days to hold enough moisture (30-forty%) and blanketed with dry weeds to prevent predators. Worms start depositing brown to black colored vermicast at the pinnacle of the feed within 1-2 weeks. as soon as the vermicast is detected watering is stopped for drying and clean series of vermicasts. The vermicast is harvested with the aid of scraping the top layer (6 inches) with the help of a lawn rake after every week. The conversion of wastes into vermicompost may additionally range from sixty five-80%of the raw materials in different vermi-gadgets. In a span of 60-70 days from the release of earthworms, the entire semi decomposed residue will be transformed into vermicasts. The cloth may be harvested and sieved via cord mesh to separate earthworms and cocoons for reusing. The

Figure 1. Seriwaste as vermicompost

Eisenia foetida

Mulberry waste

cycle can be repeated2-3 instances in a yr with the new feed (Dandin et al., 2006).The nutrient reputation of vermicompost is Nitrogen-1.5 – 2.five % Calcium-0.five – 1.0%, Phosphorus-0.9 – 1.7% Magnessium-zero.2 – zero. three %, Potash-1.five – 2.4% Sulphur-zero.4 - zero. Five % and other micronutrients with vitamins, enzymes and plant boom regulators.

MUSHROOM CULTURE FROM SERI-WASTE

The cultivation of mushroom has increased highly in recent years. A mushroom is a macrofungus with a unique fruiting frame, which may be determined either above ground (epigeous) or under floor (hypogeous), huge enough to be seen with the naked eye and to be picked by means of hand. Cultivation of mushrooms starts from spawn, which isused to inoculate substrate. Underneath most effective situations, mushrooms can be harvested from 10 days to 6 months after spawning depending upon the strains of species used. nearly sincerely prehistoric people used mushrooms as meals. The excellent early civilizations of the Greeks, Egyptians, Romans, Chinese language and Mexicans prized mushrooms as a delicacy, appreciated their therapeutic nature and, in a few cases, used them in non-secular rites. Of the ten 000 or greater species of macro fungi (mushrooms), approximately 5000 are considered to possess varying degrees of edibility and extra than 2000species from about 31 genera are regarded as top fit to be eaten mushrooms. of these, most effective about100 are experimentally grown, 50 economically cultivated, and around 30 commercially cultivated; and best 5 or six species are cultivated on an business scale.

Silkworm muddle (or silkworm dregs), is a combination of excreta (the digestibility of mulberry leaves is about fifty five%; sloughs (moulting residues) and mulberry leaves. because of natural nature, Seriwaste can be used for mushroom subculture includes one of a kind levels viz., making compost from silkworm excreta, completing the compost, spawning, casing, pinning and cropping.

PHARMACEUTICAL USES OF SILKWORM EXCRETA

Silkworm excreta or feces are excreted from silkworm (Bombyx mori) and considered to be a first-rate waste product of sericulture (silkworm cultivation). but, silkworm excreta have both pharmaceutical and food industrial uses. In traditional medicinal drug, silkworm feces have been used as a therapeutic agent in China, Korea and some eastern Asian countries to treat infectious diseases, headache and stomach pain (Tulp & Bohlin, 2004). Moreover, silkworm excreta are a good supply of herbal colorant for the meals industry. Silkworm excreta are also available in the form of an instantaneous tea, which is claimed to be a nutraceutical product. Despite the fact that silkworm feces have had medicinal makes use of for a long time, the published reports at the isolation of the chemical compounds and their shape elucidation are limited. Silkworm fed on mulberry leaves were found to excrete greater than half of the leaves without digestion, suggesting that some of the compounds in the excreta are from mulberry leaves and some are bio transformed in the silkworm intestine. physically, silkworm excreta have a cylindrical shape of 2-three mm in period with a deep green color. The chemical components of silkworm excreta that have basically been reported are chlorophyll and chlorophyll derivatives, xanthophyll, carotenoid, and flavonoids (Park et al. 2011). The lipid profile of silkworm excreta has additionally been investigated. institution compositions of the lipids were reported in silkworm excreta in which there were focused liposoluble compounds such as phytosterol, unsaturated fatty acid, and fatty alcohol. however, identification of liposoluble compounds in the excreta is very confined. only triacontanol, a very long chain fatty alcohol, andβ-sitosterol, a phytosterol, had been suggested in silkworm excreta. Three compounds 1-tritriacontanol,lupeol, and β-sitosterol were remoted. of these, 1-tritriacontanol and lupeol have not earlier been located in silkworm excreta. it is probably that lupeol and β-sitosterol are derived from the mulberry leaves on which the silkworm fed and are excreted in an unchanged shape at the same time as 1-tritriacontanol is likely synthesized within the silkworm gut. Identification of lupeol and β-sitosterol IDS the conventional medicinal use of silkworm excreta in treating inflammation. Further, the cytotoxicity of the crude extract and 1-tritriacontanol were tested and found to be non-toxic to all cell sorts tested in this take a look at. in step with the lipid profile and fatty alcohols said, silkworm excreta are rich in liposoluble compounds mainly fatty alcohols and fatty acids. Moreover, silkworm excreta are very reasonably-priced and easily harvested due to the fact they are unavoidably and abundantly produced as a primary waste product of sericulture.

SERICULTURE WASTE AS ANIMAL/POULTRY FEED

The pressing need to harness the potentials of the agro-industrial by product usually known as "Wastes" as element substitute for the greater expensive traditional feed components had been critically expressed (Onifade & Babatunde,1998). This need from the increasing demand for and supply deficit of conventional feed resources with a concomitant sharp rise in their cost prices. The net impact of growing unit price of the conventional feed resources is increased fee of the compound rations which by means of extension gives rise to multiplied fee of meat and animal merchandise. It turns into definitely imperative that other assets for speedy cattle output to satisfy the growing human call for animal protein foods to be sourced. Such other assets have to be cheap and nutritionally good enough for feeding birds / animals with the aim of lowering fee of producing meat. One of such measures is the recycling of predominant reeling and rearing waste of sericulture industry i.e. silkworm pupae and silkworm excreta as element substitute for conventional and steeply-priced grains used within the cattle nutrition. Right utilization of waste products of sericulture industry also can generate a further earnings in addition to silk, the principle output, which can also be taken into consideration as an opportunity way to clear up in part the hassle with the decline of silk production and the efficient usage of the present sericulture human capability, studies and production centers. Dried silkworm meal could totally update soybean meal in balanced diets for growing rabbits without any problem. In China silkworm meal is a conventional factor in rabbit feeds.

UTILIZATION OF SILKWORM LITTER AND PUPAL WASTEAN ECO-FRIENDLY APPROACHFOR MASS PRODUCTION OF BACILLUS THURINGIENSIS

Bt is a Gram-high-quality, spore-forming soil bacterium that produces parasporal crystal inclusions which might be poisonous to the larval forms of a extensive variety of insects together with nematodes. Bt has been remoted from exclusive environments inclusive of soil, stored grains and sericulture environments (Xavier et al., 2007). More than ninety species of naturally occurring, insect specific (Entomopathogenic) bacteria had been identified from bugs, plants and the soil, however just a few have been studied intensively. plenty attention has been given to Bt, bought underneath numerous exchange names like Dipel, Javelin, Thuricide, trojan horse attack, Caterpillar Killer, Bactospeine, and SOK-Bt. The Bt biopesticide preparations are primarily based on endotoxin proteins collected as parasporal crystals produced by the bacterial cells. Diverse agricultural and industrial by means of-merchandise may be used as raw-materials in biopesticide industry, along with maize glucose, soybean flour, peanuts, sugarcane molasses, and liquid swine manure. It is argued that mass production of Bt making use of wheat bran, soybean, defatted floor nut cake, defatted milk powder, rice, pupal waste by submerged and semi-stable-country fermentation strategies are extensively to be had (Capalbo et al.,2001; Karthikeyan and Sivakumar, 2007). there may be awesome scope of converting sericulture via-products into fee delivered merchandise like biopesticides. As has been found by the studies that the clutter is generated in substantial portions; its usage for cultivation of Bt is a cost-effective method. Tirado-Montiel et al.(2001) first examined the usage of wastewater sludge for the productionof Bt primarily based biopesticide. Mathematical relationships among spore concentrations, delta-endotoxin concentrations and entomotoxicity (Tx) of Bacillus thuringiensis var. kurstaki HD-1 (Btk HD-1)preparations produced in six special media were analyzed through Vuet al. (2012). Such relationships indicate that delta-endotoxin concentration may be used as a hallmark for tracking the fermentation process and for predicting the viable entomotoxicity of cultured media of Btk HD-1 (Vu et al., 2012).Use of such substrates represents a higher alternative for disposal and/or recycling wastes and minimizes pollutants.

CONCLUSION

Silk industry has played a very important role in the development of rural population. It has provided employment to a good chunk of people. With the concern towards environment the silk industry is also focusing on environmental friendly technologies. The useful conversion of by-products through indigenously available processing techniques brings additional income lead to socio-economic advantage of rural people. The need for introduction of integrated processing complexes with redesigning of present practices deserves special mention to make the practice more attractive and people to participate passionately. The hybrid silk, net raw silk, silk tow and silk waves were converted as high valued fancy jackets, carpets and furnishings also used as silkworm cocoon, nutritive value of pharmaceutical, bio-medical, cosmetic, animal feed, chemical industry, pupae, silk through methodical diversified dispensation certainly boosts up the Sericulture. There is an urgent need to involve the expertise from silk industry to utilize the waste from this industry and search for the opportunities to convert the same into more value added services. The production of biogas, vermicomposting, and animal feed should further involve the R&D to make them more efficient processes.

REFERENCES

Bansal, R. C., & Goyal, M. (2005). *Activated Carbon Adsorption*. New York: CRC Press. doi:10.1201/9781420028812

Bansode, R. R., Losso, J. N., Marshall, W. E., Rao, R. M., & Portier, R. J. (2003). Adsorption of volatile organic compounds by pecan shell- and almond shell-based granular activated carbons. *Bioresource Technology*, *90*(2), 175–184. doi:10.1016/S0960-8524(03)00117-2 PMID:12895561

Bergmann, W. C. (1934). Contributions to the study of marine products: II. The sterols of mollusks. *The Journal of Biological Chemistry*, *104*, 317–328.

Blakeman, J. W. (2004). Recycling. *Coloring of Plastics*, *353–357*, 353–357. doi:10.1002/0471721581.ch24

Capalbo, D.M.F., & Valicente, F.H., Oliveira Moraes, I., & Pelizer, L.H. (2001). Solid-statefermentation of Bacillus thuringiensis tolworthi to control fall armyworm inmaize. *Electronic Journal of Biotechnology*, 4.

Dandin, S. B., Das, P. K., & Bhogesha, K. (2006). *National Seminar on soil health and water management of sustainable sericulture*. Central Silk Board.

Das, P. K. (2002). *Advancein Indian Sericultural Research*. Mysore: CSRTI.

Daud, W. M. A. W., & Ali, W. S. W. (2004). Comparison on pore development of activated carbon produced from palm shell and coconut shell. *Bioresource Technology*, *93*(1), 63–69. doi:10.1016/j.biortech.2003.09.015 PMID:14987722

Dobre, P., Nicolae, F., & Matei, F. (2014). Main factors affecting biogas production – an overview. *Romanian Biotechnological Letters*, *19*, 9283–9296.

Ellabban, O., Abu-Rub, H., & Blaabjerg, F. (2014). Renewable energy resources: Current status, future prospects and their enabling technology. *Renewable & Sustainable Energy Reviews*, *39*, 748–764. doi:10.1016/j.rser.2014.07.113

ElShafei, G. M. S., ElSherbiny, I. M. A., & Philip, C. A. (2013). Silkworms' feces-based activated carbons as cheap adsorbents for removal of Cadmium and Methylene Blue from aqueous solutions. *Chemical Engineering Research & Design*. doi:10.1016/j.cherd.2013.09.004

Haimour, N. M., & Emeish, S. (2006). Utilization of date stones for production of activated carbon using phosphoric acid. *Waste Management (New York, N.Y.)*, *26*(6), 651–660. doi:10.1016/j.wasman.2005.08.004 PMID:16256326

Hastik, R., Basso, S., Geitner, C., Haida, C., Poljanec, A., Portaccio, A., ... Walzer, C. (2015). Renewable energies and ecosystem service impacts. *Renewable & Sustainable Energy Reviews*, *48*, 608–623. doi:10.1016/j.rser.2015.04.004

Hiraki, A., Ikuno, Y., Kim, J., & Ueda, S. (1996). Suppression of enveloped virus production with a substance from silk worm fecal matter cell. *Struc. Func.*, *21*(6), 501–514. doi:10.1247/csf.21.501 PMID:9078408

Ioannidou, O., & Zabaniotou, A. (2007). Agricultural residues as precursors for activated carbon production—A review. *Renewable & Sustainable Energy Reviews*, *11*(9), 1966–2005. doi:10.1016/j.rser.2006.03.013

Jensen, P. (2014). Behavior genetics and the domestication of animals. *Annual Review of Animal Biosciences*, *2*(1), 85–104. doi:10.1146/annurev-animal-022513-114135 PMID:25384136

Karthikeyan, A., & Sivakumar, N. (2007). Sericulture pupal waste – a new productionmedium for mass cultivation of Bacillus thuringiensis. *Indian Journal of Biotechnology*, *6*, 557–559.

Kim, K. Y., Kang, P. D., Lee, K. G., Hyung, K., Kim, M. J., Kim, K. H., & ... . (2010). Microsatellite analysis of the silkworm strains (Bombyx mori): High variability and potential markers for strain identification. *Genes & Genomics*, *32*(6), 532–543. doi:10.100713258-010-0066-x

Larson, G., & Fuller, D. Q. (2014). The evolution of animal domestication. *Annual Review of Ecology Evolution and Systematics*, *45*(1), 115–136. doi:10.1146/annurev-ecolsys-110512-135813

Lee, L. H., Wu, T. Y., Shak, K. P. Y., Lim, S. L., Ng, K. Y., Nguyen, M. N., & Teoh, W. H. (2018). Sustainable approach to biotransform industrial sludge into organic fertilizer via vermicomposting: A mini-review. *Journal of Chemical Technology and Biotechnology (Oxford, Oxfordshire)*, *93*(4), 925–935. doi:10.1002/jctb.5490

Lim, S. L., Lee, L. H., & Wu, T. Y. (2016). Sustainability of using composting and vermicomposting technologies for organic solid waste biotransformation: Recent overview, greenhouse gases emissions and economic analysis. *Journal of Cleaner Production*, *111*, 262–278. doi:10.1016/j.jclepro.2015.08.083

Łochyńska, M., & Frankowski, J. (2018). The biogas production potential from silkworm waste. *Waste Management (New York, N.Y.)*, *79*, 564–570. doi:10.1016/j.wasman.2018.08.019 PMID:30343788

Lyytimäki, J. (2018). Renewable energy in the news: Environmental, economic, policy and technology discussion of biogas. *Sustain. Prod. Consum.*, *15*, 65–73. doi:10.1016/j.spc.2018.04.004

Majumder, S. K. (1997). Scope for new commercial products from sericulture. *Indian Silk.*, *35*(12), 13–18.

Mao, C., Feng, Y., Wang, X., & Ren, G. (2015). Review on research achievements of biogas from anaerobic digestion. *Renewable & Sustainable Energy Reviews*, *45*, 540–555. doi:10.1016/j.rser.2015.02.032

Medipally, S. R., Yusoff, F. M., Banerjee, S., & Shariff, M. (2015). Microalgae as sustainable renewable energy feedstock for biofuel production. *BioMed Research International*. PMID:25874216

Meidl, J. A. (1997). Responding to changing conditions: How powdered activated carbon systems can provide the operational flexibility necessary to treat contaminated groundwater and industrial wastes. *Carbon*, *35*(9), 1207–1216. doi:10.1016/S0008-6223(97)00093-6

Neelagund, S. E., Ingalhalli, S. S., Savanurmath, C. J., Hinchigeri, S. B., & Hiremath, M. B. (2007). Purification and characterization of antiviral protein from fecal matter of silkworm. *Journal of Environmental Sciences (China)*, *5*, 77–85.

Onifade, A. A., & Babatunde, G. M. (1998). Comparison of theutilization of palm kernel meal, brewer's dried grain andmaize offals by broiler chicks. *British Poultry Science*, *39*(2), 245–250. doi:10.1080/00071669889204 PMID:9649879

Paolini, V. (2018). Environmental impact of biogas: A short review of current knowledge. *J. Environ. Sci. Heal. Part A*, 1–8, doi:10.1080/10934529.2018.1459076

Park, J. H., Lee, D. Y., Yun, P., Yeon, S. W., Ko, J. H., Kim, Y. S., & Baek, N. I. (2011). Flavonoids from silkworm droppingsand their promotional activities on hemeoxygenase-1. *Journal of Asian Natural Products Research*, *13*(4), 377–382. doi:10.1080/10286020.2011.555331 PMID:21462044

Park, Y. J., Kim, W. S., Ko, S. H., Lim, D. S., Lee, H. J., Lee, W. Y., & Lee, D. W. (2003). Separation and characterization ofchlorophyll degradation products insilkworm using HPLC-UV-APCI-MS. *Journal of Liquid Chromatography & Related Technologies*, *26*(19), 3183–3197. doi:10.1081/JLC-120025517

Patil, B. R., Singh, K. K., Pawar, S. E., Maarse, L., & Otte, J. (2009). *Sericulture: an alternative source of income to enhance the livelihoods of small-scale farmers and tribal communities*. Pro-Poor Livestock Policy Initiative. A Living from Livestock Research Report 09-03.

Patil, S. R., Amena, S., Vikas, A., Rahul, P., Jagadeesh, K., & Praveen, K. (2013). Utilization of silkworm litter and pupal waste-an eco-friendly approach for mass production of Bacillus thuringiensis. *Bioresource Technology*, *131*, 545–547. doi:10.1016/j.biortech.2012.12.153 PMID:23403062

Raju, S. (1996). Utilization of sericultural byproducts: A Chinese example. *IndianSilk*, *35*, 19–20.

Scarlat, N., Dallemand, J. F., & Fahl, F. (2018). Biogas: Developments and perspectives in Europe. *Renewable Energy*, *129*, 457–472. doi:10.1016/j.renene.2018.03.006

Singhal, B. K., Dhar, A., Khan, M. A., & Bindroo, B. B. (2005). Utilization of sericulturalbyproducts as urgent need for sustainable sericulture. In Progress of Research in Organic Sericulture and Seri byproduct Utilization. Seri Scientific Publishers.

Singhal, B. K., Dhar, A., Khan, M. A., Sengupta, D., & Dhar, S. L. (2005). Mulberry by-products utilization for sustenance of sericulture industry of Jammu and Kashmir. *Proc. The 20th Congress of the International Sericultural Commission*, 152-156.

Tirado-Montiel, M. I., Tyagi, R. D., & Valero, J. R. (2001). Wastewater treatment sludge asa raw material for the production of Bacillus thuringiensis based biopesticides. *Water Res. N.Y.*, *35*(16), 3807–3816. doi:10.1016/S0043-1354(01)00103-8

Toles, C. A., Marshall, W. E., & Johns, M. M. (1997). Granular activated carbons from nutshells for the uptake of metals and organic compounds. *Carbon*, *35*(9), 1407–1414. doi:10.1016/S0008-6223(97)00073-0

Tulp, M., & Bohlin, L. (2004). Unconventionalnatural sources for future drug discovery. *Drug Discovery Today*, *9*(10), 450–458. doi:10.1016/S1359-6446(04)03066-1 PMID:15109950

Viswanath, P., & Nand, K. (1994). Anaerobic digestion of silk industry wastes. *Bioresource Technology*, *49*(3), 273–276. doi:10.1016/0960-8524(94)90052-3

Vu, K. D., Tyagi, R. D., Surampalli, R. Y., & Valéro, J. R. (2012). Mathematical relationshipsbetween spore concentrations, delta-endotoxin levels, and entomotoxicity ofBacillus thuringiensis preparations produced in different fermentation media. *Bioresource Technology*, *123*, 303–311. doi:10.1016/j.biortech.2012.07.110 PMID:22940334

Wenhua, L. (2001). *Agro-ecological Farming Systems in China (Man and the Biosphere Series). Chinese Academy of Sciences*. Beijing, China: UNESCO.

Xavier, R., Nagarathinam, P., Gopalakrishnan, M. V., & Jayaraman, K. (2007). Isolation ofLepidopteran active native Bacillus thuringiensis strains through PCR Panning. *Asia-Pacific Journal of Molecular Biology and Biotechnology*, *15*, 61–67.

Chapter 8
Zero Waste:
A Sustainable Approach for Waste Management

Saima Hamid
University of Kashmir Srina, India

Bhat Mohd Skinder
https://orcid.org/0000-0001-5038-7125
University of Kashmir, India

Muzaffar Ahmad Bhat
Barkatullah University, India

ABSTRACT

Zero waste management means the holistic concept of waste management which recognizes waste as a resource produced during the interim phase of the process of resource consumption. Zero waste strategies may be applied to companies, to communities, industrial sectors, to schools, and homes since they include many stakeholders, not only those of the environment, but also technological aspects. Sustainability is also strongly supported by environmental protection, cost reduction, and additional jobs when it comes to waste management and handling back into the industrial cycle. Lowering global resource requirements force us to consider resource management and product management. The management of zero waste is therefore a holistic view of the sustainable avoidance and management of waste and resources. Although there are many null practice approaches and null waste approaches in the modern world, zero waste is a very complex system, and in the future, there are still many works.

INTRODUCTION

In 1973, Dr. Paul Palmer first applied the term "zero waste" to recover chemicals (Palmer, 2004). Zero waste approach is one of the most widely studied concept but still the most controversial topic in recent decades in waste management research (LaBrecque et al., 2015; Greyson et al., 2007). Zero Waste (ZW) is defined as "a system to design and management of products and processes for waste removal and

DOI: 10.4018/978-1-7998-0031-6.ch008

material retention and rehabilitation, not burning or burying" (Zero Waste International Alliance, ZWIA, 2004). ZW is therefore concerned with waste prevention through sustainable design and consumption practices, optimal waste recovery and not waste management by landfill or incineration (Zaman, 2015). ZW supports waste prevention and avoidance rather than waste treatment and disposal strongly. It is understandable; It may however not be feasible, under the existing resource consumption and waste management systems, to achieve nil incineration and nil landfill targets. Waste was dealt with as a burden and social problem and was therefore largely managed by "end-of-pipe" solutions such as waste disposal systems (Zaman et al., 2015). The traditional waste management system, which mainly relies on sites, significantly pollutes our environment, thus requiring an improved and efficient waste management system with a limited exception in developed countries. Thus the zero-waste objective, using industrial symbiosis, recycling or "up cycling," is to use and consume resources within a circular economic model with minimum environmental degradation, based on the "no-waste" principle from nature. Local governments and business organizations frequently use strategic waste management plans to manage waste issues (Liao et al., 2011). For a successful implementation of a waste management plan it is essential to establish an effective planning process (King et al., 2016). Several studies on the development of waste management frameworks have been conducted, including decision frameworks, legislative frameworks and hierarchical frameworks (Sentime et al., 2013). A framework assists decision-makers in understanding, improving, assessing and guiding waste management systems. The 3R principles (reduction, recycling and reuse) are among the top three in the waste hierarchy and are regarded as the founding principles of the sustainable waste management system (Murphy & Pincetl, 2013; Mason et al., 2003; Colon & Fawcett, 2006). The "3R" principles were extending to five steps in the waste hierarchy in the European Union Waste Framework Directive 2008, including prevention (avoidance), recycling (including recovery of energy) and disposal. Waste prevention is a major problem of zero waste and calls for collective social sensibility and knowledge of waste, innovation in production and business models (Cox et al., 2010). Awareness and transformative knowledge in relation to the choice of pro-environmental lifestyle are often believed to motivate behavior change (Jackson et al., 2005). Waste management and processing technologies have been used over centuries to solve waste problems (Greyson, 2007; Matete & Trois, 2008). Zero waste believes that technology alone cannot solve waste problems on the basis of community participation, service infrastructure, regulatory policy and ecological treatment technology in a sustainable manner. Nevertheless, it limits the application of waste energy (WTE), which consumes waste to make electricity (heating and electricity) and waste disposal in an "ideal" zero waste environment, between traditional waste handling and non-residual waste management. Figure 1 shows the symbolic material flux of a circular waste system, when the end-of-life product or output waste is treated and used as metabolism process (Curran & Williams, 2012; Matete & Trois, 2008).

Zero Waste Strategy and Solid Waste Management

Continuing population growth, booming economy, rapid urbanization and rising standards of community living have greatly accelerated the generation of solid waste worldwide, particularly from developing countries (Guerrero et al., 2013). Solid waste is now a global ecological issue (Seng et al., 2010) as global volume of solid waste is estimated to be around 11 billion tons per year (2.5 tons of trucks can rotate around 300 circles around the equator) in 2011, with solid waste production per capita at about 1.74 tons / year worldwide. The large volume of waste has also created enormous pressure on the waste management authority to be more sustainable way to achieve sustainability (Cheng & Hu, 2010). Cur-

Figure 1. Cyclic and linear resource flow

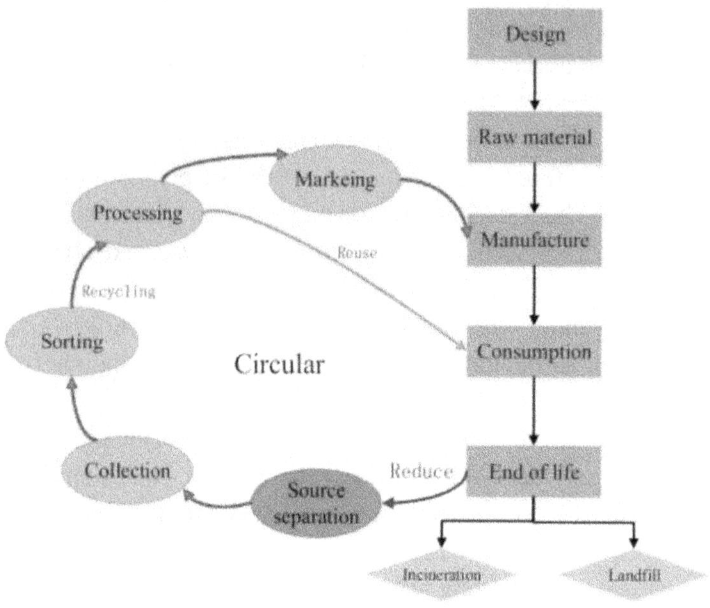

rent society is driven towards greater sustainability by global climate changes and their diverse effects on human life. Lowering global finite resources also forces us to take resource and product management into account. One approach to zero waste was therefore proposed to address these concerns (Phillips et al., 2011). A state of zero wastes may finally become a necessity in the world with limited resources because of the great environmental pressures. Currently some good zero waste practices in cities, businesses, individuals and waste recycling industries have been proposed and implemented (GAIA, 2013). Following case-studies should inspire the urban, the private and the waste recycling industries to pursue their own zero waste efforts and the development of the new zero waste investment and implementation plans for the organizations and individuals concerned.

ZERO WASTE CITIES

Adelaide, Australia

Adelaide is South Australia's capital city and includes 19 municipalities in an urban area of 8415 km^2 with population 1,089,728 (UNHABITAT, 2010). The City Council of Adelaide (ACC) is responsible in Adelaide for waste management. Zero Waste SA (ZWSA) is a government organization created under the Zero Waste SA Act of the South Australian state. ZWSA allows people to improve recycling and waste management practices in their homes, workplaces and industry (ZWSA, 2011). Adelaide has a high percentage of waste collection systems in relation to other capital cities in Australia in waste management systems. The legislation on the deposit of containers has been adopted in 1977; therefore, it is more than thirty years since recycling of various containers. Zero Waste SA works in Southern Australia to achieve a zero-waste area (Zaman, 2013).

Community Action, Taiwan, China

Taiwan was confronted by a waste crisis in the 1980s because there was no room to increase its site deposits. When the government turned to mass incineration it not only stopped dozens of burners being made by the vicious opposition of the community, it also pushed the Government to adopt waste prevention and recycling targets and programs (Allen, 2012b). These policies and programs were so effective that waste volumes decreased significantly, even as both population and gross domestic product grew. Its waste disposal rate was around 48.82 percent, and waste per capita expenditures were 25.40 dollars a year. The government, however, has cut down on the potential of waste prevention strategies through the maintenance of both pro-incinerator policies and waste prevention policies, because large investments in incinerator drain resources could otherwise be used for their improvement and expansion.

INNOVATIVE APPROACH FOR ZERO CONSTRUCTION WASTE

The building, one of mankind's most ancient activities, places an important influence on socio-economic development and also has an indelible influence on the environment and the surroundings. It affects the economic dynamics of society and has a significant environmental and environmental impact. Construction activities have long-term consequences both for the change of the appearance of the region, as well as for the management and management of natural resources. Innovation is of vital importance not only to those in a given region (region, state and like) that want to increase or sustain economic growth, but also to those who benefit directly. Production is therefore no longer a core problem that should influence or modify the economic course of development or improve quality of life as much as possible. Since the majority of natural resources are non-limited and renewable, we can guarantee equality of opportunity only if we are responsible for the management of resources. The EU has set one of its main targets, in its sustainable development strategy, the breaking down of links between economic growth, natural resource utilization and waste production (Hart, 2007). The recycling model for lightweight aggregate construction waste is developed to include environmental considerations.

Model Approach for Zero Construction Waste

In order to include environmental performance in designing building operations and minimizing construction waste, the model for recycling construction waste from lightweight aggregates containing expanded-glass was created. The quantity of waste lightweight concrete (LWC) made of expanded glass is currently less important, and this raises the question of whether such waste can be collected separately and economically. But the recycling or re-use of a building is usually better from an environmental point of view than the demolition, because of the environmental cost of energy, water and renovation and reuse materials are fewer. Life cycle assessment (LCA) has been used frequently during recent years to assess environmental issues linked to the management of solid waste (Grant et al., 2003). The main benefit of using LCA for solid waste management systems is that the approach systematically covers all the impacts of waste management, including all processes of solid waste systems and of waste management systems upstream / downstream (Kirkeby et al., 2006). In the last decade, several models for environmental evaluation of municipal solid waste treatment and disposal have been presented. The IWM (McDougall et al., 2001) model reflects life cycle thoughts on city waste management, first with a chart and then with a more advanced model for calculating the inventories of life cycles. We started with LCA in our research. LCA is a standardized tool designed to minimize potential environmental, human health and resource impacts. Therefore, limits are defined more narrowly as the economy is not covered (Hansen & Gilberg, 2003; Kirkebyet al., 2006). Creative cooperation and the new philosophy of waste management have led to a new perspective: resting (or residing) and waste (building) material is the material raw (building). The production of cement from lightweight aggregates is made from a number of raw materials. One of these is LWC with expanded glass aggregates. The raw material for LWC with expanded glass or recycled glass added aggregate which employs - to all intents and purposes a perfect recycling system every year in the Federal Republic of Germany for millions of tonnes. LWC's expanded glass aggregates make use of only the precious raw material, which the glass industry is unable to use for technical reasons for the production of new, e.g., fine glass shards. This makes a significant contribution to the development of the recycling process of glass with aggregates containing expanded glass while at the same time protecting natural resources. In addition to its areas of application for the classical and other construction materials industries, LWC with expanded glass aggregates is also becoming increasingly popular for special applications. The LWC waste from aggregates containing expanded glass can be seen in Figures 2a to 2c.

Figure 2. (a) LWC waste of LWC from aggregates containing expanded glass. (b) Crushed LWC waste from LWC with aggregates containing expanded glass. (c) Recycled "new material from LWC with aggregates containing expanded glass
(Kralj, 2011).

Zero Waste Approach Via Bio-Refinery Using Sustainable Material: Lignocellulose

Diverse aspects of biorefinery as a sustainable technology for the processing of' materials' of lignocelluloses into value-added products have been examined. There have been considerable explicit efforts to convert the lignocellulosic into value-added products, including composites, fine chemical, animal feeding stuffs, pulp and paper, biofuels and enzymes (Figure 3). In many processes relating to lignocellular biotechnology, it has shown significant improvements over the past few years and has triggered broad and innovative studies of lignocellulose, various fungal consortia and ligninolytic enzymes, including lignin peroxidase (LiP), manganese peroxidase (MnP), and laccase (Lac), their purification and immobilization to present their potential for a wider spectrum of biotechnological applications (Iqbal et al., 2013; Asgher et al., 2017). Innovations in the world of bio-refiners are offering a portfolio of sustainable and eco-efficient products to compete in the petroleum-based products market which is currently dominated by intensive research. It is interesting that a bio-refinery has enormous potential for using / capitalizing all kinds of biomass sources, including agricultural, agro-industrial, algae and municipal resources. Historically speaking, bio-based producers have targeted value-added products, chemical platforms and specialty markets, where multi-functionality often plays a critical role to justify the objective of the end product. This integrated transition from the petrochemical approaches mentioned above to a sustainable bio-refinery has numerous justifications (Langeveld et al., 2012; McCormick et al., 2013).

Figure 3. Lignocellulose bio-conversions into value-added bio-products
(Iqbal et al., 2013)

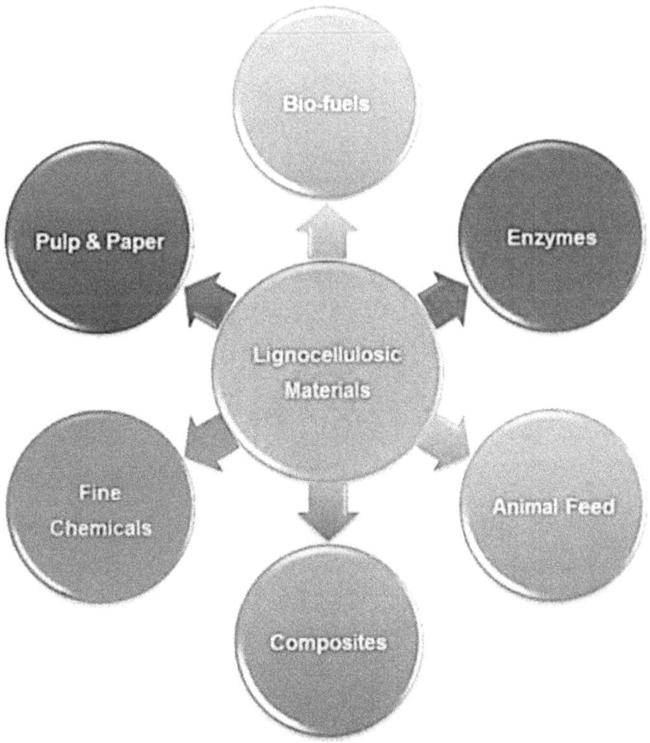

- To prevent excess dependence on petrochemical products
- To avoid price increases
- To avoid over - consumption of petroleum, gas, coal and other potential minerals
- To enhance and diversify bio - renewable sources
- In addressing global climate issues
- Greenhouse gas emissions
- To safeguard the natural ecosystem and
- To boost regional and rural greener growth

The schematic representation of the above-mentioned concept of biorefinery platforms is shown in Figure 4. Bio-refinery is normally based on sugar fermentation, e.g. C5/C6, extracted from many feedstocks based on natural polysaccharides. Different types of bio-refineries were proposed or developed by researchers based on the raw materials and processing techniques (Parajuli et al., 2015). The dimethyl ether was used for intermediate assessment of a thermochemical bio-refinery consisting of 12 procedures for the manufacture of ethanol, methyl acetate, hydrogen and electricity, as up to 50.20% was achieved with energy efficiency. For this process, an economic analysis was carried out, which concluded that there is a greater profitability for a thermochemical bio-refinery than for normal thermochemical processes, which adds more than one product to produce (Haro et al., 2013).

Green Bio-Refinery (GBR)

A green bio-refinery system is also considered to be a complex, yet multi-functional and fully integrated system which have great potential to use green biomass, such as grazing, to produce various products, like fibers, fuel, animal feed, or as a component of the bio-refinery of lignocellulose. Furthermore, new

Figure 4. Schematic representation of bio-refinery platform concept
(Iqbal et al., 2013)

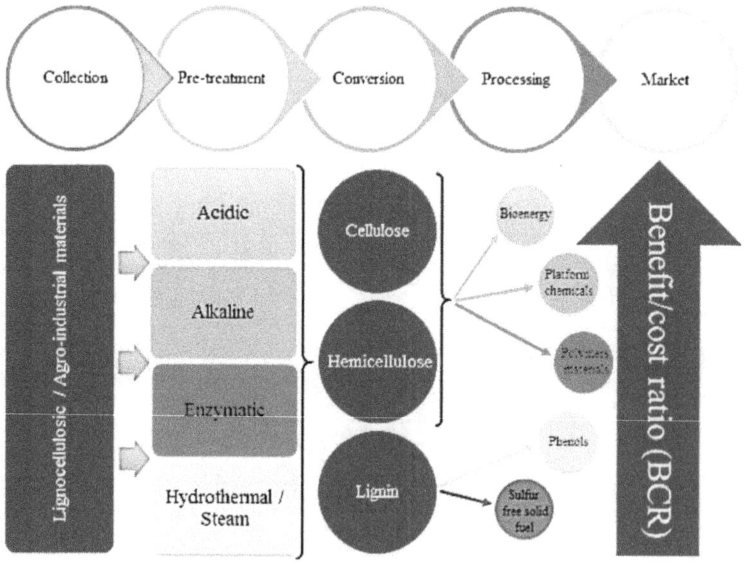

resource protection machinery for the full use of residual biomass was also considered. The production of ethanol is one of the main aims of a green bio-refinery. The following ethanol bio-refineries have been investigated from various sources: from starch via grinding (wet or dry) and fermentation, which is integrated into animal feed production (Scholey et al., 2016). For the production of ethanol and also as a typical bio-refinery for the production of animal feed the use of *Faba beans* was also proposed, as a green bio-refinery (Karlsson et al., 2015). A different kind of beans were used in an integrated bio-refinery with field bean seeds for the production of ethanol, feed components and the bonus of an edible and ethanol-producing fungal biomass (N. *intermedia*) (Pietrzak et al., 2016). In terms of industry and biotechnology, a broader spectrum of natural, agro-industrial and lignocellulosic polysaccharide substrates is available for the (bio) process into high value and industrial products. In the last decade, a major development has been achieved in many integrated bio-refinery approach procedures for the future. Due to increasing socioeconomic concern, targeted markets have been identified in recent years for possible applications such as alternative types of energy, industrially relevant enzymes, platform chemicals etc. The (bio) conversion of such natural substances to auxiliary resources to satisfy the demands of both industry and modern society has been the most striking (Sakamoto et al., 2012; Shahsavarani et al., 2013).

Zero-Waste Design for Sustainable Fashion

Zero-waste design is popular as a sustainable manufacturing way. In many other manufacturing regions Subaru, Proctor & Gamble, DuPont and Caterpillar are just some of the businesses that boast of their zero-waste-to-landfill facilities. The result of this design can be sustainability, but null-waste design can also be a tremendously creative challenge. The designer does not only create a creative reflection on model shapes of the cloth, but also needs innovative sewing, seam closings and seam construction. The concept of zero-waste clothing therefore seeks to eliminate any wasteful clothing from the production process by developing designs using 100% of the length of the cloth, so that no waste hits 'the ground for the cutting room.' The most common method of manufacturing apparel in cutting clothing production is that of cutting material waste by average 15 percent (Rissanen, 2011; Townsend & Mills, 2013). The reason is that design parts for most clothing items have irregular shapes which do not combine perfectly like pieces in a puzzle. The fashion industry uses different methods for minimizing this waste, such as software for planning the introduction of model cutting parts (which is similar in part to resolving the puzzle on the computer) or cutting multiple sizes and styles together. However, these methods do not eliminate the waste material that ends in the cutting room floor in its entirety. The focus of the paper is on the challenge of designing garments that do not produce waste when cutting while the scraps are recycled (denim insulation) or recycled (appliqués, pillow stuffing). Indigenous garments have used zero waste for centuries The Indian sari, a cloth long, is often remembered as draped and wrapped around the body without being cut or stitched. The Japanese kimono is also a null-waste, but is made of cut and sewn fabric pieces. Therefore the challenge for designers of a sustainable nature is to create styles that could be adopted in west mode within zero-fabric waste parameters. Cutting waste generated is important as 'the clothing industry accounts for 7% of global carbon emissions, Ericson (2010) estimates. The trend of eco-mode (Thomas, 2008) has led to sustainability talks that have often focused more on organic textiles and fair labour practices instead of minimising waste. Designers like Yeohlee Teng and Tara St. James include zero-waste practices, but they not only design zero-waste solutions. The emerging designer Titania Inglis used zero waste design practices to design collections and received the 2012 Sustainable Design award of the Ecco Domani Fashion Foundation (Swanson, 2012). Certainly, a great deal contributes to sustainable fashion design practices; the Textiles Environment Design Project (2012) outlines 10 strategies for sustainable design, with' design to minimize waste' as first in its list. In comparison with other sustainable practices like recycling materials and using organic dyes, while minimizing waste involves not just the cutting of zero wastes, literature is very lacking in designing textile clothes for zero waste cutting. Indeed, in the recent article of Townsend and Mills (2013), the limited information on the use of (zero-waste pattern cutting) was mainly provided by contemporary designers, including Holly McQuillan and Julia Lumsden. The authors began a study of the design and literature of zero-waste clothing by examining current and historical examples. Thankfully, the designer will study the practice for you the authors started a study of the design of non-discharge clothing by examining contemporary and historical examples and the literature available. Fortunately, designers frequently share the design pattern of their non-waste clothing for a study of the practice, including the YIELD catalog (McQuillan & Rissanen, 2011).

Jigsaw and Embedded Jigsaw Design Practices

In terms of the embedded puzzle practice, the design process is the same for embedded designs as for traditional puzzles. The built-in puzzle has become more effective when multiple clothe materials are manufactured and less of a new way of designing null-waste patterns. By integrating a traditional pattern into a fabric length, McQuillan suggests that what is waste from the traditional pattern can be used to produce zero waste in a different garments design. The study focused on puzzle practice and ignored the embedded since the authors practise individual clothing. In contrast, both puzzles and embedded puzzle methods allow designers to use different pattern forms, much like traditional design patterns. However, the design forms in the puzzle are manipulated to lock each other and share the cut edges to achieve zero waste. Timo Rissanen and McQuillan are both well-known for their workmanship. The practice of puzzles provides a way of working which is comfortable for seasonal designers since it allows the designer to build a set design, whether it is the sleeves, pant legs or any other established pattern piece. The little black PARTY dress (Figure 5) in which the raglan sleeved is the shape of a cursive r (r) lower case with the original idea that the pattern forms would form the letters P-A-R-T-Y. The rest of the models were developed in the form of drafts and designs at the same moment, and a possible decision was reached to abandon the hope that the word ' party' remained intact within the design.

Multiple Cloth Approach

McQuillan describes this practice as "a way to design two or more designs for various fabrics simultaneously." This could indeed be the most similar to mass production methods for marking patterns by combining several styles into one layout. While the integrated puzzle practice can also produce multiple garments in one marker, McQuillan's approach to the multi clothes shows results with only illustrated examples of two hooded jackets and two t-shirts. When designing for mass consumption, multiple clothing linked to fabric is essential, since the repeat of textiles contributes towards esthetic cohesion. However, once again, because the studios of the authors are not willing to produce multiple clothes, they have not taken up the task of designing in several clothes. The combination of cutting several garment patterns from the same cloth is shrewd and functional, and zero waste can be achieved. However, this study focuses on exploring methods of zero waste on individual clothing. During zero residue design surveys, McQuillan's view of zero-waste design practice was adopted by the authors as one of' the cre-

Figure 5. Design and layout of little black PARTY dress
(Melanie et al., 2013)

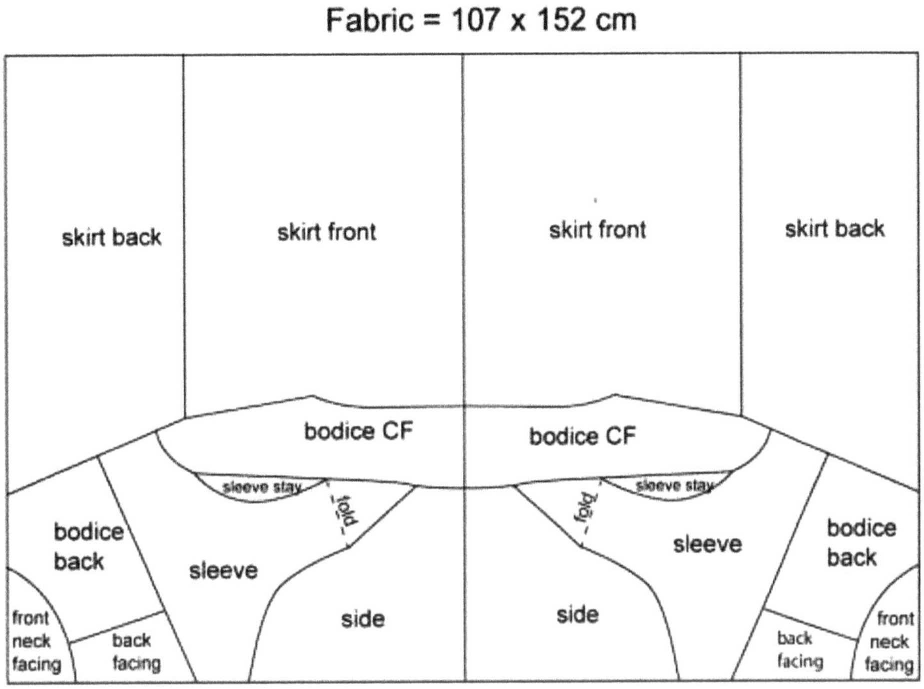

ative advantage of unpredictable design processes as a creative advantage. Hopefully, zero waste clothing design will thrive as more designers study practices, be they based on sustainable interests or just on a new creative pattern challenge. There are some obstacles to overcoming practices that are feasible in mass production. More experiments are necessary to find ways of making the zero-waste patterns the world's second largest polluting sector, as fashion industry.

Economic Growth, Sustainability, or Zero Waste

Numerous provinces and nations, including: DuPont, Fuji Xerosox, Collins Pine, Ricoh, Konika-Minolta, NEC, Toyota, Hewlett Packard, Epson and Interface have made "zero waste" commitments, with the following companies increasingly being: South Australia, Victoria, Western Australia, New Zealand, Lebanon, Taiwan and China. This shows widespread progress in preventive thought, technical performance and exemplary initiatives. The zero waste objectives, however, is often pursued gradually: no waste is reduced in reality. More than just a preventive objective, a prevention approach is needed. Common terms for describing social objectives are not ideal. The traditional aim of economic growth is to undermine sustainability, while sustainable economic growth, which is traditionally sought in ways that undermine economic growth (by over-regulation, but not by impact), is undermined. The public make little use of' sustainability' and' sustainable development' and are regularly abused by organizational and governmental bodies seeking approval for gradual improvements. Zero waste' is often misinterpreted as unrealistic, because it cannot be achieved with the economic signals of today but with different signals on a global scale. The potential for preventive approach is apparent to zero waste and sustainable development. Economic growth could also be compatible with a preventive approach is less well recognized. The counterproductive competition between economic, social and environmental objectives demonstrates the need for a review and adaptation of societal objectives (rather than balance and compromise). Renewed society objectives invite the most powerful intervention in the complex system to renewed thinking and international views (Meadows et al., 1999; Rob 1999).

A Goal-Set and Actions to Achieve the Goal-Set

The incremental approach frequently makes plans with weak connections to long-term objectives. Current trends are identified and measures to improve the situation are chosen. This method of planning is called' foresight.' Backcasting' (Robinson et al., 1990) is a preventive approach whereby planning is based on objectives instead of the present. Backcasting can be likened to traveling to a preferred destination rather than to the unwanted environment. Backcast does not hinder actions to meet longer-term goals by reducing current impacts. Backcasting helps to ensure that measures to minimize current impacts are not hindering measures to achieve long-term targets. The economist Kenneth Boulding described the 1966 goal proposed: a' cercular economy' is an economic growth, sustainability and no waste compatible long-term goal. The actions taken to implement a circular economy are appropriately described in a word first used in 1988 by Maureen O'Rorke, the social marketing manager. The combination of objectives and actions is the basis for society, economic renovation, the planning of projects and the daily making of decisions.

Action to Prevent a Range of Problems

The scope of precycling knowledge has expanded over time to include product-driven measures by households, municipalities, retailers, businesses and industry (Baldwin et al. 1997). Precycling can now be understood to include all the actions which create new resources for old products, in industrial, social, environmental and economic circumstances. Reduction, reuse, recycling, recomposting, composting and pyrolysis infrastructure can be extended to enhance product capacity. Industrial independence from known accumulated substances can be enhanced for this purpose (persistent synthetics, heavy metals, fossil fuels, radiation) as they too can be elevated so that substitution can become precycled. Environmental habitats can be wider and diversified so that more emissions and effluents can become natural resources. The ability of society to satisfy every need can be increased so that everybody can precycle. The economy can become more circular, enabling economic, environmental and social objectives to be achieved in parallel. New resources for people and for nature could be created on a' recycled planet' at a rate similar to that of today's problem waste. Forms of precycling are many actions that minimize impacts. For example, recycling is one way to precycle as action is taken prior to the release of materials into the environment. But precycling is ambitious rather than recycling. Each product is recyclable today and any error can only be corrected by pre-cycling it (Boulding et al., 1996).

From Recycling Insurance to Precycling Insurance

The practice of recycling insurance can be converted into an economic instrument for market correction for the implementation of a circular economy:

- The full spectrum of products can also be covered.
- The full range of producers may require participation, which maintains equity and encourages economic improvements.
- The whole range of waste prevention and problem prevention opportunities can be financed. Recycling funding is only one opportunity for precycling. Other opportunities include restoration, redesign, substitution, phasing out, and more effective or alternative ways to meet needs, local repair / reuse, expansion of environmentally friendly habitats and biodegradation of substances.

A generalized form of reclassification, known as "precycling insurance" (Wright, 1982), could support all forms of precycling. Government would be required, just like any economic instrument, to legislate and monitor a precycling insurance scheme. If government also collects and invests premiums, however, additional taxation will inevitably be perceived. Clearly precycling insurance is not a tax, as it is reasonably expected that producers are responsible for their own products and not to charge society with avoidable costs. The costs of premiums are shared by the market and not by government throughout society. Insurance undertakings or non-profit organizations could operate a scheme. Insurance companies have an expertise in risk calculation, investment management, and problem prevention. Even if precycling insurance reduces claims for other businesses, the diversification of the insurance companies into management consequences remains clear before it happens (Holmberg, 2000).

Precycling Insurance

Precycling insurance will then establish premiums according to the risk that a product will become a waste instead of a new resource for people or nature. Prime funding would fund pre-cycling measures. Substantial products, such as soap, gold, silver jewellery and aluminum beer kegs, with insignificant risks would be awarded low fees (Gillilan, 1996). The waste risk would be small if: products can be easily recycled or biodegraded; the manufacturer has invested in end time treatment; and the product does not accumulate in the land, atmosphere or in the sea. This last factor is concerned with the acute effects of known substances such as heavy metals, mineral fuels and persistent synthetic compounds (CFCs, PCBs, DDT, etc). Pre-cycling insurance and normal market forces could compatible with a circular economy by using fuels, heavy metals, pesticides, automobiles, houses and sofas (and anything else). Larger economic, social and environmental damage can be avoided through the inclusion of lower prevention costs in the purchase price while offering manufacturers new market opportunities and maximum freedom of choice. In 2003, the recycling assurance was first sold as a service to manufacturers with products required for end-of-life recycling by Swedish insurer Lansforsakringar Miljo (Miljo, 2002). This insurance, guaranteeing payment of all future recycling charges can be purchased by producers instead of arranging recycling themselves.

Zero Waste to Industrial Networks

Project ZeroWIN

ZeroWIN, a five-year project, running between 2009-2014 which was funded by the EC under the Seventh Framework Programme, established by the Towards Zero Waste in Industrial Networks (www.zerowin.eu). It consists of 30 academic and industry partners throughout Europe (and one in Taiwan) that integrated their skills to allow the strategies chosen to be studied in real-life case studies. The Consortium investigates and demonstrates, by adopting a network approach and combining methods & tools, using technology and design and innovations and policy measures, how the closed-loop philosophy can help to achieve zero waste. The ZeroWIN partners have elaborated the draft proposal from the common perspective that the zero waste objective has been necessary and achievable, that there has been a clear and determined need for research in this field, and that such an ambitious study has required a consortium of research institutions and industries with the necessary research expertise and practical ability to achieve the project objectives. To develop the zero waste vision before decide which new technologies and methods for waste prevention to apply, and what tools to use, before developing the manufacturing model for optimisation of resources and prevention of waste. Quantitative evaluation of the success and the policy implications of the ZeroWIN approach was reported throughout the project and extensive dissemination of results was planned. Overall management and coordination have been formally integrated into a work package to ensure efficiency of the project. In four industries, ZeroWIN focused on two key waste types:

- Building and demolition waste
- Electronic / high-tech waste; from three areas:
- Electrical and Electronic Facilities (EEF)
- Car industry
- Photovoltaic sector (PV)

The ZeroWIN project has identified the best way to improve and bring together existing approaches and tools in an industrial system, as well as how to use innovative technologies to achieve a zero waste vision. The specific environmental targets were:

- 30% reduction in greenhouse gas emissions
- 70% overall reuse and waste recycling
- 75% decrease in freshwater utilization

Key Strategies Identified for Applying Zero Waste in Industry

A network approach and everyone involved will require a shared vision in order to achieve this, make best use of a number of existing waste prevention approaches as well as tools and technologies and develop new ones to improve the whole system.

1. **Eco-Design**: The design of system waste is especially important for real improvements in system-wide waste reduction (Figure 6). For ZeroWIN this principle includes extended use, de-materialisation and green chemistry concepts (Anastas et al, 1998; BSI, 2008).
2. **Industrial Symbiosis:** It's part of the concept of industrial ecology. This is particularly important for ZeroWIN, which is aimed at reaching zero waste by promoting industrial networks. The theoretical rationale and methods of information and resources sharing in different industries constitute industrial symbiosis; eco-industry parks are real world industrial symbiosis applications (Chertow, 2000, EPA, 2009; Baas et al., 2008, for example).
3. **Closed Log Supply Chains:** ZeroWIN developed a version of an established management strategy with a view to incorporating loop-closure reverse logistics and optimizing the supply, distribution and recycling flows of materials, produce and waste. The efficiency and sustainable performance of the industrial network were improved (Rogers et al., 2001; Dekker et al., 2003; Lambert, 2004).
4. **Uses of New Technologies**: Innovative technology can have powerful environmental and economic implications for new products and industrial processes. With ZeroWIN RFID and the D4R-Re-use, Recycle, Repair-Laptop and grid-connected and stand-alone PV systems, the investigation, design and testing of these systems was carried out.
5. **Management of Products:** Extended manufacturer's liability methods, and in particular individual manufacturer's liability (IPR), are viewed as important tools to enforce producers ' full responsibil-

Figure 6. Showing zero waste system proposed for ZeroWIN
(Curran et al., 2013)

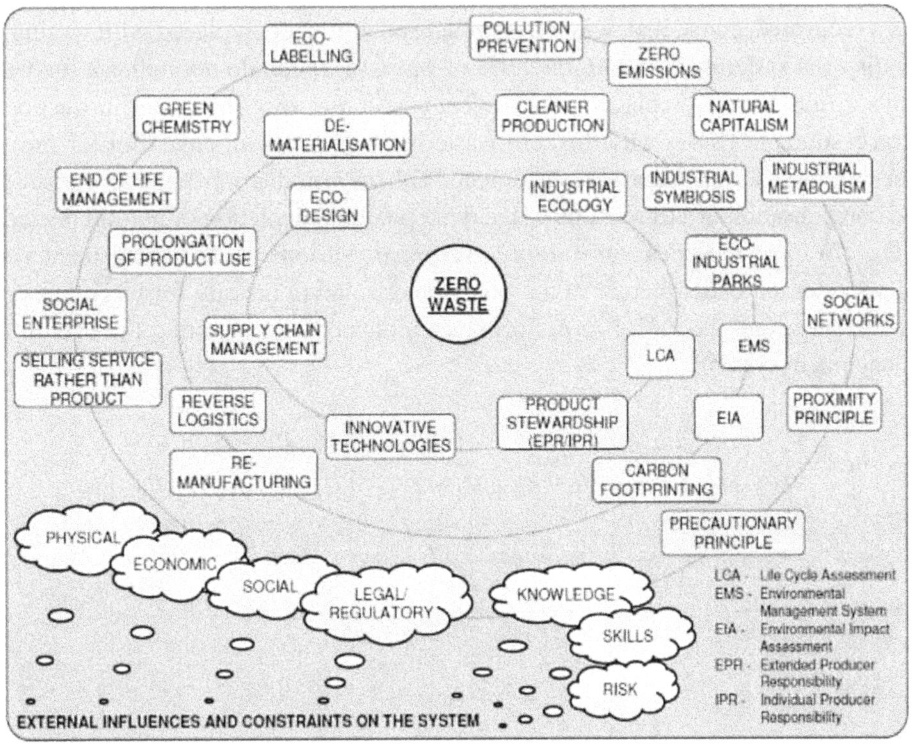

ity, especially at the end of their lifetimes for their products (OECD, 2001 ; Tojo, 2004; Rossem, 2008).

6. **Life Cycle Assessment**: It is now recognized that measuring impacts from raw materials through the production and utilization phases to end-of-life throughout the physical life cycle of real product environmental impacts can lead to more sustainable production and consumption patterns) The measurement of the carbon footprint was also identified as important for further increased use of carbon objectives and budgets in Europe. LCA and carbon footprinting methods are used to evaluate the success of ZeroWIN's various waste prevention and case-studies in relation to greenhouse gas emission targets, waste and water use (Rex 2008; Borghi 2009 and JRC-IES 2009).
7. An Environmental Management System (EMS) plans, schedules, implements and oversees those environmental performance improvement activities. It provides organizations with a method to consistently manage their environmental activities, products and services and helps them meet their environmental obligations and achieve their performance goals. Many standards are available to model EMS. There are also standards available. ISO14001 and the eco-management and audit (EMAS) system at EU level are the most widely distributed on the international stage.
8. EMS is moderate to ZeroWIN because it can be regarded as an adequate support to manage sustainable industrial activities, but does not constitute a key issue for its development itself.

Zero Waste Index

The Zero Waste Index is a tool for measuring the potential of virgin materials for zero waste systems to be offset. One of the key objectives of the concept of null waste is the zero depletion of natural resources. The measurement of efficiency of the city of zero waste would therefore eventually measure resources that have been recovered, consumed, wasted, recycled and ultimately replaced with virgin materials and offset waste disposal systems. However, the rates of waste diversion do not indicate the waste management system's virgin material replacement efficiency which is very important in the conservation of global natural resources. This is why the zero waste index is a cutting-edge tool for measuring waste management systems in virgin material replacement. With the introduction of the zero waste global index, the potential compensation of virgin materials and the potential depletion of natural resources could be measured. The ZWI also provides a useful tool to compare various waste management systems across the city and provides a broader picture of the potential demand in the city for virgin materials, energy, CO_2 and water. Therefore, the ZWI is a performance indicator for evaluating the overall performance of waste management systems.

$$Zero\,waste\,index = \sum \begin{matrix} potential\,amount\,of\,waste\,managed\,by\,the\,city \\ *substitution\,for\,the\,systems\,/\,Total\,generated\,by\,the\,city \end{matrix}$$

$$Diversion\,rate = \frac{Weight\,of\,recyclables}{Weight\,of\,garbage + Weight\,of\,recyclables} \times 100$$

Recyclables = waste that is reused, recycled, composted or digested
Garbage = waste that is landfilled or incinerated (City of Toronto, 2012).

Many towns such as Adelaide, San Francisco and Stockholm currently try to be zero-waste towns by getting waste to 100% off site. However, for zero waste initiatives, diversion from waste and recycling are not sufficient. The diversion rate as indicated above does not include waste avoidance by industrial design, effective policies and behavior changes; therefore, waste diversion rate is not enough to measure the city's nil waste performance (Zaman et al., 2013).

CONCLUSION

The rapid economic growth and globalization produced a massive amount of waste and attracted the global attention due to potential environmental impact and resource waste, such as illegal dumping and the cross-border movement of industrial waste, informal e-waste recycling, food loss and the release of greenhouse gas, resource consumption. The idea of' zero waste' is therefore an efficient means to solve the problems of solid waste. Zero waste is to promote the redevelopment of resource life cycles in order to reuse all products. There have now been significant efforts in the cities, companies and individuals that offer many good suggestions for the future production of zero waste. In order to measure the performance of waste management systems, decision makers and waste experts use various indicators. In the past decade, waste diversion, green town and zero waste indexes were used to measure the performance of a town as a key indicator. In order to produce zero waste, certain key strategies, such as Eco-Design and Eco-Labeling should be applied from the extraction of raw materials to the end disposal. The zero waste management for solid waste, industrial waste, building waste and electronic waste must move beyond recycling, to the largely unknown territory of the higher end of the waste management, such as eco - design, eco - labeling, closed loop chain and clean production. The world's efforts at reducing its problems can ironically block a preventive approach. As currently practiced, zero waste, sustainability and sustained economic growth may not be achieved. The alternative suggested seems viable and feasible; it does, however, call for a change of thinking. People have always been inventive and have changed the face of the world, but how easy minds can change remains to be seen. Most people in the world already suffer from an increasing number of worse effects. All of these impacts are not guaranteed to remain tolerable until everyone is ready.

REFERENCES

Allen, C. (2012). *Taiwan: Community Action Leads Government Towards Zero Waste*. Available at http://no-burn.org/downloads/ZW%20Taiwan.pdf

Anastas, P., & Warner, J. (1998). *Green Chemistry: Theory and Practice*. New York: Oxford University Press.

Asgher, M., Ahmad, Z., & Iqbal, H. M. N. (2017). Bacterial cellulose-assisted de-lignified wheat straw-PVA based bio-composites with novel characteristics. *Carbohydrate Polymers*, *161*, 244–252. doi:10.1016/j.carbpol.2017.01.032 PMID:28189235

Baas, L. W., & Huisingh, D. (2008). The synergistic role of embeddedness and capabilities in industrial symbiosis: Illustration based upon 12 years of experiences in the Rotterdam Harbour and Industry Complex, Progress in Industrial Ecology. *International Journal (Toronto, Ont.)*, *5*(5–6), 399–421.

Baldwin, J. H. (1997). Describes industrial design and reuse as precycling, Conservation. *Encarta Online Encyclopedia.* Retrieved from: http://www.encarta.msn.com

Borghi, A. D., Gallo, M., & Borghi, M. D. (2009). A survey of life cycle approaches in waste management. *The International Journal of Life Cycle Assessment*, *14*(7), 597–610. doi:10.100711367-009-0111-7

Boulding, K. (1966). The economics of the coming spaceship Earth. In H. Jarrett (Ed.), *Environmental quality in a growing economy, resources for the future* (pp. 3–14). Baltimore, MD: Johns Hopkins University Press.

Brezet, H., & Hemel, C. V. (1997). *ECODESIGN – A Promising Approach to Sustainable Production and Consumption*. Paris: UNEP.

BSI. (2008). *PAS 2050:2008, Specification for the Assessment of the Life Cycle Greenhouse Gas Emissions of Goods and Services*. Available at: http://www.bsigroup.com/en/Standards-andPublications/Industry-Sectors/Energy/PAS-2050/PAS-2050-Form-page/

Cheng, H., & Hu, Y. (2010). Municipal solid waste (MSW) as a renewable source of energy: Current and future practices in China. *Bioresource Technology*, *101*(11), 3816–3824. doi:10.1016/j.biortech.2010.01.040 PMID:20137912

Chertow, M. R. (2000). Industrial symbiosis: Literature and taxonomy. *Annual Review of Energy and the Environment*, *25*(1), 313–337. doi:10.1146/annurev.energy.25.1.313

City of Toronto. (2012). *How We Calculate Waste Diversion Rates*. Available: http://www.toronto.ca/environment/initiatives/wdt.htm

Colon, M., & Fawcett, B. (2006). Community-based household waste management: Lessons learnt from EXNORA's 'zero waste management' scheme in two South Indian cities. *Habitat International*, *30*(4), 916–931. doi:10.1016/j.habitatint.2005.04.006

Cox, J., Giorgi, S., Sharp, V., Strange, K., Wilson, D. C., & Blakey, N. (2010). Household waste prevention—A review of evidence. *Waste Management & Research*, *28*(3), 193–219. doi:10.1177/0734242X10361506 PMID:20215491

Curran, T., & Williams, I. D. (2012). A zero waste vision for industrial networks in Europe. *Journal of Hazardous Materials*, *207-208*, 3–7. doi:10.1016/j.jhazmat.2011.07.122 PMID:21899950

Davorin, K. (2011). Innovative systemic approach for promoting sustainable innovation for zero construction waste. *Kybernetes*, *40*(1/2), 275–289. doi:10.1108/03684921111118040

Dekker, R., Fleischmann, M., Inderfurth, K., & Wassenhove, V. (2003). *Reverse Logistics – Quantitative Models for Closed-Loop Supply Chains*. Berlin: Springer.

Ericson, S. (2010). No waste allowed. *Threads*, *149*, 60–63.

GAIA. (2013). *Individual Zero Waste Case Studies*. Available at http://no-burn.org/-1-18

Gillilan, S., Werner, C. M., Olson, L., & Adams, D. (1996). Teaching the concept of precycling: A campaign and evaluation. *The Journal of Environmental Education*, *28*(1), 11–18. doi:10.1080/00958964.1996.9942810

Grant, T., James, K.L., & Partl, H. (2003). *Life Cycle Assessment Of Waste Andresource Recovery*. Academic Press.

Greyson, J. (2007). An economic instrument for zero waste, economic growth and sustainability. *Journal of Cleaner Production*, *15*(13-14), 1382–1390. doi:10.1016/j.jclepro.2006.07.019

Guerrero, L. A., Maas, G., & Hogland, W. (2013). Solid waste management challenges for cities in developing countries. *Waste Management (New York, N.Y.)*, *33*(1), 220–232. doi:10.1016/j.wasman.2012.09.008 PMID:23098815

Hansen, M. S., & Gilberg, U. M. (2003). *Cost-benefit analysis and LCA in environmental decision-making – potentials and limitations* (Master thesis). E&R Technical University of Denmark, Kongens Lyngby.

Haro, P., Ollero, P., & Villanueva, A. L. (2013). Thermochemical biorefinery based on dimethyl ether as intermediate: Technoeconomic assessment. *Applied Energy*, *102*, 950–961. doi:10.1016/j.apenergy.2012.09.051

Hill, M. H., & Bucknell, P. A. (1967). *The evolution of fashion: Pattern and cut from 1066 to 1930*. London: Batsford.

Holmberg, J., & Robert, K. H. (2000). Backcasting from non-overlapping sustainability principles: A framework for strategic planning. *International Journal of Sustainable Development and World Ecology*, *7*, 291–308. doi:10.1080/13504500009470049

Iqbal, H. M. N., & Asgher, M. (2013). Characterization and decolorization applicability of xerogel matrix immobilized manganese peroxidase produced from Trametes versicolor IBL-04. *Protein and Peptide Letters*, *20*(5), 591–600. doi:10.2174/0929866511320050013 PMID:23016633

Iqbal, H. M. N., Kyazze, G., & Keshavarz, T. (2013). Advances in the valorization of lignocellulosic materials by biotechnology: An overview. *BioResources*, *8*(2), 3157–3176. doi:10.15376/biores.8.2.3157-3176

Jackson, T. (2005). *Motivating Sustainable Consumption: A Review of Evidence on Consumer Behaviour and Behavioural Change* (T. Jackson, Ed.). Surrey, UK: Sustainable Development Research Network.

JRC-IES. (2010). *International Reference Life Cycle Data System (ILCD) Handbook: General Guide for Life Cycle Assessment – Detailed Guidance (Plus 6 Supporting Documents)*. Author.

Karlsson, H., Ahlgren, S., Strid, I., & Hansson, P. A. (2015). Faba beans for biorefinery feedstock or feed? Greenhouse gas and energy balances of different applications. *Agricultural Systems*, *141*, 138–148. doi:10.1016/j.agsy.2015.10.004

King, S. (2004). *Strategic Framework—The Foundation of an Effective Planning Process*. Available online: http://www.infomanagementcenter.com/enewsletter/200411/third.htm

Kirkeby, J. T., Birgisdottir, H., Hansen, T. L., Christensen, T. H., Bhander, G. S., & Hauschild, M. (2006). Environmental assessment of solid waste systems and technologies: EASEWASTE. *Waste Management & Research*, *24*(1), 3–15. doi:10.1177/0734242X06062580 PMID:16496866

LaBrecque, S. (2015). What does zero waste mean to you? *The Guardian*.

Lambert, D. M. (2004). *Supply Chain Management: Processes, Partnerships, Performance* (1st ed.). Sarasota, FL: Supply Chain Management Institute.

Langeveld, H., Sanders, J., & Meeusen, M. (2012). *The Biobased Economy: Biofuels*. Materials, and Chemicals in the Post-oil Era, Earthscan.

Lansforsakringar Miljo. (2002). *Recycling Insurance*. Retrieved from http://www.lansforsakringar.se

Liao, C. H., & Chiu, A. S. F. (2011). Evaluate municipal solid waste management problems using hierarchical framework. *Procedia: Social and Behavioral Sciences*, *25*, 353–362. doi:10.1016/j.sbspro.2011.10.554

Matete, N., & Trois, C. (2008). Towards zero waste in emerging countrie Sea South African experience. *Waste Management (New York, N.Y.)*, *28*(8), 1480–1492. doi:10.1016/j.wasman.2007.06.006 PMID:17714928

McCormick, K., & Kautto, N. (2013). The bioeconomy in Europe: An overview. *Sustainability*, *5*(6), 2589–2608. doi:10.3390u5062589

McDougall, F. R., White, P., Francke, M., & Hindle, P. (2001). *Integrated Solid Waste Management – A Lifecycle Inventory 2*. London: Blackwell.

McQuillan, H. (2011). Zero-waste design practice: Strategies and risk taking for garment design. In A. Gwilt & T. Rissanen (Eds.), Shaping sustainable fashion: Changing the way we make and use clothes. London: Earthscan.

McQuillan, H., & Rissanen, T. (2011). *YIELD exhibition catalogue*. Brooklyn, NY: Textile Arts Center. Retrieved August 20, 2012, from http://yieldexhibition.com/yieldexhibitioncatalogue.pdf

Meadows, D. (1999). *Leverage points: places to intervene in a system*. Sustainability Institute. Retrieved from http://www.sustainabilityinstitute.org

Melanie, C., & Victoria, K. (2014). Expanding zero-waste design practices: A discussion paper, International Journal of Fashion Design. *Technology and Education*, *7*(1), 58–64. doi:10.1080/17543266.2013.837967

Murphy, S., & Pincetl, S. (2013). Zero waste in Los Angeles: Is the emperor wearing any clothes? *Resources, Conservation and Recycling*, *81*, 40–51. doi:10.1016/j.resconrec.2013.09.012

O'Rorke, M. (1988). *Public information campaign on precycling*. Prepared for City of Berkeley.

OECD. (2001). Extended Producer Responsibility: A Guidance Manual for Governments. Organization for Economic Cooperation and Development.

Palmer, P. (2004). *Getting to Zero Waste*. Sebastopol, CA: Purple Sky Press.

Parajuli, R., Dalgaard, T., Jørgensen, U., Adamsen, A. P. S., Knudsen, M. T., Birkved, M., & Schjørring, J. K. (2015). Biorefining in the prevailing energy and materials crisis: A review of sustainable pathways for biorefinery value chains and sustainability assessment methodologies. *Renewable & Sustainable Energy Reviews*, *43*, 244–263. doi:10.1016/j.rser.2014.11.041

Phillips, P. S., Tudor, T., Bird, H., & Bates, M. (2011). A critical review of a key waste strategy initiative in England: Zero waste places projects 2008-2009. *Resources, Conservation and Recycling*, *55*(3), 335–343. doi:10.1016/j.resconrec.2010.10.006

Pietrzak, W., Kawa-Rygielska, J., Król, B., Lennartsson, P. R., & Taherzadeh, M. J. (2016). Ethanol, feed components and fungal biomass production from field bean (Vicia faba var. equina) seeds in an integrated process. *Bioresource Technology*, *216*, 69–76. doi:10.1016/j.biortech.2016.05.055 PMID:27233099

Rex, E., & Baumann, H. (2008). Implications of an interpretive understanding of LCA practice. *Business Strategy and the Environment*, *17*(7), 420–430. doi:10.1002/bse.633

Rissanen, T. (2011). *Zero waste fashion repatterned exhibition catalogue*. Chicago, IL: Columbia College Chicago. Retrieved July 5, 2013, fromhttp://www.colum.edu/ADGallery/ Exhibits/2010_zerowaste/zerowastecatalog_final.pdf

Robinson, J. B. (1990, October). Future under glassda recipe for people who hate to predict. *Futures*.

Rogers, D. S., & Tibben-Lembke, R. S. (2001). An examination of reverse logistics practices. *Journal of Business Logistics*, *22*(2), 129–148. doi:10.1002/j.2158-1592.2001.tb00007.x

Rossem, C. V. (2008). *Individual producer responsibility in the WEEE Directive: from theory to practice?* (Ph.D. Dissertation). IIIEE, Lund University, Sweden.

Sakamoto, T., Hasunuma, T., Hori, Y., Yamada, R., & Kondo, A. (2012). Direct ethanol production from hemicellulosic materials of rice straw by use of an engineered yeast strain codisplaying three types of hemicellulolytic enzyme on the surface of xylose-utilizing Saccharomyces cerevisiae cells. *Journal of Biotechnology*, *158*(4), 4203–4210. doi:10.1016/j.jbiotec.2011.06.025 PMID:21741417

Scholey, D. V., Burton, E. J., & Williams, P. E. V. (2016). The bio refinery; producing feed and fuel from grain. *Food Chemistry*, *197*, 937–942. doi:10.1016/j.foodchem.2015.11.063 PMID:26617037

Seltenrich, N. (2017). *Incineration versus Recycling: In Europe, a Debate over Trash*. Available online:http://360.yale.edu/feature/incineration_versus_recycling__in_europe_a_debate_over_trash/2686/

Seng, B., Kaneko, H., Hirayama, K., & Katayama-Hirayama, K. (2010). Municipal solid waste management in Phnom Penh, capital city of Cambodia. *Waste Management & Research, 29*(5), 491–500. doi:10.1177/0734242X10380994 PMID:20813763

Sentime, K. (2013). The impact of legislative framework governing waste management and collection in South Africa. *Afr. Geogr. Rev., 33*, 81–93.

Shahsavarani, H., Hasegawa, D., Yokota, D., Sugiyama, M., Kaneko, Y., Boonchird, C. H., & Harashima, S. (2013). Enhanced bio-ethanol production from cellulosic materials by semi-simultaneous saccharification and fermentation using high temperature resistant Saccharomyces cerevisiae TJ14. *Journal of Bioscience and Bioengineering, 115*(1), 20–23. doi:10.1016/j.jbiosc.2012.07.018 PMID:22925900

Textiles Environment Design. (2012). *TED's Ten: 1 – design to minimise waste*. Retrieved October 26, 2012, from http://www.tedresearch.net/1-minimise-waste

The Carbon Trust. (2006). *Carbon Footprint in the Supply Chain: The Next Step for Business*. Author.

Thomas, S. (2008). From 'Green Blur' to ecofashion: Fashioning an eco-lexicon. *Fashion Theory, 12*(4), 525–539. doi:10.2752/175174108X346977

Tojo, N. (2004). *Extended producer responsibility as a driver for design change – utopia or reality?* (Ph.D. Dissertation). IIIEE, Lund University, Sweden.

Townsend, K., & Mills, F. (2013). Mastering zero: How the pursuit of less waste leads to more creative pattern cutting. *International Journal of Fashion Design, Technology, and Education, 6*(2), 104–111. doi:10.1080/17543266.2013.793746

UN-HABITAT. (2010). *Solid Waste Management in the World's Cities. Water and Sanitation in the World's Cities Report 2010*. Earthscan.

US EPA. (2008). *Introduction to the Concept of Green Chemistry*. Retrieved from http://www.epa.gov/oppt/greenchemistry/pubs/aboutgc.html

WRI-WBCSD. (2007). *The GHG Protocol for Project Accounting*. Author.

Wright, B. (1982). *The British Fire Mark 1680-1879*. Cambridge, UK: WoodheadFaulkner. Retrieved from http://www.firemarkcircle.fsnet.co.uk/history3.htm

Zaman, A. U. (2015). A comprehensive review of the development of zero waste management: Lessons learned and guidelines. *Journal of Cleaner Production, 91*, 12–25. doi:10.1016/j.jclepro.2014.12.013

Zaman, A. U. (2015). A Strategic Waste Management Framework and Tool for the Development of Zero Waste Cities. School of Art, Architecture and Design; University of South Australia.

Zaman, A. U., & Lehmann, S. (2011). Challenges and Opportunities in Transforming a City into a "Zero Waste City". *Challenges, 2*(4), 73–93. doi:10.3390/challe2040073

Zaman, A. U., & Lehmann, S. (2013). The zero waste index: A performance measurement tool for waste management systems in a'zero waste city'. *Journal of Cleaner Production, 50*, 123–132. doi:10.1016/j.jclepro.2012.11.041

Zero Waste International Alliance. (2004). *Zero Waste Definition Adopted by Zero Waste Planning Group*. Retrieved from http://www.zwia.org/main/index.php?option=com_content&view=article&id=49&Itemid=37

Zero Waste SA. (2013). *South Australia's Waste Strategy 2011-2015*. Available at Error! Hyperlink reference not valid.

Zero Waste South Australia. (2013). *South Australia's Kerbside Three-Bin System Waste Report*. Retrieved from http://www.zerowaste.sa.gov.au/upload/resource-centre/publications/localgovernment/ZWSA%20Kerbside%20report%202015%20DE_02.pdf

ZWSA. (2011). *Zero Waste SA. About Us*. Available at http://www.zerowaste.sa.gov.au/about-us

Chapter 9
Waste Management Technology for Sustainable Agriculture:
Waste Management

Muzaffar Ahmad Bhat
Barkatullah University, India

A. Wani Adil
Sagar University, India

Bhat Mohammad Sikander
University of Kashmir, India

Yaqoob Lone
Sagar University, India

Junaid Ahmad. Malik
PG College Videysha, India

ABSTRACT

The process of collection, transport, disposal, recycling, and monitoring of wastes is called waste management. The waste management is undertaken to recycle the wastes so as to reduce the ill effects of wastes on environment, health, and aesthetics. There are several kinds of wastes produced such as agricultural wastes, municipal wastes, industrial waste, mining waste. Some wastes are more hazardous such as medical wastes and nuclear wastes. Various techniques are used for the management of wastes which includes landfilling, incineration, anaerobic digestion, pyrolysis, plasma gasification, recycling, composting. Anaerobic digestion produces biofuel in the form of biogas. Plasma gasification results in the generation of electricity from wastes. Recycling of wastes involves the collection, sorting, and reprocessing of wastes into new products. Vermicomposting is the preferred form of composting as it results in the formation of vermicompost called black gold due to the presence of rich nutrients and growth promoting factors in it.

DOI: 10.4018/978-1-7998-0031-6.ch009

INTRODUCTION

Due to industrialization, urbanization and increase in population density, there has been an increase in the accumulation of wastes. The wastes include radioactive substances, agricultural wastes, food wastes, industrial wastes, municipal wastes, garbage, paper waste etc. There has been decrease in the availability of open land space for the disposal of wastes due to intensive use of land for residential, industrial and commercial purposes. In order to reduce the load of wastes on earth, the collection, transport, recycling and disposal of wastes needs to be improvised (AshaAalok et al., 2008).

The process of collection, transport, disposal, recycling and monitoring of wastes is called as waste management (Burke et al., 2005). Waste management can be costly so it is important to understand the various effective, sustainable and safe means of its management (El-Haggar, 2007).The three R`s "reduce, reuse, and recycle have become the basic tenet in waste management due increase in generation of wastes, increase in processing costs and decrease in available landfill space (El-Haggar, 2007; Seadon, 2006; Suttibak & Nitivattananon, 2008; Tudor et al., 2011). There should be flexibility in waste management systems in light of changing environmental, social and economic conditions (McDougall et al., 2001; Scharfe, 2010).To optimize, evaluate, adapt and define waste management systems, the information and feedback can be obtained from system analysis (Pires et al., 2010).

The reduction of wastes is placed at the top of waste management hierarchies because the best means of waste management is to reduce waste by not creating it in first place (USEPA, 2010). The reduction of wastes can also be achieved through the reuse of products. The reduction and reuse of wastes saves natural resources, reduces generation of wastes and reduces the costs associated with waste disposal (USEPA, 2010).

The waste management is undertaken to recycle the wastes so as to reduce the ill effects of wastes on environment, health and aesthetics (Gajalakshmi & Abbasi, 2003). The wastes may be either in solid, liquid or gaseous form. The process of waste management varies for rural and urban areas, for municipal and industrial wastes, for developed and developing nations. The management of municipal wastes is the responsibility of local government while as the management of industrial waste is the responsibility of generator (Verdone & De Filippis, 2004). The developed nations use various novel technologies to reduce the negative impacts of waste or use an effective management to exploit it (Henry & Heinke, 1989; Cunningham et al., 2007).

WASTE CATEGORIES

There are several kinds of wastes produced; some are more hazardous such as medical wastes and nuclear wastes (British Medical Association, 1999). Wastes are released into air, water or land. Following are the main categories of wastes:

Solid Waste

Solid wastes are the wastes that are that are deposited at the site where they are produced. They are further categorized into following types:

Agricultural Waste

It includes sugarcane trash, rice straw, leaves of trees and crops, rotten vegetables, fruits and flowers, animal wastes such as cattle dung, pig manure, poultry manure etc. Most of the agricultural wastes are organic and hence can be converted to biofertilizer; vermicompost through vermicomposting (Venkatachalaiah, 1996; Garg et al., 2006; Muthukumaravel et al., 2008; Reddy et al., 2009; Ansari & Rajpersaud, 2012). Rest of the agricultural wastes can be incinerated or disposed in landfills. Air pollutants are released from incineration (Lai et al., 2009). Landfilling results in the emission of greenhouse gases such as carbon dioxide and methane which result in global climate change (Hurst et al., 2005).

Food Waste

Food waste includes uneaten food and left-over food from residences, cafeteria and restaurants. These wastes are either disposed in landfills or incinerated, which result in the loss of nutrients of soil and pollute environment. Recycling of food waste reduces the cost of waste treatments, air pollutants from incineration and refuses from landfills (Lai et.al. 2009). Landfill gas includes greenhouse gases such as carbon dioxide and methane which result in global climate change (Hurst et.al. 2005). Therefore, Vermicomposting of food wastes is an alternative for the management of food wastes as it is cost effective, eco-friendly and hygienic (Kaviraj & Sharma 2003). Othman et.al. (2012) used African night crawler *Eudrilus Eugeniae* for the vermicomposting of food waste and obtained vermicompost in 14 days.

Industrial Waste

It includes the wastes from food industry, paper industry, herbal industry, pharmaceutical industry, agro industry etc. These wastes can be disposed in landfills, incinerated or composted. Various researchers have been successful on the conversion of wastes generated from above mentioned industries into vermicompost with the use of earthworm through vermicomposting (Suthar & Singh, 2008; Garg et al., 2011; Kumari et al., 2011; Singh & Suthar, 2012; Basheer & Agrawal, 2013).

Municipal Solid Waste

It includes the household wastes, wastes from commercial establishments and institutions. These wastes are usually dumped at low-lying areas that cause soil and water pollution, emission of methane, communicable diseases and production of obnoxious gases (Rao & Shantaram, 1996). Further pathogens are found in municipal solid waste which comes from feces of animals, putrifying food or dirty clothes (Pehren & Clark, 1987). These wastes are managed by landfill, incineration, recycling or composting. In India municipal solid waste is composed of more than 40% vegetative matter which includes vegetable waste, paper waste, fruit waste, food waste etc., this is termed as compostable municipal solid waste (Central Pollution Control Board, 2000). There is risk in the use of unhygienic compost prepared from municipal solid waste (Engeliet al., 1993; Aranconet al., 2005). Kaviraj et al.(2011),used bioinoculants*Penicilliumchrysogenum, P. funiculosum and Azotobacterchroococcum* along with earthworm *Eiseniafetida*for the transformation of municipal solid waste into vermicompost.

Mining Waste

The mining operations release large quantity of waste in the form of rocks, tailings, chemicals involved in the processing of ore. These wastes are either dumped in the land or in a water body. This results in water pollution and soil pollution. The mining waste contains hazardous substances such as radioactive substances, arsenic, lead and asbestos. Mining disturbs the natural vegetation of an area and causes siltation in water bodies. Further large quantity of water is utilized for mining operations and makes it unsuitable for drinking, irrigation and recreation.

Bio-Waste

It includes biodegradable green waste from parks and kitchen waste (Enger et al., 2006; Cunningham et al., 2007). It is regarded as wet waste because it contains 50-80% water. The management of bio-waste includes incineration, landfilling, anaerobic digestion and composting.

Hazardous Waste

It includes radioactive wastes, medical wastes, and explosives, volatile and flammable wastes. Special precautionary measures are adopted for the collection, storage, transport and disposal, of these wastes to avoid damage to life and property. Environment hazards may occur due to disposal of the hazardous wastes generated by industrial society (British Medical Association, 1999; Communication from the Commission Europe, 2010)

EFFECT OF WASTE ON HEALTH AND ENVIRONMENT

According to AshaAalok et al.(2008), Gajalakshmi and Abbasi (2003), wastes generated from industries and urban sources contain toxic substances which when disposed in open land, cause health hazards like respiratory infections, intestinal disorders, cholera, typhoid, plague etc. These toxic substances get deposited in the soil; thereby reduce the fertility of land. There is an important need for the recycling of these wastes so as to prevent the deterioration of the soil and also to reduce the level of pollution caused by the disposal of these wastes (Gajalakshmi & Abbasi, 2003). The wastes are either disposed in landfills or incinerated, which result in the loss of nutrients of soil and pollute environment. Landfill gas includes greenhouse gases such as carbon dioxide and methane which result in global climate change (Hurst et al., 2005).

WASTE MANAGEMENT TECHNIQUES

Landfilling

This method is still the favored means of disposal worldwide although being the worst means of waste disposal Figure 1. In this method wastes are dumped in open land and wetlands. This affects the vegetation of an area and releases greenhouse gases methane and carbon-dioxide which cause global climate change (Hurst et al., 2005). A well designed modern landfill is given in Figure 2.

Incineration

It includes burning of wastes in open areas or in an incinerator. Nowadays incineration in most of the developed countries is carried out in an incinerator that results in the conversion of waste to energy in the form of heat, electricity or production of steam. Incineration releases pollutants like acid gases, certain metals and chemicals dioxins and furans into the atmosphere which cause cancer and birth defects. The ash collected from incinerator may contain heavy metals such as mercury, lead, cadmium and arsenic. The dumping of ash in land can result in the seepage of these heavy metals into ground water and water bodies. This will affect vegetation, biota of water bodies and can cause various health problems. The good combustion of the wastes in an incinerator depends upon Temperature, Time and Turbulence. Optimum temperature causes adequate combustion of wastes. Adequate residence time of waste in an incinerator causes proper combustion of waste. Turbulence refers to the mixing of gases through incinerator. For good combustion proper turbulence is required. Incineration flow chart is given Figure 3.

Figure 1. Changes in waste disposal methods (Enger et al., 2006)

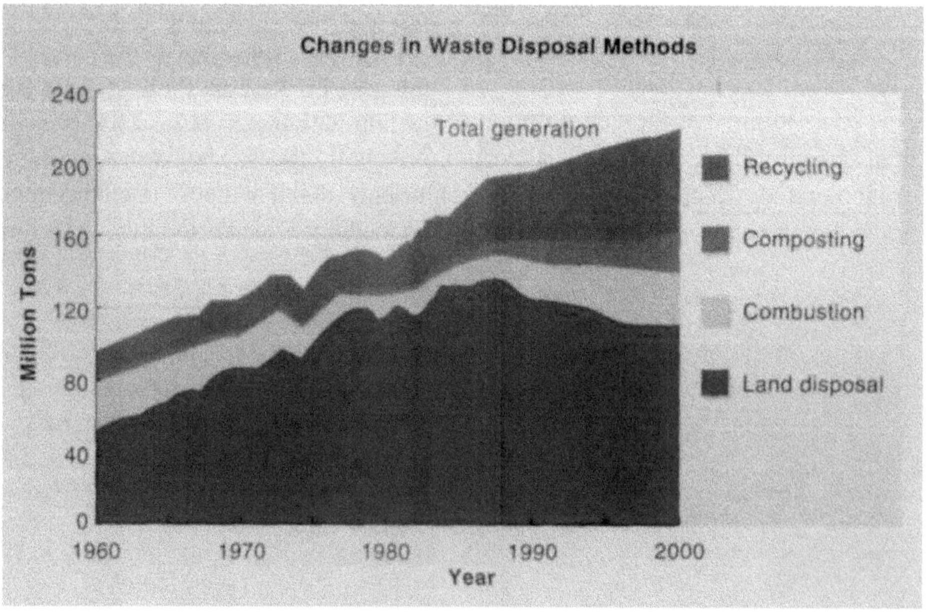

Figure 2. A well-designed modern landfill (Enger et al., 2006)

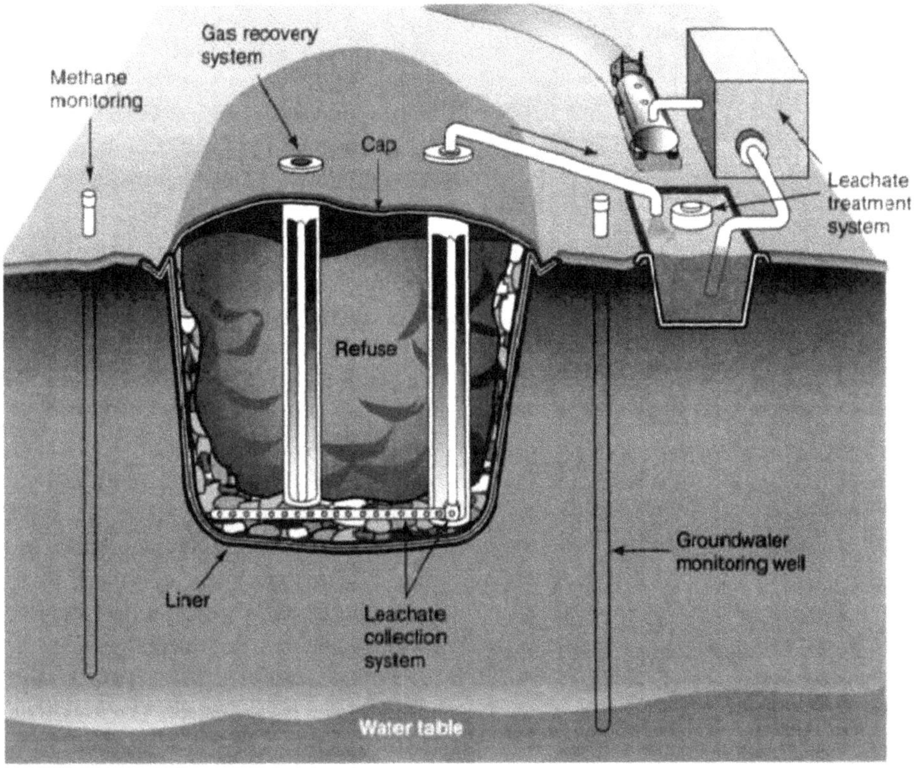

Anaerobic Digestion of Wastes (Bio-Gas Production)

Under anaerobic conditions in controlled reactors the organic wastes produce biogas which contains mainly methane 50-70% by volume and carbon-dioxide. It reduces greenhouse gas emission and bio-gas can be used as biofuel for transport or it can be directly injected into the gas distribution grid. Bio-gas production flow chart is given in Figure 4.

Pyrolysis

It involves the thermal treatment of wastes materials in a sealed vessel under high temperature and pressure, with limited availability of oxygen to produce energy, activated carbon and other refined products. Pyrolysis of plastic waste is given in Figure 5.

Plasma Gasification

It offers new opportunities for waste disposal. It involves the thermal treatment of waste materials in a sealed vessel under high temperature and pressure, with limited availability of oxygen to produce highly ionized and electrically charged syngas for the generation of electricity and heat. Syngas is composed of carbon-monoxide and hydrogen. Plasma gasification technique is given in Figure 6.

Figure 3. Incineration flow chart (Nidoni, 2017)

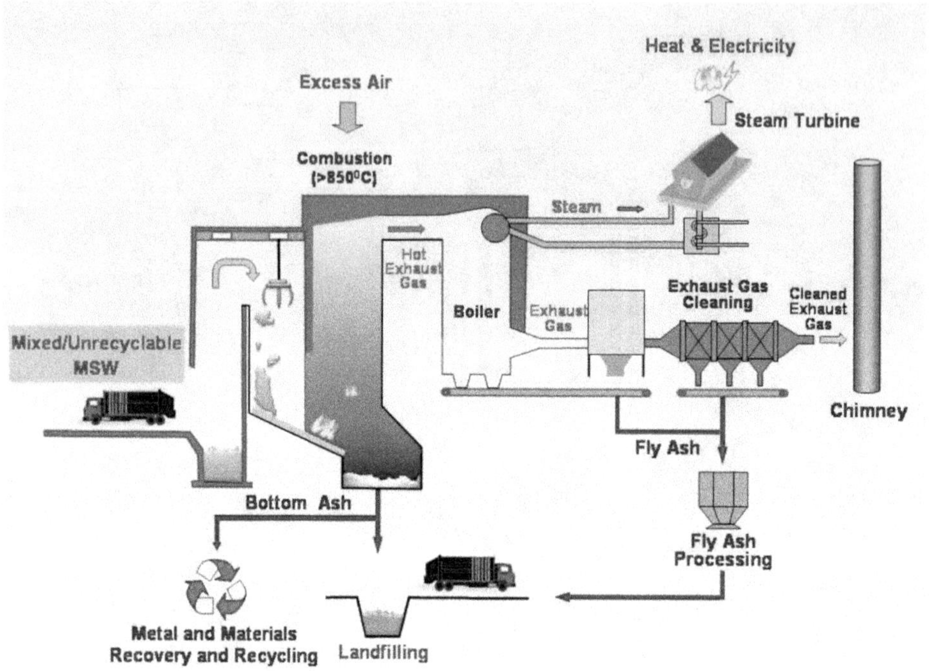

Figure 4. Bio-gas production flow chart (Flotats & Gibert, 2002)

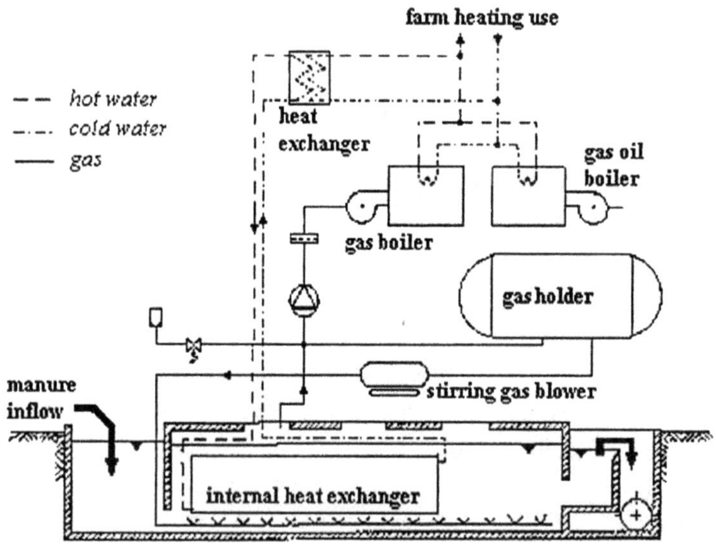

Figure 5. Pyrolysis of plastic waste (Kalargariset al., 2017)

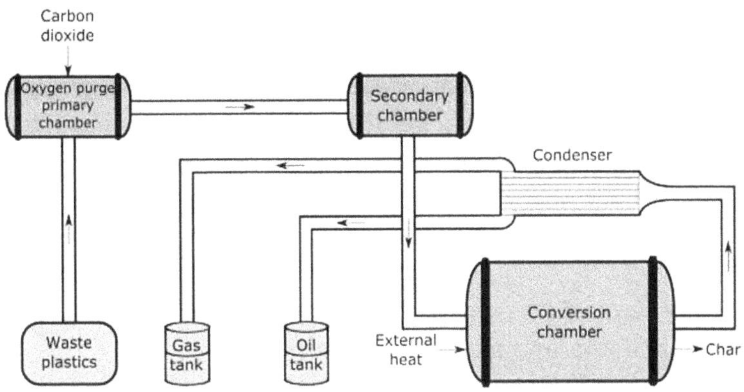

Figure 6. Plasma gasification (Westinghouse Plasma Corporation WPC, 2010)

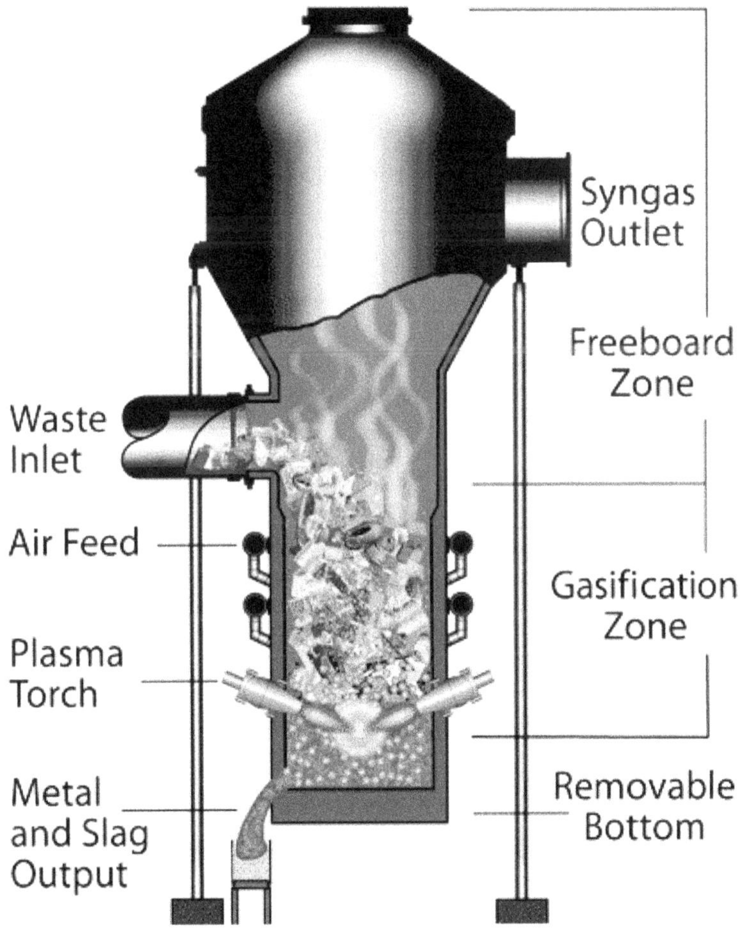

Recycling

It involves the collection, sorting and reprocessing of wastes into new products. The wastes that are commonly recycled includes plastic bottles, jars and glasses, newspapers, magazines, aluminum tins and cans, steel food cans, paperboard cartons etc. Recycling reduces need for incineration and landfilling. It helps in pollutant reduction, energy saving and resource conservation.

Composting

Composting is a process where solid waste is digested aerobically or anaerobically to produce compost; bio fertilizer. Following are the various methods of composting:

Passive Windrow

It is one of the simplest methods of composting. In this method composting is done for organic matter in windrows. It is cheapest method but produces poorest quality of compost.

Turned Windrow

It is similar to passive windrow but in this method frequent turning of organic matter is done. It produces good quality compost then passive windrow.

Aerated Static Pile Composting

In this method processed organic matter is placed over exhaust piping or grid of aeration which results in the formation of compost over a period of time. It requires frequent and close monitoring.

In-Vessel Composting

In this method composting is done in a closed vessel or container. It can be done either using only biological treatment or both biological and mechanical treatment (sorting). The compost produced by this method is of good quality, but it requires large area, skilled labor and heavy investment.

Vermicomposting

The process of preparation of organic fertilizers with the use of earthworms is called as vermicomposting. It is the technique of breeding and raising earthworms for the production of vermicompost. Vermicompost is popularly called as Black gold due to the presence of rich nutrients and growth promoting factors. It is an eco-friendly technology that helps in waste management. It involves biochemical action of earthworms on organic wastes that produces earth worm biomass and vermicompost. Vermicompost being organic manure improves the soil quality through alteration in various biochemical parameters of soil and hence it must be promoted (Datta et al., 2016). It is natural manure which contains nutrients and growth hormones and is powerful growth promoter than organic fertilizers and chemical fertilizers (Narkhede et al., 2011; Attarde et al., 2012). It promotes 30-40 per cent growth more than chemical fertilizers (Sinha et al., 2010). It improves water retention and aeration of soil, increases the nutrient uptake by plants and promotes better root growth (Punjab State Council for Science and Technology, 2010).It adds beneficial microorganisms to the soil (Shrivastava & Singh, 2013) and the soil with vermicompost contains roughly 1000 times more beneficial bacteria than the soil without it (Pathma & Sakthivel, 2012; Munroe, 2004). It promotes better plant growth and improves crop yield (Shrivastava & Singh, 2013). It gives protection to plants against pests, plant pathogens and nematodes (Pathma & Sakthivel, 2012).

The earthworm's breakdown the organic matter they eat and leave behind castings that act as valuable fertilizer (Bureau of Plant Industry (BPI), Department of Agriculture Philippines, 2010). The earthworms can be used to produce vermicompost from sewage sludge, animal wastes, industrial wastes and crop residues (Hartenstein & Bisesi, 1988; Van Gestelet al., 1992; Dominguez & Edwards, 1997). The importance of earthworm in waste management and organic farming has been highlighted by several workers (Senapati, 1992; Bhawalkar, 1993; Ghatnekaret al., 1998; Talashikar & Powar, 1998). Vermicomposting provides quality products through organic farming system (Yadav et al., 2013). It is economical, non-polluting and helps in waste disposal and recycling of organic waste (Narkhede et al., 2011; Attarde et al., 2012).It is eco-friendly technology and helps in the disposal of solid wastes such as garbage, farm waste, used papers and grasses (Bureau of Plant Industry (BPI), Department of Agriculture Philippines, 2010).Its aim is to convert the substrate material into vermicompost in shortest time period with best possible recovery (Rupani et al., 2013). It provides two-fold benefit through disposal of waste and its conversion into organic fertilizer. It degrades wastes 75% faster than traditional systems and enriches soil with nutrients and beneficial microbes (Sinha et al., 2010). To increase the efficiency of vermicomposting the worms should be provided favorable bedding, optimum temperature, moisture, aeration and food source (Munroe, 2004). The success of vermicomposting relies on the selection of the bedding material and good quality bedding material provides optimum temperature, moisture and aeration to the worms (Latifah et al., 2009).

Mechanism of Vermicomposting

Raw Materials Required for Vermicomposting

The raw material for vermicomposting includes plant wastes such as agricultural wastes; sugarcane trash, rice straw, leaves of trees and crops, rotten vegetables, fruits and flowers, animal wastes such as cattle dung, pig manure, poultry manure, food waste, municipal solid waste, paper waste, sugar factory effluent and garbage. Venkatachalaiah, 1996developed method for composting fruit and vegetable wastes. The urban waste which includes discarded vegetables, fruits and flowers can be converted to vermicompost (Vikram Reddy et al., 2009). The organic waste vermicomposed contains high carbon nitrogen ratio (Karthikeyanet al., 2007). The management of kitchen waste, industrial wastes and agricultural wastes through decomposition by earthworm *Eiseniafetida* into valuable vermicompost has been studied by (Garg et al., 2006).

Types of Earthworms Used for Vermicomposting

There are two types of earthworms; burrowing and non-burrowing earthworms. The burrowing earthworms e.g. *Pheretima elongate* and *Pheretimaasiatica* ingest 90% soil and 10% organic waste while as non-burrowing earthworms e.g. *Eiseniafetida*and *Eudriluseugenae* ingest 90% organic waste and 10%

soil and produce vermicompost at faster rate than burrowing earthworms. Further the non-burrowing earthworms live in the upper layer of soil surface while as the burrowing earthworms make holes in the soil upto depth of 3.5 m and come onto the surface of the soil only at night. Keeping the above factors under consideration, the non-burrowing earthworms are preferred for vermicomposting than burrowing earthworms. The compositing ability of three species of earthworms namely *Eiseniafetida, Eudriluseuginae and Perionyx excavates* on food waste, garden waste and cattle dung was compared and it was found that, *Eudriluseuginae* was best followed by *Eiseniafetida* (Sinha et al., 2002). The fast growing earthworm in animal waste is *Eudriluseuginae* (Vasanthi et al., 2013).

Methods of Vermicomposting

Pits Below the Ground

In this method the pits below the ground are made for vermicomposting.The pits made for vermicomposting are 1 m deep and 1.5 m wide.

Heaping Above the Ground

In this method the waste material is spread over the polythene sheet and covered with cattle dung. The efficacy of heap method is better than pit below the ground method in terms of biodegradation of waste, production of biomass and vermicompost (Sunitha et al., 1997).

Tanks Above the Ground

Tanks above the ground of various dimensions are used for vermicomposting. These tanks are constructed using different materials such as bricks, asbestos sheets, stones and rocks. The partition wall between the tanks contains small holes for easy movement of earthworms between tanks.

Cement Rings

Cement rings above the ground of various dimensions are used for vermicomposting.

Process of Vermicomposting

- Cover the bottom of the pit/tank/cement ring with coconut husk or polythene or sand and broken bricks. Spread the layer of 15-20 cm organic waste material on the polythene sheet if heap method is used.
- Sprinkle rock phosphate powder if available and then cow dung slurry over the layer of organic waste material.
- Fill the pit/tank/ring with various layers of organic waste material using above described procedure. Paste the top layer with soil or cow dung.
- Allow the material to decompose for 15-20 days.
- After 15-20 days release earthworms through the cracks developed on the top layer

Figure 7. Process of vermicomposting (Ansari, 2012)

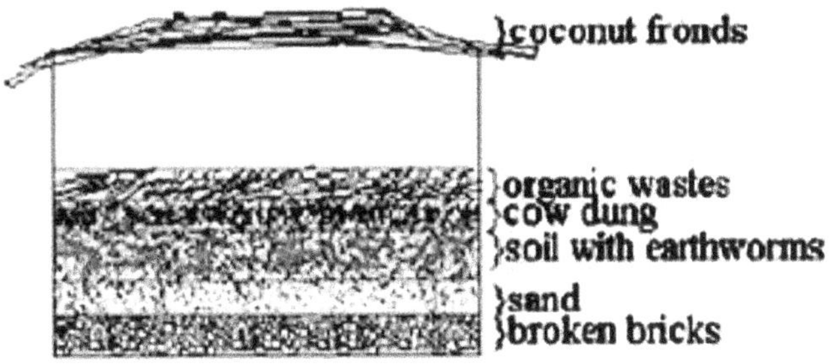

- Cover the top layer with gunny bag,coconut fronds or wire mesh to prevent picking of earthworms by birds.
- Sprinkle water every three days to provide adequate moisture and temperature to the earthworms.
- Vermicompost is formed within 2 months if agricultural waste is used as substrate.
- Pile the compost in small heaps and leave for few hours so that earthworms move down the heap. Separate the upper portion of the compost and sieve the lower portion to separate the earthworms from it.
- Pack the compost in bags and store at cool place.

Role of Vermicomposting in Waste Management

In India nearly 700-million tons organic waste which includes agricultural and domestic wastes, is produced annually which is either burnt or land filled (Bhiday, 1994). Such huge quantity of organic waste creates a problem for safe disposal. Vermicomposting is an important alternative for this problem which provides the most important nutrients for the sustainable productivity (Wani, 2002). The food waste generated from uneaten food can be recycled using earthworms (George, 2001). Paper sludge can be used as good source of carbon or as bulking agent in vermicomposting (Suriyanayanamet al., 2010). The management of kitchen waste, industrial wastes and agricultural wastes through decomposition by earthworm *Eiseniafetida* into valuable vermicompost has been studied by (Garg et al., 2006). Vermicomposting has been used for various household solid waste, animal wastes, pharmaceutical waste, vegetable waste, agricultural wastes, industrial waste, food waste, sewage, paper waste, garbage using earthworms such as *Eiseniafetida*, *Eudriluseuginae*, and *Drawidawillis*, for the bio-conversion of waste into bio-fertilizer; vermicompost (Table 1). The vermicomposting period for above mentioned wastes ranged between 28-120 days using these earthworms (Table 2). The NPK composition of vermicompost ranged from 0.15-3%, 0.0285-6% and 0.0136-4.98% respectively (Table 3).

WASTE MANAGEMENT TECHNOLOGIES USED BY DEVELOPING COUNTRIES

In developing countries, recycling of wastes such as paper, plastic, glass and metal is regarded as the effective means to prevent generation of wastes. Landfilling of wastes in open land and wetlands is commonly practiced in developing countries. Organic wastes are either composted to obtain organic compost

Table 1. The species of earthworms used for vermicomposting of various types of wastes

S.No.	Type of Waste used	Specie of earthworm	Reference
1.	Waste coffee	*Eisenia fetida*	Liu and Prince, 2011
2.	Food Industry waste	*Eisenia fetida*	Garg et al., 2011
3.	Herbal pharmaceutical waste	*Eisenia fetida*	Singh and Suthar, 2012
4.	Animal waste	*Eisenia fetida*	Garg et al., 2005
5.	Vegetable waste	*Eisenia fetida*	Suthar, 2009
6.	Sewage sludge and paper waste	*Eisenia fetida*	Ndegwa et al., 2000
7.	Bio solids	*Eisenia fetida*	Ndegwa and Thompson,2001
8.	Soybean husk	*Eudrilus eugeniae*	Lim et al., 2011
9.	Rice husk	*Eudrilus eugeniae*	Lim et al., 2012
10.	Herbal industrial waste and cow dung	*Eudrilus eugeniae*	Kumari et al., 2011
11.	Temple waste	*Eudrilus eugeniae*	Gurav and Pathade, 2011
12.	Sugarcane bagasse	*Eudrilus eugeniae*	Palsania et al., 2008
13.	Wood waste	*Drawidawillis*	Shweta, 2011
14.	Household solid waste and cow dung	*Eisenia fetida*	Garg and Gupta, 2011

Table 2. The vermicomposting period for various vermicomposted wastes

S.No.	Waste Vermicomposted	Vermicomposting Period in days	Reference
1.	Rice husk	63	Lim et al., 2012
2.	Grass and Cow dung	45	Ansari and Sukhraj, 2010
3.	Leaves and Cow dung	30	Thamaraj et al., 2011
4.	Sugarcane bagasse	48	Palsania et al., 2008
5.	Bio Solids	25	Ndegwa and Thompson,2001
6.	Kitchen waste and Agro waste	50-60	Suthar, 2009
7.	Animal waste	50-70	Garg et al., 2005
8.	Municipal waste	60	Narkhede et al., 2011
9.	Food industry waste	117	Garg et al., 2011
10.	Sewage sludge and Waste paper	28	Ndegwa et al., 2000
11.	Farmyard garbage	120	Indrajeet et al., 2010
12.	Herbal industry wastes and Cow dung	62	Kumari et al., 2011
13.	Leaf litters	60	Sundaravadivelan et al., 2011
14.	Wood waste	40	Shweta, 2011

Table 3. Quality of vermicompost produced from various vermicomposted wastes

S.No.	Waste Vermicomposted	N (%)	P (%)	K (%)	References
1.	Grass and Cow dung	1-3	2-6	1-3	Ansari and Rajpersaud, 2012
2.	Municipal waste	1.2	0.9	1.4	Narkhede et al., 2011
3.	Domestic waste	2.040	0.635	0.96	Suthar and Singh, 2007
4.	Food waste	0.197	0.0285	0.0136	Othman et al., 2012
5.	Paper waste	0.15	0.78	0.090	Basheer and Agrawal, 2013
6.	Cow dung and vegetable waste	1.45-1.76	0.57-1.60	1.98-4.98	Muthukumaravel et al., 2008

rich in plant nutrients like Nitrogen, Potassium and Phosphorus and other essential micronutrients or they are used to produce bio-gas under anaerobic conditions. Incineration in most of the developed countries is carried out in an incinerator that results in the conversion of waste to energy in the form of heat, electricity or production of steam.

CONCLUSION

Due to industrialization, urbanization and increase in population density, there has been an increase in the accumulation of wastes. In order to reduce the load of wastes on earth, the collection, transport, recycling and disposal of wastes needs to be improvised. Various techniques are used for the management of wastes which includes landfilling, incineration, anaerobic digestion, pyrolysis, plasma gasification, recycling, composting. Landfilling and incineration pollute the atmosphere by releasing toxic and greenhouse gases into air. Anaerobic digestion produces biofuel in the form of biogas. Plasma gasification results in the generation of electricity from wastes. Recycling of wastes involves the collection, sorting and reprocessing of wastes into new products. Vermicomposting is the preferred form of composting as it results in the formation of vermicompost called Black gold due to the presence of rich nutrients and growth promoting factors in it. It degrades wastes 75% faster than traditional systems and enriches soil with nutrients and beneficial microbes.

REFERENCES

Abdullah Adil Ansari. (2012). *Vermicomposting-permutations and combinations of organic waste recycling*. Lambert Academic Publishing.

Ansari, A. A., & Rajpersaud, J. (2012). Physicochemical changes during vermicomposting of water hyacinth (Eichhorniacrassipes) and grass clippings. International Scholarly Resource Network, Soil Science.

Ansari, A. A., & Sukhraj, K. (2010). Effect of vermiwash and vermicompost on soil parameters and productivity of okra (*Abelmoschusesculentus*) in Guyana. *African Journal of Agricultural Research, 5*(14), 1794–1798.

Arancon, N. Q., Edwards, C. A., Bierman, P., Metzger, J. D., & Lucht, C. (2005). Effects of vermicomposts produced from cattle manure, food waste and paper waste on the growth and yield of peppers in the field. *Pedobiologia, 49*(4), 297–306. doi:10.1016/j.pedobi.2005.02.001

AshaAalok, Tripathi, A. K., & Soni, P. (2008). Vermicomposting a better option for Organic. *Journal of Ecology, 24*(1), 59–64.

Attarde, S. B., Narkhede, S. D., Patil, R. P., & Ingle, S. T. (2012). Effect of organic and inorganic fertilizers on the growth and nutrient content of *Abelmoschusesculentus* (okra crop). *International Journal of Current Research, 4*(10), 137–140.

Basheer & Agrawal. (2013). Management of paper waste by vermicomposting using epigeic earthworm, *Eudriluseugeniae* in Gwalior, India. International. *Journal of Current Microbiology and Applied Science, 2*(4), 42–47.

Bhawalkar, U. (1993). *Turning garbage into gold: An introduction to vermiculture biotechnology*. Bhawalkar Earthworm Research Institute Pune.

Bhiday, M. R. (1994). Earthworms in agriculture. *Indian Farming, 43*(12), 31–34.

British Medical Association. (1999). *Health and Environmental impact assessment, an integrated Approach*. Earthcan Publications Ltd.

Bureau of Plant Industry (BPI), Department of Agriculture Philippines. (2010). Retrieved on September 29, 2010 from http://www.bpi.da.gov.ph/Services/vv.htm

Burke, G., Singh, B. R., & Theodore, L. (2005). *Handbook of environmental management and technology*. John Wiley & Sons.

Central Pollution Control Board. (2000). *Management of Municipal Solid Waste*. Delhi Centre Pollution Control Board. Available online: www.cpcb.nic.in

Communication from the Commission Europe. (2010). *A strategy for smart, sustainable and inclusive growth*. Author.

Cunningham, W., Cunningham, M. A., & Saigo, B. W. (2007). Environmental Science, A global concern. *Mc. Graw Hill. Higher Education.*

Datta, S., Singh, J., Singh, S., & Singh, J. (2016). Earthworms, pesticides and sustainable agriculture: A review. *Environmental Science and Pollution Research International, 23*(9), 8227–8243. doi:10.100711356-016-6375-0 PMID:26951221

Domínguez, J., & Edwards, C. A. (1997). Effects of stocking rate and moisture content on the growth and maturation of *Eiseniaandrei* (Oligochaeta) in pig manure. *Soil Biology & Biochemistry, 29*(3-4), 743–746. doi:10.1016/S0038-0717(96)00276-3

El-Haggar, S. M. (2007). *Sustainable industrial design and waste management: Cradle-to-cradle for sustainable development*. Oxford, UK: Elsevier/Academic Press.

Engeli, H., Edelmann, W., Fuchs, J., & Rottermann, K. (1993). Survival of plant pathogens and weed seeds during anaerobic digestion. *Water Science and Technology*, *27*(2), 69–76. doi:10.2166/wst.1993.0079

Enger, E., Smith, B., & Bockarie, A. T. (2006). Environmental Science. A study ofinterrelationships. *McGraw Hill. Higher Education.*

Flotats & Gibert. (2002). Mas El Cros Biogas Plant. Evaluation of 18 years in Operation. *7th FAO/SREN Workshop.Anaerobic digestion for sustainability in waste (water) treatment and re-use*, 1-8.

Gajalakshmi, V., & Abbasi, S. A. (2003). High rate vermicomposting system for recycling paper waste. *Indian Journal of Biotechnology*, *2*, 613–615.

Garg, P., Gupta, A., & Satya, S. (2006). Vermicomposting of different types of waste using *Eiseniafetida*: A comparative study. *Bioresource Technology*, *97*(3), 391–395. doi:10.1016/j.biortech.2005.03.009 PMID:16168639

Garg, V. K., Chand, S. A., & Yadav, A. (2005). Growth and reproduction of *Eiseniafetida* in various animal wastes during vermicomposting. *Applied Ecology and Environmental Research*, *3*(2), 51–59. doi:10.15666/aeer/0302_051059

Garg, V. K., & Gupta, R. (2011). Effect of temperature variations on vermicomposting of household solid waste and fecundity of *Eiseniafetida*. *Bioremediation Journal*, *15*(3), 165–172. doi:10.1080/10889868.2011.598487

Garg, V. K., Suthar, S., & Yadav, A. (2011). Management of food industry waste employing vermicompost technology. *Bioresource Technology*, *10*, 1016. PMID:22197330

George. (2001). *Vermicomposting*. Retrieved at July 22th, 2010, from http://aces.nmsu.edu/pubs/_h/h-164.pdf

Ghatnekar, S. D., Mahavash, F. K., & Ghatnegar, G. S. (1998). Management of solid waste through vermiculture biotechnology. In R. K. Trivedy & A. Kumar (Eds.), *Ecotechnology for pollution control and environmental management* (pp. 58–67). EnviromediaKarad.

Gurav, M. V., & Pathade, G. R. (2011). Production of vermicompost from temple waste (Nirmalya): A case study. *Universal Journal of Environmental Research and Technology*, *1*(2), 182–192.

Hartenstein, R., & Bisesi, M. S. (1988). Use of earthworm biotechnology for the management of effluents from intensively housed livestock. *Outlook on Agriculture*, *18*(2), 72–76. doi:10.1177/003072708901800205

Henry, J. G., & Heinke, G. W. (1989). *Environmental Science and Engineering*. Prentice Hall.

Hurst, C., Longhurst, P., Pollard, S., Smith, R., Jefferson, B., & Gronow, J. (2005). Assesment of municipal waste compost as daily cover material for odour at landfill sites. *Environmental Pollution*, *135*(1), 171–177. doi:10.1016/j.envpol.2004.09.004 PMID:15701404

Indrajeet, S. N., & Singh, J. (2010). Vermicomposting of farm garbage in different combination. *Journal of Recent Advances in Applied Science*, *25*, 15–18.

Kalargaris, Tian, & Gu. (2017). The utilisation of oils produced from plastic waste at different pyrolysis temperatures in a DI diesel engine. *Energy, 131*, 179-185. Retrieved from http://www.ipcc.ch/pdf/assessmentreport/ar4/syr/ar4_syr.pdf

Karthikeyan, V., Sathymoorthy, G. L., & Murugesan, R. (2007).Vermicomposting of market waste in Salem, Tamil Nadu, India. *Proceedings of the international conference on sustainable solid waste management*, 276-281.

Kaviraj & Sharma. (2003). Municipal solid waste management through vermicomposting employing exotic and local species of earthworms. *Biosource Tech*nology, *90*(2), 169-173.

Kumari, M., Kumar, S., Chauhan, R. S., & Ravikanth, K. (2011). Bioconversion of herbal industry waste into vermicompost using an epigeic earthworm *EudrilusEugeniae. Waste Management & Research, 29*(11), 1205–1212. doi:10.1177/0734242X10384432 PMID:20952444

Lai, C. M., Ke, G. R., & Chung, M. Y. (2009). Potential of food wastes for power generation and energy conservation in Taiwan. *Renewable Energy, 34*(8), 1913–1915. doi:10.1016/j.renene.2008.12.007

Latifah, A. M., Mohd, L. C., Mohd, K. Y., Tengku, H., Tengku, I., Rosta, H., ... Kamaruzaman, J. (2009). Influences of bedding material in vermicomposting process. *International Journal of Biology, 1*(1), 81–91.

Lim, P. N., Wu, T. Y., Sim, E. Y. S., & Lim, S. L. (2011). The potential reuse of soybean husk as feedstock of *EudrilusEugeniae* in vermicomposting. *Journal of the Science of Food and Agriculture, 91*(14), 2637–2642. doi:10.1002/jsfa.4504 PMID:21725978

Lim, S. L., Wu, T. Y., Sim, E. Y. S., Lim, P. N., & Clarke, C. (2012). Biotransformation of rice husk into organic fertilizer through vermicomposting. *Ecological Engineering, 41*, 60–64. doi:10.1016/j.ecoleng.2012.01.011

Liu, K., & Price, G. W. (2011). Evaluation of three composting systems for the management of spent coffee grounds. *Bioresource Technology, 102*(17), 7966–7947. doi:10.1016/j.biortech.2011.05.073 PMID:21704514

McDougall, F. R., White, P. R., Franke, M., & Hindle, P. (2001). *Integrated solid waste management: a life cycle inventory* (2nd ed.). Blackwell. doi:10.1002/9780470999677

Munroe, G. (2004). *Manual of on-farm vermicomposting and vermiculture*. Academic Press.

Muthukumaravel, K., Amsath, A., & Sukumaran, M. (2008). Vermicomposting of vegetable wastes using cow dung. *E-Journal of Chemistry, 5*(4), 810–813. doi:10.1155/2008/572431

Narkhede, S. D., Attarde, S. B., & Ingle, S. T. (2011). Study on effect of chemical fertilizer and vermicompost on growth of chilli pepper plant (*Capsicum annum*). *Journal of Applied Science in Environmental Sanitation, 6*(3), 327–332.

Ndegwa, P. M., & Thompson, S. A. (2001). Integrating composting and vermicomposting in the treatment and bioconversion of bio solids. *Bioresource Technology, 76*(2), 107–112. doi:10.1016/S0960-8524(00)00104-8 PMID:11131792

Ndegwa, P. M., Thompson, S. A., & Das, K. C. (2000). Effect of stocking density and feeding rate on vermicomposting of bio solids. *Bioresource Technology*, *7*(1), 5–12. doi:10.1016/S0960-8524(99)00055-3

Nidoni, P. G. (2017). Incineration Process for Solid Waste Management and Effective Utilization of By Products. *International Research Journal of Engineering and Technology*, *4*(12), 378–382.

Othman, N., Irwan, J. M., & Roslan, M. A. (2012). Vermicomposting of food waste. *InternationalJournal of Integrated Engineering*, *4*(2), 39–48.

Palsania, J., Sharma, R., Srivastava, J. K., & Sharma, D. (2008). Effect of moisture content variation over kinetic reaction rate during vermicomposting process. *Applied Ecology and Environmental Research*, *6*(2), 49–61. doi:10.15666/aeer/0602_049061

Pathma, J., & Sakthivel, N. (2012). Microbial diversity of vermicompost bacteria that exhibit useful agricultural traits and waste management potential. *SpringerPlus*, *1*(26), 2–20. PMID:23961356

Pehren, H. R., & Clark, C. S. (1987). Microorganisms in municipal solid waste and public health implications. *Critical Reviews in Environmental Control*, *17*(3), 187–228. doi:10.1080/10643388709388334

Pires, A., Martinho, G., & Chang, N.-B. (2010). Solid waste management in European countries: A review of systems analysis techniques. *Journal of Environmental Management*, *92*(4), 1033–1050. doi:10.1016/j.jenvman.2010.11.024 PMID:21194829

Punjab State Council for Science and Technology Chandigarh. (2010). Retrieved on September 30, 2010 from http://agri.and.nic.in/vermi_culture.htm

Rao, K. J., & Shantaram, M. V. (1996). Soil pollution due to disposal of urban solid wastes at landfill site, Hyderabad. *Indian Journal of Environment Protection*, *16*(5), 373–385.

Rupani, P. F., Mohd, F., Md, D., & Zelam, D., & Kanhere. (2013). Sustainable management of different organic waste by vermicomposting technology. *Agricultural Technology (Thailand)*, *9*(3), 529–539.

Scharfe, D. (2010). *Integrated Waste Management Plan*. Centre & South Hastings Waste Services Board/ Waste Diversion Ontario and Stewardship Ontario.

Seadon, J. K. (2006). Integrated waste management--looking beyond the solid waste horizon. *Waste Management (New York, N.Y.)*, *26*(12), 1327–1336. doi:10.1016/j.wasman.2006.04.009 PMID:16765038

Senapati, B. K. (1992). Biotic interactions between soil nematodes and earthworms. *Soil Biology & Biochemistry*, *24*(12), 1441–1444. doi:10.1016/0038-0717(92)90130-P

Shrivastava, S., & Singh, K. (2013). Vermicompost to save our agricultural land. *Research Journal of Agriculture and Forestry Science*, *1*(4), 18–20.

Shweta, R. K. (2011). Enhancement of wood waste decomposition by microbial inoculation prior to vermicomposting. *Bioresource Technology*, *102*(2), 1475–1480. doi:10.1016/j.biortech.2010.09.090 PMID:21036037

Singh, & Sharma, Teixeira da Silva, Satya, & Arora. (2011). Vermicomposting of municipal solid waste employing *Eiseniafetida* together with *Penicilliumspp.*and*AzotobacterBioinoculants*. *Dynamic Soil. Dynamic Plant*, *1*, 89–95.

Singh, D., & Suthar, S. (2012). Vermicomposting of herbal pharmaceutical industry solid wastes. *Ecological Engineering*, *39*, 1–6. doi:10.1016/j.ecoleng.2011.10.015

Sinha, R. K., Agarwal, S., Chauhan, K., & Valani, D. (2010). The wonders of earthworms and its vermicompost in farm production: Charles Darwin's 'friends of farmers', with potential to replace destructive chemical fertilizers. *Agricultural Science*, *1*(2), 76–94. doi:10.4236/as.2010.12011

Sinha, R. K., Heart, A. S., Asadi, R., & Carretero, E. (2002). Vermiculture technology for environmental management: Study of action of earthworms *Eiseniafetida, Eudriluseuginiae* and *Perionyx excavates* on biodegradation of some community wastes in India and Australia. *The Environmentalist*, *22*(2), 261–268. doi:10.1023/A:1016583929723

Sundaravadivelan, C., Isaiarasu, L., Manimuthu, M., Kumar, P., Kuberan, T., & Anburaj, J. (2011). Impact analysis and confirmative study of physicochemical, nutritional and biochemical parameters of vermiwash produced from different leaf liters by using two earthworm species. *Agricultural Technology (Thailand)*, *7*(5), 1443–1457.

Sunitha, N. D., Giraddi, R. S., Kulkarni, K. A., & Lingappa, S. (1997). Evaluation methods of vermicomposting under open field conditions. *Karnataka Journal of Agricultural Sciences*, *10*(4), 987–990.

Suriyanarayanan, S., Mailappa, A. S., Jayakumar, D., Nanthakumar, K., Karthikeyan, K., & Balasubramania, S. (2010). Studies on the characterization and possibilities of reutilization of solid wastes from a waste paper based paper industry. *Global Journal Environmental Research*, *4*(1), 18–22.

Suthar, S. (2009). Vermicomposting of vegetable-market solid waste using *Eiseniafetida*: Impact of bulking material on earthworm growth and decomposition rate. *Ecological Engineering*, *35*(5), 914–920. doi:10.1016/j.ecoleng.2008.12.019

Suthar, S., & Singh, S. (2007). Vermicomposting of domestic waste by using two epigeic earthworms. *International Journal of Environmental Science and Technology*, *5*(1), 99–106. doi:10.1007/BF03326002

Suthar, S., & Singh, S. (2008). Feasibility of vermicomposting in bio stabilization of sludge from a distillery industry. *The Science of the Total Environment*, *394*(2-3), 237–243. doi:10.1016/j.scitotenv.2008.02.005 PMID:18313726

Suttibak, S., & Nitivattananon, V. (2008). Resources, Conservation and Recycling Assessment of factors influencing the performance of solid waste recycling programs. *Conservation And Recycling*, *53*(1-2), 45–56. doi:10.1016/j.resconrec.2008.09.004

Talashikar, S. C., & Powar, A. G. (1998). Vermibiotechnology for eco-friendly disposal of wastes. Ecotechnology for pollution control and environment management. *Indian Journalof Environmental-Ecoplanning*, *7*, 535–538.

Thamaraj, K., Ganesh, P., Kolanjinathan, K., Suresh, K. R., & Anandan, A. (2011). Influence of vermicompost and vermiwash on physicochemical properties of rice cultivated soil. *Current Biotica*, *2*(3), 18–21.

Tudor, T., Robinson, G., Riley, M., Guilbert, S., & Barr, S. (2011). Challenges facing the sustainable consumption and waste management agendas: Perspectives on UK households. *Local Environment*, *16*(1), 51–66. doi:10.1080/13549839.2010.548372

U.S Environmental Protection Agency. (2010). Reduce, Reuse, Recycle. *Wastes - Resource Conservation*. Retrieved January 5, 2011, from http://www.epa.gov/wastes/conserve/rrr/

Van Gestel, C. A. M., Dirven-van Breemen, E. M., & Baerselman, R. (1992). Influence of environmental conditions on the growth and reproduction of the earthworm *Eiseniaandrei* in an artificial soil substrate. *Pedobiologia*, *36*, 109–120.

Vasanthi, K., Chairman, K., & Ranjit Singh, A. J. A. (2013). Influence of animal wastes on growth and reproduction of the African earthworm species *Eudriluseugeniae* (Kinberg). *Pollution Research*, *32*(2), 337–342.

Venkatachalaiah, B. K. (1996). The role of Karnataka compost development corporation Ltd., In: Organic farming. Abstract. *National Seminar on Organic Farming and Sustainable Agriculture*, 24-25.

Verdone, N., & De Filippis, P. (2004). Thermodynamic behaviour of sodium and calcium based sorbents in the emission control of waste incinerators. *Chemosphere*, *54*(7), 975–985. doi:10.1016/j.chemosphere.2003.09.041 PMID:14637355

Vikram Reddy, Gath, & Rossi. (2009). Fertilizers from vermiculture as an option for organic wastes recovery. *Agrochemical*, *27*(3), 244–251.

Wani, S. P. (2002). *Improving the livelihoods: New partnerships for win-win solutions for natural resource management.* Paper submitted in the 2nd International Agronomy Congress, New Delhi, India.

Westinghouse Plasma Corporation (WPC). (2010). *Technology and solutions-Plasma Gasification Vitrification Reactor*. Retrieved 22 July 2010, from http://www.westinghouse-plasma.com/

Yadav, S. K., Babu, S., Yadav, M. K., Singh, K., Yadav, G. S., & Pal, S. (2013). A review of organic farming for sustainable agriculture in Northern India. International Journal of Agronomy.

KEY TERMS AND DEFINITIONS

Incineration: It includes burning of wastes in open areas or in an incinerator. Nowadays incineration in most of the developed countries is carried out in an incinerator that results in the conversion of waste to energy in the form of heat, electricity, or production of steam.

Landfilling: In this method wastes are dumped in open land and wetlands.

Plasma Gasification: It involves the thermal treatment of waste materials in a sealed vessel under high temperature and pressure, with limited availability of oxygen to produce highly ionized and electrically charged syngas for the generation of electricity and heat.

Pyrolysis: It involves the thermal treatment of wastes materials in a sealed vessel under high temperature and pressure, with limited availability of oxygen to produce energy, activated carbon and other refined products.

Vermicomposting: The process of preparation of organic fertilizers with the use of earthworms is called as vermicomposting. It is the technique of breeding and raising earthworms for the production of vermicompost.

Waste Management: The process of collection, transport, disposal, recycling and monitoring of wastes is called as waste management.

Chapter 10
Innovations in Recycling for Sustainable Management of Solid Wastes

Nazia Parveen
Baba Saheb Bhimrao Ambedkar University Lucknow, India

Dig Vijay Singh
Babasaheb Bhimrao Ambedkar University Lucknow, India

Rifat Azam
Babasaheb Bhimrao Ambedkar University Lucknow, India

ABSTRACT

Rapid generation and accumulation of waste in developing countries is due to the increase in development, urbanization, industrialization, poor government policies, and population explosion. Various kinds of waste are produced in the developing countries but due to non-availability of ecofriendly environmental techniques, lack of resources, or the technical expertise, it is becoming difficult to treat the waste at the disposal sites. Most of the waste produced in developing countries is directly disposed to the landfills without any proper sorting and segregation, where it produces greenhouse gases, thus results in global warming. Recycling and composting is helpful in reducing the volume of the waste and producing valuable products which can have multiple applications. Thus, the requirement is to manage the waste by implementing strict laws, increasing awareness, utilization of innovative, as well as latest techniques (global system of mobile, geographical information system) in order to control the growing menace of the solid waste in developing countries.

DOI: 10.4018/978-1-7998-0031-6.ch010

INTRODUCTION

Human activities are responsible for the creation of the solid waste which degrade quality of the environment as well as the health of organisms living in that environment (Saxena et al., 2010; Zhu et al., 2008). Solid waste is the waste that has been rejected for further use (Nandan et al., 2017) and include industrial, mining, municipal and agricultural wastes (Sharholy et al., 2007). The enormous quantity of the waste generated is due to rapid increase in development, urbanization, rapid industrialization, population explosion, poor collection, inadequate transportation (Bundela et al., 2010; Sharholy et al., 2007) and improvement in the living standards of the people (Gidde et al., 2008; Rathi, 2007). Municipal solid waste (MSW) is generally a combination of household and commercial refuse which is generated from the living community (Rajkumar et al., 2010). The waste generated mainly consists of organic matter (51%) (Jha et al., 2011) recyclables (17.5%) and inert material (31%) (Annepu, 2012). Metropolitan cities (Delhi, Mumbai, Kolkata and Chennai) account for more than 42% of India's urban population (Ghosh & Kansal 2014) and has been a top producer of MSW in India due to its high occupancy. Municipal solid waste (MSW) production has rapidly enhanced since 1960 and has drastically increased by a factor of 2.6 (Tozlu et al., 2016). According to World Bank, world cities generates about 1.3 billion tonnes of MSW and the amount is expected to reach 2.2 billion tonnes by the end of 2025 (World Bank, 2012). It has been noticed that with economic improvement, the generation of waste has increased drastically (Turan et al., 2009) and most of waste is contributed by Industries and urban cities (Nandan et al., 2017). The per capita waste generation varies with countries population size and living standards (Table 1) and excessive use of resources generates immense quantity of solid waste which is challenging global sustainability (Ziadat & Mott, 2005). All these consequences lead to severe human health hazards which can only be minimized by implementing cost effective technical and policy measures (Saleem et al., 2016).

MSW generation is influenced by economic conditions, living standards, urbanization (Liu & Wu, 2010), and population (Chiemchaisri et al., 2007). The main objectives of solid waste management are to use the waste collection and disposal techniques in economical way so that environment and human health can be easily protected (Nandan et al., 2017). Developed countries like Italy, Japan, USA, and UK are practicing zero waste concept of municipal waste management (Zia & Devadas, 2008; Mickael, 2016; Jones & Harrison, 2016; Kothari et al., 2010; Heberlein & Murphy, 2008; Tsang and Apps, 2005) and are strictly following the concept of reduce, reuse and recycle (Gutberlet, 2010). Due to improper solid waste management, waste has become one of the pollution sources that have caused diverse environment impacts as well as detrimental to human health (Mohanty et al., 2014; Das & Bhattacharya, 2013). The management of solid waste is big problem in urban areas of developing nations as continuous establishment of industries and good economic growth has resulted in increased municipal solid waste generation per person (PPCB, 2010). Current SWM systems are inefficient and emission of greenhouse gases is common problem related to these methods of waste management (Biswas, 2010). Developing countries are lacking far behind than developed countries in using latest technology for the management of the waste (Saleem et al., 2016).

Waste Generation in India

The estimation of the quantity and characteristics of MSW in India and forecasting future waste generation is fundamental to successful waste management planning (Rana et al., 2014). The quantity of MSW generated depends on living standards, the extent and type of commercial activity, eating habits

Table 1. Per capita MSW waste generation in different countries of the world

Country	MSW generation per capita	References
United States	1.99	United States Environmental Protection Agency (2014)
Jordan	1.04	(Hoornweg & Bhada-Tata, 2012
Algeria	1.21	(Hoornweg & Bhada-Tata, 2012
Qatar	1.33	(Hoornweg & Bhada-Tata, 2012
Bahrain	1.10	(Hoornweg & Bhada-Tata, 2012
Kuwait	1.01	Al-Jarallah & Aleisa, 2014
Srilanka	0.89	World Bank, 1999
India	0.37	Annepu, 2012
Bangladesh	0.41	Enayetullah et al., 2005
Nepal	0.32	Asian Development Bank, 2013
Maldives	2.80	Peterson, 2013
Pakistan	0.65	Masood et al., 2014

and season (Kaushal et al., 2012). MSW generation per capita in India ranges from approximately 0.17 kg per person per day in small towns to approximately 0.62 kg per person per day in cities (Kumar et al., 2009). Megacities are a relatively recent phenomenon associated with globalization of the economy, culture and technology (ISWA, 2012). There is a noticeable contrast between the type of waste generated in developed and developing countries as developed countries generate more recyclable wastes whereas developing countries generate more biodegradable wastes (UNEP, 2005).

Composition and Characteristics of Waste

Waste composition depends on a wide range of factors such as food habits, cultural traditions, climate and income (Srivastava et al., 2014; Patle et al., 2014; Naveen et al., 2013; Gupta et al., 2013; Kumar et al., 2009). Different categories of waste like food waste (Bhat et al., 2013), rubbish, commercial waste, construction and demolition waste are included in municipal solid waste (Gupta et al., 2015) and also contains compostable organic matter (fruit and vegetable peels, food waste) recyclables (paper, plastic, glass, metals, etc.) and toxic substances (Upadhyay et al., 2012). The composition of municipal solid waste like paper, plastic and glass are having the increasing trend from 4.1%, 0.7% and 0.4% respectively in 1971 to 8.18%, 9.22% and 1.01% respectively in 2005 in India and metals are also having the increasing trend during the same period while inert materials and compostable matter are having the decreasing trend from 49.2% and 41.3% respectively in 1971 to 25.61% and 40% in 2005 (Gupta et al., 2015). The average composition of MSW produced by Indian cities is approximately 41 wt % organic, approximately 40 wt % inert, with approximately 19 wt % potentially recyclable materials (Sharholy et al., 2008). The generation of waste in India has increased from 6 million tons (1947) to 48 million tons (1997) and to 90 million tons (2009) and to reach 300 million tons in 2047 (Teddy, 2010; Sharholy et al., 2006). The quantity and characteristics of solid waste vary from place to place (Kumar & Goel, 2009) and factors that influence the quantity and composition are the average income level, the sources, the population, social behaviour, climate, industrial production and the market for waste materials (Late & Mule, 2013; Yadav & Devi, 2009).

Predictions on Future Waste Growth

The per day waste production of the whole world in 2050 is estimated to be approximately 27 billion tonnes per year and almost 33% of which will be contributed by Asian countries with major proportion of waste from China and India (Modak et al., 2010). Waste generation in urban areas of India will be 0.7 kg per person per day in 2025, approximately four to six times higher than in 1999 (Kumar et al., 2017). The problems associated with waste become more acute as the size of communities increase and this provides opportunities for decentralized waste management by self-help groups and NGOs (Kumar et al., 2014). The waste produced in urban areas of India is approximately 170 000 tonnes per day equivalent to about 62 million tonnes per year and this is expected to increase by 5% per year owing to increases in population and changing lifestyles (Planning Commission, 2014).

Municipal solid waste management (MSWM) is network of activities from storage to disposal and to avoid the environmental contamination and human health risks due to the poor waste management, developing countries should shift towards latest technologies of MSWM (Saleem et al., 2016). The technologies of waste management from storage, collection, recycling, processing, energy recovery and final disposal are as follows

COLLECTION AND STORAGE OF MSW

According to Municipal Waste Management Rules (2000), it is the sole responsibility of municipalities to collect and prevent littering of solid waste in cities, towns and in urban areas (Nandan et al., 2017). It is essential to collect waste regularly from community dust bins or by house to house collection in order to prevent the accumulation of the waste at every nook and corner of the society (Khan, 2014). The integration of collection methods are necessary to increase the efficiency of waste collection (Talyan et al., 2008) and with support of NGOs and local communities, waste collection efficiency has increased remarkably in few rural areas (Nandan et al., 2017). Many local bodies have taken initiative for efficient waste collection along with certain NGOs having expertise in this sector of solid waste management (Nandan et al., 2017). The waste collection in India is much unorganized and most of the urban areas are lacking in MSW storage at the source (Gupta et al., 2015). The collection bins used in various cities are neither properly designed nor properly located and average collection efficiency for MSW in Indian cities and states is about 70% (Saxena et al., 2010; Rathi, 2006; Siddiqui et al., 2006). Storage bins can be movable or fixed bins (Gupta et al., 2015) and the fixed bins are more durable, but their positions cannot be changed while the movable bins are flexible in transportation but less durable (Sharholy et al., 2008; Nema, 2004; Malviya et al., 2002). Local bodies spend around Rs. 500–1000 per tonne on SWM with 70% of this amount spent on collection and 20% spent on transport (Kumar et al., 2017) but improvements in waste collection and transport infrastructure in India will create jobs, improve public health and also the increase tourism in the country (USEPA, 2002).

There is no organized and scientifically planned segregation of MSW either at household level or at community bin (Joshi & Ahmed, 2016) as segregation and sorting of solid waste takes places in unscientific manner (Sharholy et al., 2008) and the efficiency of segregation is reasonably low (Goswami,

2018). India most often collect MSW in a mixed form (Nandan et al., 2017) and only papers and plastics are separately collected by waste pickers at source level (Marshall & Farahbakhsh, 2013). Source separation increases recycling efficiency and also improves performance of waste treatment units due to good quality of feed and lesser amount of impurities (Nandan et al., 2017).

Waste Disposal

There is a need to develop facilities to treat and dispose of increasing amounts of MSW (Sharholy et al., 2007). It is estimated that approximately 1400 km^2 was occupied by waste dumps in 1997 and this is expected to increase in the future (Singhal & Pandey, 2001). Stationary compactors, mobile compactors/closed tempos, and tarpaulin-covered vehicles are used in the transportation of MSW and about 65, 15, and 20% of waste is transported through these compacters, respectively (Goswami, 2018). The common and old methods of municipal solid waste disposal were open dumping, burning and incineration (Feng, 2010). The process of open dumping leads to water and air pollution in the form of land litter, particulates and toxic gases (Saleem et al., 2016). Disposal is the last step of MSWM where the remaining trash after recycling, processing, and WTE are disposed (Feng, 2010). Most of the developing countries are disposing waste inopen without any proper leachate treatment system and collection of landfill gas (Al-Awadhi & Shuaibi, 2012). Methods like open dumping were responsible for causing much aesthetic and other environmental issue (Stone, 1978). Indian cities with containment landfill sites include Mumbai, Kolkata, Chennai, Nashik, Vadodara, Jamshedpur, Allahabad, Amritsar, Rajkot, Shimla, Thiruvananthapuram and Dehradun (Kumar et al., 2009).

India is facing the lack of resources or the technical expertise necessary to deal with the disposal of municipal solid waste (Kaushal et al., 2012). The two leading innovative mechanisms of waste disposal being adopted in India include composting (aerobic composting and vermi-composting) and waste-to-energy (WTE) (incineration, biomethanation) (Gupta et al., 2015). These mechanisms have shown good result in developed countries but in India, these mechanisms have not been used on wide scale (Sharholy et al., 2008). An incineration and biomethanation type of waste-to-energy system of solid waste disposal has also been introduced in India but contributes at minor level in present scenario (Gupta et al., 2015).

MODERN TECHNOLOGY FOR COLLECTION AND TRANSPORT OF WASTE

Latest technologies including underground collection system, Web based GIS technology, Waste bin monitoring system using GSM, and Waste compactors are being used which are discussed further:

Underground Collection System

New technologies for waste storage are underground and semi underground storing systems (Saleem et al., 2016). In these technologies waste bins or containers are being replaced by underground collection points (Kum et al., 2005). The underground surface utilization is key for attaining sustainable development in the urban area (Dixit & Rastogi, 2016). This includes the placement of plastic container in excavation of 2-3m with only inlet in environment (Sexton & Spelman, 2003). Use of underground collection system helps to improve living condition in the urban areas and also provide opportunity of utilizing above ground surfaces for other purposes (Dixit & Rastogi, 2016). These types of containers are being used worldwide since years. They have quite extensive range of application from recyclables, to organic waste and oils (Saleem et al., 2016). The technology is beneficial for the regions with extremely hot climatic conditions as the waste would be stored underground in relatively low temperature and this technology require less maintenance and is more aesthetically acceptable (Zia & Devadas, 2008).

Web Based GIS (Geographic Information System) Technology

Geographical Information System (GIS) is the innovative and best tool to know about the optimum distances and shortest routes for waste collection and management (Gilberto et al., 2008; Dene, 2013; Christos & Lasaridi, 2009). Apart from spatial modelling techniques, GIS may provide large economic and environmental savings through the reduction of travel time, distance, fuel consumption and pollutants emissions (Gilberto et al., 2008). GIS technology is playing a crucial role in various fields like urban utilities planning, transportation, natural resources, health sciences, forestry, geology, natural disasters prevention and relief (Christos & Lasaridi, 2009). The utilization of the GIS technology in combination with waste collection has become very popular in developed countries over the last few years (Chang et al., 2011; Rada et al., 2013)). Now, the efficiency of waste management from source of production to the disposal site of municipalities has improved with the use of such innovative and automated system at every step of waste management (Idowu et al., 2012). With the advancement of Information and Communication technology (ICTs), the simple and integrated systems are being developed which can easily meet the requirements of the waste management in economical way (Saleem et al., 2016). This automatic and integrated system of waste management from collection onward to recycling and then disposing to the landfill site can be used for single municipality as well as the multiple one (Saleem et al., 2016). As per Italian and European case studies, the collection and segregation of waste at source level has efficiency of 80% by using web based GIS (Rada et al., 2013). As GIS can model the world landmarks and streets and can play an important role in waste collection sector (Saleem et al., 2016). GIS in combination of other software can also give information regarding the most reliable routes, number of residents, number of contracts and their validation (Rada et al., 2013).

Waste Bin Monitoring Technology Using Global System of Mobile (GSM)

The precise identification of location on the earth surface can be easily provided by global positioning system as it works by measuring distance from satellites (International Engineering Consortium, 2015). In this technology, sensors are placed in public garbage bins to detect a certain optimum level of waste (Saleem et al., 2016). The location of dustbin is provided by global positioning system and when the dustbin is 90% filled, the message is forwarded to concerned authority by GSM (Latifah et al., 2009). As the garbage reaches the threshold level, indication will be transferred to the controller which will further give indication to driver of collection truck by sending SMS using GSM for emptying the bin urgently (Arebey et al., 2010).

Compact Garbage Collection Trucks

In many developing countries, because of narrow and congested roads small garbage collection trucks are used (Ogwueleka, 2009). Latest technologies introduced garbage compactors in collector trucks in order to increase the collection capacity of vehicles (Saleem et al., 2016). With continuous modification, these trucks have achieved high compression rate as they can carry 1.5 times more waste as compare to flat pile trucks (Sembiring & Nitivattananon, 2010). The technology does not only increase collection capacity but also increases the fuel efficiency which is more environmentally and economically feasible. Compaction in waste management has traditionally been utilized during transport and final disposal phases through the use of compactors in trucks and landfills (Emery et al., 2007). Researchers are been working to introduce electric motor drive and hybrid type collection trucks to overcome problems like greenhouse gas emissions and air pollution (Sembiring & Nitivattananon, 2010).

Segregation and Sorting

After collection, the second step involved in MSW management system is sorting or segregation of different types of wastes for further processing (Mickael, 2016). Latest automatic technologies for municipal waste sorting include optical sorting, eddy current sorting, multi compartment bins, and optical sensor-based sorting technologies (Christensen, 2011; Metcalfe et al., 2012; Wangyao et al., 2009). Recently, developed countries including Sweden are using multi compartment bins for source segregation of waste (Saleem et al., 2016). These types of waste bins have separate compartments for different types of waste (Birpınar et al., 209). Through this organic, paper waste and recyclables can be segregated on the spot of generation, while recyclable can be recycled or reused (Metcalfe et al., 2012).

Automated Waste Sorting

Automated waste sorting techniques can be categorized into two types: direct sorting and indirect sorting (Gundupalli et al., 2017). Direct sorting techniques is based on utilization of properties of the material by applying magnetic, eddy current and gravity as external fields (Gaustad et al., 2012; Mesina et al., 2003; Svoboda, 2004) while indirect sorting utilizes sensors to detect the presence of recyclable material in the waste (Gundupalli et al., 2017). Different types of plastic, composites, and other wastes are being sorted with the help of colour sensitive cameras, UV sensors and infrared spectroscopy (Saleem et al., 2016). With the help of sensors the position of different waste components are identified and this type of sorting is most commonly used for glass waste (Erdogan & Baris, 2007). The latest technology included optical sorting with laser which is relatively expensive (Schott, 2001) but this technology can reach a purity of 99.7% for flint glass (Christensen, 2011). Modern sorting plants are converting to sensor-based sorting systems to improve sorting efficiency (Saleem et al., 2016). This technology is beneficial because it has high recovery rate, low operation cost and high reorganization capability and can convert useless garbage to highly useful product output which can reduce carbon footprint and emissions (Wangyao et al., 2009).

Automatic Bottle Sorting System

Automated sorting for plastic recyclables has been seen as the way forward in the plastic recycling industry and provides significant improvements in terms of efficiency and consistency in the sorting process (Wahab et al., 2006). This technology is widely used in Japan in recent years. It is comprised on sizing, aligning and clearing machine along with colour identification sensors (Christensen, 2011). Optical Sorting Systems use visible light to separate plastic bottles by colour and use black and white cameras for single colour separation (Wahab et al., 2006). Automated sorting systems employ a detection system or a combination of detection systems to identify the different types of plastic recyclable and these detection systems to automatically sort plastics either by resin type, shape, colour or both (Wahab et al., 2006). The IR sensors are clamped at specific height to detect the bottles according to their heights and when the sensor detects the bottle, it gives signal to the controller (Elahi et al., 2017). Optical sorting systems that provide multiple colour separations use a colour camera and the bottles of each colour are shredded and cullet is prepared which can be further used in different fields like as substitute in building material and as raw material in insulation (Christensen, 2011).

Mechanical Biological Treatment (MBT)

The term MBT is used for the combination of biological and mechanical processes for the transportation and segregation of waste into various outputs (Velis et al., 2010). In the mechanical part, waste is treated physically by processes such as crushing, grinding, milling, and similar physical treatment processes (Plavac et al., 2017). In biological treatment, biological processes occur in the waste which is biodegraded by drying out and converted into compost or secondary source of energy (fuel) (Burnley et al., 2011). MBT is considered as pretreatment and is mainly used to recover energy from the waste (Saleem et al., 2016). The biological processes are designed to minimize the content of water and the mechanical process aims to separate metals and glass from the waste (Velis et al., 2010). MBT had the highest energy efficiency when connected with biogas purification systems (38.5% energy recovery), and could significantly avoid pollution and is well-adapted to changing municipal solid waste composition making it a promising supplementary to solid waste source separation (Fei et al., 2018).

WASTE RECYCLING

3R Concept

The term 'Reduce' can be defined as a reduction in the amount and/or toxicity of waste entering the waste stream (Glavic & Lukman, 2007) and is the top ranking component of solid waste management hierarchy because it represents most effective means of reducing economical costs and environmental impacts associated with handling waste (Wilson et al., 2006). The term 'Reuse' means usage (or utilization) of a product in the same application for which it was originally used (Nandan et al., 2017), while repair means only those parts that have failed are replaced (Nandan et al., 2017).

Recycling

Recycling is the process of changing waste into new product and reducing the consumption of fresh raw materials (Mehta, 2019). Recycling is a key component of modern waste reduction and is the third component of waste hierarchy (VanEwijk & Stegemann, 2014). Materials to be recycled are brought to collection centres or picked, sorted, cleaned and reprocessed into a new material (Gencer, 2009). Recycling of material would produce a fresh supply of the same material like used office paper would be conveyed into new office paper or used foamed polystyrene into new polystyrene (Pang et al., 2010).

Recycling of plastic waste is used in our daily lives has number of applications (Al-Salem et al., 2009; Goto, 2006) like greenhouses, mulches, packaging, films, covers, bags and containers (Safavi et al., 2010). There are different types of recycling which include:

Waste Paper Recycling

The recycling of the paper consumes less energy, conserves the natural resources and decreases the environmental pollution (Cabalova et al., 2011). Recycling can become useful only when the final product obtained finds its use in the society so it is need of the hour to aware people to use recycled products which can be helpful in changing trends in the society (Laroche et al., 2001). Hence, it become important to maximize the wastepaper collection at the source level so that natural resource as well as environment can be conserved (Pati et al., 2008).

Plastic Recycling

Plastic is less expensive and can be moulded to form new products that have wide range of beneficial applications in day to day life (Al-Salem et al., 2009). Most plastic products carry a code indicating what type of plastic resin are used in the product thus facilitating separation and recycling of single-resin plastic products (Singh et al., 2017). The plastics industry is working to develop an infrastructure to make recycling of more complex plastic containers a viable option (Kang & Schoenung, 2005). In plastics the recycling process consists of following steps which include: primary, secondary and tertiary recycling (Grigore, 2017).

Primary Recycling

Primary recycling is the most popular process due to simplicity and low cost. This process refers to the reuse of products in their original structure (Hopewell et al., 2009). Primary recycling means that the recyclable material is recovered and reused without being changed in any way and usually used for the very same purpose (Singh et al., 2017). In order to maintain the balance of cost in the recycling industry, sorting and identification of waste to be recycled must be done within short time (Ribeiro et al., 2016).

Secondary Recycling

Secondary recycling was promoted and commercialized all over the world back in the 1970s as mechanical recycling of plastic solid waste can only be performed on single-polymer plastic (Singh et al., 2016). The mechanical recycling does not involve the alteration of the polymer during the process (Siracusa et al., 2008) but in physical method the plastic wastes will be formed by cutting, shredding or washing into granulates, flakes or pellets of appropriate quality for manufacturing (Al-Salem et al., 2009) and then melted to make the new product by extrusion (Poskrobko et al., 2012). Separation, washing and preparation of plastic solid waste are all essential to produce high quality, clear, clean and homogenous end-products (Brems et al., 2012).

Tertiary Recycling

Tertiary recycling includes chemical recycling and in this process solid plastic materials are converted into smaller molecules through the use of heat and/or chemical treatment (Kumar et al., 2011). This method is able to transform the plastic material into smaller molecules suitable for use as feedstock material starting with monomers, oligomers, or mixtures of other hydrocarbon compounds (Francis, 2016; Olah et al., 2008). The chemical reactions used for decomposition of polymers into monomers are: hydrogenation, glycolysis, gasification, hydrolysis, pyrolysis, methanolysis, chemical depolymerisation, thermal cracking, catalytic cracking and reforming, photodegradation and ultrasound degradation (Grigore, 2017). The gasification and pyrolysis method is under extensive research to establish the suitable conditions for their use and the processes that have reached commercial maturity at this moment are glycolysis and methanolysis (Francis, 2016; Achilias et al., 2007). Hydrolysis is a recycling method that involves a reaction of PET (Polyethylene terephthalate) with water in an acid, alkaline or neutral environment leading to total depolymerization into its monomers (Grigore, 2017). The original material of plastic is obtained by the hydrolysis in which linkage points are attacked by water in order to recover the original raw material (Karayannidis & Achilias, 2007).

WASTE TREATMENT

The objective of treatment is to improve physical and/or chemical characteristics of waste, reduce toxicity and its final volume (Misra & Pandey, 2005). Different methods of treatment (Figure 2.) are used in India which totally depends upon the type of the waste to be treated (Gupta et al., 2012).

Autoclaving

Autoclaving process involves the high-pressure sterilization of waste by steam (Hossain et al., 2012) and destroys all microorganism contained in the waste (Friends of the Earth, 2008). This technology involves treating the waste with steam at 140-160° C for 30-40 minutes (Christensen, 2011). During autoclaving process, different type of waste will have different physical reaction to the treatment process. Biodegradable waste such as paper is broken into a fibre like material while plastics are softened or deformed into unrecognisable state (DEFRA, 2007; Papadimitriou et al., 2008). Metals and plastics is send for recycling (Saleem et al., 2016) and the residue of the process is then sent for disposal to the landfill site (Vehlow, 1996).

Composting

Composting is a microbial technology that reduces waste volume by 40-50% (Hargreaves et al., 2008) and metabolically destroys pathogens in the thermophilic phase (Singh et al., 2010). The main aim of composting is the decomposition of organic materials to produce an end-product (compost) with high nutrients content that can be used as soil conditioner (Liang et al., 2003) but only 6%–7% of the MSW is converted into compost in India (Bhat et al., 2018; Sharholy et al., 2008). The benefits of composting biogenic waste are replacement of synthetic fertilizers and reduction in greenhouse gas emissions (Couth & Trois, 2012). Composting is the decomposition of organic matter by microorganism in warm, moist, aerobic and anaerobic environment (Gupta et al., 2015). Composting of MSW is the simplest and cost effective technology for treating the organic fraction of MSW (Asnani, 2006). It is suitable for organic biodegradable fraction of MSW or waste containing high proportion of lignocelluloses materials which do not readily degrade under anaerobic conditions (Gupta et al., 2015). Aerobic composting means bacterial conversion of organics in presence of air and final product (Khan, 1994) is free from odour, pathogens and be extensively used as fertilizer (Nandan et al., 2017) while anaerobic microorganisms by the process of reduction metabolize the nutrients by breaking down the organic compound (Khalid et al., 2011) and the final product on minor oxidation can be easily applied to the field (CPHEEO, 2000).

Biomethanation

Biomethanation is the process of conversion of organic matter into stable form by microorganisms in the oxygen-free environment (Nandan et al., 2017). The solid wastes from agro-based industries have high organic content and hence its treatment by the process of biomethanation is most viable as it produces useful products like biogas and enriched manure (Gupta et al., 2015). The major advantage of biomethanation is the reduction in the release of potent greenhouse gases like methane (Nandan et al., 2017). The biomethanation process is a two stage process consisting of acidification and methanation (CPCB, 2007) and methanogens (Nandan et al., 2017) are used to produce methane-rich biogas which can be used for generation of electricity and remained residue can be used as manure in the agriculture field (Yuan & Zhu, 2016).

Vermi-Composting

Vermi-composting is a process for turning organic waste into nutrient rich soil as it is processed by worms (Gupta et al., 2015). In this process, specific species of earthworms are fed on waste which converts organic waste into nutrient rich manure (Nagavallemma et al., 2004). The processing period ranges from 28-120 days and temperature ranges between 18-67° C with pH between 5.9-8.3 and moisture content of 80% (Gupta & Garg, 2011). Vermi-compost is an odourless, dark brown bio-fertilizer (Manyuchi & Phiri, 2013; Manyuchi et al., 2013; Abbasi et al., 2009), natural organic manure produced from the excreta of earthworms fed on scientifically semi-decomposed organic waste (Asnani, 2006). In this process animal waste, pharmaceutical waste, food and sewage waste is processed through earthworms to give output known as vermin-wash which is very rich in nitrogen, phosphate and potassium (Saleem et al., 2016). Nutrients in vermin-compost are often much higher than traditional garden compost (Datar & More, 2006) and the vermin-wash produced during composting is used as biofertilizer for crops like maize, soyabean, marigold and cowpea (Gupta & Garg, 2011). Vermin-composting is totally different from composting as worms grow and reproduce easily thus helps to convert enormous organic waste into the compost (Gupta et al., 2015). Vermi-composting facilities have already entered domestic and industrial marketing in countries like Canada, USA, Italy and Japan but have not been commercially used in India (Aalok et al., 2008).

Bio-Conversion

Almost 33% of the food produced all over the world for humans is wasted along the supply chain (Dasgupta & Mondal, 2012). The large amount of waste generated from the industries leads to the loss of the valuable material and also become the concern from environmental point of view (Mirabella et al., 2014). Currently food waste along with waste are collectively incinerated and landfilled for recovering energy (Saleem et al., 2016). Food waste is rich in carbohydrate, proteins, lipids and organic acids (Kiran et al., 2014) and the waste rich in proteins and lipids through different fermentation processes can be used for production of biofuel (Sarkar et al., 2012).

Thermal Conversion

This is a technology utilize water, heat or pressure to convert organic and inorganic waste to useful products (Saleem et al., 2016). Raw materials like plastics, computer cases, tyres and crop residue are subjected to thermal processing system which converts them to useful molecules of fuel gas, oil and other beneficial products (Marculescu, 2011) and even heavy metals are converted to harmless oxides (Price, 2005). Thermal technologies like incineration and combustion have been used for the conversion of waste to energy (WTE) (Saleem et al., 2016). With the advancement in technology, many thermal treatment technologies have been introduced some of them are discussed further.

Advance Thermal Treatment Technologies

Advance thermal treatment technologies (pyrolysis, gasification and incineration) have been utilized for efficient and pollution free waste to energy conversion (WTE) (Saleem et al., 2016). Pyrolysis and gasification have been used for the production of charcoal, coke and producer gas (Balat et al., 2009). Thermal processes (incineration, pyrolysis, melting or vitrification) are used for treating hazardous and municipal solid waste (MSW) prior to disposal (Dave & Joshi, 2010). Gasification process, which takes place in substoichiometric quantity of oxygen converts carbonaceous materials to a combustible or synthetic gas (H_2, CO, CO_2, CH_4) (Dave & Joshi, 2010). Gasification is the reaction of gases at higher temperature (704°C or higher) in order to produce gaseous product which can be used as a source of electric power and heat or hydrogen fuel (Rezaiyan & Cheremisinoff, 2005). Combustion that takes place in excess oxygen releases gases like CO_2, NOx, SOx, H_2O and particulate matter while pyrolysis results in formation of syngas (H_2 + CO), lighter hydrocarbons and soot particles (Huang & Tang 2007) whereas gasification results in higher syngas generation and soot particle conversion (Lemmens et al., 2007).

Incineration

Incineration is combustion process which simply means drying and burning of waste (Nandan et al., 2017) and is mostly used to reduce volume of waste and recovering energy from the waste (Singh et al., 2014). A thermal waste treatment process in which the unprocessed waste is burned at high temperature is commonly known as incineration (Saleem et al., 2016). The incineration is burning of solid waste in an oxygen-rich environment at temperature higher than 850°C (Zaman, 2010). This process releases large amount of heat which if recovered properly can turn out to be a potential source of energy generation (Tan et al., 2015). Bio-medical waste and certain toxic industrial waste are also treated by this method having small capacity incinerators (Sharholy et al., 2005). Incineration is less common in India due to the composition of municipal solid waste (Sharholy et al., 2008) as solid waste in India has high organic matter (40% to 60%), high moisture content (40% to 60%) and high inert material (30% to 50%) and low calorific value (800 to 1100 kcal/kg) (Nandan et al., 2017) and for good energy generation, calorific value must fall between 1200-1400 kcal/kg (Kansal, 2002; Joardar, 2000; Rajput et al., 2009).

Incineration is a widely used waste management technology (Ezechi et al., 2017) and involves combustion of waste at high temperatures (Linak & Wendt, 1993). Incineration has the capability to reduce total organic matter content, destroy organic pathogens, significantly reduce waste volume, and preserve raw materials and resources (Sabbas et al., 2003). However, this process does not totally eliminate the waste generated but reduce the volume of the waste significantly and the remained waste also requires the landfill for its disposal (Narayana, 2009).

Pyrolysis

Pyrolysis is an effective waste-to-energy concept refers to destructive distillation of the solid waste to recover its energy in absence of air (Nandan et al., 2017). This process yield various products like oil, char which are recyclable (Chen et al., 2014) and is preferred for low moisture containing waste as the more heat is required for the treatment of high moisture waste (Nandan et al., 2017). Pyrolysis is a conversion of waste to liquid or gaseous fuels along with residue char which is a mixture of non-combustible material and carbon (Kothari et al., 2010). The reactions taking place initially are decomposition ones where organic components of low volatility are converted into other more volatile ones (Moustakas & Loizidou, 2010). The volatilize gases and volatilize liquids are efficiently used to run a steam engine (Saleem et al., 2016). The bottom of the reactor has ash, carbon char, and metals in which carbon char and metals can be recycled and ash remained is disposed of in a landfill (Young, 2010).

Gasification

Gasification lies between combustion and pyrolysis (Arena, 2012) and involve heating of the waste which result in the production of syngas (H_2 and CO) along with minor amounts of CH_4, N_2, CO_2 and H_2O (Maya et al., 2016). The feedstock is fed into gasifiers along with limited amount of air. The products of gasification are steam, electricity, hydrogen, fertilizers, and natural gas (Arena et al., 2010). The advantage of gasification is reduction in quantity of residue (about 70-80%), volume (about 80-90%) (Arena, 2012), land use, destruction of organic pollutants, reduction in emissions of greenhouse gases (Kothari et al., 2010) and helpful in generating renewable energy from solid waste (Rebitanim et al., 2013).

Melting Technology

Melting technology involves melting of the waste with the help of electricity or fuel (1400° C) (Christensen, 2011). This technology is helpful in reducing the volume the waste and the byproduct obtained from this process is stable slag (Saleem et al., 2016). The solidified residue has much application in construction industry and in land reclamation (Lam et al., 2010). This technology has many advantages as it overcomes the problem of fly ash and is helpful in avoiding water contamination due to the stability of solidified slag (Christensen, 2011).

Landfills

In most of the Indian cities, waste is thrown outskirts of the city area without any prior treatment (Figure 1.) which leads to environmental deterioration (Nandan et al., 2017). Due to rapid industrialization and urbanization, the land availability is decreasing day by day (Chen, 2007) and it became important to reduce volume of waste by different methods before ultimate disposal to the landfills (Mor et al., 2006). Landfill should be provided with composite liners to restrict leachate percolation to underground water level (Nandan et al.,2017) and should also contain proper gas collecting and ventilating system (Pacey et al., 1999; Talyan et al., 2008). The other main problem in the waste disposal to the landfill sites is the release of leachate (Alekhya et al., 2013) into the soil, surface water and ground water (Smahi et al., 2013; Nagarajan et al., 2012; Aljaradin & Persson, 2012; Akinbile & Yusoff, 2011; Mor et al., 2006) which deteriorate the quality of the environment in totality.

Engineered Landfills in India

The UN Environmental Programme defines landfill as the controlled disposal of MSW on land in such a way that contact between waste and the environment is significantly reduced (Kumar et al., 2017). The safe disposal of the waste is possible only by having engineered landfill as these prevent the contamination of groundwater and protects the surrounding environment (MoEF, 2015). Engineered landfills reduces the impact of waste on the environment by replacing the dumps of waste from the land (UNEP, 2005). India has many landfill sites but their efficiency is very low and thus become the matter of concern for the whole country (Sharholy et al., 208). The generation of the waste in different cities of India and amount of waste disposed to landfill varies from place to place and state to state (Table 2).

Sanitary Landfill

Landfill is a professionally engineered depression in low population area for the final disposal of waste (Cheremisinoff, 2003) and is known as the most economical method especially in developing countries (Gupta et al., 2015). Landfill is mostly preferred because it has the widest range of capabilities and is least expensive method of waste disposal (Saleem et al., 2016). The term 'sanitary landfill' refers to sites allocated for managing municipal solid wastes (Masters & Ela, 2008). Improvement in design and

Figure 1. Glimpse of crude dumping of waste in India

Table 2. Waste generation in different states of India and quantity of waste disposed to landfill site

State	Solid Waste Generation (TPD)	Collected (TPD)	Treated (TPD)	Landfilled (TPD)
Andhra Pradesh	6470.0	6396.0	1623.5	233.0
Arunachal Pradesh	16.20	11.76	0	0
Bihar	8110	7200	200	0
Karnataka	11186	9706	3475	5170
Madhya Pradesh	6773	5480	1141	4339
Punjab	4544.35	4520.3	39.17	3278.6
Maharastra	23449.66	23079.5	7543.1	1553
Tamilnadu	14658.5	14416.6	4776.2	7336.9
Haryana	4514	3159.8	188	2371.8
Nagaland	337	255	28	7
Tripura	428.09	379.2	134.4	244.8
Uttarakhand	1180	1180	0	0
Uttar Pradesh	15500	12000	3115	0
West Bengal	14000	12600	830	515

Source: Central Pollution Control Board (2016-17)

planning is necessary as waste remained after different treatments are disposed to the sanitary landfill site (Sharholy et al., 2007; Kansal, 2002). The landfill site which is managed in scientific way results in less effect on the adjoining environment (Misra & Pandey, 2005) and properly managed landfills should replace dumps in India and thus can reduce the environmental impact of waste (UNEP, 2005). Modern landfills have microturbines for generation of electricity from landfill gas (Saleem et al., 2016) and is helpful in resolving the issue of air pollution and global warming due to the emission of landfill gases (He et al., 2007).

The use of landfill for waste management has become less sustainable due to the rising cost of construction, operation and environmental consequences (Akor et al., 2013; Narayana, 2009). The landfill gases (LFGs) which are product of anaerobic decomposition of complex molecules contribute to urban ozone problems and global warming (Nkwachukwu et al., 2010). The methane (CH4) emission from landfill is about 13% of global CH4 emission and is about 818 million metric tons per annum in terms of CO2 equivalent (Rachel et al., 2009). The biotic and abiotic factors lead to generation of gas at landfill which is the combination of CH_4 and CO_2 (Saleem et al., 2016). If mismanaged this gas can cause explosion of landfill or gradual leaking can cause global warming as both the gases are greenhouse gases (He et al., 2007).

Figure 2. Strategies for recycling of waste

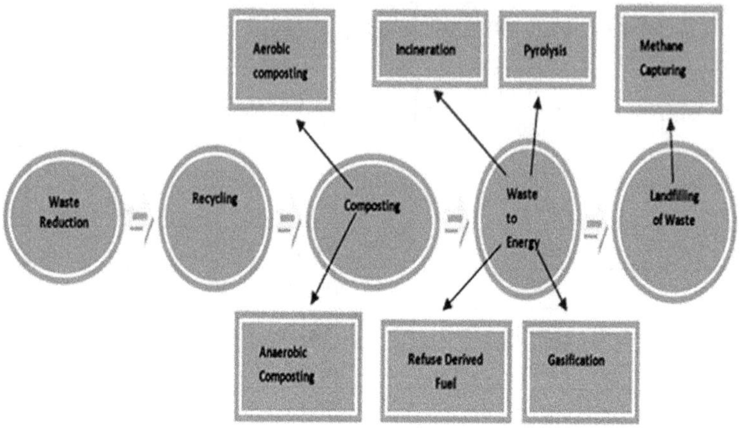

PROCESSING OF WASTE

The next step in integrated waste management is processing of collected waste. Processing can helpful in decreasing the waste volume and recovering many productive outputs including compost, steam and electricity (Kanat, 2010).

Fluffing

Waste to energy is one of the greatest sources of energy especially municipal solid waste (Cheng & Hu, 2010) as it has high heat recovery rate and high energy density (Park et al., 2016). Recently a processing method has been evolved in which the solid waste is separated, sterilize and the organic portion is processed to form pulp like material known as fluff (Saleem et al., 2016) and this technology is currently adopted by western countries where 95% recycling rate has been achieved (Narayana, 2009). High temperature steam is then introduced for further breaking of molecular bonds which destroys pathogens and the product is further grinded, dewatered and separated from other types of waste (Saleem et al., 2016). The high lignin and cellulose content of fluff have similar dust control properties with commercially produced lignosulfonates (Park et al., 2016). Additionally, the textural characteristics and pore space of fluff make it an ideal for use as a dust control agent alone and in combination with other dust control compounds (Torbert et al., 2011). The Fluff by product is important to agriculture and found to have significant nutrient concentrations that could serve as an organic fertilizer source (Busby et al., 2006; Busby, 2003). Utilization of non-composted waste material such as fluff provides significant benefits over handling costs associated with composting (Park et al., 2016).

Energy Recovery

Energy recovery from waste is the conversion of non-recyclable waste materials into usable heat, electricity, or fuel (Al-Salem et al., 2009) through a variety of processes including combustion, gasification, anaerobic digestion, and landfill gas (LFG) recovery and these process are often called waste-to-energy process (WTE) (Chinwan & Pant, 2014). Waste to Energy (WTE) is a proven environmentally sound process that provides reliable electricity generation and is extensively used in Europe and other developed nations in Asia (Kumar et al., 2010). The entire waste residue after sorting, reuse, recycle and processing is further utilized for energy recovery (Saleem et al., 2016). In 2009-2010, UK generates 32 million of waste from which 48% was returned to landfills, 39% was recycled and energy was produced from 13% of MSW and energy from waste could account for 17% of UK's electricity by 2020 (Brennand, 2004). Waste to Energy (WTE) provides a renewable alternative of energy in the world where we have limited fossil reservoirs (Saleem et al., 2016). Source separation of inert and high moisture content fractions would maximize the potential for thermal recovery and other treatment options in India (Kumar et al., 2017). There is also a lack of accountability in current SWM systems throughout India (Khajuria et al., 2010) but has good potential to produce energy from the waste (Table 3). Waste-to-energy technologies help to produce energy, recover materials and free land that would otherwise be used for dumping (Kumar et al., 2017). Converting non-recyclable waste materials into electricity and heat generates a renewable energy source and reduces carbon emissions (Chinwan & Pant, 2014). The thermal treatment of MSW results in the generation of 500-600 Kwh of electricity per ton of MSW combusted (Kumar et al., 2010). Planning Commission Report (2014) indicated that 62 million tons of annual MSW generated in urban area can produce 439 MW of power from combustible component and refuse derived fuel (RDF), 72 MW of electricity from landfill gas and 5.4 million metric tons of compost for agriculture use.

Refuse Derived Fuel (RDF)

The commercialization of waste using RDF has been utilized to make wastes useful. RDF generally refers to the processed MSW which is basically the fraction of segregated high calorific waste (Gallardo et al., 2014). In the European Union, it is estimated that the total amount of RDF produced from MSW is about 4-5 million tonnes annually (Grau and Farre, 2011). Solid recovered fuel is a new term which is becoming famous in Europe (Hernandez-Atonal et al., 2007). In this method, the size of solid waste is reduced by the shredder and the material is transferred to digestion tank (6-8 days) for preservation (Saleem et al., 2016). The temperature of 60 and 65° C is maintained for four days and the increased in last two days upto 70° C (Hernandez-Atonal et al., 2007). Removal of steel cans and magnetic things from the mainstream is done through a magnetic separator (Saleem et al., 2016). The waste is separated

Table 3. Waste to energy potential of different states in India

State	Capacity
Tamilnadu	151 MW
Gujarat	112 MW
Karnataka	450 MW
West Bengal	148 MW
Maharashtra	287 MW
Delhi	131 MW
Andhra Pradesh	123 MW
Bihar	73 MW
Rajasthan	62 MW

Source: MNRE Annual Report, 2016–2017

into different range on the basis of size like smaller than 8 mm, 8-16 mm and greater than 16 mm (Sarc & Lorber, 2013). The main advantages of using RDF as a fuel are an important reduction in the volume of waste and the possibility of energy recovery (Gallardo et al., 2014). It is predicted that use of fuel derived from solid fuel can decrease global warming (Anurag et al., 2007) and recovering energy from the solid waste also results in reduction of waste, transportation cost as well as also help to free up the earth surface and use it for other purposes (Singh et al., 2011).

BARRIERS TO IMPROVED WASTE MANAGEMENT IN INDIA

The current status of SWM in India is poor because the best and most appropriate methods from waste collection to disposal are not being used (Sharholy et al., 2008). There is a lack of training in SWM and the availability of qualified waste management professionals is limited (Kumar et al., 2017). Components of a waste management system include the facilities and equipment used to temporarily store waste (collection bins) or transfer collected waste to its final disposal site (Klundert & Anschutz 2001). Poor collection and inadequate transportation are responsible for the accumulation of MSW at every nook and corner (Sharholy et al., 2008). In India, it is the responsibility of municipal authorities to manage municipal solid waste but due to low financial resource, fewer amounts is utilised for the treatment of the waste (Guerrero et al., 213). About 90% of MSW is disposed of unscientifically in open dumps and landfills creating problems to public health and the environment (Sharholy et al., 2008). Due to unscientific management of landfills in developing countries, it is not feasible to collect methane and use is as the source of energy (Singh et al., 2011). The lack of strategic MSW plans, waste collection/ segregation and a government finance regulatory framework are major barriers to achieving effective SWM in India (Kumar et al., 2017). The management of MSW is going through a critical phase due to the unavailability of suitable facilities to treat and dispose of the larger amount of MSW generated daily in metropolitan cities (Sharholy et al., 2008) Limited environmental awareness combined with low motivation has inhibited innovation and the adoption of new technologies that could transform waste management in India (Kumar et al., 2017).

CONCLUSION

Unorganised planning and non-availability of scientific techniques is major concern in developing countries like India in which the waste generation is increasing at rapid rate and poor waste disposal is adding more problems to the environment. Most of the waste is disposed to the landfills without any proper sorting and segregation of the waste. Passing only laws for waste management and disposal without strict implementation is increasing the generation of the waste in India. Composting facility at the source of the waste generation is helpful in reducing the volume of the waste and help to produce valuable product which can have multiple uses in the field. Recycling is also the best tool for the management of waste as it saves material, reduces the need to landfill and incinerate, cuts down pollution, and helps to protect the environment. Recycling also helps to create jobs and also help to conserve resource as well as the environment. It is important to take help from developed countries which are using innovative techniques from waste collection to waste processing in order to solve the waste accumulation menace.

REFERENCES

Aalok, A., Tripathi, A. K., & Soni, P. (2008). Vermi-composting: A better option for organic solid waste management. *Journal of Human Ecology (Delhi, India)*, *24*(1), 59–64.

Abbasi, T., Gajalakshmi, S., & Abbasi, S. A. (2009). Towards modeling and design of vermi-composting systems: Mechanisms of composting/vermi-composting and their implications India. *Journal of Biotechnology*, *8*, 177–182.

Achilias, D. S., Roupakias, C., Megalokonomos, P., Lappas, A. A., & Antonakou, E. V. (2007). Chemical recycling of plastic wastes made from polyethylene (LDPE and HDPE) and polypropylene (PP). *Journal of Hazardous Materials*, *149*, 536–542. PMID:17681427

Akor, A. J., Ayotamuno, M. J., Aman, L. I., & Enokela, S. O. (2013). Assessment of domestic solid waste generation in PortHarcourt by separator – Receptacle technology. *International Journal of Scientific Engineering and Research*, *4*, 1–7.

Al-Awadhi, J. A., & Shuaibi, A. (2012). Potentiality of Zubair International Journal of Advanced Science and Research 29 Formation for deep slurry injection in Kuwait. *Environmental Earth Sciences*, *67*(5), 1269–1280.

Al-Jarallah, R., & Aleisa, E. (2014). A baseline study characterizing the municipal solid waste in the State of Kuwait. *Waste Management (New York, N.Y.)*, *34*, 952–960. PMID:24656421

Al-Salem, S. M., Lettieri, P., & Baeyens, J. (2009). Recycling and recovery routes of plastic solid waste (PSW). *Waste Management (New York, N.Y.)*, *29*, 2625–2643. PMID:19577459

Alekhya, M., Divya, N., Jyothirmai, G., & Reddy, K. R. (2013). Secured landfills for disposal of municipal solid waste. *International Journal of Engineering Research and General Science*, *1*(1).

Aljaradin, M., & Persson, K. M. (2012). Environmental impact of municipal solid waste landfills in semi-arid climates – case study – Jordan. *Open Waste Management Journal*, *5*, 28–39.

Annepu, R. K. (2012). *Sustainable Solid Waste Management in India* (M.Sc. Thesis). Earth Engineering Center, Columbia University, New York, NY.

Annepu, R. K. (2012). Report on sustainable solid waste management in India. *Waste-to-Energy Research and Technology Council*. Retrieved from http:// swmindia.blogspot.in

Anurag, G., Smith, R., & Hill, D. (2007). Wastes as co-fuels: The policy framework for solid recovered fuel (SRF) in Europe, with UK implications. *Environmental Science & Technology*, *41*, 4868–4874. PMID:17711195

Arebey, M., Hannan, M., Basri, H., Begum, R., & Abdullah, H. (2010). Integrated technologies for solid waste bin monitoring system. *Environmental Monitoring and Assessment*, *177*(1-4), 399–408. PMID:20703798

Arena, U. (2012). Process and technological aspects of municipal solid waste gasification. A review. *Waste Management (New York, N.Y.)*, *32*(4), 625–639. PMID:22035903

Arena, U., Zaccariello, L., & Mastellone, M. (2010). Fluidized bed gasification of waste-derived fuels. *Waste Management (New York, N.Y.), 30*(7), 1212–1219. PMID:20172708

Asian Development Bank. (2013). *Solid waste management in Nepal: current status and policy recommendations*. Asian Development Bank.

Asnani, P. U. (2006). *Solid Waste Management*. India Infrastructure Report, 160–189.

Asnani, P. U. (2007). Methane to markets and landfill gas energy in India. *Proceeding of the International Conference on Sustainable Solid Waste Management*, 519–525.

Balat, M., Balat, M., Kırtay, E., & Balat, H. (2009). Main routes for the thermo-conversion of biomass into fuels and chemicals. Part 2: Gasification systems. *Energy Conversion and Management, 50*(12), 3158–3168.

Bhat, R. A., Dar, S. A., Dar, D. A., & Dar, G. H. (2018). Municipal Solid Waste Generationand current Scenario of its Management in India. *International Journal of Advance Research in Science and Engineering, 07*(2), 419–431.

Bhat, R. A., Kamili, A. N., & Bandh, S. A. (2013). Characterisation and composition of Municipal Solid Waste (MSW) generated in Yusmarg: A health resort of Kashmir valley: A Glance at the World. *Waste Management (New York, N.Y.), 33*, 774–777.

Birpınar, M. E., Bilgili, M. S., & Erdoğan, T. (2009). Medical waste management in Turkey: A case study of Istanbul. *Waste Management (New York, N.Y.), 29*(1), 445–448. PMID:18550354

Biswas, A. K., Kumar, S., Babu, S. S., Bhattacharyya, J. K., & Chakrabarti, T. (2010). Studies on environmental quality in and around municipal solid waste dumpsite. *Resources, Conservation and Recycling, 55*, 129–134.

Brems, A., Baeyens, J., & Dewil, R. (2012). Recycling and recovery of post-consumer plastic solid waste in a European context. *Thermal Science, 16*(3), 669–685.

Brennand, T. (2004). Renewable energy in the United Kingdom: Policies and prospects. *Energy for Sustainable Development, 8*(1), 82–92.

Bundela, P. S., Gautam, S. P., Pandey, A. K., Awasthi, M. K., & Sarsaiya, S. (2010). Municipal solid waste management in Indian cities. *International Journal of Environmental Science and Technology, 1*(4), 591–605.

Burnley, S., Phillips, R., Coleman, T., & Rampling, T. (2011). Energy implications of the thermal recovery of biodegradable municipal waste materials in the United Kingdom. *Waste Management (New York, N.Y.), 31*, 1949–1959. PMID:21600755

Busby, R., Gebhart, D. L., & Torbert, H. A. (2006). Effects of an uncomposted municipal waste processing by product on prairie grass establishment. *Agronomy Journal, 98*, 1073–1080.

Busby, R. R. (2003). *Suitability of a municipal solid waste byproduct as a soil amendment for reestablishing native grasses on disturbed Army training lands* (M.Sc. Thesis). University of Illinois, Urbana-Champaign, IL.

Cabalova, I., Kacik, F., Geffert, A., & Kacikova, D. (2011). *The effects of paper recycling and its environmental impact*. Environmental Management in Practice; doi:10.5772/23110.2011

Central Pollution Control Board. (2016-17). *Consolidated annual report (For the year 2016-2017)*. Author.

Central Pollution Control Board (CPCB). (2007). *Bio-methanation potential of solid wastes from agro-based industries*. Delhi, India: Ministry of Environment and Forests.

Chang, N. B., Pires, A., & Martinho, G. (2011). Empowering systems analysis for solid waste management: Challenges, trends, and perspectives. *Critical Reviews in Environmental Science and Technology*, *41*(16), 1449–1530.

Chen, D., Yina, L., Wanga, H., & He, P. (2014). Pyrolysis technologies for municipal solid waste. *Waste Management*, 1-21.

Chen, J. (2007). Rapid urbanization in China: A real challenge to soil protection and food security. *Catena*, *69*(1), 1–15.

Cheng, H., & Hu, Y. (2010). Municipal solid waste (MSW) as a renewable source of energy: Current and future practices in China. *Bioresource Technology*, *101*(11), 3816–3824. PMID:20137912

Cheremisinoff, N. P. (2003). *Handbook of solid waste management and waste minimization technologies*. Elsevier Science.

Chiemchaisri, C., Juanga, J. P., & Visvanathan, C. (2007). Municipal solid waste management in Thailand and disposal emission inventory. *Environmental Monitoring and Assessment*, *135*, 13–20. PMID:17492361

Chinwan, D., & Pant, S. (2014). Waste to Energy in India and its Management. *Journal of Basic and Applied Engineering Research*, *1*(10), 89–94.

Christensen, T. (2011). *Solid waste technology & management. In International Journal of Advanced Science and Research Chichester* (p. 108). West Sussex, UK: Wiley.

Christos, C., & Lasaridi, K. (2009). A GIS based model for the optimization of municipal solid waste collection: The case study of Nikea, Athens, Greece. *WSEAS Transactions on Environment and Development*, *5*(10), 640–650.

Couth, R., & Trois, C. (2012). Cost effective waste management through composting in Africa. *Waste Management (New York, N.Y.)*, *32*, 2518–2525. PMID:22857939

CPHEEO — Central Public Health and Environmental Engineering Organization. (2000). Manual on Municipal Solid Waste Management. Ministry of Urban Development, Govt. of India.

Das, S., & Bhattarcharya, B. K. (2014). Estimation of municipal solid waste generation and future trends in greater metropolitan regions of Kolkata, India. *Journal of Industrial Engineering and Management Innovation*, *1*(1), 31–38.

Dasgupta, B., & Mondal, M. (2012). Bio Energy Conversion of Organic Fraction of Varanasi's Municipal Solid Waste. *Energy Procedia*, *14*, 1931–1938.

Datar, M. T., & More, A. B. (2006). *Vermi-composting – A Technological Option for Solid Waste Management.* Available at:http://www. bvucoepune.edu.in/pdf's/Research%20and%20Publication Research%20Publications_2006-07/International%20Conference_2006-07/Vermi-composting%20Mr%20 AB% 20More.pdf

Dave, P. N., & Joshi, A. K. (2010). Plasma pyrolysis and gasification of plastics waste – a review. *Journal of Scientific and Industrial Research, 177*(69), 177–179.

DEFRA (Department for Environment, Food and Rural Affairs). (2007). *Incineration of municipal solid waste.* Waste Management Technology Brief, the New Technologies Work Stream of the DEFRA Waste Implementation Programme.

Dene, L. O. (2013). *Solid Waste Collection Vehicle Route Optimization for the City of Redlands.* University of Redlands.

Elahi, H., Rehman, W., Swati, R. F., Khan, M. H., Mushtaq, T., Haider, Z., ... Riaz, S. (2017). Automatic sorting, counting and bottle filling robotic statio. *International Journal of Scientific Engineering and Applied Science, 3*, 5.

Emery, A., Davies, A., Griffiths, A., & Williams, K. (2007). Environmental and economic modelling: A case study of municipal solid waste management scenarios in Wales. *Resources, Conservation and Recycling, 49*, 244–263.

Enayetullah, I., Sinha, A. H. M. M., & Khan, S. S. A. (2005). *Urban solid waste management scenario of Bangladesh: problems and prospects.* Waste Concern Technical Documentation. Retrieved from http://www.wasteconcern.org/Publication/Waste%20Survey_05.pdf

Erdogan, N., & Baris, E. (2007). Environmental protection programs and conservation practices of hotels in Ankara, Turkey. *Tourism Management, 28*(2), 604–614.

Ezechi, E. H., Isa, M. H., Kutty, S. R., & Sapari, N. B. (2011). Boron recovery, application and economic significance: A review. In *National Postgraduate Conference (NPC).* Kuala Lumpur, Malaysia. Institute of Electrical and Electronic Engineering.

Ezechi, E. H., Nwabuko, C. G., Enyinnaya, O. C., & Babington, C. J. (2017). Municipal solid waste management in Aba, Nigeria. *Challenges and Prospects Environmental Engineering and Resources, 22*(3), 231-236.

Fang, M., Wong, J., Ma, K., & Wong, M. (1999). Co-composting of sewage sludge and coal fly ash: Nutrient transformations. *Bioresource Technology, 67*, 19–24.

Fei, F., Wen, Z., Huang, S., & Clercq, D. (2018). Mechanical biological treatment of municipal solid waste: Energy efficiency, environmental impact and economic feasibility analysis. *Journal of Cleaner Production, 178*, 731–739.

Feng Li, R. (2010). The composting treatment of residue from biological municipal waste anaerobic digestion. *Advanced Materials Research, 129*(131), 703–707.

Francis, R. (2016). *Recycling of polymers: methods, characterization and applications.* Hoboken, NJ: John Wiley & Sons.

Friends of the Earth. (2008). Brief Autoclaving. Author.

Gallardo, A., Carlos, M., Bovea, M. D., Colomer, F. J., & Albarrán, F. (2014). Analysis of refuse-derived fuel from the municipal solid waste reject fraction and its compliance with quality standards. *Journal of Cleaner Production*, *83*, 118–125.

Gaustad, G., Olivetti, E., & Kirchain, R. (2012). Improving aluminum recycling: A survey of sorting and impurity removal technologies. *Resources, Conservation and Recycling*, *58*, 79–87.

Gencer, Y. G. (2016). *Mystery of Recycling: Glass and Aluminum Examples*. Doi:10.4018/978-1-4666-9723-2.ch009

Ghosh, R., & Kansal, A. (2014). Urban challenges in India and the mission for a sustainable habitat. *Interdisciplina*, *2*(2), 281–304.

Gidde, M. R., Todkar, V. V., & Kokate, K. K. (2008). Municipal solid waste management in emerging mega cities: a case study of Pune city. *Proceedings of Indo Italian Conference on Green and Clean Environment*.

Gilberto, T. Z. Z., Maria, V. S., & Carvalho, D. G. (2008). A case study of fuel savings through optimization of MSW transportation routes. *Management of Environmental Quality*, *19*(4), 444–454.

Glavic, P., & Lukman, R. (2007). Review of sustainability terms and their definitions. *Journal of Cleaner Production*, *15*(18), 1875–1885.

Goswami, K. B. (2018). Challenges and opportunities associated with waste management in India. *International Journal of Research in Social Sciences*, *8*(2).

Goto, M., Sasaki, M., & Hirose, T. (2006). Reactions of polymers in supercritical fluids for chemical recycling of waste plastics. *Journal of Materials Science*, *41*, 1509–1515.

Grau, A., & Farre, O. (2011). *Situation and Potential of the DirectWaste Energy Recovery. Technical study. PER 2011-2020*. IDAE.

Grigore, M. E. (2017). Methods of recycling, properties and applications of recycled thermoplastic polymers. *Recycling*, *2*, 24.

Guerrero, L. A., Maas, G., & Hogland, W. (2013). Solid waste management challenges for cities in developing countries. *Waste Management, 33*(1), 220-232.

Gundupalli, S. P., Haitb, S., & Thakur, A. (2017). A review on automated sorting of source-separated municipal solid waste for recycling. *Waste Management*, 1-39.

Gupta, N., Yadav, K. K., & Kumar, V. (2015). A review on current status of municipal solid waste management in India. *Journal of Environmental Sciences (China)*, *37*, 206–217. PMID:26574106

Gupta, R., & Garg, V. (2011). Potential and possibilities of vermi-composting in sustainable solid waste management: A review. *International Journal of Environment and Waste Management*, *7*(3/4), 210.

Gupta, S. K., Pandey, S. K., Singh, N. B., Singh, A., Singh, V. K., & Verma, T. (2013). Municipal solid waste characterizations and management strategies for the Lucknow city, India. *International Journal of Engine Research, 8*(17), 2031–2036.

Gupta, V. K., Ali, I., Saleh, T. A., Nayak, A., & Agarwal, S. (2012). Chemical treatment technologies for waste-water recycling—an overview. *RSC Advances, 2*(16), 6380–6388.

Gutberlet, J. (2010). Waste, poverty and recycling. *Waste Management (New York, N.Y.), 30*(2), 171–173. PMID:19962085

Hargreaves, J. C., Adl, M. S., & Warman, P. R. (2008). A review of the use of composted municipal solid waste in agriculture. *Agriculture, Ecosystems & Environment, 123*(1-3), 1–14.

He, R., Liu, X., Zhang, Z., & Shen, D. (2007). Characteristics of the bioreactor landfill system using an anaerobic–aerobic process for nitrogen removal. *Bioresource Technology, 98*(13), 2526–2532. PMID:17071082

Heberlein, J., & Murphy, A. (2008). Thermal plasma waste treatment. *Journal of Physics. D, Applied Physics, 41*(5), 053001.

Hernandez-Atonal, F., Ryu, C., Sharifi, V., & Swithenbank, J. (2007). Combustion of refuse-derived fuel in a fluidised bed. *Chemical Engineering Science, 62*(1-2), 627–635.

Hoornweg, D., & Bhada-Tata, P. (2012). *What a waste: a global review of solid waste management.* Washington, DC: World Bank.

Hopewell, J., Dvorak, R., & Kosior, E. (2009). Plastics recycling: Challenges and opportunities. *Philosophical Transactions of the Royal Society of London. Series B, Biological Sciences, 364*, 2115–2126. PMID:19528059

Hossain, M. S., Balakrishnan, V., Rahman, N. N. N. A., Sarker, M. Z. I., & Kadir, M. O. A. (2012). Treatment of clinical solid waste using a steam autoclave as a possible alternative technology to incineration. *International Journal of Environmental Research and Public Health, 9*(3), 855–867. PMID:22690168

Huang, H., & Tang, L. (2007). Treatment of organic waste using thermal plasma pyrolysis technology. *Energy Conversion and Management, 48*, 1331–1337.

Idowu, A., Adagunodo, E., Esimai, O., & Olapade, T. (2012). Development of a web based gis waste disposal management system for Nigeria. *International Journal of Information Engineering and Electronic Business, 4*(3), 40–48.

International Engineering Consortium. (2015). Retrieved from http://www.uky.edu/~jclark/mas355/GSM.PDF

ISWA (International Solid Waste Association). (2012). *Globalization and waste management final report from the ISWA task force.* Retrieved from http://www.iswa. org/knowledgebase/tfgfinal

Jafari, A., Godini, H., & Mirhousaini, S. H. J. (2010). Municipal solid waste management in KhoramAbad city and experiences. *World Academy of Science, Engineering and Technology, 4*, 163–168.

Jenkins, R. R., Martinez, S. A., & Podolsky, P. K. (2013). The determinant of household recycling; A material-specific analysis of recycling program features and unit recycling. *Journal of Environmental Economics and Management*, *45*(2), 294–318.

Jha, M. K., Sondhi, O. A. K., & Pansare, M. (2003). Solid waste management- a case study. *Indian Journal of Environment Protection*, *23*, 1153–1160.

Joardar, S. D. (2000). Urban residential solid waste management in India. *Public Works Management & Policy*, *4*(4), 319–330.

Jones, A., & Harrison, R. (2016). Emission of ultrafine particles from the incineration of municipal solid waste: A review. *Atmospheric Environment*, *140*, 519–528.

Joshi, R., & Ahmed, S. (2016). Status and challenges of municipal solid waste management in India: A review. *Cogent Environmental Science*, *2*(1), 1139434.

Kalinauskas, Z. (2010). Interdependence between the Advance in Science and the Advance in Technology. *Evolution of Science and Technology*, *1*(2), 157–166.

Kanat, G. (2010). Municipal solid-waste management in Istanbul. *Waste Management (New York, N.Y.)*, *30*(8-9), 1737–1745. PMID:20185290

Kang, H. Y., & Schoenung, J. M. (2005). Electronic waste recycling: A review of US infrastructure and technology options. *Resources, Conservation and Recycling*, *45*(4), 368–400.

Kansal, A. (2002). Solid waste management strategies for India. *Indian Journal of Environment Protection*, *22*, 444–448.

Karayannidis, G. P., & Achilias, D. S. (2007). Chemical recycling of poly (ethylene terephthalate). *Macromolecular Materials and Engineering*, *292*(2), 128–146.

Kaushal, R. K., Varghese, G. K., & Chabukdhara, M. (2012). Municipal solid waste management in India—current state and future challenges: A review. *International Journal of Environmental Science and Technology*, *4*, 1473–1489.

Khajuria, A., Matsui, T., Machimura, T., & Morioka, T. (2010). Assessment of the challenge of sustainable recycling of municipal solid waste management in India. *International Journal of Environmental Technology and Management*, *13*, 171–187.

Khalid, A., Arshad, M., Anjum, M., Mahmood, T., & Dawson, L. (2011). The anaerobic digestion of solid organic waste. *Waste Management, 31*(8), 1737-1744.

Khan, N. A. (2014). Perspectives of transport and disposal of municipal solid waste in Srinagar city. *International Journal of Engineering Research and General Science*, *2*(4), 238–251.

Khan, R. R. (1994). Environmental management of municipal solid wastes. *Indian Journal of Environment Protection*, *14*(1), 26–30.

Kiran, E. U., Trzcinski, A., Ng, W., & Liu, Y. (2014). Bioconversion of food waste to energy: A review. *Fuel*, *134*, 389–399.

Kiran, E. U., Trzcinski, A. P., Ng, W. J., & Liu, Y. (2014). Bioconversion of food waste to energy: A review. *Fuel*, *134*, 389–399.

Klundert, A. V. D., & Anschutz, J. (2001). *Integrated Sustainable Waste Management—The Concept: Tools for Decision-Makers, Experiences from the Urban Waste Expertise Programme (1995–2001)*. Gouda, The Netherlands: WASTE.

Kothari, R., Tyagi, V., & Pathak, A. (2010). Waste-to-energy: A way from renewable energy sources to sustainable development. *Renewable & Sustainable Energy Reviews*, *14*(9), 3164–3170.

Kum, V., Sharp, A., & Harnpornchai, N. (2005). Improving the solid waste management in Phnom Penh city: a strategic approach. *Waste Management, 25*(1), 101-109.

Kumar, J. S., Subbaiah, K. V., & Rao, P. V. P. (2014). Municipal solid waste management scenario in India. *Australian Journal of Civil Engineering Research*, *2*, 1–8.

Kumar, J. S., Subbaiah, K. V., & Rao, P. V. V. P. (2010). Waste to Energy: A Case Study of Eluru, A.P, India. *International Journal of Environmental Sciences and Development*, *1*(3), 238–243.

Kumar, K. N., & Goel, S. (2009). Characterization of municipal solid waste (MSW) and a proposed management plan for Kharagpur,West Bengal, India. *Resources, Conservation and Recycling*, *53*, 166–174.

Kumar, S., Bhattacharya, J. K., Vaidya, A. N., Chakrabarti, T., Devotta, S., & Akolkar, A. B. (2009). Assessment of the status of municipal solid waste management in metro cities, state capitals, class I cities, and class II towns in India: An insight. *Journal of Waste Management*, *29*, 883–895. PMID:18595684

Kumar, S., Pandaa, A. K., & Singha, R. K. (2011). A review on tertiary recycling of high-density polyethylene to fuel. *Resources, Conservation and Recycling*, *55*, 893–910.

Kumar, S., Smith, S. R., Fowler, G., Velis, C., Kumar, S. J., Arya, S. R., ... Cheeseman, C. (2017). Challenges and opportunities associated with waste management in India. *Royal Society Open Science*, *4*, 160764. PMID:28405362

Lam, C. H., Ip, A. W., Barford, J. P., & McKay, G. (2010). Use of incineration MSW ash: A review. *Sustainability*, *2*(7), 1943–1968.

Laroche, M., Bergeron, J., & Barbaro-Forleo, G. (2001). Targeting consumers who are willing to pay more for environmentally friendly products. *Journal of Consumer Marketing*, *18*(6), 503–520.

Late, A., & Mule, M. B. (2013). Composition and characterization study of solid waste from Aurangabad city. *Universal Journal of Environmental Research and Technology*, *3*(1), 55–60.

Latifah, A., & Mohd, A. A., & NurIlyana, M. (2009). Municipal solid waste management in Malaysia: Practices and challenges. *Waste Management (New York, N.Y.)*, *29*, 2902–2906. PMID:19540745

Lemmens, B., Elslander, H., Vanderreydt, I., Peys, K., Diels, L., Oosterlinck, M., & Joos, M. (2007). Assessment of plasma gasification of high caloric waste streams. *Waste Management (New York, N.Y.)*, *27*, 1562–1569. PMID:17134888

Liang, C., Das, K. C., & Clendon, R. (2003). The influence of temperature and moisture contents regimes on the aerobic microbial activity of a biosolids composting blend. *Bioresource Technology*, *86*, 131–137. PMID:12653277

Linak, W. P., & Wendt, J. O. (1993). Toxic metal emissions from incineration: Mechanisms and control. *Progress in Energy and Combustion Science*, *19*(2), 145–185.

Liu, C., & Wu, X. W. (2010). Factors influencing municipal solid waste generation in China: A multiple statistical analysis study. *Waste Management & Research*, *29*, 371–378. PMID:20699292

Ma, Y., Wang, L., & Sun, X. (2011). Deinking technology and deinking agent of waste paper. *Advanced Materials Research*, *335-336*, 1340–1343.

Malviya, R., Chaudhary, R., & Buddhi, D. (2002). Study on solid waste assessment and management -Indore city. *Indian Journal of Environmental Protection*, *22*(8), 841–846.

Manyuchi, M. M., & Phiri, A. (2013). Vermi-composting in solid waste management: A review. *International Journal of Engineering Science and Technology*, *2*(12), 1234–1242.

Manyuchi, M. M., Phiri, A., Muredzi, P., & Chirinda, N. (2013). Effect of drying on vermicompost macronutrient composition. *Internatioanl Journal of Engineering Science and Invention*, *1*(10), 1–3.

Marculescu, C. (2011). Thermal-chemical treatment of solid waste mixtures. *Energy Procedia*, *6*, 558–564.

Marshall, R. E., & Farahbakhsh, K. (2013). Systems approaches to integrated solid waste management in developing countries. *Waste Management (New York, N.Y.)*, *33*(4), 988–1003. PMID:23360772

Masood, M., Barlow, C. Y., & Wilson, D. C. (2014). An assessment of the current municipal solid waste management system in Lahore, Pakistan. *Waste Management & Research*, *32*, 834–847. PMID:25106534

Masters, G. M., & Ela, W. P. (2008). *Environmental engineering and science* (3rd ed.). Prentice Hall.

Maya, D. M. Y., Sarmiento, A. L. E., Oliveira, C. A. V. B. S., Lora, E. E. S., & Andrade, R. V. (2016). Gasification of municipal solid waste for power generation in Brazil, a review of available technologies and their environmental benefits. *Journal of Chemistry and Chemical Engineering*, *10*, 249–255.

Mehta, P. (2019). E-Waste, Chemical Toxicity, and Legislation in India. In *Advanced Methodologies and Technologies in Engineering and Environmental Science* (pp. 144–156). IGI Global.

Mesina, M. B., De Jong, T. P. R., & Dalmijn, W. L. (2003). Improvements in separation of non-ferrous scrap metals using an electromagnetic sensor. *Physical Separation in Science and Engineering*, *12*(2), 87–101.

Metcalfe, A., Riley, M., Barr, S., Tudor, T., Robinson, G., & Guilbert, S. (2012). Food waste bins: Bridging infrastructures and practices. *The Sociological Review*, *60*, 135–155.

Mickael, D. (2016). Categorization and Sorting for Waste Management. *International Journal of Waste Resources*, *6*(2).

Ministry of Environment and Forests (MoEF). (2015). Municipal solid waste (Management and Handling) rules. The Gazette of India.

Ministry of New and Renewable Energy (MNRE). (2016-17). *Annual report*. Author.

Mirabella, N., Castellani, V., & Sala, S. (2014). Current options for the valorization of food manufacturing waste: A review. *Journal of Cleaner Production*, *65*, 28–41.

Misra, V., & Pandey, S. D. (2005). Hazardous waste, impact on health and environment for development of better waste management strategies in future in India. *Environment International*, *31*, 417–431. PMID:15734194

Modak, P., Jiemian, Y., Hongyuan, Yu., & Mohanty, C. R. (2010). Municipal solid waste management: turning waste into resources. In Shanghai manual: a guide for sustainable urban development in the 21st century, (pp. 1–36). Academic Press.

Mohanty, C. R., Mishra, U., & Beuria, P. R. (2014). Municipal solid waste management in Bhubaneswar, India — a review. *International Journal of Latest Trends in Engineering & Technology*, *3*(3), 303–312.

Mor, S., Ravindra, K., Dahiya, R. P., & Chandra, A. (2006). Leachate characterization and assessment of groundwater pollution near municipal solid waste landfill site. *Environmental Monitoring and Assessment*, *118*(1-3), 435–456. PMID:16897556

Moustakas, K., & Loizidou, M. (2010). Solid waste management through the application of thermal methods. *Waste Manage*. Retrieved from http://www. intechopen.com/books/waste-management/solid-wastemanagement-through-the-application-of-thermal-methods

Nagarajan, R., & Thirumalaisaumalaisamy, S., Lakshumanan, & Eakshumanan, E. (2012). Impact of leachate on groundwater pollution due to non-engineered municipal solid waste landfill sites of erode city, Tamil Nadu, India. *Iranian Journal of Environmental Health Sciences & Engineering*, *9*(1), 1–12. PMID:23369323

Nagavallemma, K. P., Wani, S. P., Lacroix, S., Padmaja, V. V., Vineela, C., Rao, M. B., & Sahrawat, K. L. (2004). *Vermi-composting: Recycling wastes into valuable organic fertilizer*. Global Theme on Agroecosystems Report no. 8.

Nandan, A., Yadav, B. P., Baksi, S., & Bose, D. (2017). Recent Scenario of Solid Waste Management in India. *World Scientific News*, *66*, 56–74.

Narayana, T. (2009). Municipal solid waste management in India: From waste disposal to recovery of resources. *Waste Management (New York, N.Y.)*, *29*(3), 1163–1166. PMID:18829290

Naveen, B. P., Sivapullaiah, P. V., & Sitharam, T. G. (2013). Disposal options for solid waste of Bangalore city base on its characteristics. *International Journal of Environment and Waste Management*, *12*(1), 77–88.

Nema, A. K. (2004). *Collection and transport of municipal solid waste. Training Program on Solid Waste Management*. Delhi, India: Springer.

Nkwachukwu, O. I., Chidi, N. I., & Charles, K. O. (2010). Issues of roadside disposal habit of municipal solid waste, environmental impacts and implementation of sound management practices in developing country "Nigeria." *International Journal of Environmental Sciences and Development*, *5*, 409–418.

Ogwueleka, T. C. (2009). Municipal solid waste characteristics and management in Nigeria. *Iranian Journal of Environmental Health Sciences & Engineering*, *6*(3).

Olah, G. A., Goeppert, A., & Prakash, G. S. (2008). Chemical recycling of carbon dioxide to methanol and dimethyl ether: From greenhouse gas to renewable, environmentally carbon neutral fuels and synthetic hydrocarbons. *The Journal of Organic Chemistry*, *74*, 487–498. PMID:19063591

Pacey, J., Augenstein, D., Morck, R., Reinhart, D., & Yazdani, R. (1999). The bioreactor landfill-an innovation in solid waste management. *Municipal Solid Waste Management*, 53-60.

Pang, L. I., Scorpio, D. E., & Kibert, C. J. (2010). Strategies for successful construction and demolition waste recycling operations. *Journal of Construction Management and Economics*, *15*, 49–50.

Papadimitriou, E. K., Barton, J. R., & Stentiford, E. I. (2008). Sources and levels of potentially toxic elements in the biodegradable fraction of autoclaved non-segregated household waste and its compost/digestate. *Waste Management & Research*, *26*(5), 419. PMID:18927061

Park, S. W., Lee, J. S., Yang, W. S., Kang, J. J., Sung, J. H., Alam, M. T., ... Oh, J. H. (2016). For Waste to Energy, Assessment of Fluff Type Solid Refuse Fuel by Thermal Characteristics Analyses International Conference on Solid Waste Management, 5IconSWM 2015. *Procedia Environmental Sciences*, *35*, 498–505.

Pati, R. K., Vrat, P., & Kumar, P. (2008). A goal programming model for paper recycling system. The international journal of management science. *Omega*, *36*, 405–417.

Patle, A. V., Williams, S. P. M. P., Gabhane, J., Dhar, H., & Nagamaik, P. B. (2014). Microbial assisted rapid composting of agriculture residues. *International Journal of Scientific and Engineering Research*, *5*(5).

Peterson, C. (2013). *Assessment of solid waste management practices and its vulnerability to climate risks in Maldives Tourism Sector*. Retrieved from http://tourism.gov.mv/downloads/tap/2014/ Solid_Waste.pdf

Planning Commission, Government of India. (2014). *Report of the task Force on waste to energy (Volume I) in the context of integrated municipal solid waste management*. Retrieved from http://planningcommission.nic.in/reports/genrep/rep_ wte1205.pdf

Plavac, B., Filipan, V., Sutlovic, I., & Sveticic, J. (2017). Sustainable waste management with mechanical biological treatment and energy utilization. *Technical Gazette*, *24*(4), 1283–1289.

Poskrobko, B., & Poskrobko, T. (2012). *Environment management in Poland*. Bialystok: Bialystok University Press.

PPCB (Punjab Pollution Control Board). (2010). *Status report on municipal solid waste in Punjab*. Punjab Pollution Control Board, Patiala. Retrieved from http://www. ppcb.gov.in/Attachments/Annual%20 Reports/AR.pdf

Price, A. (2005). UK power: An ever changing challenge for civil engineers. *Proceedings of the Institution of Civil Engineers. Civil Engineering*, *158*(6), 4–11.

Rachel, O. A., Komine, H., Yasuhara, K., & Murakami, S. (2009). *Municipal solid waste management in developed and developing countries-Japan and Nigeria as case studies*. Available at: http://wwwgeo.civil.ibaraki.ac.jp/komine/ mypapers/JGSPaper/JGS2009(973)Rachel.pdf

Rada, E., Ragazzi, M., & Fedrizzi, P. (2013). Web-GIS oriented systems viability for municipal solid waste selective collection optimization in developed and transient economies. *Waste Management (New York, N.Y.)*, *33*(4), 785–792. PMID:23402896

Rajkumar, N., Subramani, T., & Elango, L. (2010). Groundwater contamination due to municipal solid waste diposal — a GIS based study in Erode City. *International Journal of Environmental Science*, *1*(1), 39–55.

Rajput, R., Prasad, G., & Chopra, A. K. (2009). Scenario of solid waste management in present Indian context. *Journal of Environmental Sciences (China)*, *7*(1), 45–53.

Rana, P. R., Yadav, D., Ayub, S., & Siddiqui, A. A. (2014). Status and challenges in solid waste management: A case study of Aligarh city. *Journal of Civil Engineering & Environmental Technology*, *1*, 19–24.

Rana, R., Ganguly, R., & Gupta, A. K. (2015). An assessment of solid waste management system in Chandigarh city India. *The Electronic Journal of Geotechnical Engineering*, *20*, 1547–1572.

Rathi, S. (2006). Alternative approaches for better municipal solid waste management in Mumbai, India. *Waste Management (New York, N.Y.)*, *26*, 1192–1200. PMID:16288861

Rathi, S. (2007). Optimization model for integrated municipal solid waste management in Mumbai, India. *Environment and Development Economics*, *12*, 105–121.

Rebitanim, N. Z., Karim, G. W. A., Rebitanim, N. A., & Amran, M. S. M. (2013). Potential applications of wastes from energy generation particularly biochar in Malaysia. *Renewable & Sustainable Energy Reviews*, *21*, 694–702.

Rezaiyan, J., & Cheremisinoff, N. P. (2005). *Gasification technologies: a primer for engineers and scientists*. CRC Press.

Ribeiro, M. C. S., Fiúza, A., Ferreira, A., Dinis, M. D. L., Castro, M. A. C., Meixedo, J. P., & Alvim, M. R. (2016). Recycling approach towards sustainability advance of composite materials industry. *Recycling*, *1*, 178.

Sabbas, T., Polettini, A., & Pomib, R. (2003). Management of municipal solid waste incineration residues. *Waste Management (New York, N.Y.)*, *23*, 61–88. PMID:12623102

Safavi, S. M., Masoumi, H., Mirian, S. S., & Tabrizchi, M. (2010). Sorting of polypropylene resins by color in MSW using visible reflectance spectroscopy. *Waste Management (New York, N.Y.)*, *30*, 2216–2222. PMID:20674323

Saleem, W., Zulfiqar, A., Tahir, M., Asif, F., & Yaqub, G. (2016). Latest technologies of municipal solid waste management in developed and developing countries: A review. *International Journal of Advanced Science and Research*, *1*(10), 22–29.

Sarc, R., & Lorber, K. (2013). Production, quality and quality assurance of Refuse Derived Fuels (RDFs). *Waste Management (New York, N.Y.), 33*(9), 1825–1834. PMID:23746983

Sarkar, N., Ghosh, S. K., Bannerjee, S., & Aikat, K. (2012). Bioethanol production from agricultural wastes: An overview. *Renewable Energy, 37*(1), 19–27.

Saxena, S., Srivastava, R. K., & Samaddar, A. B. (2010). Sustainablewaste management issues in India. *Journal of Soil & Water Science, 3*(1), 72–90.

Schott, A. (2001). *Modern solid waste management in practice*. Academic Press.

Sembiring, E., & Nitivattananon, V. (2010). Sustainable solid waste management toward an inclusive society: Integration of the informal sector. *Resources, Conservation and Recycling, 54*(11), 802–809.

Sexton, D., & Spelman, D. (2003). Current best practices and guidelines. *Cardiology Clinics, 21*(2), 273–282. PMID:12874898

Sharholy, M., Ahmad, K., Mahmood, G., & Trivedi, R. C. (2005). Analysis of municipal solid waste management systems in Delhi – a review. *Proc. for the second International Congress of Chemistry and Environment*, 773–777.

Sharholy, M., Ahmad, K., Mahmood, G., & Trivedi, R. C. (2006). Development of prediction models for municipal solid waste generation for Delhi city. *Proceedings of National Conference of Advances in Mechanical Engineering*.

Sharholy, M., Ahmad, K., Mahmood, G., & Trivedi, R. C. (2008). Municipal solid waste management in Indian cities – A review. *Waste Management (New York, N.Y.), 28*, 459–467. PMID:17433664

Sharholy, M., Ahmad, K., Vaishya, R. C., & Gupta, R. D. (2007). Municipal solid waste characteristics and management in Allahabad India. *Journal of Waste Management, 27*, 490–496. PMID:16766176

Siddiqui, T. Z., Siddiqui, F. Z., & Khan, E. (2006). Sustainable development through integrated municipal solid waste management (MSWM) approach — a case study of Aligarh District. *Proceedings of National Conference of Advances in Mechanical Engineering*.

Singh, G. K., Gupta, K., & Chaudhary, S. (2014). Solid Waste Management: Its Sources, Collection, Transportation and Recycling. *International Journal of Environmental Sciences and Development, 5*, 4.

Singh, N., Hui, D., Singh, R., Ahuja, I., Feo, L., & Fraternali, F. (2017). Recycling of plastic solid waste: A state of art review and future applications. *Composites Part B Engineering Journal, 115*, 409–422.

Singh, N., Hui, D., Singh, R., Ahuja, I. P. S., Feo, L., & Fraternali, F. (2016). *Recycling of plastic solid waste: A state of art review and future applications*. Academic Press.

Singh, R. P., Ibrahim, M. H., Esa, N., & Iliyana, M. (2010). Composting of waste from palm oil mill: A sustainable waste management practice. *Reviews in Environmental Science and Biotechnology, 9*, 331–344.

Singh, R. P., Tyagi, V., Allen, T., Ibrahim, M. H., & Kothari, R. (2011). An overview for exploring the possibilities of energy generation from municipal solid waste (MSW) in Indian scenario. *Renewable & Sustainable Energy Reviews, 15*, 4797–4808.

Singhal, S., & Pandey, S. (2001). Solid waste management in India: Status and future directions. *TERI Information Monitoring on Environment Science*, *6*(1), 1–4.

Siracusa, V., Rocculi, P., Romani, S., & Dalla Rosa, M. (2008). Biodegradable polymers for food packaging: A review. *Trends in Food Science & Technology*, *19*(12), 634–643.

Smahi, D., Hammoumi, O. E., & Fekri, A. (2013). Assessment of the impact of the landfill on groundwater quality: A case study of Mediouna Site, Casablanca, Morocco. *Journal of Water Resource and Protection*, *5*, 440–445.

Srivastava, R., Krishna, V., & Sonkar, I. (2014). Characterization and management of municipal solid waste: A case study of Varanasi city, India. *International Journal of Current Research and Academic Reviews*, *2*(8), 10–16.

Stone, R. (1978). Municipal solid wastes and their disposal. *Environmental Health Perspectives*, *27*, 239–244. PMID:738240

Svoboda, J. (2004). *Magnetic Techniques for the Treatment of Materials*. Springer.

Talyan, V., Dahiya, R. P., & Sreekrishnan, T. R. (2008). State of municipal solid waste management in Delhi, the capital of India. *Waste Management (New York, N.Y.)*, *28*, 1276–1287. PMID:17692510

Tan, S. T., Ho, W. S., Hashim, H., Lee, C. T., Taib, M. R., & Ho, C. S. (2015). Energy, economic and environmental (3E) analysis of waste-to-energy (WTE) strategies for municipal solid waste (MSW) management in Malaysia. *Energy Conversion and Management*, *102*, 111–120.

TERI Energy Data Directory & Yearbook (TEDDY). (2010). New Delhi, India: The Energy and Resource Institute.

Teuten, E. L. (2009). Transport and release of chemicals from plastics to the environment and to wildlife. *Philosophical Transactions of the Royal Sociey B*, *364*, 2027–2045. PMID:19528054

Torbert, H. A., Gebhart, D. L., & Busby, R. R. (2011). New municipal solid waste processing technology reduces volume and provides beneficial reuse applications for soil improvement and dust control. In Integrated Waste Management (vol. 1). In Tech.

Tozlu, A., Özahi, E., & Abuşoglu, A. (2016). Waste to energy technologies for municipal solid waste management in Gaziantep. *Renewable & Sustainable Energy Reviews*, *54*, 809–815.

Tsang, C., & Apps, J. (2005). *Underground injection science and technology*. Amsterdam: Elsevier.

Turan, N., Çoruh, S., Akdemir, A., & Ergun, O. (2009). Municipal solid waste management strategies in Turkey. *Waste Management (New York, N.Y.)*, *29*(1), 465–469. PMID:18644708

UNEP (United Nations Environment Programme). (2005). *Solid waste management*. Retrieved from http://www.unep.org/ietc/Portals/136/SWM-Vol1-Part1-Chapters1to3.pdf

Upadhyay, V., Jethoo, A. S., & Poonia, M. P. (2012). Solid waste collection and segregation: A case study of MNIT campus, Jaipur. *International Journal of Engineering and Innovative Technology*, *1*(3), 144–149.

USEPA (United States Environmental Protection Agency). (2002). *Solid waste management: a local challenge with global impacts.* Retrieved from http://www. epa.gov/osw/nonhaz/municipal/pubs/ghg/f02026.pdf

USEPA (United States Environmental Protection Agency). (2014). *Municipal solid waste generation, recycling, and disposal in the United States: Facts and figures for 2012.* Washington, DC: United States Environmental Protection Agency.

VanEwijk, S., & Stegemann, J. A. (2014). Limitations of the WH for achieving absolute reductions in material throughput. *Journal of Cleaner Production.*

Vehlow, J. (1996). Municipal solid waste management in Germany. *Waste Management (New York, N.Y.), 16*(5-6), 367–374.

Velis, C. A., Longhurst, P. J., Drew, G. H., Smith, R., & Pollard, S. J. (2010). Production and quality assurance of solid recovered fuels using mechanical—biological treatment (MBT) of waste: A comprehensive assessment. *Critical Reviews in Environmental Science and Technology, 40*(12), 979–1105.

Wahab, D. A., Hussain, A., Scavino, E., Mustafa, M. M., & Basri, H. (2006). Development of a Prototype Automated Sorting System for Plastic Recycling. *American Journal of Applied Sciences, 3*(7), 1924–1928.

Wangyao, K., Towprayoon, S., Chiemchaisri, C., Gheewala, S., & Nopharatana, A. (2009). Application of the IPCC Waste Model to solid waste disposal sites in tropical countries: Case study of Thailand. *Environmental Monitoring and Assessment, 164*(1-4), 249–261. PMID:19415519

Wilson, D. C., Velis, C., & Cheeseman, C. (2006). Role of informal sector recycling in waste management in developing countries. *Habitat International, 30*(4), 797–808.

World Bank. (1999). *What a waste: solid waste management in Asia.* Retrieved from http://web.mit.edu/urbanupgrading/urbanenvironment/resources/references/pdfs/WhatAWasteAsia.pdf

World Bank. (2012). *What a Waste: Global Review of Solid Waste Management.* Author.

Yadav, I. C., & Devi, N. L. (2009). Studies on municipal solid waste management in Mysore city — a case study. *Report and Opinion, 1*(3), 15–21.

Young, G. C. (2010). *Municipal Solid Waste to Energy Conversion Processes: Economic, Technical and Renewable Comparison.* John Wiley & Sons, Inc.

Yuan, H., & Zhu, N. (2016). Progress in inhibition mechanisms and process control of intermediates and by-products in sewage sludge anaerobic digestion. *Renewable & Sustainable Energy Reviews, 58,* 429–438.

Zaman, A. U. (2010). Comparative study of municipal solid waste treatment technologies using life cycle assessment method. *International Journal of Environmental Science and Technology, 7*(2), 225–234.

Zhang, S., & Herbell, J. D., Gaye, & Haake, B. (2004). Biodegradable organic matter in municipal solid waste incineration bottom ash. *Waste Management (New York, N.Y.), 24,* 673–679. PMID:15288299

Zhu, D., Asnani, P. U., Zurbrugg, C., Anapolsky, S., & Mani, S. (2008). *Improving solid waste management in India. A Source book for Policy Makers and Practitioners, World Bank Institute, WBI Development Studies*. The World Bank.

Zia, H., & Devadas, V. (2008). Urban solid waste management in Kanpur: Opportunities and perspectives. *Habitat International*, *32*(1), 58–73.

Ziadat, A., & Mott, H. (2005). Assessing solid waste recycling opportunities for closed campuses. *Management of Environmental Quality*, *16*(3), 250–256.

KEY TERMS AND DEFINITIONS

Cellulose: Cellulose is the organic compound consisting of hundreds and sometime thousands of Carbon, Hydrogen and Oxygen atoms. Being the main construing, it helps plant to remain stiff and upright.

Gasification: It is the burning of solid fuel in limited air supply for the combustion of the waste.

Incineration: It is the process of burning waste at high temperature and converting waste into ash, flue gas, and heat.

Leachate: Leachate is the highly toxic wastewater released due to the precipitation or decomposition of the waste from the landfill site.

Recycling: It is the process of recovering the product from the waste and reusing it again and again.

Reduce: Reduce means to lessen the volume and size of the waste.

Reuse: It means using of the object or material again and again for the same purpose or for different purpose without altering the form of the product.

Solid Waste: The useless and unwanted products in the solid state derived from the activities of and discarded by society. It is produced either by-product of production processes or arise from the domestic or commercial sector when objects or materials are discarded after use.

Chapter 11
Recovery and Regeneration of Energy From Wastes

Syed Maqbool Geelani
Sher-e-Kashmir University of Agricultural Sciences and Technology, India

Moonisa Aslam Dervash
Sher-e-Kashmir University of Agricultural Sciences and Technology, India

S. J. A. Bhat
Sher-e-Kashmir University of Agricultural Sciences and Technology, India

ABSTRACT

Management of domestic and industrial wastes is of great concern to all sections of society. Huge quantities of solid wastes are generated from cities, industries, agricultural activities, markets, and hotels on a daily basis. Inadequate management of these wastes poses a serious risk to environment. Generation of energy from these wastes could be helpful for proper management of waste. Recovery and generation of energy from wastes isn't only of economic importance but also could be a boon for conserving natural resources.

INTRODUCTION

The energy recovery from waste, energy from waste or waste to energy is used to describe a number of technologies and treatment processes adopted for generating a usable form of energy from discarded or waste materials. The energy generated is usually in the form of electricity, heat and fuels (WEC, 2013). The technologies of waste processing for energy recovery develop a fuel, gas which can be combusted to generate heat or electricity. The material generated can also be converted to transport fuels, synthetic natural gas, or other products (Reinhard, 2015). The fuel replacement or co-combustion involves the opportunities for adapting industrial and power generation to use fossil fuels (coal, oil or gas) and accept a proportion of alternative fuels derived from wastes (Bogner *et al.,* 2007). This facility occurs in power plants, brick works and cement kilns (Ali, 2012). The other indirect energy from waste pathways include the processing of waste to produce combustible Refused derived fuels (RDF), which is achieved using

DOI: 10.4018/978-1-7998-0031-6.ch011

mechanical sorting and processing techniques like screening, crushing and grinding facilities known as Materials recovery facilities (MRF), Mechanical Heat Treatment (MHT) if it involves a thermal pre-treatment, mechanical heat treatment. Mechanical Biological Treatment (MBT, separates organic material for biological processing like anaerobic digestion (EPA, 2013). The other technologies which fall outside the scope of energy recovery includes, thermal combustion of waste for disposal, or flaring of biogas from landfill without energy recovery (DOELWP, 2016).Aerobic conversion of waste to create outputs without energy recovery like composting, bioenergy where purpose grown energy-crops (non-waste materials) are used as feedstock and treatment of contaminated soil or other hazardous waste without energy recovery (EPA, 2017).

The waste hierarchy (EPA, 2017b):

- **Avoidance**: "Practices which prevent the generation of waste all together"
- **Reuse**: "Direct reuse of materials without additional processing"
- **Recycling**: "Using valuable components of waste in other processes"
- **Recovery of Energy**: "Extraction of calorific value to create usable energy"
- **Treatment**: "Reduce volume or change composition to reduce hazard or nuisance"
- **Containment**: "Long-term storage of wastes requiring a high degree of control to prevent contamination"
- **Disposal**: "Deposit of materials, typically into landfill"

The energy from waste is divided in two broad categories:

1. Biological processing of biodegradable waste (anaerobic digestion or fermentation to produce biogas or alcohol).
2. Thermal treatment of Residual waste which include direct combustion, gasification and Pyrolysis (Direct combustion of waste to create heat, which can be used directly or to generate electricity).

The technologies adopted for the generation of energy usually include:

- Combustion for producing heat.
- Gasification, which includes plasma gasification or plasma assisted gasification for producing a combustible syngas
- Pyrolysis for producing syngas, oil or char.

TECHNIQUES FOR WASTE TO ENERGY GENERATION

The classification and methodology of waste to energy generation are discussed as:"

Thermo-Chemical Conversion

It is a thermal treatment process in which the energy content of waste is extracted at high temperatures and utilized. The process depends on the choice of the fuel used.

Table 1. Different types of waste to energy technologies (Funk et al., 2013)

S N	Technology	Waste constituents	Products	Residue
1	"Combustion"	"Mixed residual MSW, mixed C&I and C&D wastes, refuse derived fuels (e.g. Pellets manufactured from residual materials)"	"Heat, electricity, (Bottom ash residues may in some circumstances have value as road base or as an additive in building blocks)"	"Bottom ash, fly ash, air pollution control residues, metals"
2	"Gasification"	"Typically refuse derived fuel (RDF) prepared from mixed residual MSW, mixed C&I and C&D wastes, organic waste".	Heat, electricity, syngas,	"Bottom ash, air pollution control residues"
3	"Pyrolysis"	"Homogenous feedstocks from sorted C&I + C&D waste (eg., wood, tyres), sorted residual MSW (eg., plastics), organic waste".	"Syngas, biochar (when biomass used as feedstock), pyrolysis oil to make liquid fuels".	"Air pollution control residues".
4	"Mechanical Biological Treatment"	"Residual MSW, C&I waste, organic waste".	"Biogas, electricity, refuse derived fuels, separated recyclables (eg., plastics, paper, glass, metals), compost-like material – depending on technology configuration that is adopted".	"Process water, air pollution control residues, inert materials and residual materials that have no economic value"
5	"Anaerobic Digestion"	"Bio-solids, food waste, green waste, crop residues"	"Digestate, compost, heat, electricity, biogas"	"Liquid residues, wastewater, inert and non-compostable material, contaminants" (e.g. plastics)
6	"Fermentation"	"Organic waste high in sugar" (eg., corn, beetroot, sugarcane crop waste)	"Alcohols" (e.g. ethanol for fuel), digestate	"Liquid residues, wastewater"

Incineration

A simple incineration is utilized by way of combined heat and power (CHP) or cogeneration plant technology with mixed waste input (Figure 1). CHP uses a variety of fuels both fossil and renewable based, which can be deployed cost effectively and quickly. This kind of technology is has been employed for many years mostly in large commercial, industrial and institutional applications. This technology is providing highly efficient electricity and process heat to most vital urban centers, largest employers, industries and campuses mostly in United States.

Figure 1. Combined heat and power technology with mixed waste input

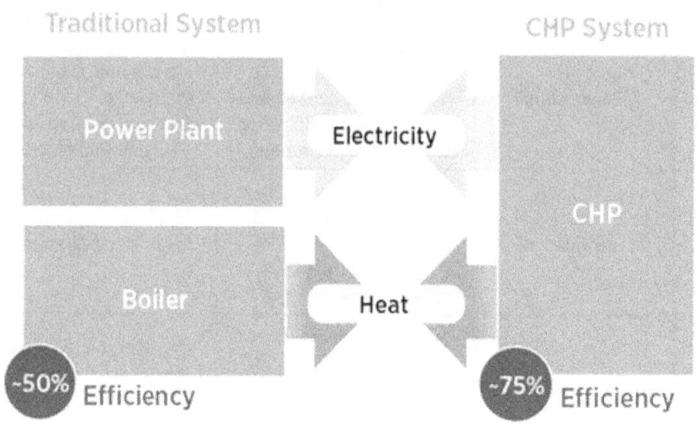

Co-Combustion

In terms of the low cost and low emission levels, co combustion is an attractive alternative to simple coal combustion. Combustion with another fuel (Co-combustion) typically coal and biomass is an alternative which makes thermal properties of the fuel easier in particular the Lower Heating Value (Rechberger, 2011).

Residual Derived Fuel (RDF) Plant

Residual Derived Fuel (RDF) is a kind of energy recovery that after different processes produces as residues fuel (Table 2). The residues are dried by hot air passing or by sun, after separating and selecting suitable residues. The process is followed by screening for separating residues in different sizes, which is followed by size reduction and magnetic separation. The final RDF produced by 3 methods includes Bricks, Pellets and Fluff. The large size materials separated by screening include plastic bags etc. produce Bricks. Reduction and Screening also develops small size materials which produce Fluff. The small residues like plants mix finally produce Pellets (Ahmad Hajinezhad, 2016).The possibility of achieving higher energy contents is the main advantage of Refuse-Derived Fuel (RDF), which can be achieved from different kinds of waste fractions. The high and uniform energy content makes it attractive for energy production, both by mono-combustion and co-combustion with MSW or coal (Rotter, 2011).

Thermal Gasification

The process of Gasification was originally developed in the 1800s and was also used to make town gas for lighting and cooking. The conversion of organic or fossil fuel based carbonaceous materials into Hydrogen, Carbon monoxide and carbon is called Gasification. The technologies of gasification process are based on the type of reactor, contact of fuel and gasification media along with physical properties. The three main technologies adopted are fixed bed, fluidized bed and entrained flow.

Table 2. Specifications of residual derived fuel for energy recovery

S.No.	RDF type	Specifications
1	"RDF_1"	"Raw waste or with minimal processing"
2	"RDF_2 (Coarse RDF) or C-RDF"	"Waste processed into coarse particles with no Separation metals in a way that 95% by weight of 6-inch square mesh sieve pass"
3	"RDF_3 (fluff RDF)"	"Processed fuel derived from waste by separating the metals, glass, and other inorganic materials. The material in a way that a 2 inch square mesh sieve passes".
4	"RDF_4 (Power RDF or dust RDF)"	"Combustible waste components in powder form and 95% by weight of 0.035 inch square mesh sieve pass".
5	"RDF–5 (Densified RDF or D-RDF)"	"Flammable waste extruded sections (compressed) in the form of pellets, cubes, briquettes, or similar forms. Due to the numerous advantages of portability and storage, and the ability to coordinate with a variety of combustion systems in developing".
6	"RDF-6"	"The combustible waste in liquid form is processed".
7	"RDF-7"	"The combustible waste gas is processed to form".

Nowadays, biomass gasification is performed in small, medium and large-scale gasification units. Thermochemical Gasification produces a fuel gas at high temperature, which after cleaning provides a good environmental performance and high flexibility in applications. The process converts biomass (solid biomass, wastes) into a combustible gas also used for different purposes. Cellulosic biomass like wood chips, pellets or wood powder or agricultural byproducts like straw or husks acts as feedstock for gasification. The gas produced is called producer gas or synthetic gas (syngas). The gasification of the feedstock takes place at 700°–1600°C in presence of the gasification medium (Figures 2-5). The media

Figure 2. Schematic of a small-scale fixed bed updraft and downdraft gasifier (WBA, 2015)

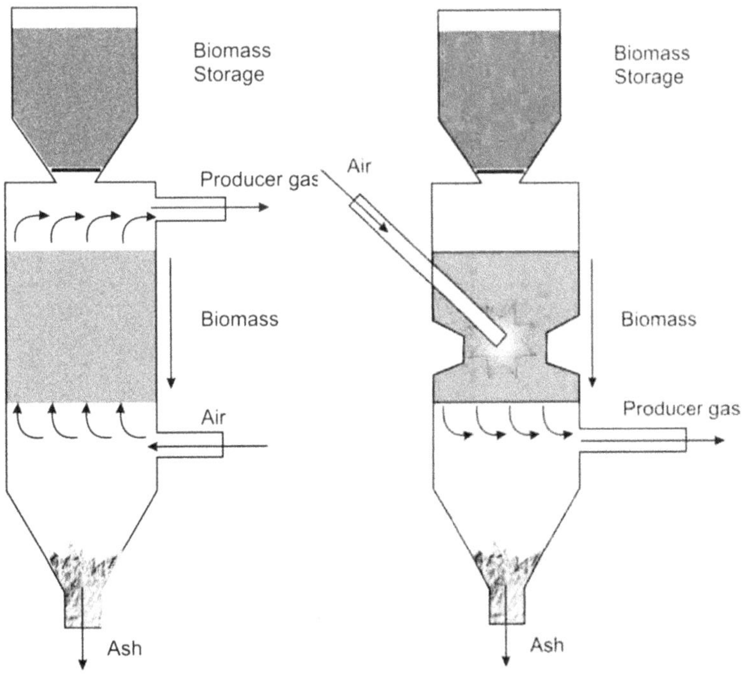

Figure 3. Schematic of a bubbling fluidized-bed gasifier (WBA, 2015)

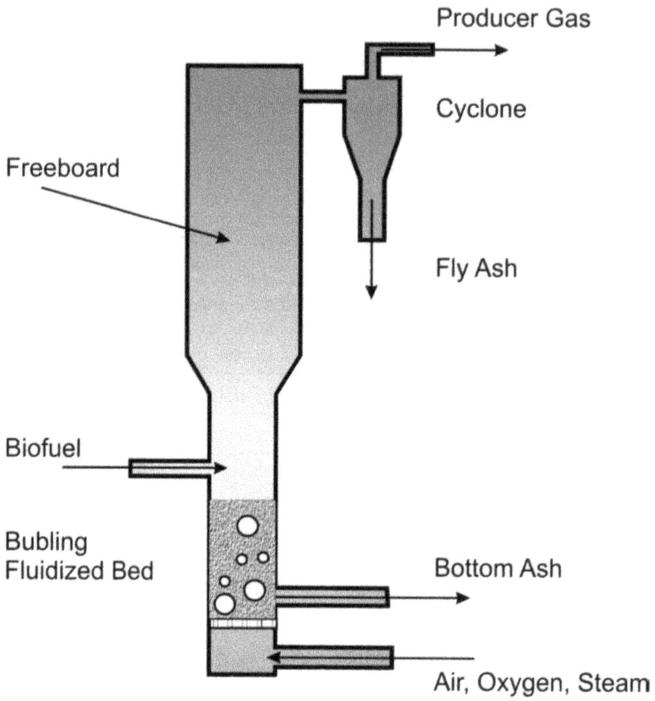

Figure 4. Schematic of a Circulating fluidized-bed gasifier (WBA, 2015)

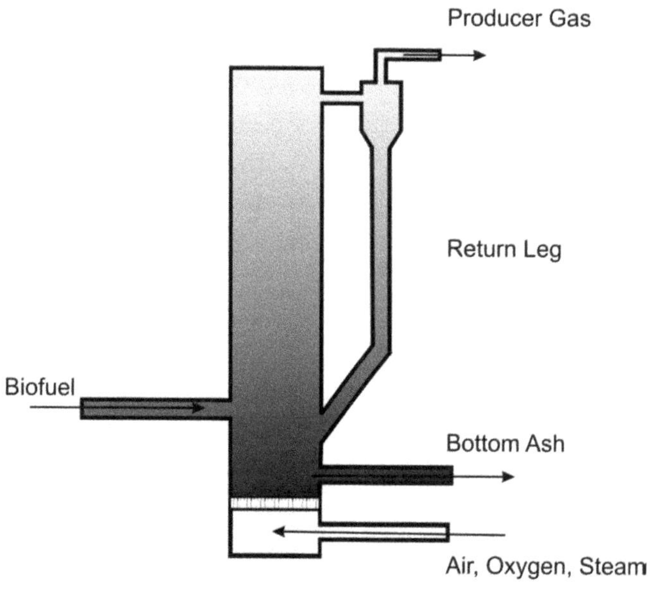

Figure 5. Schematic of two entrained-flow gasifiers operating under slagging conditions (WBA, 2015)

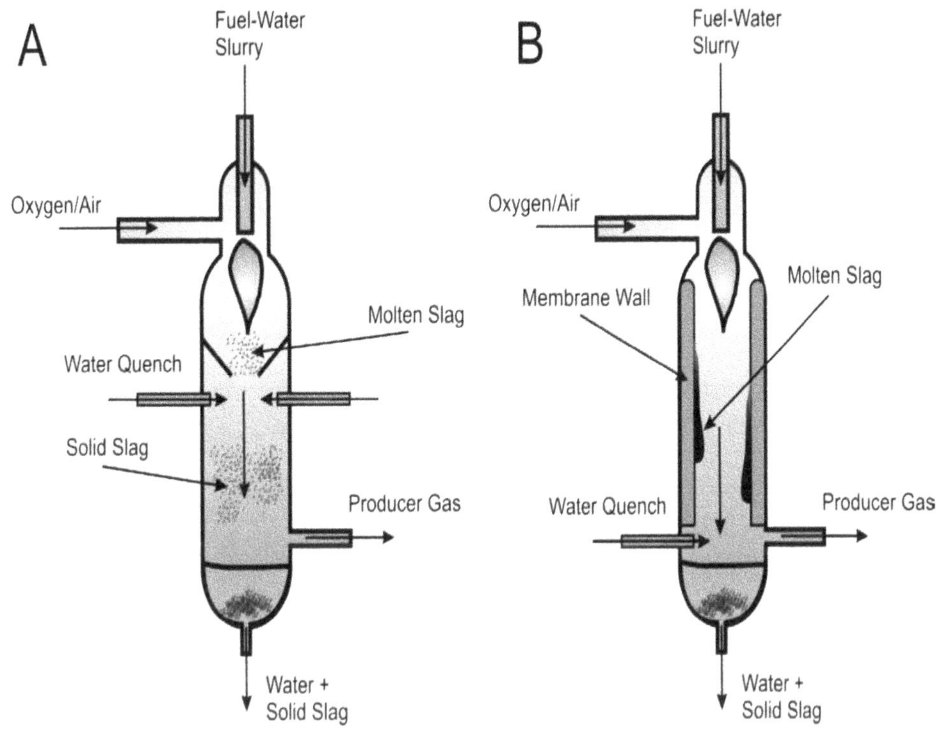

used in gasification are air, oxygen, steam or a mixture of these. Thermal gasification process is an another process, where a carbon rich material such as biomass (or coal) is broken down into a gas consisting of consisting of Carbon monoxide (CO) and Hydrogen (H_2) with Carbon dioxide (CO_2), Methane (CH_4) and other gaseous hydrocarbons along with contaminants like tars, char, ash etc. Thermal gasification is able to convert carbonaceous materials into an energy rich gas (WBA, 2015). This process is also having favorable characteristics of overall higher efficiency, better quality of gaseous outputs and of solid residues and potentially lower facility costs (Astrup, 2011). The gasification with proper future technology developments is a valuable alternative to combustion of waste (Figure 6).

Figure 6. Gasification process including four stage process (WBA, 2015)

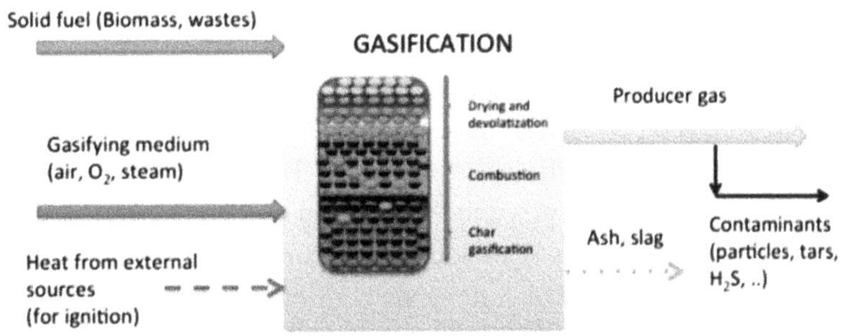

Table 3.

S.NO.	Phases	Process
1	"Feedstock supply"	"The feedstock are selected for optimum size, composition and quality. The biochemical processing are essential to cost effective, high-yield operations".
2	"Pretreatment"	"The Biomass after heating is combined with an acid or base for breaking the tough, fibrous cell walls down and make the cellulose and hemicellulose easier to hydrolyze".
3	"Hydrolysis"	"The Enzymes (or other catalysts) facilitate the sugars within cellulose and hemicellulose in the pretreated material to be separated and released over a period of several days".
4	"Biological conversion"	"The Microorganisms are added, which then use the sugars for generating other molecules suitable for use as fuels or building-block chemicals".
5	"Chemical conversion"	"The sugars are alternatively converted to fuels or an entire suite of other useful products using chemical catalysis".
6	"Product recovery"	"In this process, Products are separated from water, solvents, and any residual solids".
7	"Product Distribution"	"The Fuels are transported to blending facilities and other products and intermediates are sent to traditional refineries or processing facilities for usage in a diverse slate of consumer products".
8	"Heat and Power"	"The remaining solids are composed primarily of lignin, which are burned for heat and power". (USDOE, 2013)

BIO-CHEMICAL CONVERSION

Biochemical conversion involves the breaking down of the biomass to make carbohydrates available for processing into sugars, which can be converted into biofuels and bio products by the use of the biocatalyst and microorganisms. The potential fuel bend stocks include Renewable chemical products, Renewable diesel, Renewable gasoline and Ethanol and other alcohols. The usage of biocatalysts such as enzymes in addition to heat and other chemicals for converting a carbohydrate portion of biomass (hemicellulose and cellulose) into an intermediate sugar stream, which can act as intermediate building blocks that can be fermented or chemically catalyzed into a range of advanced biofuels and value added chemicals (USDOE, 2013). The process of Bio conversion involves the following essential steps:

BIO-ETHANOL PRODUCTION

Bioethanol is produced by treating a range of certain organic fractions of waste. Different technologies are involved, which includes separate stages for fermentation (using microorganisms), hydrolysis (by enzymatic treatment) and distillation. The adoption of these technologies are can also be used to obtain hydrogen, which is a potential and promising energy carrier (Karakashev & Angelidaki, 2011). Bioethanol is widely used as biofuel for transportation worldwide (Balat, 2011).The biochemical production involves fermentation, distillation, hydrolysis and dehydration (Scott et al. 2013). The biomass degradation begins during the hydrolysis of complex molecules into simple compounds readily available for the fermentation step. The natural pathway which depends both on operating conditions and raw materials is carried by microorganisms that ferment sugars into lactic acid, alcohol or other end products. The obtained bioethanol from fermentation process requires further treatment. The fractional distillation is typically implemented to separate ethanol from water based on their different volatilities (Alessandra & Vincenzo, 2015).

DARK FERMENTATION AND PHOTO-FERMENTATION PRODUCING BIO-HYDROGEN

Dark fermentation and Photo-fermentation techniques involves the conversion of organic substrates into hydrogen in the presence or absence of light, which becomes possible only due to the processing and functioning of diverse groups of bacteria's. This technology is a valuable option for the wastewater treatment processes (Angenent et al., 2004).Bio-hydrogen and bio-methane plays an important role for future energy economy as clean, CO_2 neutral and environmental friendly energy (Figure 7). The most capable and eco-friendly method seems to be dark fermentation from organic wastes, which combines the hydrogen generation with waste treatment (Benemann, 1996). The anaerobic digestion process includes methanogenesis and hydrolysis/acidogenesis. Hydrogen gas and organic acids are produced by Hydrolysis and acidogenesis, which can be further used to produce methane in methanogenesis. The step of hydrogen production requires 1-2 days hydraulic retention time (HRT) and step of methane production requires longer HRT (12-20 days).When methane is not further used and hydrogen gas is not harvested, it is called one-stage fermentation process. Otherwise it is called two-stage fermentation process.

Figure 7. Two-stage process for hydrogen and CH_4 production

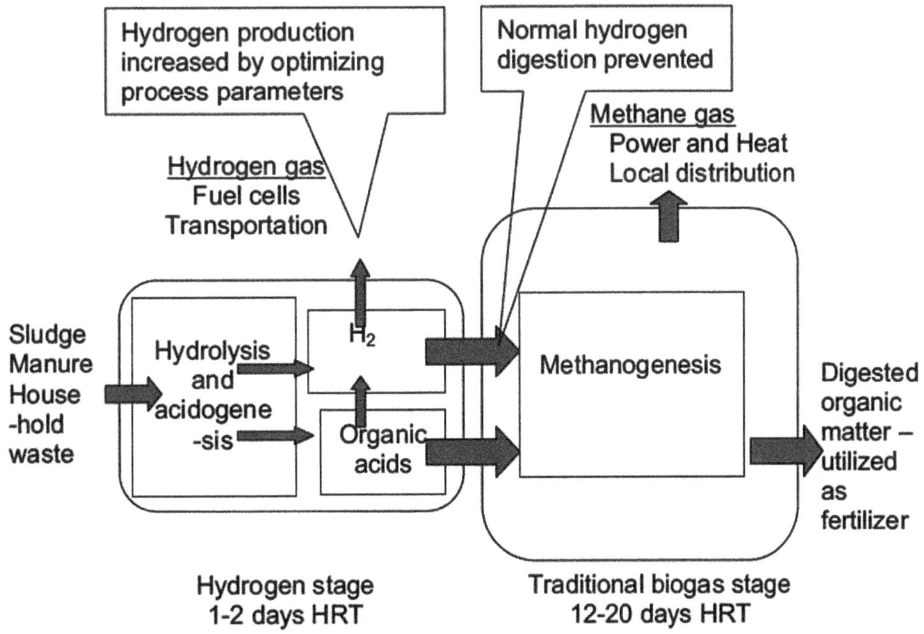

BIOGAS PRODUCTION FROM ANAEROBIC DIGESTION

A biological conversion process which is carried out in the absence of an electron acceptor like oxygen is called anaerobic digestion (Liu, 2008). This technology has been traditionally viewed as a symbol of "energy from waste" and provides varied benefits apart from valuable renewable energy from biogas like waste treatment and reduce diseases, odors and pollution (Angelidaki 2002).The main products of the process are an effluent (or digest) residue and energy rich biogas. The different group of microorganisms drives the required chemical reactions in the entire conversion chain broken into several stages. The biogas obtained can be used to generate power, heat and produce biofuels. The digest, based on the composition can be utilized in different ways. This kind of process is utilized in several technologies developed throughout the year and are still considered as immature and non-economical in comparison to other waste to energy technologies.

BIOGAS PRODUCTION FROM LANDFILLS

The biogas can be extracted directly from landfill site, Other than in an anaerobic digester (Tchobanoglous *et al.*, 2002). This can be done by constructing appropriate collecting systems for the produced gas. The biogas in landfills is produced by means of complex conversion processes, which usually includes different phases like Acid phase, Transition phase, Initial adjustment, Maturation phase and Methane fermentation (Zaman, 2009).The organics in a Municipal solid waste (MSW) contains a significant fraction (30-50%). The organics are useful resource, when it is used for power generation. The exhaustion of the conventional energy sources has necessitated the search for alternative energy source (Bhattacharyya, 2007). Biogas and leachate are usually generated from present municipal solid waste landfill sites. Inorganic contaminants like mercury are also reported to be present in landfill gas. The origin of source like amount of hydrogen sulphide in the landfill gas and in the farm biogas varies as per the composition of biogas, while as hydrogen sulphide is not detected from the sewage digester (Ras, 2007). The sewage digesters usually contain 55- 65% methane, 35-45% CO_2 and less than 1% nitrogen. The organic waste digesters biogas contain methane (60-70%), CO_2 (30-40%), Nitrogen less than 1%. However, methane content in landfill varies from 45-55%, 30- 40% CO_2, 5-15% Nitrogen (Ras, 2007). The solid residuals of fermentation might be also reused as fertilizers (Mohammad et al. 2011). The production of Biogas deals with the problems of pollution and waste treatment.

Microbial Fuel Cell

Microbial fuel cells are directly used to generate electricity from the oxidation of dissolved organic matter. The conversion of the chemical energy content of organic matter to produce electricity is carried out in a microbial fuel cell. This is done by the catalytic reaction of microorganisms and bacteria present in nature. This technology could be used for power generation in combination with a waste water treatment facility (Min *et al.* 2005).

CHEMICAL CONVERSION (ESTERIFICATION)

The chemical process of esterification occurs when an alcohol and an acid react to form anester. If applying this process to WtE treatment, it is possible to obtain various types of biofuels from waste (Nic *et al.*, 2006).In Biochemical conversion, fermentation and anaerobic digestions are used (McKendry, 2002).

Fermentation

This method is commercially utilized on a large scale in various countries to produce ethanol from sugar crops (sugar beet, sugar cane) and starch crops (wheat, maize etc.). The starch is converted to sugars by enzymes then converting the sugars to ethanol with yeast. The residue of fermentation are used in cattle feed in sugar cane and the bagasse is used as fuel for boilers or subjected to further gasification (Coombs, 1996).

Anaerobic Digestion

The process of the conversion of organic material directly into gas (biogas), which is a mixture of methane and carbon dioxide with small quantities of hydrogen sulphide (EU,1999). The production of gas from biomass is carried in anaerobic environment by bacteria. The treatment of the high moisture content organic wastes has proved an efficient technology commercially. The higher quality natural gas of biogas is made by the removal of CO_2. The biogas or the quality gas can be used directly to produce electricity.

WASTE TO ENERGY OPPORTUNITIES

The generation of energy from waste offers a range of opportunities worldwide. It could help us to make better use of our residual waste by generating a valuable product - energy. It could help us to tackle with the problems of water, soil and air problems. Waste to energy technologies offer opportunities to extract more value from residual wastes that cannot be reused or recycled and would otherwise be landfilling. It could help to reduce greenhouse gas emissions and achieve its emissions reduction. Waste to energy facilities could help reduce greenhouse gas emissions from the waste sector by diverting organic waste from landfills. They could also reduce emissions from the energy sector by displacing some energy derived from fossil fuels with energy derived from organic sources. It could help to achieve its renewable energy targets. The waste sector generates greenhouse gas emissions from the disposal of waste to land and through the collection, sorting, transporting, processing and reprocessing of waste and recovered resources. The greenhouse gas emissions from waste could be reduced by diverting more organic waste from landfill. The investment in waste to energy projects can surely increase the economic development and jobs all over the world.

CONCLUSION

The waste to energy is emphasizes on the technologies and treatment processes used to generate energy usually electricity, heat and fuels from discarded or waste materials. These technologies encourage the use of alternative fuels derived from wastes and reduce pressure on existing conventional resources like coal, oil and gas. The main process involved is to concert waste to Refused derived fuels (RDF), which further rely on techniques like screening, crushing and grinding facilities known as Materials recovery facilities (MRF) and Mechanical Heat Treatment (MHT). Aerobic conversion processes like composting and other bioenergy forms are very much useful in agriculture and to decontaminate the degraded soils and sites.

REFERENCES

Angelidaki I. & Batstone D.J. (2011). Anaerobic digestion: process. *Solid Waste Technology & Management.*

Angenent, L. T., Karim, K., Al-Dahhan, M. H., Wrenn, B. A., & Domíguez-Espinosa, R. (2004). Production of bioenergy and biochemicals from industrial and agricultural wastewater. *Trends in Biotechnology, 22*(9), 477–485. doi:10.1016/j.tibtech.2004.07.001 PMID:15331229

Asgari, Safavi, & Mortazaeinezahad. (2011). Landfill Biogas production process. *International Conference on Food Engineering and Biotechnology, 9*, 208-212.

Astrup T. (2011). Pyrolysis and Gasification. *Solid Waste Technology & Management.*

Balat, M. (2011). Production of bioethanol from lignocellulosic materials via the biochemical pathway: A review. *Energy Conversion and Management, 52*(2), 858–875. doi:10.1016/j.enconman.2010.08.013

Bhattacharyya, J. K., Kumar, S., & Devotta, S. (2007). Studies on acidification in two-phase biomethanation process of municipal solid waste. *Waste Management (New York, N.Y.), 28*(1), 164–169. doi:10.1016/j.wasman.2006.11.014 PMID:17276666

Bogner, J., Abdelrafie Ahmed, M., Diaz, C., Faaij, A., Gao, Q., Hashimoto, S., ... Zhang, T. (2007). Waste Management. In *Climate Change Mitigation. Contribution of Working Group III to the Fourth Assessment Report of the Intergovernmental Panel on Climate Change.* Cambridge, UK: Cambridge University Press.

CEFC. (2015). *Clean Energy Finance Corporation: The Australian bioenergy and energy from waste market November 2015.* Sydney: CEFC.

Cesaro, A., & Belgiorno, V. (2015). Combined Biogas and Bioethanol Production: Opportunities and Challenges for Industrial Application. *Energies, 8*(8), 8121–8144. doi:10.3390/en8088121

Coombs, J. (1996). *Bioconversion Assessment Study.* CPL Scientific Ltd.

Department of Environment, Land, Water, and Planning. (2016). *Victoria's Climate Change Framework.* Retrieved from:https://www.climatechange.vic.gov.au/__data/assets/pdf_file/0021/552 54/DELWPClimateChange_Framework.pdf

Department of Environment and Energy. (2015). *Australian National Greenhouse Gas Accounts – State and Territory Greenhouse Gas Inventories: Table 3*. Author.

Department of Environment and Energy. (2016). *Quarterly Update of Australia's National Greenhouse Gas Inventory: June 2016*. Commonwealth of Australia.

EU. (1999). Biomass Conversion Technologies. EUR 18029 EN.

Funk, K., Milfold, J., & Simpkins, T. (2013). *Technology for Site-Specific Optimization of Renewable Energy Options*. Technical Report. NREL/TP-6A50-52829. Retrieved from http://www.osti.gov/bridge

Hajinezhad, A., Halimehjani, E. Z., & Tahani, M. (2016). Utilization of Refuse-Derived Fuel (RDF) from Urban Waste as an Alternative Fuel for Cement Factory: A Case Study. *International Journal of Renewable Energy Research.*, *6*(2), 702–714.

Hasanbeigi, A., Lu, H., Williams, C., & Price, L. (2012). International Best Practices for Pre-Processing and Co Processing Municipal Solid Waste and Sewage Sludge in the Cement Industry. China Energy Group Energy Analysis and Environmental Impacts Department Environmental Energy Technologies Division Lawrence Berkeley National Laboratory. *LBNL, 5581E*, 1–115.

Karakashev D. & Angelidaki I. (2011). Emerging biological technologies: biofuels and biochemicals. *Solid Waste Technology & Management.*

Liu, D. (2008). *Bio-hydrogen production by dark fermentation from organic wastes and residues. Kgs.* Lyngby: DTU Environment.

McKendry, P. (2002). Energy production from biomass (part 2): Conversion technologies. *Bioresource Technology*, *83*(1), 47–54. doi:10.1016/S0960-8524(01)00119-5 PMID:12058830

Min, B., Cheng, S., & Logan, B. E. (2005). Electricity generation using membrane and salt bridge microbial fuel cells. *Water Research*, *39*(9), 1675–1686. doi:10.1016/j.watres.2005.02.002 PMID:15899266

Nic, M., Jirat, J., & Kosata, B. (Eds.). (2006). esters. IUPAC Compendium of Chemical Terminology. doi:10.1351/goldbook.E02219

Ras, A. V., & Rintala, J. (2007). Trace compounds of biogas from different biogas production plants. *Journal of Energy*, *32*(8), 1375–1380. doi:10.1016/j.energy.2006.10.018

Rauch, Hrbek, & Hofbauer. (2013). Biomass gasification for synthesis gas production and applications of the syngas. *WIREs Energy Environ.* doi:10.1002/wene.97

Rechberger, H. (2011). *Incineration: Co-combustion in Solid waste technology & management*. Academic Press.

World Energy Resources. (2013). *Waste to Energy*. World Energy Council.

Rotter, S. (2011). *Incineration: RDF and SRF - Solid waste from fuels in Solid waste technology & management*. Academic Press.

Scott, F., Venturini, F., Aroca, G., & Conejeros, R. (2013). Selection of process alternatives for lignocellulosic bioethanol production using a MILP approach. *Bioresource Technology*, *148*, 525–534. doi:10.1016/j.biortech.2013.09.008 PMID:24080291

Tchobanoglous, G., & Kreith, F. (2002). *Handbook of Solid Waste Management* (2nd ed.). McGraw-Hill.

Tchobanoglous, G., & Kreith, F. (2002). *Handbook of Solid Waste Management* (2nd ed.). McGraw-Hill.

The State of Victoria Department of Environment, Land, Water, and Planning. (2017). *Turning waste into energy*. Retrieved from http://creativecommons.org/licenses/by/4.0/

United States Department of Energy. (2013). *Energy Efficiency and Renewable Energy*. DOE/EE-0948. Bioenergy Energy Gov.

Victoria, E. P. A. (2013). *Energy from waste*. EPA Victoria. Retrieved from http://www.sustainability.vic.gov.au/

Victoria E. P. A. (2017). *Environment Protection Authority Victoria (2017b), Your environment – Waste webpage*. EPA Victoria. Retrieved from http://www.epa.vic.gov.au/your-environment/waste

Victoria, S. (2017). *Statewide Waste and Resource Recovery Infrastructure plan Victoria 2017-2046 (amendment consultation draft)*. Sustainability Victoria.

WEC. (2013). *World Energy Resources: Waste to Energy*. Retrieved from https://www.worldenergy.org/wp

World Bioenergy Association (WBA). (2015). Thermochemical Gasification of Biomass. *Holländargatan, 17*, 1-8.

World Bioenergy Association (WBA). (2015). Thermochemical gasification of Biomass. *Holländargatan, 17*. Retrieved from www.worldbioenergy.org

Zaman, A. U. (2009). Life Cycle Environmental Assessment of Municipal Solid Waste to Energy Technologies. EESI, School of Architecture and the Built Environment, KTH Royal Institute of Technology.

KEY TERMS AND DEFINITIONS

AD: Anaerobic digestion.
BioSNG: Synthetic natural gas/alternative name for biomethane.
C&D: Commercial and domestic.
C&I: Commercial and industrial.
CEWEP: Confederation of European Waste-to-Energy Plants.
CHP: Combined heat and power.
DECC: Department of Energy and Climate Change.
DEFRA: Department of Environment, Food, and Rural Affairs.
DS: Dry solids.
EfW: Energy from waste.
EMR: Electricity market reform. After 2017 this will result in new form of contract for low carbon electricity, such as renewables.
EPA: Environment Protection Authority.
FIT: Feed in tariff.
GGCS: Green gas certification scheme.
I&CW: Industrial and commercial waste.
kWh Kilowatt Hour: A unit of electricity.
MBT: Mechanical biological treatment.
MHT: Mechanical heat treatment.
MSW: Municipal solid waste.
MWh: Megawatt hour – 1000 units of electricity.
NPPF: National planning policy framework.
RDF: Refuse derived fuels. A fuel made from Municipal Solid Waste, also called SRF.
REA: Renewable Energy Association.
RED: Renewable Energy Directive (2009/28/EC).
RO: Renewables obligation.
ROC: Renewable obligation certificate.
RTFO: Renewable transport fuel obligation.
SRF: Solid recovered fuel.
TS: Total solids.
TVS: Total volatile solids.

Chapter 12
Impacts of Food Industrial Wastes on Soil and Its Utilization as Novel Approach for Value Addition

Nowsheeba Rashid
Amity University, India

Ifra Ashraf
Skuast-K, India

Shazia Ramzan
University of Kashmir, India

ABSTRACT

Among the various agro-industries, food processing industries are the second prime generator of wastes after domestic sewage. In the current epoch of the rapid budding world, the wastes are mounting, which robustly sway the health of ecosystems and eventually the human population. For that reason, each agro-industrial sector has critical stipulation toward the secure utilization of agro-materials all the way through recycling of wastes. A crude disposal and littering of these waste materials frequently signifies a problem that is additionally provoked by different legal restrictions. Inadequate management of these solid waste constituents could lead to drastic change in physico-chemical properties of soils. The waste product, which is discarded into the environment, is loaded with valuable compounds. They are new, innate, and monetary sources of colorants, protein, dietary fiber, flavoring, antimicrobials, and antioxidants, which can be utilized in the food industry as a basis of natural food additives.

DOI: 10.4018/978-1-7998-0031-6.ch012

INTRODUCTION

India is a largely populated country which exists as the major reason for massive waste generation created frequently out of domestic & industrial actions which includes removal of peel followed by cutting of raw fruits and vegetables former to processing, eating and cooking (William, 2005). FAO revealed that every year, about one-third of all the food produced for the purpose of human consumption globally is lost or wasted. Large amounts of food processing by-product wastes are generated all over the world. It is estimated that about 5 million tons of sugar beet pulp and 3.5 million tons of brewers grain and almost half a million tons of onion peeling waste are created annually (Awarenet, 2004).This food wastage generation predicts a huge missed opportunity to enhance global food security, but also to alleviate environmental impacts and exhaustive resource use from food chains. According to the various quantitative food losses and waste estimations globally per year are roughly 40-50% for root crops, fruits and vegetables, 20% for oil seeds, meat and dairy plus 35% for fish and 30% for cereals (Gowe, 2015).

At the present epoch, ample quantity of waste generated marks food industry. As per the latest research carried out by FAO, almost 1.3 billion tons of food has been exhausted globally per year, which symbolizes nearly one third of the overall production of food industry (Gustafssonet al., 2013). Referring to the individual supplement fruit processing industries contribute more than 0.5 billion tons of waste. Thus, providing globally the accessibility of feed stock and encouraging the researchers, scientists and other authorities to exercise comprehensive studies on the various value added potential of fruit processing waste (FPW). On the other hand, vegetables are essential but uneconomical i.e. they produce ample waste concentration about 25%- 30% of inedible products (Ajila et al., 2010). Around twenty peculiar kinds of plants are generally developed for vegetables in the United States (US). Including this in each State these plants are grown on commercial basis out of which maximum population resides in New York, Texas, California, and Wisconsin. The profits from them to cultivars approach approximately to 300 million dollars annually despite of the fact not more than 20 to 30% of the crop is consumed. In this perspective, the utilization of the whole plant tissue could have cost-effective benefits to cultivars and a positive brunt on the environment, leading to a superior multiplicity of products (Cerezal & Duarte, 2005). Out of the total wastes 4 million tons of them are generally leaves. Several wastes are left as such on the soil to be plowed underneath. Few of them are fed, some are discarded in dumps and some are a simple nuisance a small portion is synthetically dehydrated for feed. The most prominent component of the wastes is water which accounts nearly about 75 to 90% (Willaman & Eskew, 1943). On a comparison fruit processing waste are originated to be selective and concentrated in nature as compared to other biomass derived waste. Besides, the greatest contribution provided is the utilization of peels, pomace and seed fractions as an excellent feedstock for recovery of bioactive compounds which include flavonoids, lipids, dietary fibers, pectin, etc (Kowalska et al., 2017; Banerjee et al., 2017).

A novel bio-refinery method would aim to manufacture a wider variety of important chemicals from fruit and vegetable processing waste. The wastes from bulk of the withdrawal processes may supplementary be used as recycle sources for creation of bio-fuels. These all benefits will open up as a scope for future utilization of fruit and vegetable waste for therapeutic and nutraceutical purpose as well as a great source for value addition of the end products.

According to the data acquired on global trend of fruit and vegetable production, the total of residues with prospective consumption after processing has been anticipated in millions tons every year. This demands the use of different forms of energy, water and other factors providing a by-product potential as the cardinal significance. This comes into being due to the presence of biocomponents, which may

Table 1. Percentage of food wastes and by-products in fruit and vegetable production

Production process	%age of wastes and by-products
White wine production Crow	20-30
Red wine production	20-30
Fruit and vegetable juice production	30-50
Fruit and vegetable processing and preservation	5-30
Vegetable oil production	40-70
Corn starch production	41-43
Potato starch production	80
Wheat starch production	50
Sugar production from sugar beet	85

Source: (Sadh et al. 2018)

be utilized for novel food production. This demands an appropriate measure to convert by-products into value added products for the reason of their calculus natural components.

To conquer the existing environmental condition, scientific communities have been exploiting industrial as well as agricultural wastes and effluents via process of recycling and clean technology, by modified and integrated utilization of waste or merely returning to the place of their origin and nature. Over the years pollution and organization of huge quantities of wastes generated by diverse industrial activities have been main problems practiced by developing countries. More demanding is the precarious disposal of these wastes into the ambient surroundings and biodegradation tribulations linked with it. These have resulted to substitution means of reclaiming the surroundings through bioremediation, phyto-remediation and air contamination control. Bioremediation involves the employment of microorganisms to break complex materials into simpler end products. These micro-organisms have advanced a host of enzymes that support in biodegradation of natural products contained by the ecosystem. This microbial clean-up (Bioremediation) method removes a number of supplementary destructive pollutants and conceivably is the most environmentally safe process used nowadays (Adriaens & Hickey, 1993; Beg et al, 2003; Van Hamme et al, 2003).

An additional source of industrial solid wastes is produced from pulp and paper mechanized plants. Pulp and paper are feigned from lignocellulosic raw materials from wood, recycled paper and agricultural residues. These wastes are typically disposed via landfill and incineration processes (Karthikeyan & Balasubramanian, 2010). These methods of dumping could be destructive to ecosystem and cause diseases to man. Many studies have been reported on the management of pulp and paper mill effluents via biological method like conventional aerobic, anaerobic action and use of white-rot fungi in the conduct (Wong et al, 2006; Wu et al, 2005; Dalentoft & Thulin, 1997). Recently, mushrooms are considered as the most advantageous and environment-friendly technique for recycling of the vast lignocellulosic waste substrates (Gume et al, 2013; Rai & Ahlawat, 2002) Figure1 depicts the illustration of recycled value-added applications of agro-industrial wastes. This chapter reviews numerous kinds of agro-industrial wastes and their principle recycled products which can be utilized in the factual world (Yusuf, 2017)

Figure 1. Schematic representation of recycled value-added applications of agro-industrial wastes

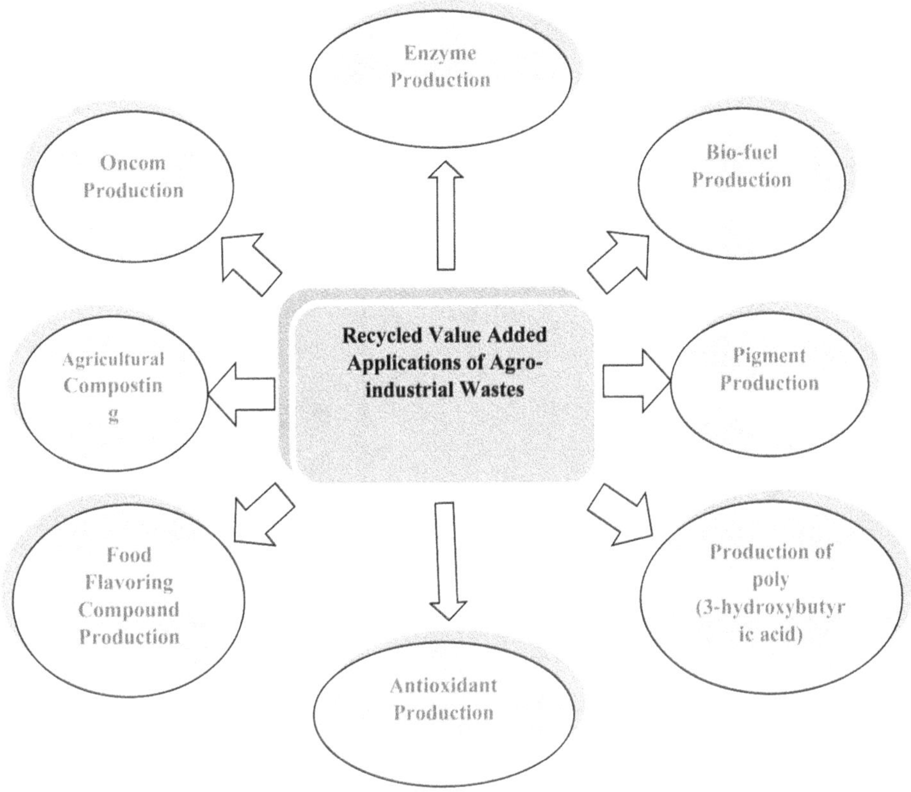

AGRO-INDUSTRIAL WASTE MATERIALS

Nowadays, one of the major sources of pollution is mostly organic waste from agro-industries. As a known fact regarding wastes mainly originating out of food industries comprise waste from pruning, manure, livestock slurry, and crop remains (Figure 2). Also, industries produce organic wastes, which constitute the by-products from agri-food industry (Table 2) like sludge from wool, coffee dregs, degummed fruits and legumes, bagasse, milk serum, cellulose, etc (Figure 3). These organic wastes goes on multiplying day by day and well thought-out to have damaging impacts on the overall environment. In this view, a number of countries prepared certain laws and rules to prevent on description of ecological concerns. Therefore, norm implies to the waste and includes the main explanations and ethics that administer waste management, highlight that waste appraisal and exclusion must be executed without generating risks for soil, water, air, or the fauna and flora to reduce such harmful effects and legalize the use of natural wastes as fertilizers in agriculture (Directive, 2008; Venglovsky et al.,2006).

In order to decrease the industrial pollution, a novel and clean technology can be executed to curtail organic waste via recycling. The clean technology assessment is a successful and efficient procedure that comprises five steps given below:

- Planning and organization
- Pre-assessment

Figure 2. Representation of agro industrial wastes (Yusuf, 2017)

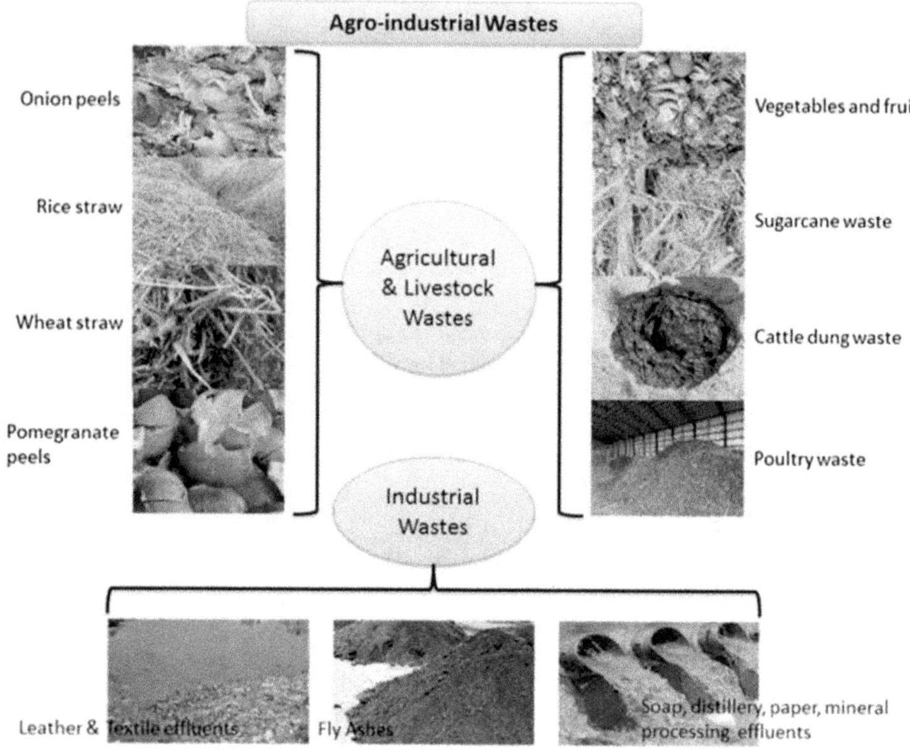

Table 2. Composition of fruit industrial waste

Fruit industrial waste	Chemical Composition (%w/w)								
	Cellulose	Hemi-cellulose	Lignin	Ash	Total solids	Moisture	Total carbon	Total nitrogen	References
Potato peel waste	2.2%	-	-	7.7%	-	9.89	1.3%	-	Al-Weshahy and Rao (2012)
Orange Peel	9.21%	10.5%	0.84%	3.5%	-	11.86	-	-	Rivas et al. (2008)
Coffee skin	23.77 (gg/100g)	16.68%	28.58 (g/100g)	5.36 (g/100g)	-	-	C/N 14.41	-	Ballesteros et al.(2014)
Pineapple Peel	18.11	-	1.37	-	93.6	91	40.8	0.99	Paepatung et al. (2009)

Figure 3. Agro-industrial wastes and their types

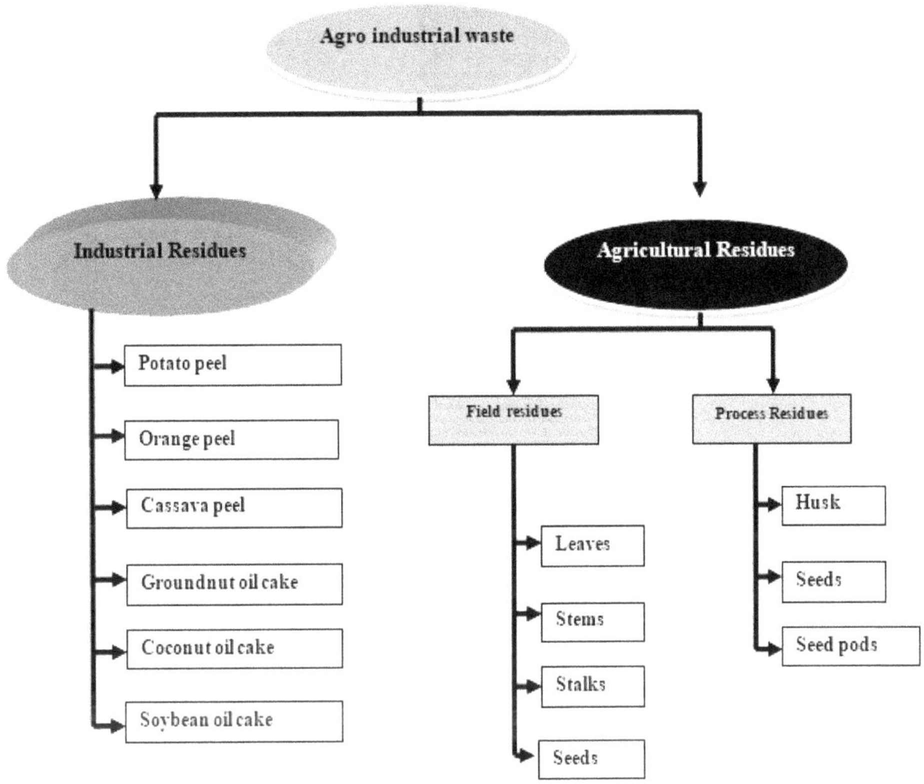

- Assessment
- Feasibility study and
- Implementation

VALUE-ADDED APPLICATIONS OF RECYCLED AGRO-INDUSTRIAL WASTES

Production of Bio Fuels

Bio-fuels continue to be important because of the fact that they are used as an alternative for fossil fuels. It is revealed from the previous studies that the bio-fuels are fabricated from affirmative agro-industrial left over's like sugar beet waste, rice straw, corn stalks,(Figure 4) sweet potato waste, potato waste, sawdust, sugarcane and bagasse (Duhan et al., 2013; Kumar et al., 2014). It is found that in the year 2011, bio-ethanol production increased all over the world as much as 85 billion liters of bio-ethanol are reported to be produced (Avci et al., 2013; Saini et al., 2014). As the main concern nowadays is to reduce industrial waste and at the same time preserving our soil and environment from these pollution calamities. Hence, it is elucidated that with the help of agricultural residues, the deforestation can be decreased thus resulting in less dependence on forest woody biomass. Despite the fact that agro-industrial

Figure 4. Positive impacts of using wastes

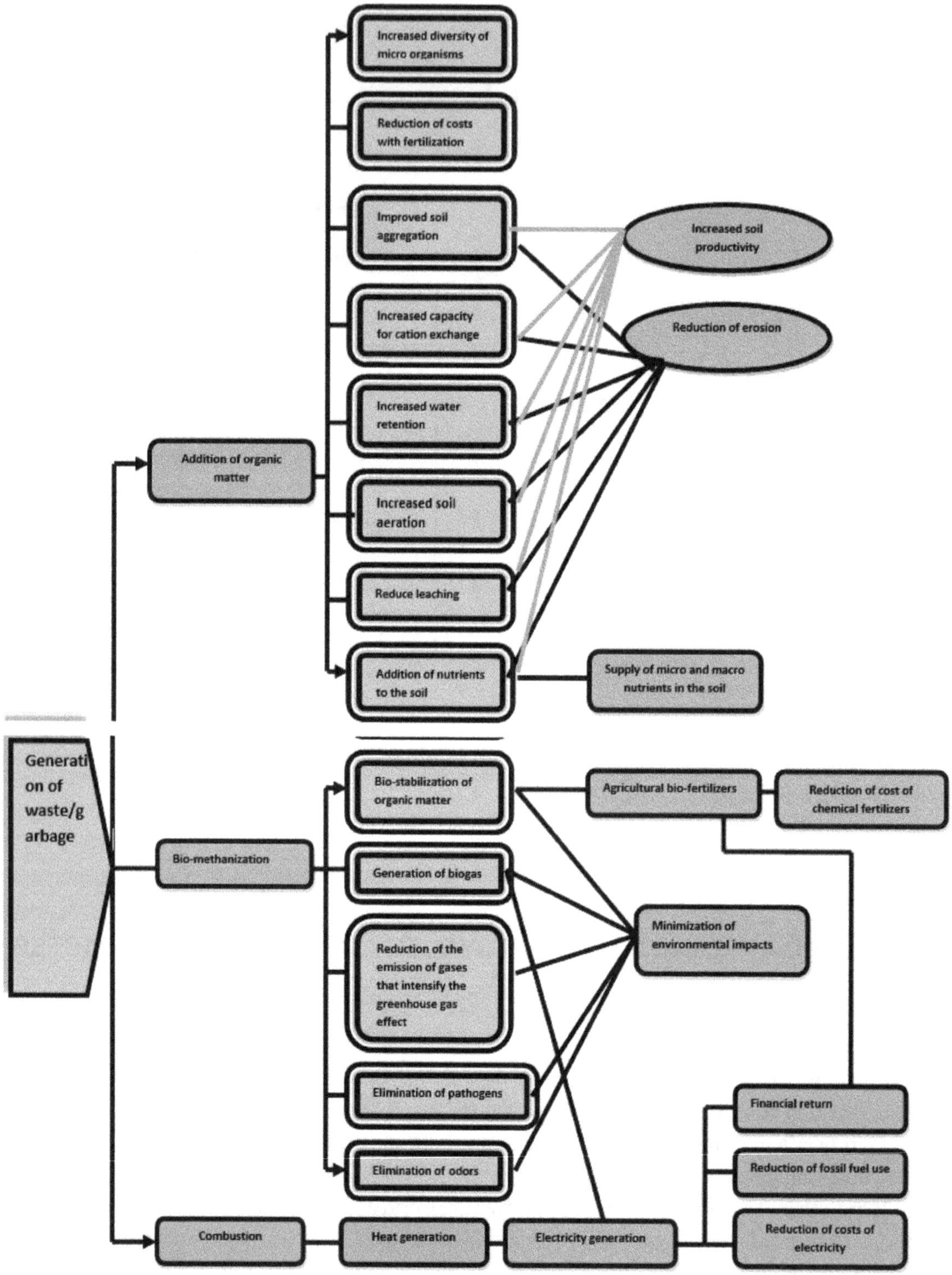

Table 3. Studies on production of enzymes by microorganisms using agro-industrial wastes

Substrates	Enzymes	Microorganisms	References
Papaya waste	α-Amylase	*A. niger*	Sharanappa et al. (2011)
Groundnut oil cake(GOC)	Lipase	*C. rugosa*	Rekha et al. (2012)
Wheat bran and orange peel	Pectin methyl esterase	*P. notatum*	Gayen and Ghosh (2011)
Linseed oil cake(LOC)	Lipase	*P. aeruginosa*	Parihar (2012)
Orange peel	α-Amylase	*A. niger*	Sindiri et al. (2013)
Coconut oil cake(COC)	α-Amylase	*A. oryzae*	Ramachandran et al. (2004)
Rice bran	α-Amylase	*Bacillus* sp.	Sodhi et al. (2005)
Corn bran	α-Amylase	*Bacillus* sp.	Sodhi et al. (2005)
(Rice, Wheat, Blackgram) bran and soyabean	α-Amylase	*A. niger*	Akpan et al. (1999)
Fruit peel waste	Invertase	*A. niger*	Mehta and Duhan (2014)

residues have short harvest time thus, reduce the extra consistently presented to bio-ethanol production (Limayema & Ricke, 2012). Several researchers have accomplished the production of ethanol from materials withlingo-cellulosic composition (Cadoche & López, 1989; Bjerre et al., 1996). Najafi et al. (2009) furthermore, premeditated the manufacture of bio-ethanol from a variety of agricultural residues attained from diverse agricultural crops. A range of agro-industrial wastes utilized for the production of bio-ethanol for next generation are discussed by Saini et al. (2014). They paid attention to the use of lingo-cellulosic composition of diverse agro-industrial wastes and accomplished the fact that the bio-fuels are valuable substitutes of different fossil fuels like diesel and petrol. On the basis of their debate and analysis of various approaches for bio-fuel creation, it is undoubtedly revealed that the lingo-cellulosic consequential bio-fuels are cost effective and also eco-friendly as well as substituent source of energy for upcoming generations. One more study for making of biogas by means of a variety of food industrial residues from diverse sources as well as two weeds namely *Eichornia crassipes* solms and *Typha angustifolia* L. were conceded out by Paepatung et al. (2009). A fast growth in population in most of the developing countries and also their fast expansion in industrialization creates the elevated demand for cheap sources of energy by means of cost-effective food industrial and other agricultural residues (Figure 4). An ample amount of vital waste is obtainable in these countries for the fabrication of bio-fuels. In a way Mushimiyimana and Tallapragada (2016) formed bio-ethanol by using wastes out of vegetable's via fermentation technique with the aid of baker's yeast *Saccharomyces cerevisiae*. In the study they made use of common vegetables residues such as carrot peel, potato peel, and onion peel. Bio-ethanol manufacture could be best substitute for the utilization of agricultural residues. Utilization of the stems of banana as a substrate for the production of bio-ethanol is an excellent alternative in India because of enormous availability of pseudo banana stem as a squander. Ingale et al. (2014) fashioned bio-ethanol by means of pseudo banana stem as a substrate with pre-treatment using microbes such as *Aspergillus fumigatus and Aspergillus ellipticus*. Furthermore, Maiti et al. (2016) utilizedfood industrial residues for the fabrication of butanol by making use of *Clostridium beijerinckii*. The highest amount of butanol (11.04 g/l) was produced subsequent to 96 hours of fermentation from the food industrial residue like SIW (starch industry wastewater). Hence, the utilization of economical and biologically agricultural waste for the creation of priceless bio-fuels is a healthier pathway for the accomplishment of necessity of energy through restricted resources and also eliminating the threat to the soil environment.

Production of Enzyme

One of the most important phenomena for the transformation of agricultural wastes into valuable products is enzymatic hydrolysis which is a valuable and significant technique. Consumption of food industrial wastes proposes enormous prospective for decreasing the production costs and mounting the usage of enzymes for industrial functions. As a known fact that agro-industrial wastes include sugarcane bagasse, wheat bran, wheat straw, corncob, rice bran, etc. are low price and have abundantly obtainable natural carbon sources which can be effectively utilized for the creation of scientifically essential enzymes (Jecu, 2000). A technique known as microbial culture-based solid state fermentation (Salim et al., 2017) is presently used to ferment various agro-industrial residual substrates and by-products for the successful production of cellulases (Salim et al., 2017), consequently, in a supplementary study explained the fabrication of enzymes by a recently secluded *Bacillus* sp. TMF-1 in solid-state fermentation on food industrial by-products. In this research, they achieved α-amylases, cellulases, pectinases and proteases (Krishna, 2005)

Antioxidant Production

Fruit peels such as skins are loaded with nutrients and include many phytochemicals that may be competently used as drugs or as food supplements or replacers (Bobinaitė et al., 2016). Antioxidants are excellent compounds as they act as additives in foodstuffs in order to enhance their nutritional value or in manufacture of fruit puree (Chacko & Estherlydia, 2014). In addition their sensory properties act as a raw material for the creation of food dyes. Very essential is the prospect of preserved and improved quality as a consequence of avoiding food oxidation (Ayala-Zavala & González-Aguilar, 2011). The probable antioxidative position and bioavailability of byproducts from creation of nectar of the pomegranate fruit were examined under a study (Surek & Nilufer-Erdil, 2016). Pomegranate seeds and the impulsive obtained by sedimentation of nectar comprise a good supply of anthocyanins. Filter cake, sediment peel contained extra phenolic compounds and were characterized by elevated antioxidant action than those extracted from the pomegranate nectar. In the certain studies the antioxidant activity and overall content of polyphenols of tomato skin were found to be 38.2 and 66.5% elevated, correspondingly, in comparison to the seeds of tomatoes (Sarkar & Kaul, 2014).

Production of Oncom

Oncom is a native fermented product of Indonesia which is prepared from a number of agricultural wastes. Basically, there are three categories of oncom. Out of these three categories, the most eminent is that prepared from peanut press-cake, waste merchandise from peanut oil processing industries. This is known as oncomkacang, it is accepted in West Java (VanVeen et al., 1968; Beuchat, 1986). Another type is oncomtahoo, it is admired in Jakarta. It is manufactured from the solid wastes out of tahoo, which is a curd of soya bean. Its preparation is analogous to that for oncomkacang. The third type is prepared from the solid wastes of starch flour (Hunkwe) and mungbean (*Phaseolus radiata*) and is popularly known by the name of oncomampashunkwe (Steinkraus, 1983).

The High-quality oncom (Figure 5), conventional Indonesian fermented food, was primed by the inoculation of pure culture of numerous Neurospora, Rhizopus, and Mucor strains. Even though the role of Mucor in oncom making was not recognized, many Mucor strains could create high-quality on-

Figure 5. Black Oncom (source: aguskrisnoblog.wordpress.com)

com. Solid inoculum includes each of 4 Mucor, 4 Rhizopus, and 3 Neurospora strains with utilization of cooked rice as a transporter were primed to manufacture high-quality oncom. All inoculum apart from one contained a sufficient amount of living cells to make oncom after storage for 180 days at room temperature. Bandung red oncom and bogor red oncom formed by the local manufacturer using the solid inoculum had very good eminence. They were brought to the neighboring market and well acknowledged by the consumers. The raw materials of oncom are chiefly the waste of agricultural goods and there are many types of oncom by the mixture of several molds and raw materials. The temperature of the solid substrate fermentation used for fermented foods in Southeast Asia is well thought-out to be mesophilic as the incubation is usually conceded out at room temperature in a very consistent steamy environment of 25° to 32°C. Oncom is the cheapest product because the raw materials for oncom are mostly the waste of agricultural products that still hold high quantity of protein appropriate for consumers with low-incomes. Thus, the enhancement of the quality of oncom ought to result not only in a deliverance of good and cheap food with high protein content, but also in the competent use of agricultural products. On the other hand, no attempt has been made to advance the quality of oncom and to deal out the enhanced products (Sastraatmadja et al,2001)

PRODUCTION OF FOOD FLAVORING COMPOUNDS

The wastes out of agro-industries are cheap and sustainable resources for the fabrication of renewable technique for value-added products. A lot of natural flavoring agents can be formed out of these wastes via microbial conversions. Because of the importance gained by potential flavorings they are being widely used. They are extensively used in various dishes as seasoning. Besides, most of them are utilized by traditional medicine, e.g. certain essential oils that are rich in terpene compounds are acknowledged for supporting in the treatment and curability of enormous health problems. An important purpose of some of them leads to various activities such as antifungal, anti viral, anti inflammatory, anti mutagenic, antibacterial, vermicide and anticancer activities (Raut & Karuppayil, 2014; Felipe et al., 2017).At the present epoch flavor companies are concerned with the aromas and flavors which are more stable and don't escape openly besides under specifically defined conditions i.e. they are mostly microscopically encapsulated (Arvanitoyannis & Varzakas, 2008).

Sometimes various factors affect the existence or availability of specific flavoring extracts such as their existence being uneconomical due to high cost or its unavailability. This leads to a great switch towards the commercial flavoring adoption. These trendy flavorings are nothing but an equal or adjacent chemical substitutes or equivalents of the existing natural flavours commonly known as the 'nature-identical (Mantzouridou et al., 2015). The production of the specific enzymatic as well as the whole-cell biocatalysis has gained a cardinal attention as well as a novelty. Their approach has been greatly observed as a substitute for the better development of several esters when compared to any naturally or chemically synthesized enzymes (Zhuang et al., 2015). These extractions of the enzymes from the particular organisms serve various vital benefits which mainly include superior productivity with respect to the elevated catalyst concentration, as well as the simpler product refinement. Some researches provide three main methods exercised for the process of aroma compounds production (Felipe et al., 2017). These processes include the method of chemical synthesis, extraction from natural sources and some of the biotechnological production processes resulting in the development of the bio-aromas.

However, among all the above-mentioned techniques majority advantages are contributed by the biotechnological process. The importance can be approved or confirmed by its natural product development as a versatile resultant. Furthermore, the most novel advantage offered is the development of the better, suitable and sustainable preservation approach towards the environment. Bioaroma production extends its approach towards the most unique renewable processing characteristic. This imparts requirement of very simple operation conditions responsible for non-toxic waste generation thus, sometimes availing the agro-industrial residues. The vital substitute is a by-product of fruit and vegetable waste which is a prospective resource of flavor production (Sadh et al., 2018).

Production of Poly (3-Hydroxybutyric Acid)

Citrus fruits are devoured throughout the world for diverse industrial purposes such as fruit juice, jams and jelly manufacture. So these categories of industries also produces an enormous quantity of waste in the form of peel residue or in supplementary form but these citrus wastes if used in fermentation process yield novel products as they contain huge quantity of carbohydrates. Various researchers obtained useful by-products by utilizing these residues besides Sukan et al. (2014) utilized orange peel waste for the creation of Poly (3-HB). Their outcome showed that orange peel is a rich and unutilized food industrial waste which can be utilized for novel approaches (Figure 6). In their research they reported first time the production of Poly (3-HB) utilizing orange peel as a solitary carbon source followed by a very simple pre-treatment method thus, preserving the soil environment and at the same time proving to be a sustainable novel product (Sadh et al., 2018).

Figure 6. Putative metabolic pathway of PHA synthesis from γ- butyrolactone in H.pseudoflava (Ahn et al., 2000)

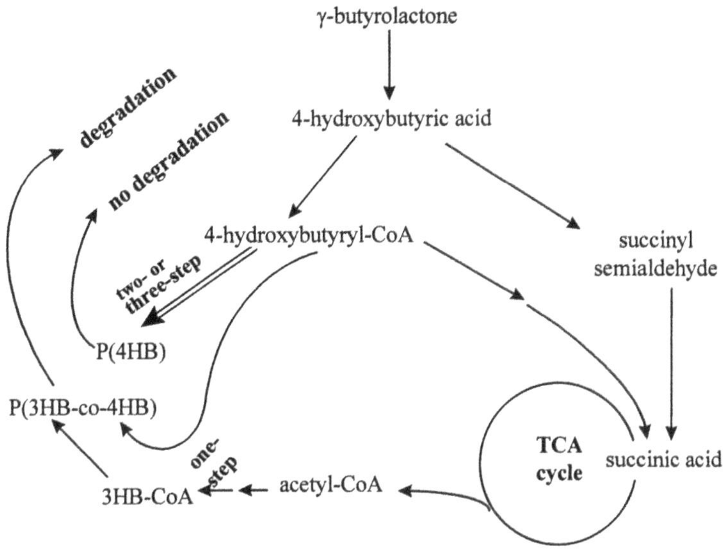

Recycled Agricultural Composting

In recent epoch, composting techniques have been reevaluated in most countries around the globe and incorporated into the novel 4-R strategy, together with countries of America, Asia, Africa and Europe. The progresses made in composting techniques in India all through the early fraction of the preceding century have led to the novel composting operations. Composting as we know is a controlled process which uses innate and natural microorganisms existing in organic matter as well as soil to decay organic wastes (Sadh et al., 2018). These microorganisms necessitate adequate vital nutrients, oxygen as well as water in order to decompose the organic matter at an accelerated speed. The raw materials which decompose after going into the compost are habitually referred as feedstock. The end product i.e. compost, has a dark brown, humus-like appearance and can be easily hence safely stored, handled, and applied to land as a priceless soil conditioner (Lopez-Real, 1996).

FUTURE PROSPECTS

Agro-industrial wastes or residues are loaded with nutrients as well as bioactive compounds. These wastes encompass inconsistency in their composition such as proteins, sugars, and minerals; therefore, they should be well thought-out to be "raw material" instead of "wastes" for several other industrial purposes. The amount of such nutrients in these wastes proposes appropriate circumstances for the productive expansion of microorganisms. The microorganisms have prospective to reprocess the waste as raw materials for their development through fermentation processes. The food industrial wastes can be used as solid support i.e. substrate in solid state fermentation technique for the manufacture of a variety of important valuable compounds. The utilization of agricultural as well as agro-based industry wastes as raw materials can facilitate to decrease the manufacturing cost and contribute in recycling of residues as well as to construct the eco-friendly environment. Currently, several studies have been performed to use solid wastes from human activities as soil conditioners and/or fertilizers for increasing crop productivity. Therefore, studies that monitor organic waste effects on agricultural soils deserve the attention of the international scientific community, as it enables increases in the productivity of agricultural crops, fiber, and biomass energy combined to reduce risks to human, plant, and animal health and environment.

CONCLUSION

There have been growing remarks and evidences of various value added food products developed from the food industrial residues. Among which the majority have been significantly implemented for the various cardinal purposes in such a way that there value is unhampered and not being adversely affected. However, their production can itself be a challenge to the manufacturer for the reason of imparting additional cost to the processing value and adding to the economical values or the overall cost. However, in the mean time value addition of products may also lead to a remarkable growth in terms of the profit, but the identification, selection, manufacturing itself pertains to a major challenge. This depends on the manufacturing practices, consumption on the agro-economic level and the slightest threats imparted to the global environment. There is a extensive dialogue regarding the suitable forms of dumping and use of the solid wastes, and their reuse on agricultural soil has been measured the most remarkable option, both from the environmental standpoint and the economic one. However, their use in agriculture should be headed by a scrutiny of environmental and economic impact and the indiscriminate utilization of wastes may lead to contamination.

REFERENCES

Adriaens, P., & Hickey, W. J. (1993). Article. In D. L. Stone (Ed.), Biotechnology for the Treatment of Hazardous Waste (pp. 97–120). Ann Arbor, MI: Lewis Publishers.

Ahn, W. S., Park, S. J., & Lee, S. Y. (2000). Production of Poly(3-Hydroxybutyrate) by Fed-Batch Culture of Recombinant Escherichia coli with a Highly Concentrated Whey Solution. *Applied and Environmental Microbiology*, *66*(8), 3624–3627. doi:10.1128/AEM.66.8.3624-3627.2000 PMID:10919830

Ajila, C. M., Aalami, M., Leelavathi, K., & Rao, U. P. (2010). Mango peel powder: A potential source of antioxidant and dietary fiber in macaroni preparations. *Innovative Food Science & Emerging Technologies*, *11*(1), 219–224. doi:10.1016/j.ifset.2009.10.004

Akpan, I., Bankole, M. O., Adesemowo, A. M., & Latunde, D. G. (1999). Production of amylase by *A. niger*in a cheap solid medium using rice bran and agricultural materials. *Tropical Science*, *39*, 77–79.

Al-Weshahy, A., & Rao, V. A. (2012). Potato peel as a source of important phytochemical antioxidant nutraceuticals and their role in human health-A review. In *Phytochemicals as Nutraceuticals-Global Approaches to Their Role in Nutrition and Health*. InTech. doi:10.5772/30459

Arvanitoyannis, I. S., & Varzakas, T. H. (2008). Vegetable waste management: Treatment methods and potential uses of treated waste. Waste Management for the Food Industries, 703-752.

Avci, A., Saha, B. C., Dien, B. S., Kennedy, G. J., & Cotta, M. A. (2013). Response surface optimization of corn stover pretreatment using dilute phosphoric acid for enzymatic hydrolysis and ethanol production. *Bioresource Technology*, *130*, 603–612. doi:10.1016/j.biortech.2012.12.104 PMID:23334017

Awarenet. (2004). *Hand book of waste management and co-products recovery in food processing*. Academic Press.

Ayala-Zavala, J. F., & González-Aguilar, G. A. (2010). Use of additives to preserve the quality of fresh-cut fruits and vegetables. In *Advances in fresh-cut fruits and vegetables processing* (pp. 233–256). CRC Press. doi:10.1201/b10263-10

Ballesteros, L. F., Teixeira, J. A., & Mussatto, S. I. (2014). Chemical, functional, and structural properties of spent coffee grounds and coffee silverskin. *Food and Bioprocess Technology*, *7*(12), 3493–3503. doi:10.100711947-014-1349-z

Banerjee, J., Singh, R., Vijayaraghavan, R., MacFarlane, D., Patti, A. F., & Arora, A. (2017). Bioactives from fruit processing wastes: Green approaches to valuable chemicals. *Food Chemistry*, *225*, 10–22. doi:10.1016/j.foodchem.2016.12.093 PMID:28193402

Beg, M. U., Saeed, T., Al-Muzaini, S., Beg, K. R., & Al-Bahloul, M. (2003). Distribution of petroleum hydrocarbon in sediment from coastal area receiving industrial effluents in Kuwait. *Ecotoxicology and Environmental Safety*, *54*(1), 47–55. doi:10.1016/S0147-6513(02)00019-2 PMID:12547635

Beuchat, L. R. (1986). Oncom (fermented peanut presscake). In N. R. Reddy, M. D. Pierson, & D. K. Salunkhe (Eds.), *Legume-based fermented foods* (pp. 135–144). Boca Raton, FL: CRC Press.

Bjerre, A. B., Olesen, A. B., Fernqvist, T., Plöger, A., & Schmidt, A. S. (1996). Pretreatment of wheat straw using combined wet oxidation and alkaline hydrolysis resulting in convertible cellulose and hemicellulose. *Biotechnology and Bioengineering*, *49*(5), 568–577. doi:10.1002/(SICI)1097-0290(19960305)49:5<568::AID-BIT10>3.0.CO;2-6 PMID:18623619

Bobinaitė, R., Viskelis, P., Bobinas, Č., Mieželienė, A., Alenčikienė, G., & Venskutonis, P. R. (2016). Raspberry marc extracts increase antioxidative potential, ellagic acid, ellagitannin and anthocyanin concentrations in fruit purees. *Lebensmittel-Wissenschaft + Technologie*, *66*, 460–467. doi:10.1016/j.lwt.2015.10.069

Cadoche, L., & López, G. D. (1989). Assessment of size reduction as a preliminary step in the production of ethanol from lignocellulosic wastes. *Biological Wastes*, *30*(2), 153–157. doi:10.1016/0269-7483(89)90069-4

Cerezal, P., & Duarte, G. (2005). Utilización de cáscaras en la elaboración de productosconcentrados de tuna (Opuntiaficus-índica (L.)Miller). *Journal of the Professional Association for Cactus Development*, 61-83.

Chacko, C. M., & Estherlydia, D. (2014). Antimicrobial evaluation of jams made from indigenous fruit peels. *International Journal of Advanced Research*, *2*(1), 202–207.

Dalentoft, E., & Thulin, P. (1997). The use of aerobic selectors in activated sludge systems for treatment of wastewater from the pulp and paper industry. *Water Science and Technology*, *35*(2-3), 181–188. doi:10.2166/wst.1997.0513

Directive, E. C. (2008). 105/EC of the European Parliament and of the Council of 16 December 2008 on environmental quality standards in the field of water policy, amending and subsequently repealing Council Directives 82/176. *EEC*, *83*(513), 84–97.

Duhan, J. S., Kumar, A., & Tanwar, S. K. (2013). Bioethanol production from starchy part of tuberous plant (potato) using Saccharomyces cerevisiae MTCC-170. *African Journal of Microbiological Research*, *7*(46), 5253–5260. doi:10.5897/AJMR2013.6122

Felipe, L., de Oliveira, A. M., & Bicas, J. L. (2017). Bioaromas–Perspectives for sustainable development. *Trends in Food Science & Technology*, *62*, 141–153. doi:10.1016/j.tifs.2017.02.005

Gayen, S., & Ghosh, U. (2011). Pectinmethylesterase Production from mixed agro- wastes by Penicilliumnotatum NCIM. 923 in Solid-State fermentation. *Journal of Bioremediation & Biodegradation*, *2*(02), 119. doi:10.4172/2155-6199.1000119

Gowe, C. (2015). Review on potential use of fruit and vegetables by-products as a valuable source of natural food additives. *Food Science and Quality Management*, *45*, 47–61.

Gume, B., Muleta, D., & Abate, D. (2013). Evaluation of locally available substrates for cultivation of oyster mushroom (Pleurotus ostreatus) in Jimma, Ethiopia. *African Journal of Microbiological Research*, *7*(20), 2228–2237. doi:10.5897/AJMR12.895

Gustafsson, J., Cederberg, C., Sonesson, U., & Emanuelsson, A. (2013). *The methodology of the FAO study: Global Food Losses and Food Waste-extent, causes and prevention*. SIK Institutetförlivsmedelochbioteknik.

Ingale, S., Joshi, S. J., & Gupte, A. (2014). Production of bioethanol using agricultural waste: Banana pseudo stem. *Brazilian Journal of Microbiology*, *45*(3), 885–892. doi:10.1590/S1517-83822014000300018 PMID:25477922

Jecu, L. (2000). Solid state fermentation of agricultural wastes for endoglucanase production. *Industrial Crops and Products*, *11*(1), 1–5. doi:10.1016/S0926-6690(99)00022-9

Karthikeyan, K., & Balasubramanian, S. (2010). Studies on the characterization and possibilities of reutilization of solid wastes from a waste paper based paper industry. *Global Journal of Environmental Research*, *4*(1), 18–22.

Kowalska, H., Czajkowska, K., Cichowska, J., & Lenart, A. (2017). What's new in biopotential of fruit and vegetable by-products applied in the food processing industry. *Trends in Food Science & Technology*, *67*, 150–159. doi:10.1016/j.tifs.2017.06.016

Krishna, C. (2005). Solid-state fermentation systems—An overview. *Critical Reviews in Biotechnology*, *25*(1-2), 1–30. doi:10.1080/07388550590925383 PMID:15999850

Kumar, A., Duhan, J. S., Gahlawat, S., & Gahlawat, S. K. (2014). Production of ethanol from tuberous plant (sweet potato) using *Saccharomyces cerevisiae* MTCC-170. *African Journal of Biotechnology*, *13*(28).

Limayem, A., & Ricke, S. C. (2012). Lignocellulosic biomass for bioethanol production: Current perspectives, potential issues and future prospects. *Progress in Energy and Combustion Science*, *38*(4), 449–467. doi:10.1016/j.pecs.2012.03.002

Lopez-Real, J. M. (1996). Composting of Agricultural Wastes. In M. de Bertoldi, P. Sequi, B. Lemmes, & T. Papi (Eds.), *The Science of Composting* (pp. 542–550). Dordrecht, The Netherlands: Springer. doi:10.1007/978-94-009-1569-5_51

Maiti, S., Sarma, S. J., Brar, S. K., Le Bihan, Y., Drogui, P., Buelna, G., & Verma, M. (2016). Agro-industrial wastes as feedstock for sustainable bio-production of butanol by *Clostridium beijerinckii*. *Food and Bioproducts Processing*, *98*, 217–226. doi:10.1016/j.fbp.2016.01.002

Mantzouridou, F. T., Paraskevopoulou, A., & Lalou, S. (2015). Yeast flavour production by solid state fermentation of orange peel waste. *Biochemical Engineering Journal*, *101*, 1–8. doi:10.1016/j.bej.2015.04.013

Mehta, K., & Duhan, J. S. (2014). Production of invertase from Aspergillusniger using fruit peel waste as a substrate. *International Journal of Pharma and Bio Sciences*, *5*(2), 353–360.

Mushimiyimana, I., &Tallapragada, P. (2016). *Bioethanol production from agro wastes by acid hydrolysis and fermentation process*. Academic Press.

Najafi, G., Ghobadian, B., Tavakoli, T., & Yusaf, T. (2009). Potential of bioethanol production from agricultural wastes in Iran. *Renewable & Sustainable Energy Reviews*, *13*(6-7), 1418–1427. doi:10.1016/j.rser.2008.08.010

Paepatung, N., Nopharatana, A., & Songkasiri, W. (2009). Bio-methane potential of biological solid materials and agricultural wastes. *Asian Journal on Energy and Environment*, *10*(1), 19–27.

Parihar, D. K. (2012). Production of lipase utilizing linseed oilcake as fermentation substrate. *Int. J. Sci. Environ. Technol*, *1*(3), 135–143.

Rai, R. D., & Ahlawat, O. P. (2002). Edible fungi: Biotechnological approaches. In *Applied Mycology and Biotechnology* (Vol. 2, pp. 87–121). Elsevier.

Ramachandran, S., Patel, A. K., Nampoothiri, K. M., Francis, F., Nagy, V., Szakacs, G., & Pandey, A. (2004). Coconut oil cake—a potential raw material for the production of α-amylase. *Bioresource Technology*, *93*(2), 169–174. doi:10.1016/j.biortech.2003.10.021 PMID:15051078

Raut, J. S., & Karuppayil, S. M. (2014). A status review on the medicinal properties of essential oils. *Industrial Crops and Products*, *62*, 250–264. doi:10.1016/j.indcrop.2014.05.055

Rekha, K. S. S., Lakshmi, M. V. C., Devi, V. S., & Siddartha Kumar, M. (2012). Production and optimization of lipase from *Candida rugosa* using groundnut oilcake under solid state fermentation. *Biosensors*, *27*, 31.

Rivas, B., Torrado, A., Torre, P., Converti, A., & Domínguez, J. M. (2008). Submerged citric acid fermentation on orange peel autohydrolysate. *Journal of Agricultural and Food Chemistry*, *56*(7), 2380–2387. doi:10.1021/jf073388r PMID:18321055

Sadh, P. K., Duhan, S., & Duhan, J. S. (2018). Agro-industrial wastes and their utilization using solid state fermentation: A review. *Bioresources and Bioprocessing*, *5*(1), 1. doi:10.118640643-017-0187-z

Saini, J. K., Saini, R., & Tewari, L. (2014). Lignocellulosic agriculture wastes as biomass feedstocks for second-generation bioethanol production: Concepts and recent developments. 3. *Biotechnology (Faisalabad)*, *5*(4), 337–353. doi:10.100713205-014-0246-5 PMID:28324547

Salim, A. A., Grbavčić, S., Šekuljica, N., Stefanović, A., Tanasković, S. J., Luković, N., & Knežević-Jugović, Z. (2017). Production of enzymes by a newly isolated Bacillus sp. TMF-1 in solid state fermentation on agricultural by-products: The evaluation of substrate pretreatment methods. *Bioresource Technology*, *228*, 193–200. doi:10.1016/j.biortech.2016.12.081 PMID:28063362

Sarkar, A., & Kaul, P. (2014). Evaluation of Tomato Processing By-Products: A Comparative Study in a Pilot Scale Setup. *Journal of Food Process Engineering*, *37*(3), 299–307. doi:10.1111/jfpe.12086

Sastraatmadja, D. D., Tomita, F., & Ksai, T. (2001). Production of high-quality oncom, a traditional Indonesian fermented food, by the inoculation with selected mold strains in the form of pure culture and solid inoculum. *Journal-Graduate School of Agriculture Hokkaido University*, *70*(2), 111–128.

Sindiri, M. K., Machavarapu, M., & Vangalapati, M. (2013). Alfa-amylase production and purification using fermented orange peel in solid state fermentation by *Aspergillusniger. Ind J Appl Res*, *3*(8), 49–5. doi:10.15373/2249555X/AUG2013/16

Sodhi, H. K., Sharma, K., Gupta, J. K., & Soni, S. K. (2005). Production of a thermostable α-amylase from Bacillus sp. PS-7 by solid state fermentation and its synergistic use in the hydrolysis of malt starch for alcohol production. *Process Biochemistry*, *40*(2), 525–534. doi:10.1016/j.procbio.2003.10.008

Steinkraus, K. H. (1983). *Handbook of indigenous fermented foods*. New York: M Dekker Inc.

Sukan, A., Roy, I., & Keshavarz, T. (2014). Agro-industrial waste materials as substrates for the production of poly (3-hydroxybutyric acid). *Journal of Biomaterials and Nanobiotechnology*, *5*(4), 229–240. doi:10.4236/jbnb.2014.54027

Surek, E., & Nilufer-Erdil, D. (2016). Phenolic contents, antioxidant activities and potential bioaccessibilities of industrial pomegranate nectar processing wastes. *International Journal of Food Science & Technology*, *51*(1), 231–239. doi:10.1111/ijfs.13000

Van Hamme, J. D., Singh, A., & Ward, O. P. (2003). Recent advances in petroleum microbiology. *Microbiology and Molecular Biology Reviews*, *67*(4), 503–549. doi:10.1128/MMBR.67.4.503-549.2003 PMID:14665675

VanVeen, A. G., Graham, D. C. W., & Steinkraus, K. H. (1968). Fermented peanut press cake. *Cereal Sci Today*, *13*, 96–99.

Venglovsky, J., Martinez, J., & Placha, I. (2006). Hygienic and ecological risks connected with utilization of animal manures and biosolids in agriculture. *Livestock Science*, *102*(3), 197–203. doi:10.1016/j.livsci.2006.03.017

Willaman, J., & Eskew, R. (1943). *Uses for vegetable wastes*. Academic Press.

William, P. T. (2005). Water treatment and disposal. John Wiley.

Wong, S. S., Teng, T. T., Ahmad, A. L., Zuhairi, A., & Najafpour, G. (2006). Treatment of pulp and paper mill wastewater by polyacrylamide (PAM) in polymer induced flocculation. *Journal of Hazardous Materials*, *135*(1-3), 378–388. doi:10.1016/j.jhazmat.2005.11.076 PMID:16431022

Wu, J., Xiao, Y. Z., & Yu, H. Q. (2005). Degradation of lignin in pulp mill wastewaters by white-rot fungi on biofilm. *Bioresource Technology*, *96*(12), 1357–1363. doi:10.1016/j.biortech.2004.11.019 PMID:15792583

Yusuf, M. (2017). Agro-Industrial Waste Materials and their Recycled Value-Added Applications. Handbook of Ecomaterials, 1-11.

Zhuang, S., Fu, J., Powell, C., Huang, J., Xia, Y., & Yan, R. (2015). Production of medium-chain volatile flavour esters in Pichiapastoris whole-cell biocatalysts with extracellular expression of Saccharomyces cerevisiae acyl-CoA: Ethanol O-acyltransferase Eht1 or Eeb1. *SpringerPlus*, *4*(1), 467. doi:10.118640064-015-1195-0 PMID:26357598

KEY TERMS AND DEFINITIONS

Agricultural Composting: Composting is the biological decomposition of organic materials by microorganisms under controlled, aerobic conditions to a relatively stable humus-like material called compost. Composting can happen in many different ways using a variety of materials, methods, equipment, and scales of operation. For agricultural operations the common materials or feedstocks that are composted are livestock manures and bedding and various residual plant materials (straw, culls, on-farm processing wastes, etc.).

Bagasse: Bagasse is the dry pulpy fibrous residue that remains after sugarcane or sorghum stalks are crushed to extract their juice. It is used as a biofuel for the production of heat, energy, and electricity, and in the manufacture of pulp and building materials.

Biofuel: Biofuel is any fuel that is derived from biomass—that is, plant or algae material or animal waste. Since such feedstock material can be replenished readily, biofuel is considered to be a source of renewable energy, unlike fossil fuels such as petroleum, coal, and natural gas.

Biomass: Biomass is organic matter, such as leaves, stems, wood. All plants collectively produce biomass that can be converted into usable energy.

Nutraceuticals: Neutraceuticals are defined as ingredients that have health benefits beyond basic nutrition. They are also defined as parts of a food or a whole food that have a medical or health benefit, including the prevention and treatment of disease.

Chapter 13
Municipal Urban Waste-to-Energy Business Model in Mexico:
A Study of Three Companies

José G. Vargas-Hernández
University of Guadalajara, Mexico

ABSTRACT

The major objective of this research is to identify the main aspects presented in three companies of energy generation from biogas produced by landfills in three Mexican cities. Analyzing these companies, their way of operating and business environment, it is possible to detect the existence of business opportunities in this area. It is concluded that there is a great potential of wasted businesses that works through market mechanisms in strategic alliance with public entities as a more viable option for the reduction of investment risk and the viability of the projects. The use of high technologies, composition of city waste, and available infrastructure are key factors. The study of three companies in operation allows the identification of factors that can help in the implementation of more companies of energy generation based on urban solid waste in more areas of the country and elsewhere with successful and profitable results for both private capital and social interest.

INTRODUCTION

The world moves with energy; transport, companies and people demand energy at all times to carry out their common activities. Society consumes energy at accelerated rates, which mostly comes from fossil fuels, mainly hydrocarbons. The vertiginous consumption of hydrocarbons for decades has made society concerned about the underlying effects it causes. The environmental impact provoked by the combustion of hydrocarbons and their characteristic feature of being a nonrenewable resource gave a sound of alarm in diverse scientific, civic and political groups arising a great interest by the renewable energy sources. Solar, wind, geothermal and bioenergy are some of the main sources of renewable energy.

DOI: 10.4018/978-1-7998-0031-6.ch013

Mexico is a country with a large population and abundant natural resources. The use of renewable energy has had importance at different levels and a legal framework has been sought for the emergence of companies that are dedicated to the production of energy from renewables. Fortunately, the use of biomass to generate energy is promising enough to pay attention to the various uses that can be given to waste. It is said that each person produces one kilo of solid urban waste (municipal solid waste - MSW) a day. Three companies are already taking advantage of this business opportunity.

The major objective of this research is to identify the main aspects presented in three companies of energy generation from biogas produced by landfills in three cities of the country: "BENLESA" in Salinas Victoria, Nuevo León; "YLEM ENERGY LIMITED" in Aguascalientes, Aguascalientes; and "Biogas Juárez" in Ciudad Juárez, Chihuahua. Analyzing these companies, their way of operating and business environment, it is possible to detect the existence of business opportunities in this area. In this research, a qualitative methodology bases on the review of the existing literature, in different countries, as well as upon the histories of three companies from the perspectives proposed in Peng's strategy tripod (Peng, 2012).

The study focuses the analysis on the potential of electric power generation plants from urban solid waste as a profitable business in the country. Identifying their strategies, it can be stipulated a serial of key factors for the opening and successful operability of this type of business in Mexico.

BACKGROUND OF THE PROBLEM

The supply of energy has been a problem that has preoccupied interest groups for decades, not only if it could satisfy the demand but also the harmful impact of its production and use on the environment through Greenhouse gases (GHG). The enormous level of consumption of hydrocarbons in the world has led to the change of paradigms to obtain the energy necessary for the operation of the economy. According to data from the World Bank (Banco Mundial, 2017), in 2014, 81% of the total energy consumed in the world came from hydrocarbons while alternative energy only had a share of around 10% globally.

In the broader definitions, alternative energies are all sources of energy no based on the combustion of the hydrocarbons and do not involve the burning of fossil fuels such as coal, gas and oil. In the most restrictive concept, and alternative energy would be equivalent to the concept of sources of renewable energy or green energy, such as wind, solar, geothermal, bioenergy and tidal. According to Masera Cerutti and others (2011), bioenergy, which is the energy obtained from biomass, represents 77% of the total renewable energy, estimating that by 2035 this type of energy source would represent 25% of the total energy required by the world.

According to Cerdá, Caparrós and Ovando (2008), renewable energy can be used for electricity generation, thermal use and biofuel production. Bioenergy is capable of being used for all three things, unlike wind and solar energy that cannot be converted into biofuel. Biomass, for its part, is a constituent part of living beings, can be divided according to its source (Monreal, 2008):

- Food source: it comes from grains or other crops, which is also used for human and animal consumption;
- Non-food source: it originates from manure, crop and forest, and food and municipal solid waste.

A shift from hydrocarbon-based to biomass-based economy would represent a significant change in technical and socioeconomic systems (Mccormick & Kauto, 2013). Although its large-scale production is possible, its effects on food security have not been satisfactorily weighed (Winchester & Reilly, 2015). Thus, biomass from food sources is often limited in many crops such as maize and beet (Becerra Pérez, 2015). This is due to the food deficit that exists in the e country, especially in maize.

Therefore, biomass from non-food sources is often seen as a better alternative. There is a significant trend towards the use of various types of waste for energy production, which led many investors to pay attention to electric power generation plants based on urban solid waste, millionaire's businesses evaluated here.

THEORETICAL-CONCEPTUAL REVIEW

Bioenergy, the energy obtained from biomass, can be found in a variety of forms. Biomass is the constitutive material of living beings, excreta or non-living remains (Masera Cerutti et al., 2011). Salinas Callejas and Gasca Quezada (2009, p. 76) extend the definition of biomass to any type of organic matter that has had its immediate origin in some biological process of organisms recently alive, classified even as follows:

- **Primary Biomass:** Vegetal biomass directly from photosynthesis, agricultural and forest residues;
- **Secondary Biomass**: Manure and meat of herbivorous animals;
- **Tertiary Biomass:** Remains and ejection of carnivorous animals;
- **Natural Biomass:** Produced in wild ecosystems;
- **Residual Biomass:** Extractable from agricultural and forestry residues and human activities;
- **Energy Crops:** Any agricultural crop whose purpose is the supply of biomass for the production of biofuels.

At the international level, circa 85% of the energy generated by biomass is considered as "traditional use", i.e., use of firewood or manure at domestic level. The other 15% is of industrial use for the creation of fuels, heat and energy processes, and electricity (Mead, 2007).

One of the many products that are the result of the processing of biomass are biofuels, such as alcohols and other chemicals that come mainly from biomass that can substitute to some extent the use of gasoline in transport or be destined for electricity production (Salinas Callejas & Gasca Quezada, 2009). On the other hand, Masera Cerutti and others (2011) classified biofuels according to their origin:

- Wood fuels from natural forests and plantations;
- Byproducts of logging and the timber industry;
- Agricultural byproducts;
- Livestock byproducts;
- Agro-industrial byproducts;
- Energy crops;
- Byproducts of municipal origin.

However, for the purpose of this work, attention is devoted to everything related to byproducts of municipal origin or also called solid urban waste or municipal solid waste (MSW). Solid waste management represents a serious administrative and environmental challenge to Mexican municipalities. According to Rodriguez and others (2015), from 1997 to 2012, the municipal solid waste generated in Mexico increased in the amount of around 44% and its composition is mostly organic matter (50%). Urban waste originated mainly from house collection (Rodriguez et al., 2015). In Mexico, according to the most recent figure published in 2015 by the Sistema Nacional de Información Ambiental y de Recursos Naturales (2018), the generation of urban solid waste reached 53.1 million tons, which represented an increase of 61.2% with respect to 2003 (10.24 million tons more generated in that period). If expressed per inhabitant, it reached 1.2 kilograms on average daily in that same year.

Municipal solid waste, trash or garbage, is defined as wastes that come from homes, schools, hospitals, restaurants, small business and other commercial units, consisting of items such as product packaging, grass clippings, furniture, clothing, bottles and cans, food scraps, newspapers, appliances, consumer electronics, and batteries (United States Environmental Protection Agency, 2018). Indeed, a solid residue is nothing more than any product, matter or substance from human activity, which can have different origins (households, industries, etc.) and diverse composition (paper, glass, plastic, etc.). The product of greatest interest in urban solid waste processing is biogas, a gaseous fuel generated by biodegradation reactions of organic matter in the absence of oxygen. The produced methane can easily be converted into electricity for use in homes or industries.

Several methods estimate the generation of methane from municipal waste. Among the most important ones is the simple Tier 1 included in the Intergovernmental Panel on Climate Change (2006) and the Corporate Accounting and Reporting (World Resources Institute; World Businesses Council for Sustainable Development, 2004). The Landfill Methane Outreach Program (LMOP) of the Unites States Environmental Protection Agency (USEPA) developed, in 2003, the first Mexican biogas estimating model in collaboration with the United States International Development Agency (USAID) and other Mexican government agencies (Global Methane Initiative, 2018). It provides an automatic tool for estimating the generation and recovery of landfill biogas in municipal sanitary landfills in Mexico. However, the Scholl Canyon model is the most representative, commonly used and accepted in North and South America (Montaño Arango, Corona Armenta & Montelongo Reyes, 2009).

Technologies for the thermal conversion of the urban waste residual include incineration boiler, pyrolysis/gasification reactor, and gasification reactor with plasma arc. One of the most appropriate methods for the generation of methane from municipal waste is anaerobic digestion (Nasir; Ghazi; Omar, 2012). For Ferrer and Pérez (2010), this process generates hydrogen sulphide and carbon dioxide in large quantities which represents a lower yield of methane quality. Even so, anaerobic digestion is still the most viable method with the current technology, because it represents a relatively lower cost and less waste results from the production process, making this technique one of the most used in the world (Morita; Sasaki, 2012).

Opportunities in this sector have been exploited by governments and companies in several countries around the world, especially in Europe (Cerdá, Caparrós & Ovando 2008). In Mexico, this potential has not been properly exploited, and three companies that have been successful in this industry are analyzed: Benlesa in Salinas Victoria, Nuevo León; Ylem Energy Limited in Aguascalientes, Aguascalientes; and Biogas Juárez in Ciudad Juárez, Chihuahua.

For the study of these three companies, three perspectives are used that are considered strategic (Peng, 2012). First, the paper undertakes an analysis of the industry in which these companies compete. Second, there is an assessment based on resources and, thirdly, an analysis from an institutionalist approach takes place.

The industry analysis corresponds to a review of the five operating forces in the well-known Porter Diamond, which emphasizes the state of five determinants in a specific industry (Porter, 1991):

- **Degree of Rivalry Between Competitors:** A high degree of rivalry implies a more intense competition among companies in the sector;
- **Threat of Substitute Products:** The ease with which a product can have substitutes poses a threat to industry´s companies;
- **Bargaining Power of Buyers:** The degree of negotiation that consumers have of the product in question will determine the production strategies of a company;
- **Threat of Entry of Competitors:** Each industry presents barriers of entry that depending on its intensity will determine the possible entry of more companies to the industry;
- **Bargaining Power of Suppliers:** A high degree of bargaining power of suppliers can bring great administrative problems to companies, being ideal a great variety of suppliers available.

From the second perspective, the resource-based view, it takes as reference the VRIO model designed by Barney and Griffin (1992). VRIO is an acronym for the words valuable, rarity, inimitable and organization (Barney & Griffin, 1992), emphasizing the heterogeneity of the resources of a company:

- **Valuable**: Resources that have or generate value for the company and the consumer;
- **Rarity**: The rarity of the resource in question makes it valuable and attractive;
- **Inimitable**: Resources that are harder to imitate by competitors give important intangible assets to the company;
- **Organization**: It is the way in which the resources are organized and how the organization aligns itself with others.

On the other hand, the analysis is done from the institutionalist perspective, i.e., all norms and rules governing human and social action, conditioning and directing any relationship that arises in it, as they say, represent the "rules of the game" (North, 1990). North (1990) distinguishes institutions in two types: formal institutions, which include laws, rules and/or regulations imposed by government entities or with formal recognition by all members of society; and informal institutions, whose main characteristic is to take into account the culture, traditions and beliefs of a given society.

REVIEW OF THE EMPIRICAL LITERATURE

For the empirical review, biomass processing plants of diverse types and in different countries or geographic areas were considered, increasingly delimiting to projects in Mexico based on solid urban waste. In the case of Europe, it had a big boost to biofuels. McCormick and Kautto (2013) noted the increase in research and development projects, especially in schemes focused on biofuels and bio refineries in countries like Germany. Cerdá, Caparrós and Ovando (2008) asserted that, since 1990, the European Union has launched a plan for the development of renewable energy to reduce climate change. They found that the development of bioenergy companies in the European Union depends largely on the existence of policies that encourage such companies and at the same time avoid a negative impact on biodiversity and food prices.

Winchester and Reilly (2015) used data from the Food and Agriculture Organization of the United Nations (FAO) to make future projections on the impact of biofuel production on food prices, concluding that there will be no significant impact as long as foreseen the growth policy of this industry. On the other hand, Razo and others (2007) argued that the best market for liquid biofuels is transportation. However, the cost of waste collection and transportation is high with current technology, especially in Latin American countries.

Many authors see forest biomass as a great business opportunity, such as Favero and Mendelsohn (2014) who argued that the production of energy from wood is cost-efficient in the presence of adequate governmental incentives. Sims (2006) placed an emphasis on the process of collecting and transporting the plant in the cheapest way possible subject to the availability of fuels and high capital costs. Virani (2011) added to this the relative price of fuels in use at that time.

There are also a large number of authors who have focused on biofuels from food sources, such as Becerra Pérez (2015) who performed simulation routines in different scenarios and concluded that ethanol production was feasible in Mexico only from sweet sorghum and sugar cane. Another conclusion was the crucial role of institutions to create long term certainty in the sector. Massieu Trigo and Acuña Rodarte (2015) added that a Law for the Promotion and Development of Biofuels exists in Mexico since 2008. However, although the sector is incipient, it has already presented problems of socio-environmental violence since some projects have already attacked peasant and indigenous peoples.

In the area of energy production based on municipal waste, Brown and Mann (2008) found great potential in companies with systems that combine heat and power (CHP) differentiating a rural market that has abundance of biomass and ease of implementation of small processing plants, and a municipal market that benefits from waste management systems and infrastructure to collect and distribute energy economically. Williams (2011) undertook a study on the effectiveness of the "Waste-to-energy" model in Sweden compared to the companies of this type in the United States, concluding that a legal framework that provides certainty in conjunction with fiscal incentives policies were the key factors for the success of these companies in Sweden.

For their part, Rafati and others (2016) indicated that in order to apply a certain technology for the conversion of urban waste to energy, must be taken into account the composition of municipal solid waste, the already functional technologies and the waste management strategies that the city has available. Bustos Flores (2009) identified a problem of waste collection and recycling in developing countries in which it suggests the need for the involvement of companies and productive associations. Similarly, Medina (2010) noted the potential for the formation of public-private partnerships among cooperatives of collectors, companies and public institutions that can treat waste at low cost. Bernache (2003) identified the lack of an efficient system of municipal waste management in Mexican cities due to a low interest on the part of the authorities.

CONTEXTUAL FRAMEWORK: THE CASE STUDIES

Some of the contextual factors are described and analyzed in specific for each one of the power generation companies based on municipal solid waste. It is important to notice that the processing of waste material is a municipal responsibility that may be transferred to private companies. In this analysis, only three companies are considered out of the unknown total operating. These companies are:

Bioenergy of Nuevo León, S.A. de C. V. (BENLESA)

BENLESA is a Private Public Association (APP) that emerged from a strategic alliance between the private company Bioeléctrica de Monterrey, S.A. de C.V., a company that is a subsidiary of Sistemas de Energía Internacional, S.A. de C.V. (SEISA) and the Government of the State of Nuevo León through the System for Ecological Management and Waste Processing (SIMEPRODE). It is responsible for the operation of the electric power generation plant based on the biogas produced by the Salinas Victoria landfill, Nuevo León. It provides electricity to 13 associated entities through a cogeneration permit by the Energy Regulatory Commission (CRE). It is recognized as the first project in Latin America to use this type of technology (Pino Pérez, 2013).

This project was sponsored by the Global Environment Facility (GEF) and the World Bank. This was due to the large amount of waste that caused pollution and public health problems. SIMEPRODE started a bidding process in 2001 in which 13 companies competed. Grupo Gentor was founded together with the Grupo Monterrey. In 1992, in its energy division, Sistemas de Energía Internacional SA de CV (SEISA) and American Gentor Corporation (AGC) were created. The combination of both constitutes Bioeléctrica S.A. de C.V. contributing 53% of the initial capital (Pino Pérez, 2013).

It was agreed that while Bioeléctrica S.A. de C. V. would be in charge of technical aspects of the project as the operation and maintenance of the biogas plant, SIMEPRODE is the owner of the site of the landfill where the plant is located. The plant was inaugurated in 2013 with a capacity of 7.42 MW, with a private investment of 5.7 million dollars and a support of 5.1 million dollars from the World Bank. After another investment, its installed capacity increased to produce 17 nominal MW and 16 MW net, over the years the plant would obtain the ISO 9001 quality management certification (SEISA, 2017).

According to Pino Pérez (2013, p. 4), from the beginning of its operation until 2013, the company has avoided some 800 thousand tons of carbon dioxide and generated a total of 202 thousand 500 MW of electricity, which are used for lighting municipalities in the metropolitan area, State Government Buildings, Water and Drainage Services of Monterrey, Paseo Santa Lucía, the public transport system and other government buildings. According to the Federal Electricity Commission (CFE) through the Institute of Electrical Research (IIE), the main benefit for the dependencies that use this energy is a saving of 10% in their electric billing (Comisión Federal de Electricidad; Instituto de Investigaciones Eléctricas, 2012, p. 67).

Ylem Energy Limited

Initially, the project was created to replace the waste dump of Cumbres by a sanitary landfill that complies with the current regulation of waste management. That landfill began to operate in the municipality of San Nicolás, Aguascalientes, in 1998. During the first years in operation, the project of a plant of generation of biogas was registered before the Mechanism of Clean Development, which arose with the signature of the Kyoto Protocol, in 1997. Then the first approaches arose of the companies Biogas Technology and ENER-G Natural Power (now called YLEM ENERGY LIMITED) presenting projects in two stages: the first would consist of the method to decompose the CH4 molecule and obtain benefits through the sale of carbon bonds; the second stage would consist of electricity generation (Comisión Federal de Electricidad; Instituto de Investigaciones Eléctricas, 2012).

The agreement reached was that the municipality would be the owner of all the land, would be in charge of closing the cells, would handle leachate, and would provide coverage and daily maintenance. In turn, the municipality of Aguascalientes would receive royalties from utilities obtained both from the sale of carbon credits and from the generation of energy. On the other hand, the companies Biogas Technology and the then, ENER-G Natural Power contributed 100% of the investment capital with a 20-year contract for the operation, maintenance, electricity production and gas rights.

In 2012, the electric power generating plant with an installed capacity of 2.7 MW came on stream. During that same year, a contract was signed among the Japanese automobile company Nissan, the City of Aguascalientes and the operating company YLEM ENERGY LIMITED in which Nissan acquired 100% of the electric energy generated by the waste processing plant, a contract that to date is still standing.

This project is not only generating benefits for private companies, it also causes savings in the electric bill of its users at a very low environmental cost, obtaining until 2012 profits of 1.7 million dollars (Comisión Federal de Electricidad; Instituto de Investigaciones Eléctricas, 2012).

Biogas of Juárez S.A of C. V.

The sanitary landfill of Ciudad Juárez, Chihuahua was opened in 1994, which has a capacity of more than 10 million tons of urban solid waste. Biogas of Juarez S.A. of C.V. is a company whose capital is totally Mexican, specialized in the extraction, burning and use of biogas generated in the sanitary landfill of Ciudad Juárez. The project was carried out jointly with the Federal Electricity Commission, registering to the Clean Development Mechanism (CDM) and generating carbon credits that same year. The operations began in 2011 with a capacity of 4.5 MW, where its clients are mainly the self-sufficient and the municipal public lighting of Ciudad Juárez and Nuevo Casas Grandes, both in the state of Chihuahua. This capacity makes it the second largest electric power plant based on municipal waste in the country, only after BENLESA in Nuevo Leon.

RESULTS

The three companies mentioned above are analyzed, taking into account the tripod of Peng's strategy (Peng, 2012), so that there are three perspectives that help to see the fundamental aspects of electricity generating plants based on waste. Government has the faculty of create autonomous bodies for this purpose or grant the management services to companies.

Results From an Industry-Based View

The Porter Diamond (Porter, 1991) is the guide to analyze the industry. Degree of rivalry between competitors: despite the existence of other possible sources of renewable energy, from the three cases analyzed, it is possible to notice that the industry mentioned in this research has a very low level of competition, where companies have a legal monopoly by being allied with public institutions of the municipalities where they are, therefore it is a market with low level of rivalry.

Threat of substitute products: this type of energy production is put aside from other sources of renewable energy or biomass, sometimes unfairly competing with each other. Among other reasons, the energy production is complicated and uncertain, because it is not a thermal conversion process, it is an alternative process of interest to the use of the organic fraction of the urban waste material in which biogas is produced as a result of bacterial decomposition of the organic matter of the urban waste, and for which separation is essential of toxic inhibitory agents such as disinfectants, insecticides and medicines, apart from metals and glass. Although in theory, the different sources of renewable energy are substitutes for hydrocarbons and not substitutes for each other, in practice it is usually chosen from one type to another. It is because of this variety of energy production techniques that electric plants based on municipal waste have substitutes, however, other substitutes would not successfully end up with the problem of overproduction of garbage in the cities, so the existing potential in landfills would remain attractive.

Buyer negotiation power: one of the great advantages of the three companies reviewed is that, as part of municipal and state projects through public private partnerships, the partner on the public sector side is usually the client, giving an enormous certainty to the business and reducing the risks that the investment entails when taking joint risks and long-term contracts. In the case of Ylem Energy Limited, the client is a large automotive company that has been committed through a long-term contract, this is also a mechanism to reduce the risk of investments.

Threat of entry of competitors: within the industry the threat of entry is practically null at the moment, this is because the operating companies won a bid and form a partnership between public and private capital that allows them to have management costs and disposal of very low resources that a privately owned company could never achieve.

Bargaining power of suppliers: much of the technology used in the plants is imported or, as in the case of Biogás de Juárez, there is an alliance with a foreign company that has the patent of the technology. However, all the necessary technology is easily adaptable in Mexico.

The waste disposal is not a problem in this industry since the integration of the private and public sectors has internalized the collection and disposal of waste without extra costs, thus ensuring the constant flow of waste to be treated.

Results From a Resource-Based View

This analysis takes as reference the VRIO model (Barney & Griffin, 1992) detailed in the conceptual section of this document.

Valuable: as stated earlier, generators used by these types of companies are easily adaptable by any company that wants to compete in this sector. The truly valuable aspects of the companies focused on the transformation of biogas into electrical energy correspond precisely to the association between the know-how belonging to private companies and the infrastructure and systems of collection and separation with which the government has.

Rare: public and private partnership is a rare resource because interested companies must compete in a public tender in which they must submit a technical and an economic proposal that correspond to the interests of the project. One of the striking aspects is the type of customers this entails, since usually the customer is the same public sector or a large company (Nissan case in Aguascalientes), the electric power needs of these entities are usually high and constant. Being the supplier of these entities represents a rare and economically attractive intangible asset for energy companies.

Inimitable: the business model presented here is and should be inimitable in any region that meets the necessary characteristics such as a composition of urban waste that facilitates the process, adequate facilities in landfills and adequate mechanisms of collection and separation of garbage. In fact, a company that is competing in the market must develop a business model as a capability to maintain their own resources and other organizational, technological and marketing capabilities.

Organization: the effect is greater when it is a large company that makes the investment. In the case of Bioeléctrica de Monterrey SA de CV, the company has great financial capacity and experience in other parts of Mexico, that together with the existence of a sanitary landfill certified in the municipality where it operates and with it an efficient collection and separation of waste. The companies of this sector have a good organization that is certified before the CDM as environmentally friendly company and with the ISO-9001 as a compliant with quality management systems.

Results From Institution-Based Vision

The role of formal institutions upon the companies studied is preponderant since, to begin with, the project would hardly have started without some incentive by a global institution such as the World Bank or the signing of the Kyoto Protocol. Based on this encouragement, the public institution prepares for a project for the use of municipal waste to generate electricity and launches a public call. Several companies respond to this call. The government entity chooses the winner of the tender and a contract is made.

In all the cases studied, the result is the creation of an association composed of private and public capital, the latter always in charge of facilitating the existing infrastructure and collection mechanisms, reducing costs and facilitating the operation. All the companies studied have negotiated with carbon credits and have benefited from monetary help from institutions that support environmental projects. For that reason, the role of formal institutions in this type of business is crucial for economic and operational viability. On the other hand, the role of informal institutions, that is, the cultural component of society, is less preponderant than that of formal institutions, but its importance lies in the consumption and recycling habits of a society.

An important point for the optimal processing of urban waste is its composition, the food habits of the inhabitants will determine the type of waste that will have and thus the type of management that they should have. Another important point is the culture of recycling, the transformation of municipal waste is not opposed to it, on the contrary, helps the separation and efficient use of waste.

DISCUSSION

This research confirms that biomass processing plants for the production of energy have a big boost in projects focused on biofuels and bio refineries in countries as McCormick and Kautto (2013) found in Germany; and Cerdá, Caparrós and Ovando (2008) uncovered in the European Union aimed to reduce climate change. Also, this research confirms their findings based on that the development of bioenergy companies depends largely on the existence of policies that encourage such companies and at the same time avoid a negative impact on biodiversity and food prices.

This study also suggests that biofuel production has an impact on food prices in Mexico, confirming the results of Winchester and Reilly (2015) and those of Razo and others (2007) who state that the cost of waste collection and transportation is high with current technology. However, this analysis reaches different conclusion of that of Sims (2006), Favero and Mendelsohn (2014) and Virani (2011) in relation to the difficulties of achieving the cheapest way of transportation subject to the availability of fuels, cost-efficiency based on adequate governmental incentives, and the relative prices of fuels.

The conclusion of Becerra Pérez (2015) limits the production of ethanol without considering other sources such as the waste of the agave. The present analysis agrees that institutions play a crucial role in long-term certainty. Indeed, we confirm the findings of Massieu Trigo and Acuña Rodarte (2015) who sustain that the Mexican Law for the Promotion and Development of Biofuels has already created problems of socio-environmental violence. The findings of the current study confirm the existing problem of waste collection and recycling (Bustos Flores, 2009) in Mexican municipalities, requiring the involvement of productive companies and associations.

This research certainly corroborates the results of Brown and Mann (2008) who foresee potential for companies in the area of energy production based on municipal waste to implement small processing plants in a municipal market that benefits from waste management systems and infrastructure to collect and distribute energy economically. In the case of Mexican municipalities, the technology used to convert urban waste to energy, contrary to the assertion of Rafati and others (2016), requires support in order to take into account the composition of municipal solid waste, and the waste management strategies of the city.

This analysis also supports the study of Williams (2011) on the effectiveness of the "Waste-to-energy" but the institutional, the availability of resources and capabilities and the creation of a competitive advantage should be considered in the case of México. However, this study found that these factors may contribute to formation of public-private partnerships among cooperatives of collectors, companies and public institutions but cannot guarantee treatment of waste at low cost as Medina (2010) suggests. Of course, this research agrees with Bernache (2003) in the identification of efficient systems of municipal waste management in Mexican municipalities.

CONCLUSION AND RECOMMENDATIONS

One of the most significant conclusions of this research is the importance of public sector involvement in this type of business and, by itself, the transformation of waste into electrical energy that can be a business with a high degree of risk begins to generate long-term gains.

For this reason, a public-private partnership has been a successful mechanism. State involvement allows to generate certainty and to provide free of charge waste and existing infrastructure.

Companies that want to enter need to adapt existing technology, as a state-of-the-art technology will improve waste performance making them cost-efficient...correct the grammar. Companies have the option of partnering with other energy companies that have technology patents to be more efficient.

There are very few companies that use waste to transform it into electrical energy, which represents a great business opportunity because it is an unsaturated market in which intangible assets can be very valuable. Of course, public institutions must first work on maintaining adequate sanitary landfills for waste management and, secondly, the initiative to carry out projects and approaches in these areas.

In this way, it is concluded that there is a wasted opportunity that can generate profits for private companies, savings and profits for government institutions and an environmentally friendly solution to deal with the problem of municipal waste abundance.

REFERENCES

Banco Mundial. (2017). *Consumo de energía procedente de combustibles fósiles (% del total)*. Disponível em: http://datos.bancomundial.org/indicador/EG.USE.COMM.FO.ZS?view=chart

Barney, J., & Griffin, R. (1992). *The management of organizations: strategy, structure behavior*. Boston, MA: Houghton Mifflin Co.

Becerra Pérez, L. (2015). *La producción de etanol en México: insumos, procesos y oxigenación de gasolina*. Culiacán, MX: Universidad Autónoma de Sinaloa.

Bernache, G. (2003). *The environmental impact of municipal waste management: the case of Guadalajara metro area* (Vol. 39). Resources, Conservation and Recycling.

Brown, E., & Mann, M. (2008). *Initial market assessment for small-scale biomass-based CHP*. National Renewable Energy Laboratory. Disponível em: http://citeseerx.ist.psu.edu/viewdoc/download?doi=10.1.1.370.1389&rep=rep1&type=pdf

Bustos Flores, C. (2009). La problemática de los desechos sólidos. *Economía*, *27*, 121-144. Disponível em: http://www.redalyc.org/articulo.oa?id=195614958006

Cerdá, E., Caparrós, A., & Ovando, P. (2017). *Bioenergía en la Unión Europea*. Madrid: Fedea. Disponível em: https://www.researchgate.net/profile/Alejandro_Caparros/publication/277259293_Bioenergia_en_la_Union_Europea/links/0f3175358c6e9ac1d8000000/Bioenergia-en-la-Union-Europea.pdf

Comisión Federal de Electricidad; Instituto de Investigaciones Eléctricas. (2012). *Guía de usuario: generación de electricidad mediante residuos sólidos urbanos*. Cuauhtémoc, MX: CFE. Disponível em: http://www.olade.org/realc/docs/doc_106229_20170612115908.pdf

Favero, A., & Mendelsohn, R. (2014). Using markets for woody biomass energy to sequester carbon in forests. *Journal of the Association of Environmental and Resource Economists*, *1*(1-2), 75-95. Disponível em: http://www.jstor.org/stable/10.1086/676033

Ferrer, Y., & Pérez, H. (2010). Los microorganismos en la digestión anaerobia y la producción de biogás. Consideraciones en la elección del inóculo para el mejoramiento de la calidad y el rendimiento. *ICIDCA sobre los derivados de la caña de azúcar*, *43*(1), 9-20. Disponível em: http://www.redalyc.org/pdf/2231/223120681002.pdf

Global Methane Initiative. (2018). *Mexico Landfill Gas Model 2.0* [Modelo de Biogas Mexicano 2.0]. Disponível em https://www.globalmethane.org/partners/country.aspx?c=mexico

Intergovernmental Panel on Climate Change. (2006). *IPCC guidelines for national greenhouse gas inventories*. Hayama, JP: IPCC. Disponível em: https://www.ipcc-nggip.iges.or.jp/public/2006gl/index.html

Masera Cerutti, O. (2011). *La bioenergía en México: situación actual y perspectivas*. Red Mexicana de Bioenergía. Disponível em: http://rembio.org.mx/wp-content/uploads/2014/12/CT4.pdf

Mccormick, K., & Kautto, N. (2017). The bioeconomy in Europe: an overview. *Sustainability*, *5*(6), 2589-2608. Disponível em: http://www.mdpi.com/2071-1050/5/6/2589/html

Mead, D. (2007). Forest for energy and the role of planted trees. *Critical Reviews in Plant Sciences*, *24*(5-6), 407-421, 2007. Disponível em: http://www.tandfonline.com/doi/abs/10.1080/07352680500316391

Medina, M. (2010). *Solid wastes, poverty and the environment in developing country cities: challenges and opportunities*. Helsinki: WIDER. Disponível em: https://www.wider.unu.edu/sites/default/files/wp2010-23.pdf

Monreal, C. (2008). *La utilización de los residuos agrícolas y otros desechos para la producción de bioenergía, biocombustibles y bioproductos*. Ottawa: ECORC.

Montaño Arango, O., Corona Armenta, J. R., & Montelongo Reyes, M. (2009). Metodología sistémica para el desarrollo de un proyecto de Biogás. *Congreso Internacional de Investigación en Ciencias Administrativas, 13*. Disponível em: https://www.uaeh.edu.mx/investigacion/productos/5173/prod_12.pdf

Morita, M., & Sasaki, K. (2012). Factors influencing the degradation of garbage in methanogenic bioreactors and impacts on biogas formation. Applied Microbiology and Biotechnology, 94(3), 575-582.

Nasir, I., Ghazi, I., & Omar, R. (2012). Production of biogas from solid organic wastes through anaerobic digestion: a review. Applied Microbiology and Biotechnology, 95(2), 321-329.

North, D. C. (1990). *Institutions, institutional change and economic performance*. Cambridge, UK: Cambridge University Press. doi:10.1017/CBO9780511808678

Peng, M. (2012). *Global business*. Dallas, TX: Cengage Learning.

Pino Pérez, F. (2013). *Asociación público-privada entre Bioeléctrica S. A. de C. V. y el gobierno del Estado de Nuevo León Bioenergía de Nuevo León S. A. de C. V.* Monterrey: Tecnológico de Monterrey. Disponível em: http://www.cca.org.mx/ps/funcionarios/muniapp/descargas/Documentos_de_apoyo/informaciontematica/capp/APP_Bioelectrica.pdf

Porter, M. E. (1991). *La ventaja competitiva de las naciones*. Barcelona: Plaza & Janés.

Rafati, L. (2016). Waste-to-energy: Challenges and opportunities in Iran. *Journal of Environmental Health and Sustainable Development, Yazd, IR*, *1*(3), 163–172.

Razo, C. (2017). *Producción de biomasa para biocombustibles liquidos: el potencial de América Latina y el Caribe*. Santiago: CEPAL. Disponível em: http://www20.iadb.org/intal/catalogo/PE/2008/02231.pdf

Rodriguez, A. (2015). Management of municipal solid waste in Mexico. *International Waste Management and Landfill Symposium, 15*. Disponível em: https://www.researchgate.net/profile/Sanchez_Salinas_Enrique/publication/292326943_management_of_municipal_solid_waste_in_mexico/links/56ad1a7a08ae19a3851408ed/management-of-municipal-solid-waste-in-mexico.pdf?origin=publication_detail

Salinas Callejas, E., & Gasca Quezada, V. (2009). Los biocombustibles. *El cotidiano, 157*, 75-82. Disponível em: http://www.redalyc.org/pdf/325/32512739009.pdf

SEISA. (2006). *Our History*. Disponível em: https://seisa.com.mx/nosotros/

Sims, R. (2006). *Energy and fuelwood*. Palmerston North, NZ: Massey University.

Sistema Nacional de información ambiental y de recursos naturales. (2018). *Sistema nacional de información ambiental y de recursos naturales: informe del medio ambiente*. Disponível em: http://apps1.semarnat.gob.mx/dgeia/informe15/tema/cap7.html#tema1

Trigo, M., & Rodarte, A. B. (2017). Biofuels: socio-environmental violence in rural Mexico. *Latin American Perspectives, 42*(5). Disponível em: http://journals.sagepub.com/doi/abs/10.1177/0094582X15594398

United States Environmental Protection Agency. (2018). *Report on environment: municipal solid waste*. Disponível em: https://cfpub.epa.gov/roe/indicator_pdf.cfm?i=53

Virani, T. (2011). *The economic feasibility of bionergy production from Miscanthus for the Ontario greenhouse industry* (Master's thesis). The Faculty of Graduate Studies, University of Guelph, Ontario, Canada.

Williams, M. (2017). *Waste-to-energy success factors in Sweden and the United States: analyzing the transferability of the swedish waste-to-energy model to the United States*. Disponível em: https://pt.scribd.com/document/122133655/Waste-to-energy-success-factors-in-Sweden-and-the-United-States

Winchester, N., & Reilly, J. (2015). The feasibility, costs, and environmental implications of large-scale biomass energy. *Energy Economics, 51*, 188-203. doi:10.1016/j.eneco.2015.06.016

World Resources Institute; World Businesses Council for Sustainable Development. (2004). *The greenhouse gas protocol: a corporate accounting and reporting standard* (Revised edition). Washington, DC: WRI.

KEY TERMS AND DEFINITIONS

Biogas: Gas produced by the decomposition of organic matter.

Business Model: It is a tool prior to the business plan that will allow you to clearly define what you are going to offer to the market, how you are going to do it, to whom you are going to sell it, how are you going to sell it and how are you going to generate income.

Energy: Capacity that has the matter of producing work in the form of movement, light, heat, etc.

Mexico: Country that is located in the Southern part of North America. Its full name is United States of Mexico. Its capital is Mexico City, and it is a democratic, representative and federal republic. It comes from the Náhuat language, where Metztli means moon and Xictli means navel.

Municipal: Territorial administrative division in which a state is organized, which is governed by a municipality.

Urban Waste: Solid waste constitutes those materials discarded after their useful life, and that usually do not have economic value alone. They consist mainly of waste from materials used in the manufacture, processing, or use of consumer goods.

Vertederos: It is called landfill space in which you can pour things. This verb (to pour), on the other hand, alludes to spilling a liquid or to throwing some type of element. The most common use of a landfill is linked to the final deposit of garbage; That is why landfills are also known as garbage dumps.

Chapter 14
Conversion of Waste Into Different By-Products of Economic Value in India

Khursheed Ahmad Wani
Government Degree College Bijbehara, India

Ashaq Ahmad Dar
Pondicherry University, India

Azad Gull
P4, India

Lutfah Ariana
Indonesian Institute of Sciences, Indonesia

ABSTRACT

The management of solid waste has become a major problem even in rural areas of India, due to shrinkage of agricultural land and depletion of forest areas. During the recent past, people in rural areas were decomposing the waste, and finally, it was used as a manure in their agricultural areas. However, the trend is completely changing the Indian scenario of converting the backyard waste into manure. Now with the help of scientific knowledge, the waste is utilized as an energy resource, and waste from the rural areas is considered a raw material for this process. Different technologies in India are available to convert waste into energy apart from the technologies that have impact on the environment.

INTRODUCTION

There has been a continuous rise in industrial and technological development since the last century which results in improved living standards and population rise all over the globe i.e., in both underdeveloped and developed nations. As it is apparent, population rise is directly correlated with the increase in demand and consumption which results in a waste generation as no procedure is 100% competent. As the humans started to settle in a permanent establishment, waste has started to settle and became a

DOI: 10.4018/978-1-7998-0031-6.ch014

problem to think off. Waste disposal is one of the major problems being faced by all nations across the world. Furthermore, quantity and composition of waste generated depend upon population, the number of people living in urban areas and level of comfort which depends upon income, Table 2 (Omran *et al.* 2009; Bogner *et al.* 2007; Hoornweg *et al.* 1999).Residents living in urban areas generate two-fold waste in contrast to individuals living in rural areas with equivalent privileged circumstances, which can grow to four-fold if they are thought-out to be more affluent. A total of 13% of the global population was residing in an urban area in year the 1900, generating 0.3 million tonnes of rubbish/day in contrast to the year 2000, 3 million tonnes of waste/day was created by 49% of the total population living in cities which are expected to be double by the end of 2025. Moreover, it was reported that solid waste was increased globally by 2.02 billion tons (7%) annually from the year 2003–2006 (UNEP, 2009) in contrast to years between 2007–2008 in which it was 8% per annum, increased by 1%. If it follows the same trend, 2.2 billion tons of solid waste will be generated by the end of 2025 (World Bank, 2017). The volume will uninterruptedly upsurge (until 2050), then drops gradually may be due to progress in scientific arenas. Similarly, waste ingredients change as the country develops more and more Table 1. More money means more expenditure on packaging (packing waste), electronics appliances (e-waste), toys (broken toys), etc. thus represents the impacts of urbanization and income on the environment. In most of the countries, management of solid waste remains a major challenge. As we know from the past there is a direct correlation between development and ecological squalor,it was reported that 1.43–2.08 kg day^{-1} $person^{-1}$ and 0.77 kg day^{-1} $person^{-1}$ of waste is generated in developed and underdeveloped countries (Troschinetz & Mihelcic, 2009). In contrast to developing nations where inadequate waste management practices cause environmental damage and pose serious hazards to human health (Moreno *et al.* 2008), developed nations generate higher quantities of waste, but due to the employment of advanced waste management technologies result in a decreased waste generation (Kathiravale & MuhdYunus, 2008). However, certain countries with a similar gross domestic product per capita generate more waste may be due to high population density.Annually, a significant amount of money has been spent on waste related challenges throughout the world. A total of $205 billion was spent in 2010 which could rise up to $375 billion per annum by the end of 2025 (World Bank, 2013). Reason for indiscriminate solid waste may be inadequate disposal facility, facilities not near residence, lack of enforcement of laws, illiteracy, no fee charged, the road is no mans land, poverty, no facilities at disposal point, attitudinal problem, carelessness, ignorance or high cost of collection (Onanuga & Odunsi 2018).

Nowadays, many people are worried about the quality of the surrounding environment as environmental management challenges are apparent in most cities of the globe. An assessment has been made that per capita waste generation is increasing by about 1.3% per year (Anonymous, 2009). Most of this is a wet waste, which can be used to produce fertilizer and generate electricity. Instead, huge mounds of it lie piled up in cities and villages, posing a serious threat to public health making them vulnerable to malaria, dengue, dysentery, typhoid, and cholera, and the environment such as attraction to insects and pests, pollutionof atmosphere, hydrosphere and lithosphere, aesthetic loss, and disagreeable odors (Onanuga & Odunsi, 2018).Moreover, the leachate containing organic as well as inorganic pollutants, generated from unmanaged landfills in the rainy season may seep belowground into the aquifers.

Thus, it must be managed appropriately with the production of byproducts such as compost, bioenergy, etc. Recycling of municipal solid waste in the united states from 1960 to 2012 had elevated from 6.4% to 34.5% (EPA, 2014). Furthermore, in the year 2012, the average recycling rate in European countries was 32% with countries such as Germany, Austria, Belgium, and Switzerland top the list with

Table 1. Composition of solid waste in global regions(Heller & Nourani, 2008)

Region	Food waste	Paper/ cardboard	Wood	Textiles	Rubber/ leather	Plastic	Metal	Glass	Other
Asia									
Eastern Asia	26.2	18.8	3.5	3.5	1.0	14.3	2.7	3.1	7.4
South-central Asia	40.3	11.3	7.9	2.5	0.8	6.4	3.8	3.5	21.9
Southeast Asia	43.5	12.9	9.9	2.7	0.9	7.2	3.3	4.0	16.3
Western Asia and Middle East	41.1	18	9.8	2.9	0.6	6.3	1.3	2.2	5.4
Africa									
Eastern Africa	53.9	7.7	7	1.7	1.1	5.5	1.8	2.3	11.6
Middle Africa	43.4	16.8	6.5	2.5		4.5	3.5	2	1.5
Northern Africa	51.1	16.5	2	2.5		4.5	3.5	2	1.5
Southern Africa	23	25	15						
Western Africa	40.4	9.8	4.4	1		3	1		
Europe									
Eastern Europe	30.1	21.8	7.5	4.7	1.4	6.2	3.6	10	14.6
Northern Europe	23.8	30.6	10	2		13	7	8	
Southern Europe	36.9	17	10.6						
Western Europe	24.2	27.5	11						
Oceania									
Australia and New Zealand	36	30	24						
Rest of Oceania	67.5	6	2.5						
America									
North America	33.9	23.2	6.2	3.9	1.4	8.5	4.6	6.5	9.8
Central America	43.8	13.7	13.5	2.6	1.8	6.7	2.6	3.7	12.3
South America	44.9	17.1	4.7	2.6	.7	10.8	2.9	3.3	13
Caribbean	46.9	17	2.4	5.1	1.9	9.9	5	5.7	3.5

a recycling rate between 50 – 64% (Eurostat, 2012). Different factors such as public awareness, lack of funding, type of waste, etc. play a significant role in material reuse and recycling apart from the progress. Robust waste recycling and recovery are expected in the upcoming years throughout the globe, as due to increasing awareness and pollution-related environmental problems. Several new techniques along with the conventional methods that are used to divert the waste include, composting, vermicomposting, incineration, landfilling, plastic roads, biogas technology, etc.

Table 2. 1995 and 2025 municipal solid waste generation in Asia (Hoornweg, 1999)

	GNP per capita (US$)		Urban population (% of total)		MSW (kg day^{-1}per capita)	
Country	**1995**	**2025**	**1995**	**2025**	**1995**	**2025**
Low income	490	1050	27.8	48.8	.64	.6–1
Nepal	200	360	13.7	34.3	.50	.6
Bangladesh	240	440	18.3	40	.49	.6
Myanmar	240	580	26.2	47.3	.45	.6
Vietnam	240	580	20.8	39	.55	.7
Mongolia	310	560	60.9	76.5	.60	.9
India	340	620	26.8	45.2	.46	.7
Laos	350	850	21.7	44.5	.69	.8
China	620	1500	30.3	54.5	.79	.9
Sri Lanka	700	1300	22.4	42.6	.89	1
Middle income	1410	3390	37.6	61.1	.73	.8–1.5
Indonesia	980	2400	35.4	60.7	.76	1
Philippines	1050	2500	54.2	74.3	.52	.8
Thailand	2740	6650	20	39.1	1.10	1.5
Malaysia	3890	9400	53.7	72.7	.81	1.4
High income	30990	41140	79.5	88.2	1.64	1.1–4.5
Republic of Korea	9700	17600	81.3	93.7	1.59	1.4
Hong Kong	22990	31000	95	97.3	5.07	4.5
Singapore	26730	36000	100	100	1.10	1.1
Japan	39640	53500	84.9	84.9	1.47	1.3

PLASTIC WASTE RECYCLING

Plastic, an astonishing, marvellous, ubiquitous petroleum product, turned out to be a key commodity, has occupied a critical part in day today life and is considered to be one of the utmost noteworthy novelties of the second half of the 20th century due to its corrosion resistance, light weight, strength, transparency, harmlessness and durability. Due to its inherent characteristics, it has found vast applications in almost every aspect of life *viz.*, in the packaging (40%) (Plastics Europe, 2018), construction, electronics, health sector, paints, and energy production, etc. Thermoplastics and thermo-setting plastics two types of plastic are mostly used. Globally, thermoplastics have a huge market because of its mouldable and re-mouldable capabilities (Akçaözoğlu*et al.* 2010).

Since the commercial production (in 1950), globally 6.3 billion tons of plastic waste has been generated (Ellen MacArthur Foundation, 2016) which grew up to 8.3 billion MT in 2017 (Geyer *et al.* 2017, Wright & Kelly 2017). A total of 2 million metric tons (Mt), 322 Mt, 330 Mt and, 380 Mt were produced in the year 1950, 2015, 2016 and 2017, respectively, (Plastics Europe, 2018; Plastics Europe, 2017, Geyer *et al.* 2017) at an annual rate of 8.4%. In developing countries are good competitors of developed countries in terms of using plastic products. In India, in the year 2008, 8 million tonnes which rose up to 12 million tonnes in 2012 of plastic products were used (Singh *et al.* 2016). As per the report of the British Plastics Federation 2008, a total of 4% of annual global petroleum production is converted into plastic products which are expected to rise up to 20% by the mid if this century (World Economic Forum, 2016). However, almost everything including plastic has its own limitations. In the 1970s the impact of plastic on the environment was revealed (Bergmann, 2015). It was reported that most of the plastic recycled, only 9% (Jambeck*et al.* 2015) while the rest is being dumped in landfills and/or result in environmental contamination, was converted into less recyclable products (Brooks *et al.* 2018). Annually about 4 million to 12 million metric tons have been dumped in oceans and are rising continuously (Schweitzer *et al.* 2018; Plastics Europe, 2017; Geyer *et al.* 2017). Plastic, a major constituent of marine litter (90%) mostly (61%) contributed by plastic packaging (Schweitzer *et al.* 2018; Lebreton*et al.* 2017). The oceans are not only distributors of plastic worldwide via ocean currents, they also act as grinders of plastic to small particles which result in 15–51 trillion particles of plastic freely moving on ocean surface (Sebille *et al.* 2015) and may bioaccumulate and biomagnify along food chain within the marine biota (Romera-Castillo *et al.* 2018).

Plastic poses a significant threat to life on the planet earth. They are non-biodegradable and can cause water pollution because of the chemical constituents, soil pollution when disposed of and air pollution when burned thus, makes its disposal a major concern for local authorities. In the absence of proper solid waste management, a small air current can spread it everywhere results in choking of the drainage system, resulting in detrimental public health and environment concerns (Lithner *et al.* 2011). Plastic waste is the most visible constituent of any dumping site as it can persist for more than 500 years (Oehlmann *et al.* 2009). According to the World Economic Forum 2016, if the current rate of plastic production will continue, greenhouse gas emission from plastic will constitute 15% of the total carbon budget by the end of this mid-century. Similarly, plastic in the marine environment can also have severe direct as well as indirect impact on sea biota which includes habitat destruction, problems related to navigation, ingestion, suffocation, etc. These problems were reported in more than 700 species of marine organisms (Wright *et al.* 2013; Gall *et al.* 2015). Moreover, microalgal exposure to plastic can lead to its 45% growth reduction (Sjollema *et al.* 2016). It was also reported that plastic debris can also act as a disease carrier as it can enhance the likelihood from 4 – 89% (Lamb *et al.* 2018). Due to these reasons, several countries have barred or are planning prohibitions or fees on plastic use.

Due to its wide use and material characteristics that limit its recycling attempts plastic can be challenging to recycle. Plastic waste recycling is the best strategy for the management of plastic waste. It is a technique by which one can diminish its detrimental influences on the ecosystem and avoid exhaustion of resources and reserves, consequently, declines energy and virgin material use per unit of yield and so product amended efficiently (Hamad *et al.* 2013; Singh *et al.* 2017). From the last two decades, researchers are focussing to find appropriate means for reusing plastic waste ingredients in an economical and eco-friendly manner. Various practical techniques have been industrialized for this purpose.

Pyrolysis which involves the combustion in limited air supply at a high temperature usually >300ºC, were used to break long polymeric chains to get the new usable products. Pyrolysis of high-density polyethylene waste with subsequent distillation leads to a hydrocarbon blend with a structure containing 96.8%saturated aliphatic paraffinic hydrogens, 2.6% aliphatic olefinic hydrogens and 0.6% aromatic hydrogens. Furthermore, the boiling point range of mixture was paralleled with diesel fuel (Sharma *et al.* 2014). The extraction of petroleum products from plastic waste was also reported by Butler *et al.* (2011). Recycling of plastic waste to concrete seems to be best due to its monetary and environmental benefits. Techniques were developed to use fine aggregates of plastic waste along with concrete mixture to get the more reliable product of high strength (Bhogayata & Arora, 2018; Saikia & de Brito, 2012). The use of plastic waste in road construction can help to reduce the immense pressure on landfills. Plastic roads mainly use plastic carry-bags, disposable cups and PET bottles that are collected from garbage dumps as an important ingredient of the construction material. Bitumen, an important road binder has low Marshall quotient, water resistance, rutting capacity and stability. Waste plastic is grounded and powder is made and 3 to 4% plastic is mixed with the bitumen. A blend of low-density polyethylene of size ranging from 0.30 – 0.92 mm to asphalt, was reported to increase the Marshall quotient by two folds and amplified the stability by 15%. Moreover, while evaluating the performance of asphalt containing plastic waste; mixture type, particle size, and nature and processing technique of waste should be considered as these can affect significantly (Huang *et al.* 2007). Also, the mixing of plastic with bitumen aids in retaining high temperatures. Plastic roads will benefit for hot and extremely humid climatic regions, where temperatures frequently cross 50°C and torrential rains create havoc, leaving most of the roads with big potholes. It is hoped that in the near future we will have strong, durable and eco-friendly roads which will relieve the earth from all types of plastic-waste. Research is also going on to use plastic waste for synthesising carbon nanotubes (Gong *et al.* 2013). The use of plastic waste in tiles is a new approach to tackle plastic waste in an eco-friendly manner. They are lightweight, corrosion and chemical resistant, low cost for production, etc (Dhawan *et al.* 2019).

VERMICOMPOSTING

In order to preserve the natural resources and also the finest use of conventional energy sources in an eco-friendly manner has stimulated the reprocessing and recovery of decomposable waste, an alternative to landfilling; as organic waste may constitute 30-70% of the municipal solid waste (Savino, 2009) which can reach up to 80% in some developing countries (Troschinetz & Mihelcic, 2009; Moreno *et al.* 2008) and is continuously rising and is predicted to constitute 75% of total municipal solid waste globally by the end of 2025 (Hoornweg, 1999), and burning. One of such method is composting, which involves the stabilisation through natural biological disintegration of organic substances in principally aerobic conditions (De Araujo *et al.* 2010) as compared to the abandoned natural decay of organic substances, the temperature can elevate to the ranges favourable for mesophilic and thermophilic organisms. The biologically stable humus-like substance is the end product of the process which can be used as a soil conditioner. Composting can be used to achieve stabilization, space reduction, aeration, removal of phytotoxic compounds and also cleanliness. Theoretically, any biodegradable waste can be treated by this technique. Composting of organic waste prior to landfilling can also diminish greenhouse gas emissions.

Vermicomposting involves the nonthermophilic breakdown of biodegradable ingredients of waste in an ecologically sound manner up to an extent so that it can be stockpiled, handled, and applied to farms without any antagonistic implication (Singh et al. 2011; Khwairakpam & Bhargava, 2009). It is a joint action of earthworms (conditioning and biological activity) and microorganisms (biological degradation) under environmental controlled conditions which coverts biodegradable material into two useful products *viz.,* earthworm biomass and humus-like product known as vermicompost (Suthar 2009; Singh et al. 2011). Among the various species of earthworms, *Eisenia fetida* is used almost all over the globe whereas as one more specie *Eudrillusfetida* is common in tropical and subtropical regions of both the hemispheres. The different types of vermicomposting systems are the windrow system, pits, tanks, rings, etc. to provide a suitable environment for worm functioning. In contrast to composting, vermicomposting is considered as one of the low-cost sustainable practices for handling harmless and organic waste, particularly when the availability of space is limited. It is not only eco-friendly but also economical and can replace and reduce the use of chemical fertilisers, thus improves the soil quality. Furthermore, it is the only pollution control technology that uses multicellular animal (earthworm) as the main agent (Abbasi *et al.* 2009). Because of high organic content in waste streams which should be returned to the soil, the technique of vermicomposting would be highly recommended. Some studies have reported that that vermicomposting can kill pathogens (Neklyudov*et al.* 2008; Dominguez *et al.* 1997). These have a conflict with other studies that reported that it lacks the ability to kill pathogens and is one of the major drawback of vermicomposting. Therefore, an integrated approach should be used composed of both the techniques i.e., composting and vermicomposting. Factors affecting vermicomposting include feeding, temperature (25–37° C), pH (4.5–9.0), moisture (50–80%), stocking density (1.60 kg worms/m^2) and C/N ratio (Ndegwa*et al.* 2000).

The suitability of garbage to earthworms can be predicted from the fact that the population of earthworms (*P. excavatus*) in the garbage was amplified. Furthermore, the C/N ratio was increased as compared to soybean, wheat and chicken pea straw, when earthworms were inoculated (Manna et al. 1997). A better quality of vermicompost from biodegradable waste can be made by using this technique (Lodha, 2007). It was reported that earthworm species *E. fetida* and *L. mauritii* can attain an extreme carbon reduction of 44.3% and 44.1% after a certain period (Kaviraj & Sharma, 2003). Vermicompost of better quality (high organic matter, total nitrogen and, nutrient availability) was generated by *E. fetida* from waste subjected to composting and shredding prior vermicomposting technique (Tognetti et al. 2007). Moreover, vermicomposting of composted organic waste can be also be done without amalgaming with the substrate using *E. fetida* if the waste is subjected to conventional composting for 21 days (Gheisari*et al.* 2010). However, Pramanik*et al.*reported a lesser nitrogen content of vermicompost prepared from municipal solid waste after70–85 days, which might be due to lower nutrient and C/N ratio of the waste. An increase in nitrate and phosphorus ion was reported by involving vermicompost technology when compared with the traditional composting (Tognetti *et al* 2007).

TECHNIQUES TO MANAGE AND CONVERT WASTE INTO ENERGY

Plastics Roads

According to recent studies, plastics can stay unchanged for as long as 4500 years on earth with increase in the global population and the rising demand for food and other essentials, there has been a rise in the amount of waste being generated daily by each household. A Bangalore-based firm and a team of engineers from R. V. College of Engineering, Bangalore, have developed a way of using plastic waste for construction of roads. Plastic roads mainly use plastic carry-bags, disposable cups and PET bottles that are collected from garbage dumps as an important ingredient of the construction material. Waste plastic is grounded and powder is made and 3 to 4% plastic is mixed with the bitumen. Plastic is used to increases the melting point of the bitumen and helps road to retain its flexibility during winters for its long life. The mixing of plastic with bitumen aids in retaining high temperature. The plastic waste is melted and mixed with bitumen in a particular ratio. The vigorous tests at the laboratory level proved that the bituminous concrete fulfilled all the specified Marshall mix design criteria for surface course of road pavement (Chavan, 2013).

Plastic roads will benefit for hot and extremely humid climatic regions, where temperatures frequently cross 50°C and torrential rains create havoc, leaving most of the roads with big potholes. It is hoped that in near future we will have strong, durable and eco-friendly roads which will relieve the earth from all type of plastic-waste.

Nisargruna biogas technology: Dr Sharad Kale (Padma Shri) Head of the Technology Transfer Division of the Bhabha Atomic Research Centre, Mumbai designed the Nisargruna biogas technology which has been used to set up several biogas plants in the country.

The waste resources which can be processed using Nisargruna Biogas plant include biodegradable kitchen waste, paper waste, green grass, leaf litter, animal remains in abattoirs, hospital waste, green plant waste, cow dung, crop residues, sugarcane, baggase, water hyacinth, etc. The waste is first segregated carefully to remove non-biodegradable material and then homogenized to make a slurry. It is then processed in a sequential manner first by aerobic and then by anaerobic process. The products of the process are biogas consisting of methane, carbon dioxide and water vapor & weed free good quality manure. About 70% of water can be recycled at the end of the process. The Bio gas plants are useful for Municipal Corporations, Hospitals, Hotels, Housing Societies, Govt. Establishments, Abattoirs, Nagar Palika, Grampanchayat, Farmers. At present, 160 Nisargruna plants are functional in the country (http://technogreen.net.in).

Figure 1.
Adopted for: technogreen.net.in

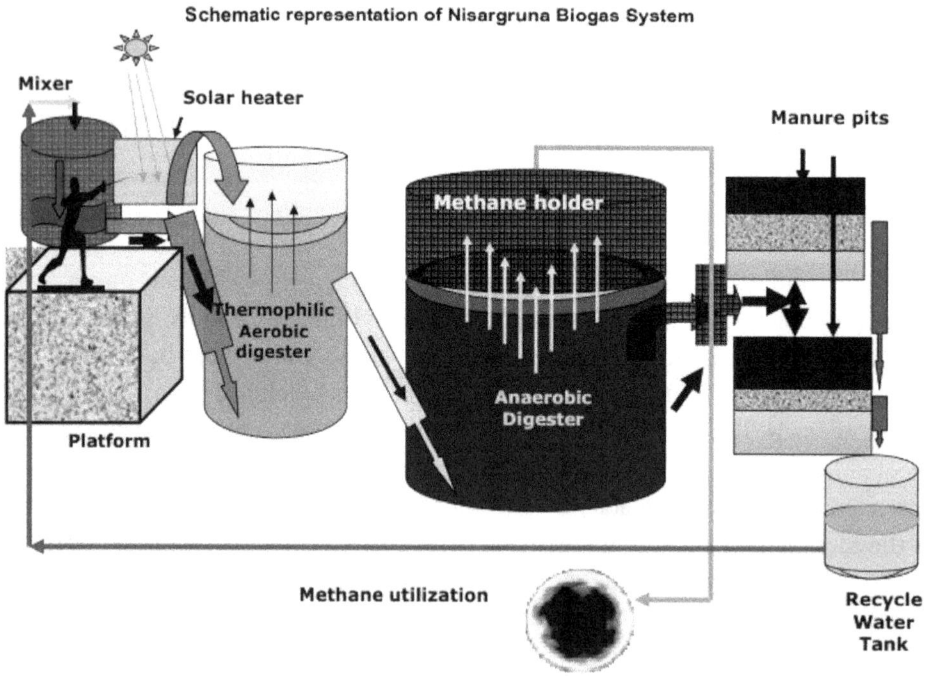

SOLID AND LIQUID RESOURCE MANAGEMENT (SLRM) AND VELLORE MODEL

The SLRM system was implemented as a pilot project in Dollars Colony in Bangalore, where a compost shed was established using locally available materials like casuarina to undertake segregation of waste and composting of the biodegradable waste. The residents and the workers of the BBMP (Bruhat Bengaluru Mahanagara Palike) or the Greater Bangalore Municipal Corporation were taught about secondary and tertiary segregation of the recyclable waste (paper, plastics, bottles, wood, etc.) and the classification of the different kinds of plastic and paper waste, the price of the various categories of recyclable waste and the buyers who purchased them. The 90% wet waste is sold after segregation and the 10% that include napkins goes to landfills. This waste is also composted which turns wet waste into manure is a great help towards realizing this zero waste management. In 60 days sweet smelling manure is produced which can be either sold to generate income for the women workers or used to grow vegetables in a terrace garden. The vegetables will be organic, free from pesticide and excellent for one's health (rural.nic.in).

Kagad Kach Patra Kashtakari Panchayat (KKPKP)

Waste picking is considered a stigmatized profession in most parts of the world. In Pune, Kagad Kach Patra Kashtakari Panchayat (KKPKP) is helping waste pickers to fight for their rights and work for the betterment of their community. The organization highlighted the fact that waste pickers contribute enormously to the city and they see their work as respectable and environmentally beneficial. KKPKP launched a pilot project in 2005 for door-to-door waste collection work that led to the formation of SWaCH (a workers' cooperative run entirely by waste pickers and for them) and the integration of waste pickers into Pune municipality's waste collection system. This organization provides a variety of waste management services to households, including door-to-door waste collection. The process of recyclables also helps them generate additional income and improve their quality of life. It has reached over 4,00,000 households in 15 municipal wards on a regular basis. All the members of this organization have taken the oath to educate their children and discourage child labour. It provides scholarships and educational loans to its members and children of waste pickers have finished school and are now studying in college (www.wastepickerscollective.org).

STREE MUKTI SANGHATANA

Mumbai-based stree Mukti Sanghatana has helped thousands of women to improve their lives and escape poverty by recycling waste and create zero-waste cities. Parisar Vikas, one of the NGO's key programmes, aims to works closely with self-employed women. Besides training these women in composting, bio-methanation, micro-saving and leadership skills, the NGO has also helped many women find alternative employment. Currently, the NGO is building an innovative facility in Navi Mumbai that will offer trainings in water recycling, rainwater harvesting, bio-methanation and composting. This centre is opened to anyone interested in learning how to manage, treat and recycle solid waste (streemuktisanghatana.org).

CONCLUSION

With rapid urbanization and incessant population growth, the amount of garbage India produces is only going to grow. The waste will turn into a resource when treated properly and to ensure that it gets the right treatment, so that any other productive resource will be obtained out of it. Some simple techniques that have been discussed seem a viable option. These techniques may reduce the waste footprint and pressure on local authorities, especially at the municipal level, to deal with waste management. However, these techniques have some political, social and scientific limitations that need to be addressed in due course of time.

REFERENCES

Abbasi, T., Gajalakshmi, S., & Abbasi, S. A. (2009). Towards modeling and design of vermicomposting systems: Mechanisms of composting/ vermicomposting and their implications. *Indian Journal of Biotechnology*, *8*, 177–182.

Akçaözoğlu, S., Atiş, C. D., & Akçaözoğlu, K. (2010). An investigation on the use of shredded waste PET bottles as aggregate in lightweight concrete. *Waste Management (New York, N.Y.)*, *30*(2), 285–290. doi:10.1016/j.wasman.2009.09.033 PMID:19853433

Ali, U., Sajid, N., Khalid, A., Riaz, L., Rabbani, M.M., Syed, J.H., & Malika, R.N. (2015). A Review on Vermicomposting of Organic Wastes. *Environmental Progress & Sustainable Energy*.

Bergmann, M., Gutow, L., & Klages, M. (Eds.). (2015). *MarineAnthropogenicLitter*. New York: Springer International Publishing.

Bhogayata, A. C., & Arora, N. K. (2018). Impact strength, permeability and chemical resistance of concrete reinforced with metalized plastic waste fibers. *Construction & Building Materials*, *161*, 254–266. doi:10.1016/j.conbuildmat.2017.11.135

Bogner, J., Abdelrafie Ahmed, M., Diaz, C., Faaij, A., Gao, Q., & Hashimoto, S. (2007). Waste management, in Climate Change: Mitigation of Climate Change. Contribution of Working Group III to the Fourth Assessment Report of the Intergovernmental Panel on Climate Change. Cambridge University Press.

British Plastics Federation. (2008). *Oil consumption*. British Plastics Federation. Retrieved from http://www.bpf.co.uk/press/Oil_Consumption

Brooks, A.M., Wang, S., & Jambeck, J.R. (2018). The Chinese import ban and its impact on globalplastic waste trade. *Science Advanced, 4*.

Butler, E., Devlin, G., & McDonnell, K. (2011). Waste Polyolefins to Liquid Fuels via Pyrolysis: Review of Commercial State-of-the-Art and Recent Laboratory Research. *Waste and Biomass Valorization*, *2*(3), 227–255. doi:10.100712649-011-9067-5

Chavan, A. (2013). Use of plastic waste in flexible pavements. *IJAIEM*, *2*(4), 540–553.

De Araujo, A. S. F., de Melo, W. J., & Singh, R. P. (2010). Municipal solid waste compost amendment in agricultural soil: Changes in soil microbial biomass. *Reviews in Environmental Science and Biotechnology*, *9*(1), 41–49. doi:10.100711157-009-9179-6

Dhameja, S. K. (2009). *Environmental Engineering and management*. Delhi: S.K. Kataria and Sons.

Dhawan, R., Bisht, B.M.S., Kumar, R., Kumari, S., & Dhawan, S.K. (2019). Recycling of plastic waste into tiles with reduced flammability and improved tensile strength. *Process Safety and Environmental Protection*. . doi:10.1016/j.psep.2019.02.018

Dominguez, J., Edwards, C., & Subler, S. (1997). Comparison of vermicomposting and composting *BioCycle*. *Journal of Composting and Organics Recycling*, *38*, 57–59.

Europe, P. (2017). *Plastics—The facts*. Plastic Europe. Retrieved from https://www.plasticseurope.org/en/resources/publications/274-plastics-facts-2017

Europe, P. (2018). *Plastics—the Facts 2017*. Retrieved from www.plasticseurope.org/application/files/5715/1717/4180/Plastics_the_facts_2017_FINAL_for_website_one_page.pdf

Gall, S. C., & Thompson, R. C. (2015). The impact of debris on marine life. *Marine Pollution Bulletin*, *92*(1-2), 170–179. doi:10.1016/j.marpolbul.2014.12.041 PMID:25680883

Geyer, R., Jambeck, J. R., & Law, K. L. (2017). Production, use, and fate of all plastics ever made. *Science Advances*, *3*(7), e1700782. doi:10.1126ciadv.1700782 PMID:28776036

Gong, J., Yao, K., Liu, J., Wen, X., Chen, X., Jiang, Z., Mijowska, E., & Tang, T. (2013). Catalytic conversion of linear low density polyethylene into carbon nanomaterials under the combined catalysis of Ni2O3 and poly(vinyl chloride). *Chemical Engineering Journal, 215-216*, 339-347.

Government of India. (2009). *The solid waste management sector in India*. Department of Economic Affairs Ministry of Finance.

Hamad, K., Kaseem, M., & Deri, F. (2013). Recycling of waste from polymer materials: An overview of the recent works. *Polymer Degradation & Stability*, *98*(12), 2801–2812. doi:10.1016/j.polymdegradstab.2013.09.025

Hoornweg, D. (1999). *What a Waste: Solid Waste Management in Asia*. Washington, DC: The International Bank for Reconstruction and Development/The World Bank.

Hoornweg, D. (1999). *What a Waste: Solid Waste Management in Asia*. The International Bank for Reconstruction and Development/The World Bank. Retrieved from http://technogreen.net.in/downloads/nisargruna-biogas-technology_concept-note.pdf[REMOVED HYPERLINK FIELD]

Huang, Y., Bird, R. N., & Heidrich, O. (2007). A review of the use of recycled solid waste materials in asphalt pavements. *Resources, Conservation and Recycling*, *52*(1), 58–73. doi:10.1016/j.resconrec.2007.02.002

Jambeck, J. R., Andrady, A., Geyer, R., Narayan, R., Perryman, M., Siegler, T., ... Law, K. L. (2015). Plastic waste inputs from land into the ocean. *Science*, *347*(6223), 768–771. doi:10.1126cience.1260352 PMID:25678662

Kathiravale, S., & Muhd Yunus, M. N. (2008). Waste to wealth. *Asia Europe Journal*, *6*(2), 359–371. doi:10.100710308-008-0179-x

Kaviraj & Sharma. (2003). Municipal solid waste management through vermicomposting employing exotic and local species of earthworms. *Bioresour Technol*, *90*, 169–173.

Lamb, J. B., Willis, B. L., Fiorenza, E. A., Couch, C. S., Howard, R., Rader, D. N., ... Harvell, C. D. (2018). Plastic waste associated with disease on coral reefs. *Science*, *359*(6374), 460–462. doi:10.1126cience.aar3320 PMID:29371469

Lebreton, L. C. M. (2017). River plastic emissions to the world's oceans. *NatureCommunication*, *8*, 15611. doi:10.1038/ncomms15611 PMID:28589961

Lithner, D., Larsson, A., & Dave, G. (2011). Environmental and health hazard ranking and assessment of plastic polymers based on chemical composition. *Science Total Environment, 409*(18).

Lodha, B. (2007). Effect of maturity and nutritional parameters during vermicomposting of municipal solid waste. *International Journal of Environmental Technology and Management*, *7*(3/4), 454–463. doi:10.1504/IJETM.2007.015157

Manna, M. C., Singh, M., Kundu, S., Tripathi, A. K., & Takkar, P. N. (1997). Growth and reproduction of the vermicomposting earthworm *Perionyx excavatus* as influenced by food materials. *Biology and Fertility of Soils*, *24*(1), 129–132. doi:10.1007/BF01420233

Moreno, E. L., Bazoglu, N., & Mboup, G. (2008). State of the World's Cities 2008/2009: Harmonious Cities. Earthscan.

Moreno, E. L., Bazoglu, N., Mboup, G., & Warah, R. (2008). *State of the World's Cities 2008/2009: Harmonious Cities*. London: Earthscan.

Ndegwa, P., Thompson, S., & Das, K. (2000). Effects of stocking density and feeding rate on vermicomposting of biosolids. *Bioresource Technology*, *71*(1), 5–12. doi:10.1016/S0960-8524(99)00055-3

Neklyudov, A., Fedotov, G., & Ivankin, A. (2008). Intensification of composting processes by aerobic microorganisms: A review. *Applied Biochemistry and Microbiology*, *44*(1), 6–18. doi:10.1134/S000368380801002X

Oehlmann, J., Schulte-Oehlmann, U., Kloas, W., Jagnytsch, O., Lutz, I., Kusk, K. O., ... Tyler, C. R. (2009). A critical analysis of the biological impacts of plasticizers on wildlife. *Philosophical Transactions of the Royal Society of London. Series B, Biological Sciences*, *364*(1526), 2047–2062. doi:10.1098/rstb.2008.0242 PMID:19528055

Omran, A., & Mahmood, A., AbdulAziz, H., & Robinson, G.M. (2009). Investigatinghouseholds attitude towards recycling of solid waste in Malaysia:a case study. *International Journal of Environmental of Research*, *3*, 275–288.

Onanuga, M. Y., & Odunsi, O. (2018). Health is wealth: Concern for households' solid waste self-disposal practices. *Environmental Quality Management*, 1–9.

Pramanik, P., Ghosh, G. K., Ghosal, P. K., & Banik, P. (2007). Changes in organic-C, N, P and K and enzyme activities in vermicompost of biodegradable organic wastes under liming and microbial inoculants. *Bioresource Technology*, *98*(13), 2485–2494. doi:10.1016/j.biortech.2006.09.017 PMID:17081750

Romera-Castillo, C., Pinto, M., Langer, T. M., Álvarez-Salgado, X. A., & Herndl, G. J. (2018). Dissolved organic carbon leaching from plastics stimulates microbial activity in the ocean. *Nature Communication*, *9*, 1430. Retrieved from rural.nic.in/sites/downloads/pura/ZWM_models

Saikia, N., & de Brito, J. (2012). Use of plastic waste as aggregate in cement mortar and concrete preparation: A review. *Construction & Building Materials*, *34*, 385–401. doi:10.1016/j.conbuildmat.2012.02.066

Savino, A. A. (2009). *Waste and Climate Change: ISWA White Paper*. International Solid Waste Association.

Schweitzer, J. P., Gionfra, S., Pantzar, M., Mottershead, D., Watkins, E., Petsinaris, F., ... Janssens, C. (2018). *Unwrapped: How Throwaway Plastic is Failing to Solve Europe's Food Waste Problem (And What We Need To Do Instead)*. Institute for European Environmental Policy.

Sharma, K., Moser, B. R., Vermillion, K. E., Doll, K. M., & Rajagopalan, N. (2014). Production, characterization and fuel properties of alternative diesel fuel from pyrolysis of waste plastic grocery bags. *Fuel Processing Technology*, *122*, 79–90. doi:10.1016/j.fuproc.2014.01.019

Singh, N., Hui, D., Singh, R., Ahuja, I. P. S., Feo, L., & Fraternali, F. (2017). Recycling of plastic solid waste: A state of art review and future applications. *Composites. Part B, Engineering*, *115*, 409–422. doi:10.1016/j.compositesb.2016.09.013

Singh, R., Embrandiri, A., Ibrahim, M., & Esa, N. (2011). Management of biomass residues generated from palm oil mill: Vermicomposting a sustainable option. *Resources, Conservation and Recycling*, *55*(4), 423–434. doi:10.1016/j.resconrec.2010.11.005

Sjollema, S. B. (2016). Do plastic particles affect microalgal photosynthesis and growth? *AquaticToxicology*, *170*, 259–261. PMID:26675372

Suthar, S. (2009). Potential of Allolobophora parva (Oligochaeta) in vermicomposting. *Bioresource Technology*, *100*(24), 6422–6427. doi:10.1016/j.biortech.2009.07.004 PMID:19648000

The New Plastics Economy: Rethinking the Future of Plastics. (n.d.). Retrieved from https://www. ellenmacarthurfoundation.org/publications/the-new-plastics-economy-rethinking-the-future-ofplastics

Tognetti, C., Mazzarino, M. J., & Laos, F. (2007). Cocomposting biosolids and municipal organic waste: Effects of process management on stabilization and quality. *Biology and Fertility of Soils*, *43*(4), 387–397. doi:10.100700374-006-0164-8

Tognetti, C., Mazzarino, M. J., & Laos, F. (2007). Improving the quality of municipal organic waste compost. *Bioresource Technology*, *98*(5), 1067–1076. doi:10.1016/j.biortech.2006.04.025 PMID:16822669

Troschinetz, A. M., & Mihelcic, J. R. (2009). Sustainable recycling of municipalsolid waste in developing countries. *Waste Management (New York, N.Y.)*, *29*(2), 915–923. doi:10.1016/j.wasman.2008.04.016 PMID:18657963

United Nations Environmental Programme. (2009). *Developing integrated solid waste management plan: Training manual: Assessment of current waste management systems and gaps therein* (Vol. 2). Osaka, Japan: Author. Retrieved from http://www.unep.or.jp/ietc/Publications/spc/ISWMPlan_Vol2.pdf

van Sebille, E., Wilcox, C., Lebreton, L., Maximenko, N., Hardesty, B. D., van Franeker, J. A., ... Law, K. L. (2015). A global inventory of small floating plastic debris. *Environmental Research Letters, 10*(12), 124006. doi:10.1088/1748-9326/10/12/124006

World Bank. (2017). *Brief: Solid waste management*. Washington, DC: Author. Retrieved from https://www.worldbank.org/en/topic/urban development/brief/solid-waste-management

World Economic Forum. (2016). *The new plastics economy. Rethinking the future of plastics*. World Economic Forum.

Wright, S. L., Thompson, R. C., & Galloway, T. S. (2013). The physical impacts of microplastics on marine organisms: A review. *Environmental Pollution, 178*, 483–492. doi:10.1016/j.envpol.2013.02.031 PMID:23545014

Chapter 15
Vermitechnology:
A Solution for Agricultural Waste

Kumar Anand
Vinoba Bhave University Hazaribagh, India

Pritam Bala Sinha
Amity University, India

ABSTRACT

Vermicomposting has been rising as an innovative ecotechnology for the conversion of various types of wastes into vermicompost. Vermicompost is humus like finely granulated and stabilized material which can be used as a soil conditioner to reintegrate the organic matter to the agricultural soils. Non-toxic and organic industrial wastes could be potential raw material for vermitechnology. The success of the vermitechnology depends upon several process parameters like quality of raw material, pH, temperature, moisture, aeration, type of vermicomposting system, and earthworm species used. Vermicomposting is a suitable means for waste remediation and organic manure production. Physical processes include substrate aeration, mixing, and grinding while biochemical processes involves decomposition of waste by various enzymes present in the gut of earthworms and is influenced by microbes present in their intestine. Earthworms have several medicinal properties and are also known to accumulate toxic residues from soil/substrates. The role of earthworms in sustainable farming is immense.

INTRODUCTION

It has been already made clear that one of the prominent problems which are being faced by mankind is solid waste management. Due to successive increase in the amount of waste being generated from various developments worldwide, there is an encroachment of fertile area and population explosion. Also due to fast urbanization, massive amount of waste is being generated (Atiyeh, et al., 1999; Kumar et al., 2017).Solid waste such as organic and inorganic waste materials are invaluable according to their owner. As a developing nation, India produces about 25 million tons of urban solid waste. Rough estimates indicate that near about 0.4kg/day of waste is being produced per capita. Approximately 50-60% of matter is compostable, i.e. 0.4kg/day. Among the most common practices related to waste production

DOI: 10.4018/978-1-7998-0031-6.ch015

are deliberate dumping of the wastes to soil and also to water which results in extensive soil and water pollution. For the disposal of solid waste finally, there is a need to incorporate several other methods as like incineration and composting, other than sanitary land filling as well as dumping. For the conversion of solid organic waste to compost, vermiculture or most often said as earthworm farming is employed (Aalok et al., 2008; Ghosh, 2004; PBCB, 2015). Vermiculture as a discipline of biotechnology involves, the breeding and propagation of earthworms. In vermiculture we use the castings of it. Vermiculture has become a significant tool of waste recycling worldwide and has also been considered as a separate and fruitful discipline of biology. "vermitechnology" involves a low cost and environmentally sound waste management practice involving the use of earthworms in the form of natural bioreactors (Atiyeh et al., 2000).

Earthworms break the organic debris present on the surface of soil and soil turn over process, as already it has been explained by Darwin (1881). With the successful use of facilities provided due to vermicomposting, the technology has explored domestic market as well as industrial marketing of it is being done in several counties like Canada, USA, Japan and Italy. First of all, the process of vermicomposting was started in Ontario (Canada) in the year 1970. At present about75 tons of refuse per week is being processed. During 1978-80, the American Earthworm Company initially prepared a farm which had the capacity to produce about 500 tonnes per month. A well-known company of Japan, Aoka Sangyo Co. Ltd., produces three 1000 tonnes per month plants processing wastes from pulp and food industries. Other than these, about 3000 other vermicomposting plants in Japan also produces 5-50 tones capacity of vermicompost per month. Similar vermicomposting plants are also in Italy and Philippines. India should think about commercialization of "vermitechnology" (Bhawalkar, 1989; Belliturk, 2016).

Being a biological process, it involves interactions between earthworms and microorganisms. Due to this there is an efficient conversion of different type of organic wastes into nutrient rich manure (Amouei et al., 2017).There is also the use of worms during vermicomposting process, due to which worms are able to transfer organic waste into a nutrient rich fertilizer. Mutual actions exist between earthworms and microorganisms which alter the physical, chemical and biological properties of waste material and convert them into vermicompost. Vermicompost can suitably be homogeneous, stabilized, peat like material, odour free which contains significant quantities of nutrients with low level of toxicants (Dominguez & Edwards, 2007; Ndegwa & Thompson, 2001).

Vermicomposting is amongst one of the healthy as well as hygienic way to eliminate wastes going into our landfills, which improves the environment. Within 2-3 months results are produced and also it only takes two or three months to produce results and is inexpensive. Not only being a cost-effective process, it is also a simple an efficient method which degrades all of the non-toxic biodegradable waste. Various types of agro industrial wastes and crop residues were successfully converted into vermicompost (Sinha et al., 2010). Chauhan and Singh (2013) worked on the effect of different types of crop residues (straw and bran) on the growth and reproduction of *E. fetida*. Successful conversion of cattle manure, orange peel and filter cake into vermicompost's has also been reported. Some valuable reports regarding the vermicomposting potential of press mud a residue of sugarcane industry and cattle dung are also available (Pigatin et al., 2016). There are some significant contributions related to vermicomposting of household and paper industry waste (Amouei et al., 2017).

IMPORTANCE OF VERMICOMPOSTING

Vermicomposting is a biotechnological process in which the organic wastes are employed to convert into humus like material by the use of earthworms is known as vermicomposting and the product is called vermicompost. Earthworms consume a tremendous range of organic wastes from sewage sludge, animal wastes, agricultural residues, domestic wastes, to industrial wastes. In vermicomposting, earthworms when provided favourable conditions of temperature and moisture maintain the aerobic conditions in the vermicomposting process. They consume organic waste materials and excrete a humus-like substance which is more homogeneous than the organic wastes or raw materials used (Arancon et al., 2004; Edwards and Burrows, 1988). Both physical and biochemical process are involved in the actions of the earthworms. Physical actions included are fragmentation, turnover and aeration. Some of the biochemical actions included in vermicomposting involves enzymatic digestion, nitrogen enrichment, transport of inorganic and organic materials (Edwards & Lofty, 1972).In biochemical processes, some of the major plant nutrients such mainly phosphorous, potassium, calcium and nitrogen which are mainly observed on waste materials are further converted into more readily available and soluble forms. During the biochemical process, important plant nutrients such as phosphorous, potassium, nitrogen and calcium present in the waste materials are converted through microbial action into such chemical forms which are much more soluble and available to the plants than those in the parent substrate (Ndegwa & Thompson, 2001).

The phenomenon of compost formation by earthworms is called as vermicomposting. Earthworms not only are important for maintaining the role in plant nutrients cycling, but also, they are equally responsible for maintenance of soil texture and turnover of organic matter, which is indeed a well-known fact.10-20% of their own biomass per day is consumed by them. In agro-ecosystems, the most important effect of earthworms in agro ecosystems is the increase in nutrient cycling, particularly nitrogen. They ingest organic matter. A relatively wide C:N ratio is used by them to convert it to earthworm tissue with a lower C:N ratio. Overall, earthworms affect the physico-chemical properties of soil. Hence, they affect the physico-chemical properties of soil. Vermicompost's can also be defined as a sustainable biofertilizer generated from organic wastes (Gajalakshmi & Abbasi, 2004).

Vermicompost's are a perfect source of sustenance not only for vegetables and ornamentals, but also for fruits and plantation crops. An increase in crop yield by about 10-15% can be obtained by the use of vermicompost, apart from the improvement in quality of the products.The first report of implementing the process of vermicomposting was in 1970 in Ontario (Canada).The leaders of vermicompost producers are Japan, U.S.A., and Philippines in recent years. India is also in the field of vermicomposting. A few government and non-government organizations are trying to popularise the vermicomposting process (Maheshwari, 1999). Earthworm can consume 1kg organic matter in a day, as it has been estimated. A most significant feature of vermicompost's is that their casting is rich in Ca, Mg, K, N and available P, are secreted by them. Vermicompost consist of 2.5-3% N, 1-1-5%P and 1.5-2%K depending upon substrate quality, useful microorganisms such as fungi, actinomycetes, bacteria, protozoa, enzymes hormones, and vitamins.

Generally, they make passage and mix them. They change of state of soil promotes the growth of actinomycetes and bacteria. The activity of soil, particularly the microbial activity is overall increased because of the increase in biological and enzymatic activities of earthworms. Various reports are available which illustrates that there are about 500 species of earthworms in India and over 3,000 in the world. Commonly used members of earthworms forvermicomposting include *Endriluseuginee, Lamitomauritii, Perionyx excavates, Dravidawillsii, Eisenia andrie, Eisenia coetida, and Lubrieusrubellus.*

Earthworms are highly used in improving soil fertility (Darwin, 1981; Kale et al., 1982), indecomposing the waste materials and also for the conversion of biodegradable matter intovermicast. In addition to froth regulators, these droppings are high in nutritive value with many micronutrients like Mn, Fe, Mo, B, Cu, Zn etc. Being converted tosoluble forms they can be readily available to crop plants (Gajalakshmi & Abbasi, 2004; Suthar, 2008). Moreover, it is rich in several microflora like *Azospirillum, Actinomycetes, Phosphobacillus* etc. It is enhanced by the digestive system ofearthworm (Hussain et al., 2011).It is mixed with the hard substance and made liquid fertilizer (Sharma, 2003). It influences the physicochemical as well as the biological properties of soil, which in turn improve the fertility. Vermicomposting is not only innovative, but also relatively a new alternative technology that makes an eco-friendly environment (Sinha et al., 2002).

PROCESS OF VERMICOMPOSTING

During vermicomposting, the procedure is performed in pits or concrete tanks, wells or wooden crates. Normally in shade a pit of 2×1×1 m^3 dimension is dug. The reason to do so is to prevent the entry of water during rain. At the bottom of the pit pebbles or broken bricks are spread. Adequate amount of coarse sand is added to it to facilitate the drainage. Further, a loamy soil layer is covered over it to keep it moistened and inoculated by earthworms. Then it is covered by small lumps or fresh or decay cattle dung followed by a layer of hay or dry leaves or agricultural wastes. Whole system should be kept moist. Regularly for about 20-25 days water is sprinkled over it. Till the pit becomes fully dry, green leaves are put into the pit each week. A sweet smelling, soft, spongy and almost dark brown coloured vermicompost appears. For future use it is harvested and kept in dark. Then it is sieved and packed in polythene, so that it can retain at least 20% moisture content.

Initially during vermicomposting, a lower or basal layer of vermi bed is prepared by adding broken bricks or pebbles of 3-4 cm^2. Then a layer of coarse sand of about 6-7 cm^2 thickness is spread over it. Then to facilitate proper drainage moist loamy soil of 15 cm^2 is added to it. About 100 earthworms are inoculated in this soil. Small lumps of cattle dung (fresh or dry) are then scattered over the soil and covered with a 10cm layer of hay. Regularly water is spread on it to keep the entire set up moist but not wet. Optimum amount of water is required and is spread, because less water kills the worms and too much water chases them away. Continued watering of the unit is done. Then it is observed regularly for almost 30 days. It is monitored for 30 days. Appearance of adolescent earthworms by this time, indicates that it's a healthy sign.

The use of "organic amendment, such as traditional thermophilic composts, has long been recognized as an effective means of improving soil structure, enhancing soil fertility" (Follet et al., 1981), "increasing microbial diversity and populations" (Barakan et al., 1995), "improving the moisture-holding capacity of soils and increasing crop yields". "Some of the marked effects on microorganisms due to vermicomposting have also been associated with their capability to suppress soil-borne plant diseases" (Hoitink & Fahy, 1986), plant parasitic nematode populations and increased crop yields.

Vermicompost contains most nutrients in plant available form such as "phosphates, nitrates and exchangeable calcium and soluble potassium" (Edwards, 1998; Orozco et al., 1996).Among so many valuable benefits of vermicompost is that, they provide many "micro sites for microbial activity and for the strong retention of nutrients are mainly due to large particulate surface areas for diverse population mainly bacteria, actinomycetes and fungi are found in vermicompost" (Grappelli, et al., 1987).One of the advantages of vermicomposting is that it consistently promotes various biological activities which are responsible for germination of plants, growth of flower and better yield than commercial container media which is not being affected by nutrient availability (Arancon et al., 2004; Atiyeh et al., 2000). Some important reports have also been observed which are concerned with the replacement of lesser amount of vermicompost's into soil-less bedding plant potting mixture, where a noteworthy increase in the germination and growth of marigolds, tomatoes as well as in peppers, during greenhouse trials, was noted when all necessary nutrients are available, even at substitution rates as low as 5-30% into the medium (Atiyeh et al., 2002).

Not only "plant growth regulators, but also vermicompost contains plant growth influencing materials produced by microorganisms" (Grappelli et al., 1987; Tomati et al., 1988, 1990; Atiyeh et al., 2002). Quite a vast amount of humic substances is present in vermicomposts (Masciandaro et al., 1997; Senesi et al., 1992).Remarked effects of these substances have on "plant growth regulators or hormones" (Muscolo et al., 1999).Most of the scientists have worked on greenhouse and some of the workers have reported on the use and effects of vermicompost's in the field.

In "vermitechnology" we use surface and subsurface local varieties of earthworms in composting and soil. It mainly comprises three processes:

- **Vermiculture**: The "cultivation of earthworms, especially in order to convert organic waste into fertilizer".
- **Vermicomposting**: Use of "earthworms to convert organic waste into fertilizer".
- **Vermiconversion**: The "product of the composting process using various species of worms, usually red wigglers, white worms, and other earthworms, to create a mixture of decomposing vegetable or food waste, bedding materials, and vermicast".

Potential environmental benefits of "vermitechnology" which are utilizable products includes; waste biomass management, animal protein production, organic pollution abatement, waste land conservation, land reclamation, soil fertility production of worm-worked manure and enhancement in plant production. Although in limited areas, but vermicomposting technology is known throughout the world. It may be considered a widely spread and efficient technology, even though it is not so much popular. It represents a different approach in waste management, as a process for handling organic residuals in as much as the material is neither landfilled nor burned. It is considered a resource that may be reprocessed. It is also a very fruitful process and is harmonious with sound environmental principles that value resource conservation and tenable practices. The process has been found to be of great value in many developing countries. It has also been observed the process may require large expenditure of labour but not so much capital when automated equipment such as front-end loaders, trommel screens, tractors, etc., are not available to handle large volumes of material. Vermicomposting may supply an opportunity for employment in areas where creation of low or prenominal jobs is of great advantage. Composting and vermicomposting offer good possibility to turn waste material into a valuable soil amendment, where accumulation of paper, cardboard, food waste, manures, agriculture waste, and biosolids are questionable. Earthworms are crucial drivers of the process, by crumbling and accustoming the substrate and dramatically altering its biological activity although microorganisms are responsible for the biochemical degradation of organic matter. By communizing the organic matter, earthworms act as mechanical blenders which modify its physical and chemical status, reducing its C:N ratio gradually, increasing the exposure of surface area to micro-organisms and making it much more favourable for microbial activity and further putrescence. Greatly during passage through the earthworm gut, they move fragments and bacteria-rich evacuation, thus systematize the organic material.

Either directly or indirectly, they are also involved in organic matter decomposition, maintenance, and nutrient turn-over. The cast of earthworms produces a special type of bio-fertilizer (vermicompost), after bioremediation of organic wastes through earthworm. The term *"Biofertilizer"* or more appropriately a *"Microbial inoculants"* can generally be defined as a substance which contains living microorganisms which, when applied to seeds, plant surfaces, or soil, colonize the rhizosphere or the interior of the plant and promotes growth by increasing the supply or availability of primary nutrients to the host plant. Biofertilisers may have live or latent cells of efficient strains of phosphate solubilising, nitrogen fixing, or cellulolytic microorganisms used for application to seeds, soil with the objective of increasing the number of such microorganisms and accelerate those microbial process which augment the availability of nutrients that can be easily assimilated by plants.

At any level vermicomposting can be done. It may be from household vermicomposting of food waste or from community or from city scale, which is also a positive aspect of it (Edwards & Lofty 1972). Following steps of waste degradation occurs by earthworms, which are given below.

- Initially ingestion of organic waste material occurs.
- Later, within the mouth of earthworms, with the aid of saliva, softening of organic waste material occurs.
- In the next step softening of organic waste and neutralization by calcium (excreted by the inner walls of oesophagus) occurs which goes to the gizzard is then for further action in the oesophagus.
- Waste is finally chopped into small particles in muscular gizzard.
- Then, digestion of organic waste occurs in stomach. For this action, proteolytic enzyme is required.
- Later, there is decomposition of pulped waste material components by various enzymes including amylases, cellulases, chitinases, proteases, lipases, etc. and it is excreted in intestine. The digested material is then absorbed in the epithelium of intestine.
- Excretion of undigested food material occurs from the castings of worm.

Success of vermicomposting process depends on a number of abiotic and biotic factors. Some of these factors are given below:

Abiotic Factors

Some of the very important abiotic factors which affect vermicomposting process include pH, temperature, moisture content, aeration, feed quality, light, C:N ratio etc.

Moisture Content

It is a necessary requirement for the working of earthworms and microorganisms in vermicomposting system. Adequate amount of moisture is required. Since, earthworms breathe through their skin; the system must have adequate moisture content. 60%-80% of moisture is ideally required for vermicomposting (Neuhauser et al., 1988; Edwards, 1998). Due to physical and chemical differences in feed stocks there may be slight variations. Even a 5% difference in moisture content significantly affects the clitellum development in *Eisenia fetida* worm species (Reinecke & Venter, 1985).

pH

pH also affects the vermicomposting. The acceptable pH range, suitable for earthworms and microorganism's activity, is 5.5–8.5. Optimum pH is neutral or near neutral, however. The pH value of feed substrates undergoes considerable changes during vermicomposting. During vermicomposting of feed substrate, at the very initial stage, a low pH is observed. The reason is that initially, during vermicomposting, there is formation of carbon dioxide and volatile fatty acids. Subsequently, with the evolution of CO_2 and utilization of volatile fatty acids, the pH begins to increase as the process progresses (Kaushik & Garg, 2004).

Temperature

Earthworms grow at an optimum temperature range of 12-28^{0}C. With the increase in temperature, there is a significant effect in the activity of earthworms. It is generally recommended that temperature should be maintained above 10 ^{0}C in winter and it should be maintained below 35 ^{0}C in summer (Ismail, 1997). Decrease in temperature results in reduced metabolic activity of earthworms and also, they are not able to reproduce. At very low temperatures, earthworms do not consume food. At higher temperature (above 35 ^{0}C) metabolic activity and reproduction of earthworms begins to decline and mortality occurs (Riggle & Holmes, 1994).Variations in preferences for temperature as well as in tolerance vary from species to species.

Aeration

Since, they are aerobic, oxygen is essential for vermicomposting. Earthworms consume oxygen, which is a function for their microbial, oxygen levels are also related to substrate temperatures. Continuous supply of moisture will result in poor aeration during vermicomposting system and can also affect the oxygen supply to worms.

Feed Quality

One of the major requirements for vermicomposting process of earthworm is suitable feed material. Variations are observed in the amount of food which can be consumed by earthworms. It is so because of a number of factors such as the state of decomposition of food, C: N ratio of food, food particle size, salt content in food, etc. Minimum particle size of feed waste ensures the worms to accelerate the process of vermicomposting. Due to small particle size of feed waste, aeration is provided properly through the pile of waste material which is later on available to worms. A worm generally consumes food of about 100 to 300 mg/g body weight/day (Edwards, 1988; Gajalakshmi & Abbasi, 2004). Nutrition is provided to earthworms from living microorganisms, organic materials, and by decomposing macro-fauna.

Light

Since earthworms are photophobic in nature, they should be kept away from light (Edwards & Lofty, 1972).

C: N Ratio

Growth as well as reproduction of earthworms is also affected to some extent by the C: N ratio of feed material. Higher C:N ratio in the feed material accelerates the growth and reproduction of worms. Waste degradation is slowed due to too high or too low C:N ratio. Plants cannot assimilate mineral nitrogen unless the C: N is in the range of 25–20:1.

BIOTIC FACTORS

A number of biotic factors also affect the vermicomposting process which includes microorganisms, enzymes earthworms stocking density, etc. A number of activities such as reproduction rate, feeding rate, respiration rate, and burrowing activity affect the earthworm population, i.e. stocking density during vermicomposting. Mortality rate increases in higher population densities. One earthworm produces reduced amount of cocoon. Also, growth rate is decreased (Uvarov & Scheu, 2004). Marked reduction in rate of growth as well as reproduction in *E. andrei* has also been observed due to increase in stocking density (Frederickson et al., 1997). A rapid turnover of fresh organic matter into earthworm casts was also seen due to high population densities of earthworms (Aira et al., 2002). While establishing a vermicomposting system, it requires to be maintained optimum earthworm density to obtain maximum population growth and reproduction in shortest possible time.

Microorganisms

The microbes present in the vermicomposting process plays a very important role in waste composition of those particular components which undergoes it. Stabilization of organic matter due to mutualistic interaction is seen between earthworms and microorganisms during vermicomposting (Edwards & Fletcher, 1988). Earthworms consume fungi with organic substrates to fulfil their protein or nitrogen requirement, fungal population in earthworm casts was almost equal or higher than that of initial substrates (Edwards & Bohlen 1996). The micro-organisms not only mineralize complex substances into plant available form, but they also synthesise biologically active substances (Pramanik, 2010; Tiunov & Scheu, 2004).

Enzymes

Complete stabilization requires enzymatic action for chemically organic wastes are very complex. The worms stimulate enzymes in their gizzard and intestine. A quick biochemical conversion, as a result, a rapid biochemical conversion of proteinaceous and cellulosic materials takes place (Hand et al., 1988). Most common enzymes required in vermicomposting are cellulases, amidohydrolase, proteases, urease, b-glucosidases, and phosphatases. Enzyme activities have often been used as indicators of microbial activity and can also be useful to interpret the intensity of microbial metabolism in soil (Schinner et al., 1996).Being a catalyst of a number of metabolic functions, they also decompose and detoxify a number of contaminants (Nannipieri & Bollag, 1991).

Status of Earthworms

All over the world, there are about 3,320 species of earthworms (Bhatnagar & Palta, 1996). Julka et al. (2009) has already mentioned that with about 590 species of earthworms, found in different ecological preferences, the functional role of a number of the species and their influence on the habitat are lacking.

Habitat of Earthworms

Species of earthworms have different ecological niches, life style habitat characteristics, and life span. The dispersal of earthworms is related to the physico-chemical characteristics of soils. Prominent among these are pH, carbon, nitrogen temperature, moisture, and C:N ratio etc. Majority of earthworm species prefer soil with a moisture of 12–34%, temperature of 10–35 ^{0}C, pH of about 7 and C:N ratio 2–8 (Edwards and Lofty, 1977; Kale and Krishnamoorthy, 1981; Lee, 1985). Earthworms are generally absent or rare in soil with a very rough texture in soil and high clay content, or soil with pH<4 (Gunathilagraj & Ravignanam, 1996).

GROWTH PARAMETERS AND VERMICOMPOST

Among so many important activities performed during vermicomposting, the most important ones included are hormones like activity and it induces increased root biomass, enhanced plant growth, greater root initiation, and development and alters the morphology of plants (Grapelli et al., 1985). Subler et al., (1998), incorporated commercial bedding plant potting media to produce a significant increase in the total biomass of tomato seedlings with the use of pig manure vermicompost. Atiyeh et al. (2000) later on confirmed that there are several beneficial effects on the increase in plant growth in the seedlings of chickpea at the lowest level of vermicompost substitution. Certain evidences are also mentioned where vermicompost promotes growth of plants (Fosgate & Babb, 1972; Reddy, 1988; Rajkhowa, 2000). Growth improvement of potato, spinach and turnip was recorded to 6t/ha vermicompost application was reported by Ansari (2008). Reports are also about the growth medium substituted with 20% pig manure vermicompost in marigold (Atiyeh et al., 2000). Tallini et al., (1991) found that vermicompost stimulates the root growth increasing proliferation of root hairs. Applications of vermicompost to field soils have also been reported to increase crop growth and yields (Venkatesh et al., 1998; Arancon et al., 2004). Remarked influence on all yield parameters of crops like paddy, sugarcane and wheat has also been observed due to the use of vermicompost (Ismail, 2005; Garg & Bhardwaj, 2000; Ansari, 2007). Ahmed et al. (2010), mentioned that total dry weight, plant height, and leaf area significantly increases on the application of biofertilizers. Plant height of wheat increased by inoculation with *Azospirillum* sp. (Mohmood et al. 2006).Certain favourable effect of a manure compost on the growth and development of plants with the use of various parameters like leaf area, rooting, time of flowering, development and lengthening of internodes has also been possible (Arancon et al., 2004; Jim et al., 2007).

With the use of 30% compost containing mixtures, in shoot dry weights of marigold plants, a significant increase was measured (Bugbee & Frink, 1989).Increased growth of corps grown in media amended with humic acid that were extracted from vermicompost, have also been reported by some workers (Atiyeh et al., 2002). With the application of farmyard manure and phosphobacteria, an increase in plant height in maize upto 40% (Nanjundappa et al., 2000), 24% in case of leaf area in soybean biofertilisers (Javid & Mohmood, 2010) and 41% in total numb of root nodules in *Vicia sativa* with organic manure has been reported (Javid & Mohmood, 2010).In rice as well as senna due to suitable application of biofertilizers there was an increase in plant height (Shanmugam & Veeraputhran, 2000; Ramaroorthy et al., 2003),an increase in number of leaves in piper beetle (Arulmozhian & Thamburaj, 1998), an increase in leaf area in maize (Nanjundappa et al., 2000),an increase in total number of root nodules in mungbean (Gomaa et al., 1999) and green gram (Rajkhowa et al., 2003).Due to the use of green manure as well as biofertilizers, the growth of plants with a greater number of tillers and broader leaves in rice was noted (Shanmugam & Veeraputhran, 2000). Rice biomass treated in soil with *Sesbania aculeate* was worked out by Azam (1990).It was found to be near maximum. Ghosal and Singh (1995) reported that there were enhancements in 33, 36 and 67% in case of rice as compared to control in biofertilizers. There was an increase in dry weight of straw berry (*Fragaria anunassa* Duch.) leaves in 7.5 t/ha vermicompost concentration (Singh et al. 2008). Dubey (1997) explained that in farmyard manure of soya bean and biofertilizers increased the total number of root nodules/plant (Dubey, 1997). Especially during non-thermophilic processes, vermicompost enhance plant growth and yields in greenhouse crops. These are stabilized organic molecules which are produced due to interactions between earthworms and microorganisms (Arancon, 2004).An increase in the growth as well as yield in case of pepper was reported by Arancon et al. (2004). Atiyeh et al. (2000, 2002) reported more marigold flower buds in Metro Mix 360 substituted with pig manure vermicompost. Suitable applications of humic substances resulted in the increased yield in case of corn, oats, soya bean, peanuts, clover, chicory plants and other tropical crops. Reports are also available which discusses about the application of vermicompost to tomatoes cultivated in the field also with an increased yield. Experiments conducted in greenhouse demonstrate that due to the application of vermicompost, an increase in germination, growth and yields of certain vegetables and ornamentals significantly and consistently (Althani et al., 1999). With some of the field experiments conducted as reported in some of the literatures, it was found that due to amendments in soils with vermicompost there was an increase in the growth and yield of some crops such as cowpea (Mba, 1983), banana (Althani et al., 1999) and straw berries (Arancon et al., 2004). Venktesh et al. (1998) reported that yields of thompson seedling grapes increased significantly in response to vermicompost application. Some other researchers have reported that vermicompost not only enhances seed germination, but also seedling growth and yield (Edwards & Burrows, 1988; Alves & Passoni, 1997). Saline compost and phosphobacteria increased the number of leaves, leaf area, plant growth, root nodules, fresh weight and dry weight in *Arachishypogaea* L (Ravindran et al., 2007). Quite a significant increase in the dry matter yield was recorded in the application of animal manure (Xie & Mackenzie, 1986). A negative affect was seen in fresh as well as dry weight cucumber seedlings because of the use of much amount of vermicompost (30%) which clearly indicated salt stress resulting in growth retardation (Parthasarathi & Ranganathan, 2002).Sufficient data is also available regarding the growth increase with several amendments, tolerant vegetables and trees treated with vermicompost have also been reported. To establish the role and growth productivity of legume plants few attempts have been made (Gupta, 2003; Sinha et al., 2010).

Eisenia Fetida

The most prevalent earthworm for vermicomposting is *Eisenia fetida.* Most popularly it is also known as red wriggler, red worm, tiger worm etc. The species has been widely used for various toxicological studies as test worm. Upto 1500 mg body weight can be acquired by mature earthworms. An adult worm on maturation produces one cocoon every third day. On hatching from each cocoon within almost 23 days 1-3 individuals emerges.

VERMICOMPOSTING BIOTECHNOLOGY

A number of various environmental problems are being created regularly and millions of tons of solid waste are being generated from the modern existing society. They end up in landfills every day and it also creates unnecessary economic and environmental problems for the local government to manage and monitor them (may be up to 30 years) for environmental safety (emission of GHG, toxic gases and leach ate discharge into ground water & soil). One of the major serious environmental concerns is the emission of greenhouse gases (GHG) methane (CH_4) & nitrous oxides (N_2O). They result in the disposal of solid waste either in the landfills or from their management by conventional composting systems. One of the major problems being faced is the escalating socio-economic and environmental cost of dealing with current and future generation of mounting solid wastes (SW). A considerable portion of SW consist of "Organic Wastes" that are "biodegradable" and can be vermicomposted into a highly "nutritive bio-fertilizer", which is 4-5-fold more powerful than conventional composts. As compared to chemical fertilizers, vermicomposting is a better and fruitful method both for better crop growth and also for safe food production. As compared to conventional microbial degradation & composting technology waste degradation, vermicomposting has been found to be an economical and environmentally preferred technology. It is a nearly odourless and rapid process. It also reduces composting time by more than half and the end product is both "detoxified" and "disinfected". It has been expected that within a year, about one million worms double every two months and their total number can be 64 million. It can also be considered that each adult worm (*Eisenia fetida)* consume organics waste equivalent to its own body weight every day. 64 million worms (weighing 64 tons) would consume 64 tons of waste every day and produce 30-32 tons of vermicompost per day at 40-50% conversion rate.

SOME IMPORTANT STUDIES ON VERMICOMPOSTING BIOTECHNOLOGY

Most earthworms consume, at the best, half their body weight of organics in the waste in a day. *Eisenia fetida* can consume organic matter at the rate equal to their body weight every day. As compared to conventional aerobic & anaerobic composting, earthworms, enhances natural biodegradation and decomposition of organic waste from 60 to 80% .In about 30 days, if provided optimum conditions of moisture (60-70%) and temperature (20-30ºC), then about 5 kg of worms (numbering approx. 10,000) can vermiprocess 1 ton of waste into vermicompost in just 30 days. If vermi-composting is done, the volume of solid waste is significantly reduced from approximately 1 cum to 0.5 cum of vermi-compost indicating 50% conversion rate. The rest is converted into worm biomass. Vermicomposting of "human excreta" was studied successfully by Bajsa et al. (2003). It was completed in six months. Anodourless and safe pathogen quality and also good physical texture was produced at the end. Among so many covering materials generally used, sawdust is considered to be the best covering material to be used in vermicomposting toilets to produce compost with a good earthy smell, a crumbly texture and dark brown colour.

APPLICATIONS OF VERMICOMPOSTING BIOTECHNOLOGY

Vermicomposting involves a great reduction in populations of human pathogenic microorganisms as like composting. Pathogenic organisms as like human pathogens do not survive vermicomposting. The thermophilic stage of the composting process eliminates pathogenic organisms, which is generally accepted. Certain evidences in support of the prolonged use of vermicomposting are well investigated. Some of them include that after 60 days of vermicomposting, amounts of faecal coliform bacteria in biosolids dropped from 39,000 MPN/g to 0 MPN/g. In that same time period, *Salmonella sp.* dropped from <3 MPN/g to <1 MPN/g. Eastman (1999) have reported similar results also for faecal coliforms, *Salmonella sp.,* enteric viruses, and for helminth ova.

Water Holding Capacity

One of the most significant reasons for losses of soil water is through the process of water loss and evaporation from the surface of soil by a process involving combination of both of these commonly known as evapotranspiration. The max dry density of compaction decreases considerably, only when, about 10% or more than that organic content is there in a given soil wastes. An increase in optimum moisture content is seen with the increase in organic content. Significant experimental effects of vermicomposting have been seen for water holding capacity also. One such observation is the addition of 200 g of vermicompost of press mud prepared by *E. fetida, E. eugeniae*and *M. megascolex* to 3 kg of soil. This showed lower densities compared to the density of soil indicating an increase in voids and water holding capacity. It can be marked that due to addition of vermicompost with aggregation properties and higher water holding capacities, they not only provide nutrients required for growth but also increase the yield of crops. Vermicompost when applied to the soil surface maintains soil evaporation to a minimum but also, they absorb atmospheric moisture as a good adsorbent and influence the energy balance.

Carbon Mineralization During Vermicomposting

Earthworms, as like most of the other members of decomposer community, assimilate carbon from the most recently deposited organic matter fractions, consisting mainly of easily-degradable substances. In all cases the degradation process resulted in carbon losses by mineralization which produced a decrease in the amounts of total organic carbon and in the carbon contributions to the organic matter, which was much higher in the final stages of decomposition when the earthworm populations were bigger and more active. Although earthworms consume and process large amounts of organic matter, their contributions to the total heterotrophic respiration is very low due to their poor assimilation efficiency and only when there are large active earthworm populations, as in vermicomposting systems, can they contribute to an appreciable extent to the heterotrophic respiration. More than 15 companies (small-medium-large sized) produce vermicompost in Thrace Region by Belliturk, (2016). Quite a number of reports are available on the vermicompost analysis regarding the results for liquids as well as solids produced in these factories in Thrace Region. Some of the results of these analyses of solid vermicompost are shown in Table 1. given below.

Role in Nitrogen Cycle

Most of the mineral nitrogen retained as nitrate when nitrogen mineralization is enhanced. Due to this, earthworms had a great impact on nitrogen transformation in pig manure.When there was greater activity of earthworms, the net total nitrogen, in all treatments and times, decreased; losses being more marked during the final stages. Various fractions of nitrogen followed trends similar to the total nitrogen.A high rate of nitrification was seen in all of the treatments applied during the final stages of the process, when the earthworm was bigger and more active. Reduced organic nitrogen content as well as high nitrification rate was also seen. As seen in this case that earthworm (*Eisenia andrei*), modified conditions in the manure that favoured nitrification. NPK and organic matter values of solid vermicompost processed by *E. fetida* from the same feedstock (cow manure) was significantly increased by 3 to 4 times and enhanced several micronutrients (Table 2).

Table 1.Chemical compositions of different solid vermicompost's (Belliturk, 2016)

Contents	"Mix food waste vermicompost"	"Pruning olive waste vermicompost"	"Cow Manure Vermicompost"	"Sheep Manure Vermicompost"
K %	1,10	1,74	1,12	0,32
N %	1,95	1,80	1,75	1,60
P %	2,75	0,23	1,29	0,45
Zn mg/kg	354,00	545,00	427,00	405,00
Organic matter %	41,00	48,00	51,00	42,00

Table 2. Organic matter and NPK values of vermicompost compared with conventional cow manure compost made from cow manure (Belliturk et al., 2015)

Nutrients	Cow Manure Compost	Vermicompost
N%	0.30-1.40	1.60-3.00
K%	0.30-1.00	1.40-2.90
P%	0.25-0.95	1.05-2.65
pH	6.50-7.05	6.80-7.89
Organic Matter%	22.00-46.00	55.00-69.00

Vermicomposting and Heavy Metal Availability

Some reports reveal that during the process of vermicomposting, changes occur in total and available contents of heavy metals in the organic matter. Due to this they may cause problems in sewage sludges, some animal manures, and industrial organic wastes. An increase in the total amounts of heavy metals (almost between 25-30%) was seen. The amounts of biologically available heavy metals resulted to decrease between 35 and 55% in the bioavailable metals in two months. Also, same type of results was seen in other studies for both composting and vermicomposting which implies a lower availability of these elements for plants from vermicompost's. Heavy metals tend to form complex aggregates with the humic acids and the most polymerized organic fractions during vermicomposting.

CONCLUSION

It has been scientifically proved that vermicomposting biotechnology can be successfully used to enhance the fertility of soil. As like biomining and other beneficial techniques, vermicomposting is like getting "gold from garbage" (solid organic waste to highly nutritive biofertilizer). *E. euginae, E. fetida,* and *P. excavatus* are the three versatile and ubiquitous earthworms which can perform wide social, economic and environmental functions. Vermi-composting not only maintains the "global human sustainability cycle" and "circular economy" using "food wastes", but also protects farm soil and improves its fertility. Vermicompost can "replace" the "chemical fertilizers" for production of "safe organic foods". This has now been proved worldwide. It will be a great step towards achieving global social, economic and environmental sustainability. Vermicomposted products are not only economical and cost effective, but also, they are highly valued end products as well as by-products. Applying vermicompost helps in plant growth. It does not cause any damage to insects as well as soil, when compared with chemical fertilizers.

REFERENCES

Aalok, A., Tripathi, A. K., & Soni, P. (2008). Vermicomposting: A better option for organic solid waste management. *Journal of Human Ecology (Delhi, India)*, *24*(1), 59–64. doi:10.1080/09709274.2008.11906100

Aira, M., Monroy, F., & Domínguez, J. (2002). *Eisenia foetida* (Oligochaeta, Lumriciadae) activates fungal growth triggering cellulose decomposition during vermicomposting. *Microbial Ecology*, *52*(4), 738–747. doi:10.100700248-006-9109-x PMID:16944346

Arancon, N. Q., Edwards, C. A., Lees, S., & Byrne, R. (2004). Effects of humic acids from vermicomposts on plant growth. *European Journal of Soil Biology*, *46*, 65–69.

Atiyeh, R. M., Edwards, C. A., Subler, S., & Metzger, J. D. (2000). Earthworm-processed organic wastes as components of horticultural potting media for growing marigold and vegetable seedlings. *Compost Science & Utilization*, *8*(3), 215–223. doi:10.1080/1065657X.2000.10701994

Atiyeh, R. M., Lee, S. S., Edwards, C. A., Arancon, N. Q., & Metzger, J. D. (2002). The influence of humic acid derived from earthworm-processed organic wastes on plant growth. *Bioresource Technology*, *84*(1), 714. doi:10.1016/S0960-8524(02)00017-2 PMID:12137272

Atiyeh, R. M., Subler, S. M., Edwards, C. A., & Metzgar, J. D. (1999). Growth of tomato plants in horticultural potting media amended with vermicompost. *Pedobiologia*, *43*, 724–728.

Bajsa, O., Nair, J., Mathew, K., & Ho, G.E. (2003). Vermiculture as a tool for Domestic Wastewater management, *Water Science and Technology*, *48*(11/12), 125-132.

Belliturk, K. (2016). Vermicomposting Technology for Solid Waste Management in Sustainable Agricultural Production. *Cukurova J Agric Food Sci*, *31*, 1–5.

Belliturk, K., Gorres, J. H., Bagdatl, M. C., Gocmez, S., & Turan, H. S. (2015). The Evaluation of Olive Pruning Waste as a Vermicompost: Micro Elements. *Agriculture Vision Agriculture*, *1*, 7–12.

Bhawalkar, U. S. (1989). Vermiculture: A promising source of biofertilizers. In National Seminars in Agriculture Biotechnology Held at Gujarat Agriculture University, Nausari, India.

Bhawalkar, U. S. (1995). *Vermiculture bioconversion of organic residues*. IIT Mumbai.

Bhiday, M. R. (1994). Earthworms in Agriculture. *Indian Farming*.

Bhoyar, R. V., & Bhide, A. D. (1996). *Vermicomposting of organic (vegetable) solid wastes*. Nagpur, India: VRCE.

Bouche, M. B., & Ferriere, G. (1986). Soil organisms as components of ecosystem. Ecological Bulletin, 25, 122-132.

Bulluck, L. R., Borsius, M., Evanylo, G. K., & Ristaino, J. B. (2002). Organic and synthetic fertility amendments influence soil microbial, physical and chemical properties on organic and conventional farms. *Applied Soil Ecology*, *19*(2), 147–160. doi:10.1016/S0929-1393(01)00187-1

Chakrabarty, D. (2008). *Vermicompost and Organic Pisciculture*. West Bengal: Akshay Krishi Vikash.

Chaoui, I., Zibiliske, M., & Ohno, T. (2003). Effects of earthworm casts and compost on soil microbial activity and plant nutrient availability. *Soil Biology & Biochemistry*, *35*(2), 295–302. doi:10.1016/S0038-0717(02)00279-1

Chattopadhyay, G. (2012). Use of vermicomposting biotechnology for recycling organic wastes in agriculture. *International Journal of Recycling of Organic Waste in Agriculture*, *1*(1), 1–6. doi:10.1186/2251-7715-1-8

Chirashree, G. (2004). Integrated Vermi-Pisciculture – an alternative option for recycling of solid municipal waste in rural India. *Bioresource Technology*, *93*(1), 71–75. doi:10.1016/j.biortech.2003.09.014 PMID:14987723

Darwin, C. (1981). *The formation of vegetable mould through the action of worms, with observations on their habitats*. London: Murray.

Dash, M. C. (1978). *Role of Earthworms in the Decomposer System*. New Delhi, India: India International Scientific Publication.

Divya, U. K. (2001). *Relevance of vermiculture in sustainable agriculture*. Agriculture General World.

Dominguez, J., Edwards, C., & Subler, S. (1997). Comparison of vermicomposting and composting. *BioCycle*, *4*, 57–59.

Dominguez, J., & Edwards, C. A. (2004). *Vermicomposting organic wastes: Soil Zoology for Sustainable Development in the 21st Century*. Cairo, Egypt: MikhaTl.

Edward, C. A. (1988). *Breakdown of animal, vegetable and industrial organic wastes by earthworms*. The Hague: SPB.

Edwards, C. A. (1998). The use of earthworms in the breakdown and management of organic wastes. In C. A. Edwards (Ed.), *Earthworm ecology* (pp. 327–354). Boca Roton, FL: CRC press.

Edwards, C. A., & Bohlen, P. J. (1996). *Biology and Ecology of Earthworms (3rd ed.)*. London: Chapman and Hall.

Edwards, C. A., & Fletcher, K. E. (1988). Interaction between Earthworms and Micro-Organisms in Organic Matter Breakdown. *Agriculture, Ecosystems & Environment*, *24*(1-3), 235–247. doi:10.1016/0167-8809(88)90069-2

Edwards, C. A., & Lofty, J. R. (1977). *Biology of Earthworms*. London: Chapman and Hall. doi:10.1007/978-1-4613-3382-1

Federico, A., Borraz, J. S., Molina, J. A. N., Nafate, C. C., Archila, M. A., Oliva, L. M., & … . (2007). Vermicompost as a soil supplement to improve growth, yield and fruit quality of tomato (*Lycopersicon escculentum*). *Bioresource Technology*, *98*(15), 2781–2786. doi:10.1016/j.biortech.2006.02.032 PMID:17400447

Follet, R., Danahue, R., & Murphy, L. (1981). *Soil and Soil Amendments*. Prentice-Hall, Inc.

Grapplli, A., Galli, E., & Tomati, U. (1987). Earthworm casting effect on *Agaricusbisporus* fructification. *Agrochimica*, *21*, 457–462.

Gupta, P. K. (2004). Vermicomposting for sustainable agriculture, *Agrobios,* India. *Indian Journal of Biotechnology*, *3*, 486–494.

Hand, P., Hayes, W. A., Satchell, J. E., Frankland, J. C., Edwards, C. A., & Neuhauser, E. F. (2004). The vermicomposting of cow slurry. *Earthworms in Waste and Environmental Management*, *31*, 49–63.

Hartenstein, R., Neuhauser, E. F., & Kaplan, D. L. (1979). Reproductive potential of the earthworm *Eisenia foetida*. *Oecologia*, *43*(3), 329–340. doi:10.1007/BF00344959 PMID:28309889

Hoitink, H. A., & Fahy, P. (1986). Basis for the control of soil borne plant pathogens with composts. *Annual Review of Phytopathology*, *24*(1), 93–114. doi:10.1146/annurev.py.24.090186.000521

Hussain, N., Abbasi, T., & Abbasi, S.A. (2005). *Vermicomposting eliminates the toxicity of Lantana (*Lantana camara*) and turns it into a plant friendly organic fertilizer*. Academic Press.

Jicong, H., Yagnum, Q., Guangging, I., & Dong, R. (2005). The influence of temperature, pH, and C/N ratio on the growth and survival of earthworms in Municipal Solid Waste. CIGR Ejournal, 7.

Kale, R. D. (1998). *Earthworm- Cinderella of Organic Farming*. Bangalore, India: Publishers- Prism Book Pvt. Ltd.

Kale, R. D., & Bano, K. (1988). Earthworm cultivation and culturing techniques for production of Vee COMP83E UAS, Vee MEAL 83P UAS. *Mysore Journal of Agricultural Sciences*, *22*, 339–344.

Kale, R. D., Bano, K., & Krishnamoorthy, R. V. (1982). Potential of *Perionyx excavatus* for utilizing organic wastes. *Pedobiologia*, *23*, 419–425.

Kale, R. D., Mallesh, B. C., Kubra, B., & Bhagyaraj, D. J. (1992). Influence of vermicompost application on available micronutrients and selected microbial populations in paddy field. *Soil Biology & Biochemistry*, *24*(12), 1317–1320. doi:10.1016/0038-0717(92)90111-A

Kaviraj, & Sharma, S. (2003). Municipal solid waste management through vermicomposting employing exotic and local species of earthworms. *Bioresource Technology*, *90*(2), 169–173. doi:10.1016/S0960-8524(03)00123-8 PMID:12895560

Krishnamoorthy, R. V., & Vajrabhiah, S. N. (1986). Biological activity of earthworm casts: An assessment of plant growth promoter levels in casts. *Proceedings of the Indiana Academy of Sciences*, *95*(3), 341–351. doi:10.1007/BF03179368

Mishra, R. K., Singh, B. K., Upadhyay, R. K., & Singh, S. (2009). Technology for vermicompost production. *Indian Farming*.

Morgan, M., & Burrows, I. (1982). *Earthworms/Microorganisms Interactions*. Rothamsted Experimental Station.

Munnoli, P. M. (2007). *Management of industrial organic solid wastes through vermiculture biotechnology with special reference to microorganisms*. Goa University.

Ndegwa, P. M., & Thompson, S. A. (2001). Integrating composting and vermicomposting in the treatment of bioconversion of biosolids. *Bioresource Technology*, *76*(2), 107–112. doi:10.1016/S0960-8524(00)00104-8 PMID:11131792

Norman, Q. (2004). Effects of vermicompost's produced from food waste on the growth and yields of greenhouse peppers. *Bioresource Technology*, *93*(2), 139–144. doi:10.1016/j.biortech.2003.10.015 PMID:15051075

Orozco, F. H., Cegarra, J., Trvjillo, L. M., & Roig, A. (1996). Vermicomposting of coffee pulp using the earthworm *Eisenia foetida*: Effects on C and N contents and the availability of nutrients. *Biology and Fertility of Soils*, *22*(1-2), 162–166. doi:10.1007/BF00384449

Pathma, J., & Sakthivel, N. (2012). Microbialdiversity of vermicompost bacteria that exhibit useful agricultural traits and waste management potential. *SpringerPlus*, 1. PMID:23961356

Piccone, G., Biosoil, B., Deluca, G., & Minelli, L. (1986). Vermicomposting of different organic wastes. Udine, Italy: Academic Press.

Rajiv, K., Sinha, Agarwal, A., Chauhan, K., Chandran, V., & Brijal Kiranbhai Soni, B.K. (2010). Vermiculture Technology: Reviving the Dreams of Sir Charles Darwin for Scientific Use of Earthworms in Sustainable Development Programs. *Technology and Investment*, 155-172.

Satchell, J. E. (1955). *Some aspects of earthworm ecology*. London: Butterworth's.

Suthar, S. (2007). Nutrients changes and biodynamic of epigeic earthworms Perionyx excavatus during recycling of some agricultural wastes. *Bioresource Technology*.

Tchobanolglous, G., Theisen, H., & Vigil, S. (1993). *Integrated Solid Waste Management*. McGraw Hill, Inc.

Thomas, S., & Trivedy, R. K. (2002). *Earthworm biotechnology for waste management and crop improvement a review of research*. Bangalore, India: Allied Publication Pvt. Ltd.

Visvanathan, C., Trankler, J., Jospeh K., & Nagendran, R. (2005). *Vermicomposting as an Eco-Tool in Sustainable Solid Waste Management*. Asian Institute of Technology, Annamalai University.

Yan, Y. W. (2013). Comparison of plant nutrient contents in vermicompost from selected plant residues. *African Journal of Biotechnology*, *12*(17), 2207–2214. doi:10.5897/AJB11.3164

KEY TERMS AND DEFINITIONS

Agricultural Waste: Agricultural waste refers to waste produced from agricultural operations, including waste from farms, poultry houses, and slaughterhouses. Soil fertility refers to the ability of soil to sustain agricultural plant growth (i.e., to provide plant habitat and result in sustained and consistent yields of high quality).

Vermicompost: The product of the composting process using various species of worms, usually red wigglers, white worms, and other earthworms, to create a mixture of decomposing vegetable or food waste, bedding materials, and vermicast.

Vermicomposting: Vermicomposting uses earthworms to turn organic wastes into very high-quality compost. This is probably the best way of composting kitchen wastes.

Compilation of References

Aalok, A., Tripathi, A. K., & Soni, P. (2008). Vermi-composting: A better option for organic solid waste management. *Journal of Human Ecology (Delhi, India)*, *24*(1), 59–64.

Abbasi, T., Gajalakshmi, S., & Abbasi, S. A. (2009). Towards modeling and design of vermicomposting systems: Mechanisms of composting/ vermicomposting and their implications. *Indian Journal of Biotechnology*, *8*, 177–182.

Abbasi, T., Gajalakshmi, S., & Abbasi, S. A. (2009). Towards modeling and design of vermi-composting systems: Mechanisms of composting/vermi-composting and their implications India. *Journal of Biotechnology*, *8*, 177–182.

Abbasi, T., Gajalakshmi, S., & Abbasi, S. A. (2009). Towards modelling and design of vermicomposting systems: Mechanisms of composting/vermicomposting and their implications. *Indian Journal of Biotechnology*, *8*, 177–182.

Abd El-Salam, M. M. (2010). Hospital waste management in El-Beheira Governorate, Egypt. *Journal of Environmental Management*, *91*(3), 618–629. doi:10.1016/j.jenvman.2009.08.012 PMID:19931245

Abdel-Shafy, I. H., & Monsour, M. S. M. (2016). A review on polycyclic aromatic hydrocarbons:Source, environmental impact, effect on human health and remediation. *Egyptian Journal of Petroleum*, *25*(1), 107–123. doi:10.1016/j.ejpe.2015.03.011

Abdulla, F., Qdais, H. A., & Rabi, A. (2008). Site investigation on medical waste management practices in northern Jordan. *Waste Management*, *28*(2), 450-458.

Abdullah Adil Ansari. (2012). *Vermicomposting-permutations and combinations of organic waste recycling*. Lambert Academic Publishing.

Abdullah, M. O., Tan, I. A. W., & Lim, L. S. (2011). Automobile adsorption air-conditioning system using oil palm biomass-based activated carbon: A review. *Renewable & Sustainable Energy Reviews*, *15*(4), 2061–2072. doi:10.1016/j.rser.2011.01.012

Abrol, V., Wali, P., Pareek, N., Mondal, A. K., & Jalali, V. K. (2003). Edaphic impact of sewage and industrial effluents on soil resources. *Indian Journal of Environmental Protection*, *23*, 741–748.

Achilias, D. S., Roupakias, C., Megalokonomos, P., Lappas, A. A., & Antonakou, E. V. (2007). Chemical recycling of plastic wastes made from polyethylene (LDPE and HDPE) and polypropylene (PP). *Journal of Hazardous Materials*, *149*, 536–542. PMID:17681427

Achilias, D., Giannoulis, A., & Papageorgiou, G. (2009). Recycling of polymers from plastic packaging materials using the dissolution–reprecipitation technique. *Polymer Bulletin*, *63*(3), 449–465. doi:10.100700289-009-0104-5

Adekomaya, O., & Ojo, K. (2016). Adaptation of plastic waste to energy development in lagos: An overview assessment. *Nigerian Journal of Technology*, *35*(4), 778–784. doi:10.4314/njt.v35i4.12

Adhikari, S., Gupta, S. K., & Banerjee, S. K. (1993). Trace metal content of city sewage and sludge. *Journal of the Indian Society of Soil Science*, *41*, 170–172.

Adhikary, S. (2012). Vermicompost, the story of organic gold: A review. *Agricultural Sciences*, *3*(7), 905–917. doi:10.4236/as.2012.37110

Adibi, J. J., Perera, F. P., Jedrychowski, W., Camann, D. E., Barr, D., Jacek, R., & Whyatt, R. M. (2003). Prenatal exposures to phthalates among women in New York City and Krakow, Poland. *Environmental Health Perspectives*, *111*(14), 1719–1722. doi:10.1289/ehp.6235 PMID:14594621

Adriaens, P., & Hickey, W. J. (1993). Article. In D. L. Stone (Ed.), Biotechnology for the Treatment of Hazardous Waste (pp. 97–120). Ann Arbor, MI: Lewis Publishers.

Agarwal, S. C., Bisht, S. S., & Sharma, C. P. (1977). Relative effectiveness of certain heavy metals in producing toxicity and symptoms of iron deficiency in barley. *Canadian Journal of Biotechnology*, *55*, 1299–1307.

Agarwal, T., Khillare, P. S., Shridhar, V., & Ray, S. (2009). Pattern, sources and toxic potential of PAHs in the agricultural soils of Delhi, India. *Journal of Hazardous Materials*, *163*(2-3), 1033–1039. doi:10.1016/j.jhazmat.2008.07.058 PMID:18757133

Ahmed, L., Bhat, F. A., & Sultan, S. (2015). Municipal Solid Waste Management in Srinagar City, Kashmir, India. *International Journal of Innovative Research in Science. Engineering and Technology*, *4*(7). doi:10.15680/IJIRSET.2015.0407148

Ahmed, S. A., & Ali, M. (2004). Partnerships for solid waste management in developing countries: Linking theories to realities. *Habitat International*, *28*(3), 467–479. doi:10.1016/S0197-3975(03)00044-4

Ahn, J. H., Shin, S. G., & Hwang, S. (2009). Effect of microwave irradiation on the disintegration and acidogenesis of municipal secondary sludge. *Chemical Engineering Journal*, *153*(1-3), 145–150. doi:10.1016/j.cej.2009.06.032

Ahn, W. S., Park, S. J., & Lee, S. Y. (2000). Production of Poly(3-Hydroxybutyrate) by Fed-Batch Culture of Recombinant Escherichia coli with a Highly Concentrated Whey Solution. *Applied and Environmental Microbiology*, *66*(8), 3624–3627. doi:10.1128/AEM.66.8.3624-3627.2000 PMID:10919830

Aira, M., Monroy, F., & Domínguez, J. (2002). *Eisenia foetida* (Oligochaeta, Lumriciadae) activates fungal growth triggering cellulose decomposition during vermicomposting. *Microbial Ecology*, *52*(4), 738–747. doi:10.100700248-006-9109-x PMID:16944346

Aira, M., Monroy, F., & Domínguez, J. (2007). Earthworms strongly modify microbial biomass and activity triggering enzymatic activities during vermicomposting independently of the application rates of pig slurry. *The Science of the Total Environment*, *385*(1-3), 252–261. doi:10.1016/j.scitotenv.2007.06.031 PMID:17628641

Aizawa, H., Yoshida, H., & Sakai, S. I. (2008). Current results and future perspectives for Japanese recycling of home electrical appliances. *Resources, Conservation and Recycling*, *52*(12), 1399–1410. doi:10.1016/j.resconrec.2008.07.013

Ajila, C. M., Aalami, M., Leelavathi, K., & Rao, U. P. (2010). Mango peel powder: A potential source of antioxidant and dietary fiber in macaroni preparations. *Innovative Food Science & Emerging Technologies*, *11*(1), 219–224. doi:10.1016/j.ifset.2009.10.004

Akçaözoğlu, S., Atiş, C. D., & Akçaözoğlu, K. (2010). An investigation on the use of shredded waste PET bottles as aggregate in lightweight concrete. *Waste Management (New York, N.Y.)*, *30*(2), 285–290. doi:10.1016/j.wasman.2009.09.033 PMID:19853433

Akor, A. J., Ayotamuno, M. J., Aman, L. I., & Enokela, S. O. (2013). Assessment of domestic solid waste generation in PortHarcourt by separator – Receptacle technology. *International Journal of Scientific Engineering and Research*, *4*, 1–7.

Akpan, I., Bankole, M. O., Adesemowo, A. M., & Latunde, D. G. (1999). Production of amylase by *A. nigerin* a cheap solid medium using rice bran and agricultural materials. *Tropical Science*, *39*, 77–79.

Alappat, B. J., & Dikshit, A. K. (1999). Management of plastic wastes. *Indian Journal of Environmental Protection*, *19*, 892–897.

Al-Awadhi, J. A., & Shuaibi, A. (2012). Potentiality of Zubair International Journal of Advanced Science and Research 29 Formation for deep slurry injection in Kuwait. *Environmental Earth Sciences*, *67*(5), 1269–1280.

Albertsson, A. C. (1980). The shape of the biodegradation curve for low and high density polyethenes in prolonged series of experiments. *European Polymer Journal*, *16*(7), 623–630. doi:10.1016/0014-3057(80)90100-7

Albrecht, M. A., Evans, C. W., & Raston, C. L. (2006). Green chemistry and the health implications of nanoparticles. *Green Chemistry*, *8*(5), 417–432. doi:10.1039/b517131h

Albrecht, R., Le Petit, J., Terrom, G., & Périssol, C. (2011). Comparison between UV spectroscopy and nirs to assess humification process during sewage sludge and green wastes co-composting. *Bioresource Technology*, *102*(6), 4495–4500. doi:10.1016/j.biortech.2010.12.053 PMID:21239169

Alekhya, M., Divya, N., Jyothirmai, G., & Reddy, K. R. (2013). Secured landfills for disposal of municipal solid waste. *International Journal of Engineering Research and General Science*, *1*(1).

Alhumoud, J. M., & Alhumoud, H. M. (2007). An analysis of trends related to hospital solid wastes management in Kuwait. *Management of Environmental Quality*, *18*(5), 502–513. doi:10.1108/14777830710778274

Ali, U., Sajid, N., Khalid, A., Riaz, L., Rabbani, M.M., Syed, J.H., & Malika, R.N. (2015). A Review on Vermicomposting of Organic Wastes. *Environmental Progress & Sustainable Energy*.

Ali, M. E., Islam, M. R., & Jahiruddin, M. (2009). Effect of integrated use of organic manures with chemical fertilizers in the rice-rice cropping system and its impact on soil health. *Bangladesh Journal of Agricultural Research*, *34*(1), 81–90. doi:10.3329/bjar.v34i1.5757

Ali, M., Wang, W., & Chaudhry, N. (2016). Management of wastes from hospitals: A case study in Pakistan. *Waste Management & Research*, *34*(1), 87–90. doi:10.1177/0734242X15616474 PMID:26628050

Ali, M., Wang, W., Chaudhry, N., & Geng, Y. (2017). Hospital waste management in developing countries: A mini review. *Waste Management & Research*, *35*(6), 581–592. doi:10.1177/0734242X17691344 PMID:28566033

Ali, S. M., Pervaiz, A., Afzal, B., Hamid, N., & Yasmin, A. (2014). Open dumping of municipal solid waste and its hazardous impacts on soil and vegetation diversity at waste dumping sites of Islamabad city. *Journal of King Saud University Science*, *26*(1), 59–65. doi:10.1016/j.jksus.2013.08.003

Aljaradin, M., & Persson, K. M. (2012). Environmental impact of municipal solid waste landfills in semi-arid climates – case study – Jordan. *Open Waste Management Journal*, *5*, 28–39.

Al-Jarallah, R., & Aleisa, E. (2014). A baseline study characterizing the municipal solid waste in the State of Kuwait. *Waste Management (New York, N.Y.)*, *34*, 952–960. PMID:24656421

Al-Khatib, I. A., Al-Qaroot, Y. S., & Ali-Shtayeh, M. S. (2009). Management of healthcare waste in circumstances of limited resources: A case study in the hospitals of Nablus city, Palestine. *Waste Management & Research*, *27*(4), 305–312. doi:10.1177/0734242X08094124 PMID:19423582

Al-Khatib, I. A., & Sato, C. (2009). Solid health care waste management status at health care centers in the West Bank–Palestinian Territory. *Waste Management (New York, N.Y.)*, *29*(8), 2398–2403. doi:10.1016/j.wasman.2009.03.014 PMID:19398317

Allen, C. (2012). *Taiwan: Community Action Leads Government Towards Zero Waste*. Available at http://no-burn.org/downloads/ZW%20Taiwan.pdf

Allen, N. S., Edge, M., Appleyardt, J. R., Jewitt, T. S., & Rorie, C. V. (1988). Degradation of Historic Cellulose Triacetate Cinematographic Film: Influence of Various Film Parameters and Prediction of Archival Life. *The Journal of Photographic Science*, *36*(6), 194–198. doi:10.1080/00223638.1988.11736999

Allen, N. S., Edge, M., Mohammadian, M., & Jones, K. (1994). Physicochemical aspects of the environmental degradation of polyethylene terephthalate. *Polymer Degradation & Stability*, *43*(2), 229–237. doi:10.1016/0141-3910(94)90074-4

Alloway, B. J. (1995). Soil Processes and the behaviour of heavy metals. In B. J. Alloway (Ed.), *Heavy Metals in Soils* (Vol. 354, pp. 11–37). London, UK: Blackie Academic and Professional. doi:10.1007/978-94-011-1344-1_2

Al-Maaded, M., Madi, N. M., Kahraman, R., Hodzic, A., & Ozerkan, N. G. (2012). An Overview of Solid Waste Management and Plastic Recycling in Qatar. *Journal of Polymers and the Environment*, *20*(1), 186–194. doi:10.100710924-011-0332-2

Al-Salem, S. M., Lettieri, P., & Baeyens, J. (2009). Recycling and recovery routes of plastic solid waste (PSW). *Waste Management (New York, N.Y.)*, *29*, 2625–2643. PMID:19577459

Altomare, C., & Tringovska, I. (2011). Beneficial soil microorganisms, an ecological alternative for soil fertility management. In *Genetics, biofuels and local farming systems* (pp. 161–214). Dordrecht: Springer. doi:10.1007/978-94-007-1521-9_6

Alvim-Ferraz, M. C. M., & Afonso, S. A. V. (2005). Incineration of healthcare wastes: Management of atmospheric emissions through waste segregation. *Waste Management (New York, N.Y.)*, *25*(6), 638–648. doi:10.1016/j.wasman.2004.07.017 PMID:15993348

Al-Weshahy, A., & Rao, V. A. (2012). Potato peel as a source of important phytochemical antioxidant nutraceuticals and their role in human health-A review. In *Phytochemicals as Nutraceuticals-Global Approaches to Their Role in Nutrition and Health*. InTech. doi:10.5772/30459

Ambade, B., Sharma, S., Sharma, Y., & Sharma, Y. (2011). Assessment of windrow composting plant's performance at Keru, Jodhpur, India. *Asian Journal of Environmental Science*, *6*(2), 107–111.

Ambulkar, A. R., & Shekdar, A. V. (2004). Prospects of biomethanation technology in the Indian context: A pragmatic approach. *Resources, Conservation and Recycling*, *40*(2), 111–128. doi:10.1016/S0921-3449(03)00037-5

American Chemistry Society. (1993). *The Bakelizer*. Washington, DC: ACS.

Amin, S., & Amin, M. (2011). Thermoplastic elastomeric (TPE) materials and their use in outdoor electrical insulation. *Reviews on Advanced Materials Science*, *29*, 15–30.

Ananth, A. P., Prashanthini, V., & Visvanathan, C. (2010). Healthcare waste management in Asia. *Waste Management (New York, N.Y.)*, *30*(1), 154–161. doi:10.1016/j.wasman.2009.07.018 PMID:19726174

Anastas, P. T., & Beach, E. S. (2007). Green chemistry: The emergence of a transformative framework. *Green Chemistry Letters and Reviews*, *1*(1), 9–24. doi:10.1080/17518250701882441

Anastas, P. T., Bickart, P. H., & Kirchhoff, M. M. (2000). *Designing safer polymers*. New York, NY: John Wiley and Sons, Wiley-Interscience.

Anastas, P. T., & Crabtree, R. H. (Eds.). (2009). Handbook of green chemistry green catalysis.: Vol. I. *Homogenous catalysis*. John Wiley & Sons.

Anastas, P. T., & Warner, J. C. (1998). *Green chemistry: theory and practice*. Oxford, UK: Oxford University Press.

Anastas, P., & Warner, J. (1998). *Green Chemistry: Theory and Practice*. New York: Oxford University Press.

Andrady, A. L. (2011). Microplastics in the marine environment. *Marine Pollution Bulletin*, *62*(8), 1596–1605. doi:10.1016/j.marpolbul.2011.05.030 PMID:21742351

Andrady, A. L., & Neal, M. A. (2009). Application and societal benefits of plastics. *Philosophical Transactions of the Royal Society of London. Series B, Biological Sciences*, *364*(1526), 1977–1984. doi:10.1098/rstb.2008.0304 PMID:19528050

Andrady, A. L., & Neal, M. A. (2009). Applications and societal benefits of plastics Philos. *Trans. R. Soc. Lond. B Biol. Sci.*, *364*, 1977–1984.

Angelidaki I. & Batstone D.J. (2011). Anaerobic digestion: process. *Solid Waste Technology & Management*.

Angelidaki, I., Alves, M., Bolzonella, D., Borzacconi, L., Campos, J. L., Guwy, A. J., ... Van Lier, J. B. (2009). Defining the biomethane potential (BMP) of solid organic wastes and energy crops: A proposed protocol for batch assays. *Water Science and Technology*, *59*(5), 927–934. doi:10.2166/wst.2009.040 PMID:19273891

Angenent, L. T., Karim, K., Al-Dahhan, M. H., Wrenn, B. A., & Domíguez-Espinosa, R. (2004). Production of bioenergy and biochemicals from industrial and agricultural wastewater. *Trends in Biotechnology*, *22*(9), 477–485. doi:10.1016/j.tibtech.2004.07.001 PMID:15331229

Anjello, R., & Ranawana, A. (1996). Death in Slow Motion- India has become the dumping ground for the West's toxic Waste. *Asiaweek*.

Annepu, R. K. (2012). Report on sustainable solid waste management in India. *Waste-to-Energy Research and Technology Council*. Retrieved from http:// swmindia.blogspot.in

Annepu, R. K. (2012). *Sustainable Solid Waste Management in India* (M.Sc. Thesis). Earth Engineering Center, Columbia University, New York, NY.

Ansari, A. A., & Rajpersaud, J. (2012). Physicochemical changes during vermicomposting of water hyacinth (Eichhorniacrassipes) and grass clippings. International Scholarly Resource Network, Soil Science.

Ansari, A. A., & Sukhraj, K. (2010). Effect of vermiwash and vermicompost on soil parameters and productivity of okra (*Abelmoschusesculentus*) in Guyana. *African Journal of Agricultural Research*, *5*(14), 1794–1798.

Antonovics, J., Bradshaw, A. D., & Turner, R. Q. (1971). Heavy metal tolerance in plants. In J. B. Cragg (Ed.), *Advances in ecological research* (Vol. 7, pp. 1–85). New York: Academic Press.

Anurag, G., Smith, R., & Hill, D. (2007). Wastes as co-fuels: The policy framework for solid recovered fuel (SRF) in Europe, with UK implications. *Environmental Science & Technology*, *41*, 4868–4874. PMID:17711195

Anyaegbunam Felix, N. C. (2013). Plasma Gasification for waste management and sustainable renewable clean energy generation. Proceedings of Nigerian Academy of Science, 6(1), 33-50.

Appleton, T. J., Colder, R. I., Kingman, S. W., Lowndes, I. S., & Read, A. G. (2005). Microwave technology for energy-efficient processing of waste. *Applied Energy*, *81*(1), 85–113. doi:10.1016/j.apenergy.2004.07.002

Arancon, N. Q., Edwards, C. A., Bierman, P., Metzger, J. D., & Lucht, C. (2005). Effects of vermicomposts produced from cattle manure, food waste and paper waste on the growth and yield of peppers in the field. *Pedobiologia, 49*(4), 297–306. doi:10.1016/j.pedobi.2005.02.001

Arancon, N. Q., Edwards, C. A., Lees, S., & Byrne, R. (2004). Effects of humic acids from vermicomposts on plant growth. *European Journal of Soil Biology, 46*, 65–69.

Arebey, M., Hannan, M., Basri, H., Begum, R., & Abdullah, H. (2010). Integrated technologies for solid waste bin monitoring system. *Environmental Monitoring and Assessment, 177*(1-4), 399–408. PMID:20703798

Arena, U. (2012). Process and technological aspects of municipal solid waste gasification. A review. *Waste Management (New York, N.Y.), 32*(4), 625–639. PMID:22035903

Arena, U., Zaccariello, L., & Mastellone, M. (2010). Fluidized bed gasification of waste-derived fuels. *Waste Management (New York, N.Y.), 30*(7), 1212–1219. PMID:20172708

Arvanitoyannis, I. S., & Varzakas, T. H. (2008). Vegetable waste management: Treatment methods and potential uses of treated waste. Waste Management for the Food Industries, 703-752.

Arvind, K., & Surat, S. (2013). Domestic Solid Waste Generation– A Case Study of Semi Urban Area of Kathua District, Jammu, J & K, India. International Journal of Scientific and Research Publications, 3(5).

Asgari, Safavi, & Mortazaeinezahad. (2011). Landfill Biogas production process. *International Conference on Food Engineering and Biotechnology, 9*, 208-212.

Asgher, M., Ahmad, Z., & Iqbal, H. M. N. (2017). Bacterial cellulose-assisted de-lignified wheat straw-PVA based bio-composites with novel characteristics. *Carbohydrate Polymers, 161*, 244–252. doi:10.1016/j.carbpol.2017.01.032 PMID:28189235

Asha Aalok, A. K. (2008). Vermi-composting: A Better Option for Organic Solid Waste Management. *Journal of Human Ecology (Delhi, India), 24*(1), 59–64. doi:10.1080/09709274.2008.11906100

AshaAalok, Tripathi, A. K., & Soni, P. (2008). Vermicomposting a better option for Organic. *Journal of Ecology, 24*(1), 59–64.

Asian Development Bank. (2013). *Solid waste management in Nepal: current status and policy recommendations*. Asian Development Bank.

Askarian, M., Vakili, M., & Kabir, G. (2004). Results of a hospital waste survey in private hospitals in Fars province, Iran. *Waste Management, 24*(4), 347-352.

Asnani, P. U. (2006). *Solid Waste Management*. India Infrastructure Report 160–189.

Asnani, P. U. (2006). *Solid Waste Management*. India Infrastructure Report, 160–189.

Asnani, P. U. (2007). Methane to markets and landfill gas energy in India. *Proceeding of the International Conference on Sustainable Solid Waste Management*, 519–525.

Astrup T. (2011). Pyrolysis and Gasification. *Solid Waste Technology & Management*.

Atiyeh, R. M., Edwards, C. A., Subler, S., & Metzger, J. D. (2000). Earthworm-processed organic wastes as components of horticultural potting media for growing marigold and vegetable seedlings. *Compost Science & Utilization, 8*(3), 215–223. doi:10.1080/1065657X.2000.10701994

Atiyeh, R. M., Lee, S. S., Edwards, C. A., Arancon, N. Q., & Metzger, J. D. (2002). The influence of humic acid derived from earthworm-processed organic wastes on plant growth. *Bioresource Technology, 84*(1), 714. doi:10.1016/S0960-8524(02)00017-2 PMID:12137272

Atiyeh, R. M., Subler, S. M., Edwards, C. A., & Metzgar, J. D. (1999). Growth of tomato plants in horticultural potting media amended with vermicompost. *Pedobiologia, 43*, 724–728.

ATSDR, Agency for Toxic Substances and Disease Registry. (1995). *Toxicological profile for polycyclic aromatic hydrocarbons (PAHs)*. Atlanta, GA: US Department of Health and Human Services, Public Health Service.

ATSDR, Agency for Toxic Substances and Disease Registry. (2003). *Toxicological profiles on CD-ROM. Agency for Toxic Substances and Disease Registry*. Public Health Service.

Attarde, S. B., Narkhede, S. D., Patil, R. P., & Ingle, S. T. (2012). Effect of organic and inorganic fertilizers on the growth and nutrient content of *Abelmoschusesculentus* (okra crop). *International Journal of Current Research, 4*(10), 137–140.

Avci, A., Saha, B. C., Dien, B. S., Kennedy, G. J., & Cotta, M. A. (2013). Response surface optimization of corn stover pretreatment using dilute phosphoric acid for enzymatic hydrolysis and ethanol production. *Bioresource Technology, 130*, 603–612. doi:10.1016/j.biortech.2012.12.104 PMID:23334017

Awarenet. (2004). *Hand book of waste management and co-products recovery in food processing*. Academic Press.

Awasthi, M. K., Pandey, A. K., Khan, J., Bundela, P. S., Wong, J. W., & Selvam, A. (2014). Evaluation of thermophilic fungal consortium for organic municipal solid waste composting. *Bioresource Technology, 168*, 214–221. doi:10.1016/j.biortech.2014.01.048 PMID:24507579

Awasthi, M. K., Wang, Q., Awasthi, S. K., Li, R., Zhao, J., Ren, X., ... Zhang, Z. (2018). Feasibility of medical stone amendment for sewage sludge co-composting and production of nutrient-rich compost. *Journal of Environmental Management, 216*, 49–61. doi:10.1016/j.jenvman.2018.01.032 PMID:29357991

Ayala-Zavala, J. F., & González-Aguilar, G. A. (2010). Use of additives to preserve the quality of fresh-cut fruits and vegetables. In *Advances in fresh-cut fruits and vegetables processing* (pp. 233–256). CRC Press. doi:10.1201/b10263-10

Baas, L. W., & Huisingh, D. (2008). The synergistic role of embeddedness and capabilities in industrial symbiosis: Illustration based upon 12 years of experiences in the Rotterdam Harbour and Industry Complex, Progress in Industrial Ecology. *International Journal (Toronto, Ont.), 5*(5–6), 399–421.

Babanyara, Y. Y., Ibrahim, D. B., Garba, T., Bogoro, A. G., & Abubakar, M. Y. (2013). Poor Medical Waste Management (MWM) practices and its risks to human health and the environment: A literature review. *International Journal of Environmental Health and Engineering, 11*(7), 1–8.

Baekeland, L. H. (1909). The synthesis, constitution, and uses of Bakelite. *Industrial & Engineering Chemistry, 1*(3), 149–161. doi:10.1021/ie50003a004

Bajsa, O., Nair, J., Mathew, K., & Ho, G.E. (2003). Vermiculture as a tool for Domestic Wastewater management, *Water Science and Technology, 48*(11/12), 125-132.

Balat, M. (2011). Production of bioethanol from lignocellulosic materials via the biochemical pathway: A review. *Energy Conversion and Management, 52*(2), 858–875. doi:10.1016/j.enconman.2010.08.013

Balat, M., Balat, M., Kırtay, E., & Balat, H. (2009). Main routes for the thermo-conversion of biomass into fuels and chemicals. Part 2: Gasification systems. *Energy Conversion and Management, 50*(12), 3158–3168.

Baldwin, J. H. (1997). Describes industrial design and reuse as precycling, Conservation. *Encarta Online Encyclopedia*. Retrieved from: http://www.encarta.msn.com

Ballesteros, L. F., Teixeira, J. A., & Mussatto, S. I. (2014). Chemical, functional, and structural properties of spent coffee grounds and coffee silverskin. *Food and Bioprocess Technology*, *7*(12), 3493–3503. doi:10.100711947-014-1349-z

Banco Mundial. (2017). *Consumo de energía procedente de combustibles fósiles (% del total)*. Disponível em: http://datos.bancomundial.org/indicador/EG.USE.COMM.FO.ZS?view=chart

Banerjee, J., Singh, R., Vijayaraghavan, R., MacFarlane, D., Patti, A. F., & Arora, A. (2017). Bioactives from fruit processing wastes: Green approaches to valuable chemicals. *Food Chemistry*, *225*, 10–22. doi:10.1016/j.foodchem.2016.12.093 PMID:28193402

Bansal, R. C., & Goyal, M. (2005). *Activated Carbon Adsorption*. New York: CRC Press. doi:10.1201/9781420028812

Bansode, R. R., Losso, J. N., Marshall, W. E., Rao, R. M., & Portier, R. J. (2003). Adsorption of volatile organic compounds by pecan shell- and almond shell-based granular activated carbons. *Bioresource Technology*, *90*(2), 175–184. doi:10.1016/S0960-8524(03)00117-2 PMID:12895561

Barnes, D. K. A., Galgani, F., Thompson, R. C., & Barlaz, M. (2009). Accumulation and fragmentation of plastic debris in global environments. *Philosophical Transactions of the Royal Society of London. Series B, Biological Sciences*, *364*(1526), 1985–1998. doi:10.1098/rstb.2008.0205 PMID:19528051

Barney, J., & Griffin, R. (1992). *The management of organizations: strategy, structure behavior*. Boston, MA: Houghton Mifflin Co.

Barthod, J., Rumpel, C., Calabi-Floody, M., Mora, M. L., Bolan, N. S., & Dignac, M. F. (2018). Adding worms during composting of organic waste with red mud and fly ash reduces CO 2 emissions and increases plant available nutrient contents. *Journal of Environmental Management*, *222*, 207–215. doi:10.1016/j.jenvman.2018.05.079 PMID:29857313

Barthod, J., Rumpel, C., Paradelo, R., & Dignac, M. F. (2016). The effects of worms, clay and biochar on CO 2 emissions during production and soil application of co-composts. *Soil (Göttingen)*, *2*(4), 673–683. doi:10.5194oil-2-673-2016

Basheer & Agrawal. (2013). Management of paper waste by vermicomposting using epigeic earthworm, *Eudriluseugeniae* in Gwalior, India. International. *Journal of Current Microbiology and Applied Science*, *2*(4), 42–47.

Bdour, A., Altrabsheh, B., Hadadin, N., & Al-Shareif, M. (2007). Assessment of medical wastes management practice: A case study of the northern part of Jordan. *Waste Management (New York, N.Y.)*, *27*(6), 746–759. doi:10.1016/j.wasman.2006.03.004 PMID:16713238

Becerra Pérez, L. (2015). *La producción de etanol en México: insumos, procesos y oxigenación de gasolina*. Culiacán, MX: Universidad Autónoma de Sinaloa.

Beg, M. U., Saeed, T., Al-Muzaini, S., Beg, K. R., & Al-Bahloul, M. (2003). Distribution of petroleum hydrocarbon in sediment from coastal area receiving industrial effluents in Kuwait. *Ecotoxicology and Environmental Safety*, *54*(1), 47–55. doi:10.1016/S0147-6513(02)00019-2 PMID:12547635

Beilenhoff, U., Neumann, C. S., Rey, J. F., Biering, H., Blum, R., Cimbro, M., ... Schmidt, V. (2008). ESGE–ESGENA guideline: Cleaning and disinfection in gastrointestinal endoscopy. *Endoscopy*, *40*(11), 939–957. doi:10.1055-2008-1077722 PMID:19009486

Belliturk, K. (2016). Vermicomposting Technology for Solid Waste Management in Sustainable Agricultural Production. *Cukurova J Agric Food Sci*, *31*, 1–5.

Belliturk, K., Gorres, J. H., Bagdatl, M. C., Gocmez, S., & Turan, H. S. (2015). The Evaluation of Olive Pruning Waste as a Vermicompost: Micro Elements. *Agriculture Vision Agriculture*, *1*, 7–12.

Bello, M., Lawan, M. K., Aluwong, T., & Sanusi, M. (2015). Management of slaughter houses in northern Nigeria and the safety of meat produced for human consumption. *Food Control*, *49*, 34–39. doi:10.1016/j.foodcont.2013.09.007

Belowska-Bien, K., & Zdrojewicz, Z. (2006). Phthalates – structure, properties, clinical significance. *Advances in Clinical and Experimental Medicine*, *15*, 677–681.

Benachour, N., & Aris, A. (2009). Toxic Effects of Low Doses of Bisphenol A on Human Placental Cells. *Toxicology and Applied Pharmacology*, *241*(3), 322–328. doi:10.1016/j.taap.2009.09.005 PMID:19769995

Benerjee, S. K., & Bagchi, K. (1999). Hospital solid waste and its management approach- A case study of a hospital in Calcutta. *Indian Journal of Environmental Protection*, *19*, 932–938.

Bera, R., Datta, A., Bose, S., Dolui, A.K., Chatterjee, A.K., Dey, G.C., … Seal, A. (2013). Comparative Evaluation of Compost Quality, Process Convenience and Cost under Different Composting Methods to assess their Large Scale Adoptability Potential as also Complemented by Compost Quality Index. *International Journal of Scientific and Research Publications, 3*(6).

Bergmann, M., Gutow, L., & Klages, M. (Eds.). (2015). *Marine AnthropogenicLitter*. New York: Springer International Publishing.

Bergmann, W. C. (1934). Contributions to the study of marine products: II. The sterols of mollusks. *The Journal of Biological Chemistry*, *104*, 317–328.

Berkun, M., Aras, E., & Nemlioglu, S. (2005). Disposal of solid waste in Istanbul and along the black sea coast of Turkey. *Waste Management (New York, N.Y.)*, *25*(8), 847–855. doi:10.1016/j.wasman.2005.04.004 PMID:15949934

Bernache, G. (2003). *The environmental impact of municipal waste management: the case of Guadalajara metro area* (Vol. 39). Resources, Conservation and Recycling.

Bernal, M. P., Alburquerque, J. A., & Moral, R. (2009). Composting of animal manures and chemical criteria for compost maturity assessment. A review. *Bioresource Technology*, *100*(22), 5444–5453. doi:10.1016/j.biortech.2008.11.027 PMID:19119002

Berry, L. L., & Parasuraman, A. (2004). *Marketing services: Competing through quality*. Simon and Schuster.

Bertram, M., Graedel, T. E., Rechberger, H., & Spatari, S. (2002). The contemporary European copper cycle: Waste management subsystem. *Ecological Economics*, *42*(1-2), 43–57. doi:10.1016/S0921-8009(02)00100-3

Beuchat, L. R. (1986). Oncom (fermented peanut presscake). In N. R. Reddy, M. D. Pierson, & D. K. Salunkhe (Eds.), *Legume-based fermented foods* (pp. 135–144). Boca Raton, FL: CRC Press.

Bhat RA, Dar GH, Jehangir A, Bhat BA, Yousuf AR (2012) Municipal solid waste generation and present scenario of waste management during Yatra season in Pahalgam: A tourist health resort of Kashmir valley. *Int. J. Cur. Res., 4*, 4-9.

Bhat, R. A., Dar, S. A., Dar, D. A., & Dar, G. H. (2018). Municipal Solid Waste Generation and current Scenario of its Management in India. *International Journal of Advance Research in Science and Engineering, 7*(2), 419-431.

Bhat, R.A., Dar, G.A., Jehangir, A., Bhat, B.A., & Yousuf, A. R. (2012). Municipal solid waste generation and present scenario of waste management during Yatra season in Pahalgam: a tourist health resort of Kashmir valley. *International Journal of Current Research 4*, 4-9.

Bhat, R.A., Dar, S.A., Dar, D.A., & Dar, G.H. (2018). Municipal Solid Waste generation and current scenario of its management in India, *International Journal of Advance Research in science and Engineering, 7*(2), 419-431.

Bhat, R. A., Dar, S. A., Dar, D. A., & Dar, G. H. (2018). Municipal Solid Waste Generationand current Scenario of its Management in India. *International Journal of Advance Research in Science and Engineering, 07*(2), 419–431.

Bhat, R. A., Dervash, M. A., Mehmood, M. A., & Hakeem, K. R. (2018). Municipal solid waste generation and its management, a growing threat to fragile ecosystem in Kashmir 756 Himalaya. *American Journal of Environmental Sciences, 2*(11), 145–167.

Bhat, R. A., Kamili, A. N., & Bandh, S. A. (2013). Characterisation and composition of municipal solid waste (MSW) generated in Yusmarg: A health resort of Kashmir valley. A Glance at the World. *Waste Management (New York, N.Y.), 33*, 774–777.

Bhat, R. A., Kamili, A. N., & Bandh, S. A. (2013). Characterisation and composition of Municipal Solid Waste (MSW) generated in Yusmarg: A health resort of Kashmir valley: A Glance at the World. *Waste Management (New York, N.Y.), 33*(3), 774–777. doi:10.1016/j.wasman.2013.01.009

Bhat, R. A., Nazir, R., Ashraf, S., Ali, M., Bandh, S. A., & Kamili, A. N. (2014). Municipal solid waste generation rates and its management at Yusmarg forest ecosystem, a tourist resort in Kashmir. *Waste Management & Research, 32*(2), 165–169. doi:10.1177/0734242X13518089 PMID:24519231

Bhattacharyya, J. K., Kumar, S., & Devotta, S. (2007). Studies on acidification in two-phase biomethanation process of municipal solid waste. *Waste Management (New York, N.Y.), 28*(1), 164–169. doi:10.1016/j.wasman.2006.11.014 PMID:17276666

Bhawalkar, U. S. (1989). Vermiculture: A promising source of biofertilizers. In National Seminars in Agriculture Biotechnology Held at Gujarat Agriculture University, Nausari, India.

Bhawalkar, U. (1993). *Turning garbage into gold: An introduction to vermiculture biotechnology*. Bhawalkar Earthworm Research Institute Pune.

Bhawalkar, U. S. (1995). *Vermiculture bioconversion of organic residues*. IIT Mumbai.

Bhiday, M. R. (1994). Earthworms in agriculture. *Indian Farming, 43*(12), 31–34.

Bhiday, M. R. (1994). Earthworms in Agriculture. *Indian Farming*.

Bhogayata, A. C., & Arora, N. K. (2018). Impact strength, permeability and chemical resistance of concrete reinforced with metalized plastic waste fibers. *Construction & Building Materials, 161*, 254–266. doi:10.1016/j.conbuildmat.2017.11.135

Bhoyar, R. V., & Bhide, A. D. (1996). *Vermicomposting of organic (vegetable) solid wastes*. Nagpur, India: VRCE.

Bianchini, F., & Hewage, K. (2012). How "green" are the green roofs? Lifecycle analysis of green roof materials. *Building and Environment, 48*, 57–65. doi:10.1016/j.buildenv.2011.08.019

Birpınar, M. E., Bilgili, M. S., & Erdoğan, T. (2009). Medical waste management in Turkey: A case study of Istanbul. *Waste Management (New York, N.Y.), 29*(1), 445–448. PMID:18550354

Biswas, A. K., Kumar, S., Babu, S. S., Bhattacharyya, J. K., & Chakrabarti, T. (2010). Studies on environmental quality in and around municipal solid waste dumpsite. *Resources, Conservation and Recycling, 55*, 129–134.

Bjerre, A. B., Olesen, A. B., Fernqvist, T., Plöger, A., & Schmidt, A. S. (1996). Pretreatment of wheat straw using combined wet oxidation and alkaline hydrolysis resulting in convertible cellulose and hemicellulose. *Biotechnology and Bioengineering, 49*(5), 568–577. doi:10.1002/(SICI)1097-0290(19960305)49:5<568::AID-BIT10>3.0.CO;2-6 PMID:18623619

Blakeman, J. W. (2004). Recycling. *Coloring of Plastics, 353–357*, 353–357. doi:10.1002/0471721581.ch24

Blanchette, R. A., Cease, K. R., Abad, A., Koestler, R. J., Simpson, E., & Sams, G. K. (1991). An evaluation of different forms of deterioration found in archaeological wood. *International Biodeterioration, 28*(1-4), 3–22. doi:10.1016/0265-3036(91)90030-U

Blowers, A. (2013). Pollution and waste-a sustainable burden? In *Planning for a Sustainable Environment* (pp. 81–104). Routledge. doi:10.4324/9781315070315

Bobinaitė, R., Viskelis, P., Bobinas, Č., Mieželienė, A., Alenčikienė, G., & Venskutonis, P. R. (2016). Raspberry marc extracts increase antioxidative potential, ellagic acid, ellagitannin and anthocyanin concentrations in fruit purees. *Lebensmittel-Wissenschaft + Technologie, 66*, 460–467. doi:10.1016/j.lwt.2015.10.069

Bogner, J., Abdelrafie Ahmed, M., Diaz, C., Faaij, A., Gao, Q., & Hashimoto, S. (2007). Waste management, in Climate Change: Mitigation of Climate Change. Contribution of Working Group III to the Fourth Assessment Report of the Intergovernmental Panel on Climate Change. Cambridge University Press.

Bogner, J., Abdelrafie Ahmed, M., Diaz, C., Faaij, A., Gao, Q., Hashimoto, S., ... Zhang, T. (2007). Waste Management. In *Climate Change Mitigation. Contribution of Working Group III to the Fourth Assessment Report of the Intergovernmental Panel on Climate Change*. Cambridge, UK: Cambridge University Press.

Bombelli, P., Howe, C. J., & Bertocchini, F. (2017). Polyethylene bio-degradation by caterpillars of the wax moth *Galleria mellonella. Current Biology, 27*(8), 292–293. doi:10.1016/j.cub.2017.02.060 PMID:28441558

Bordner, R., Winter, J. A., & Scarpino, P. (Eds.). (1978). *Microbiological methods for monitoring the environment: water and wastes*. Environmental Protection Agency, Office of Research and Development, Environmental Monitoring and Support Laboratory.

Borghi, A. D., Gallo, M., & Borghi, M. D. (2009). A survey of life cycle approaches in waste management. *The International Journal of Life Cycle Assessment, 14*(7), 597–610. doi:10.100711367-009-0111-7

Boswijk, G., Fowler, A., Lorrey, A., Palmer, J., & Ogden, J. (2006). Extension of the New Zealand kauri (*Agathis australis*) chronology to 1724 BC. *The Holocene, 16*(2), 188–199. doi:10.1191/0959683606hl919rp

Bouche, M. B., & Ferriere, G. (1986). Soil organisms as components of ecosystem. Ecological Bulletin, 25, 122-132.

Boulding, K. (1966). The economics of the coming spaceship Earth. In H. Jarrett (Ed.), *Environmental quality in a growing economy, resources for the future* (pp. 3–14). Baltimore, MD: Johns Hopkins University Press.

Boulter, J. I., Trevors, J. T., & Boland, G. J. (2002). Microbial studies of compost: Bacterial identification, and their potential for turfgrass pathogen suppression. *World Journal of Microbiology & Biotechnology, 18*(7), 661–671. doi:10.1023/A:1016827929432

Bradshaw, A. D., & Chadwich, M. J. (1981). *The restoration of land*. Oxford, UK: Blackwell.

Bradshaw, A. D., & Chadwick, M. J. (1980). *The construction of land*. Blackwell.

Brems, A., Baeyens, J., & Dewil, R. (2012). Recycling and recovery of post-consumer plastic solid waste in a European context. *Thermal Science, 16*(3), 669–685.

Brennand, T. (2004). Renewable energy in the United Kingdom: Policies and prospects. *Energy for Sustainable Development, 8*(1), 82–92.

Brezet, H., & Hemel, C. V. (1997). *ECODESIGN – A Promising Approach to Sustainable Production and Consumption*. Paris: UNEP.

Brigden, K., Labunska, I., Santillo, D., & Johnston, P. (2008). *Chemical contamination at e-waste recycling and disposal sites in Accra and Korforidua, Ghana*. Greenpeace Research Laboratories Technical Note. Greenpeace International. Retrieved from http://www.greenpeace.org/international/press/reports/chemicalcontamination-at-e-wa

Briggs, D. J., & Courtney, F. M. (1985). *Agriculture and environment*. Longman Inc.

British Medical Association. (1999). *Health and Environmental impact assessment, an integrated Approach*. Earthcan Publications Ltd.

British Plastics Federation. (2008). *Oil consumption*. British Plastics Federation. Retrieved from http://www.bpf.co.uk/press/Oil_Consumption

Brooks, A.M., Wang, S., & Jambeck, J.R. (2018). The Chinese import ban and its impact on globalplastic waste trade. *Science Advanced, 4*.

Brown, E., & Mann, M. (2008). *Initial market assessment for small-scale biomass-based CHP*. National Renewable Energy Laboratory. Disponível em: http://citeseerx.ist.psu.edu/viewdoc/download?doi=10.1.1.370.1389&rep=rep1&type=pdf

Browne, M. A., Dissanayake, A., Galloway, T. S., Lowe, D. M., & Thompson, R. C. (2008). Ingested Microscopic Plastic Translocates to the Circulatory System of the Mussel, *Mytilus edulis* (L.). *Environmental Science & Technology, 42*(13), 5026–5031. doi:10.1021/es800249a PMID:18678044

BSI. (2008). *PAS 2050:2008, Specification for the Assessment of the Life Cycle Greenhouse Gas Emissions of Goods and Services*. Available at: http://www.bsigroup.com/en/Standards-andPublications/Industry-Sectors/Energy/PAS-2050/PAS-2050-Form-page/

Buenrostro, O., & Bocco, G. (2003). Solid waste management in municipalities in Mexico: Goals and perspectives. *Resources, Conservation and Recycling, 39*(3), 251–263. doi:10.1016/S0921-3449(03)00031-4

Bulluck, L. R., Borsius, M., Evanylo, G. K., & Ristaino, J. B. (2002). Organic and synthetic fertility amendments influence soil microbial, physical and chemical properties on organic and conventional farms. *Applied Soil Ecology, 19*(2), 147–160. doi:10.1016/S0929-1393(01)00187-1

Bundela, P. S., Gautam, S. P., Pandey, A. K., Awasthi, M. K., & Sarsaiya, S. (2010). Municipal solid waste management in Indian cities. *Int. J. Environ. Sci., 1*(4), 591–605.

Bundela, P. S., Gautam, S. P., Pandey, A. K., Awasthi, M. K., & Sarsaiya, S. (2010). Municipal solid waste management in Indian cities. *International Journal of Environmental Science and Technology, 1*(4), 591–605.

Bureau of Plant Industry (BPI), Department of Agriculture Philippines. (2010). Retrieved on September 29, 2010 from http://www.bpi.da.gov.ph/Services/vv.htm

Burke, G., Singh, B. R., & Theodore, L. (2005). *Handbook of environmental management and technology*. John Wiley & Sons.

Burnley, S., Phillips, R., Coleman, T., & Rampling, T. (2011). Energy implications of the thermal recovery of biodegradable municipal waste materials in the United Kingdom. *Waste Management (New York, N.Y.), 31*, 1949–1959. PMID:21600755

Busby, R. R. (2003). *Suitability of a municipal solid waste byproduct as a soil amendment for reestablishing native grasses on disturbed Army training lands* (M.Sc. Thesis). University of Illinois, Urbana-Champaign, IL.

Busby, R., Gebhart, D. L., & Torbert, H. A. (2006). Effects of an uncomposted municipal waste processing by product on prairie grass establishment. *Agronomy Journal, 98*, 1073–1080.

Bustos Flores, C. (2009). La problemática de los desechos sólidos. *Economía*, *27*, 121-144. Disponível em: http://www.redalyc.org/articulo.oa?id=195614958006

Butler, E., Devlin, G., & McDonnell, K. (2011). Waste Polyolefins to Liquid Fuels via Pyrolysis: Review of Commercial State-of-the-Art and Recent Laboratory Research. *Waste and Biomass Valorization*, *2*(3), 227–255. doi:10.100712649-011-9067-5

Cabalova, I., Kacik, F., Geffert, A., & Kacikova, D. (2011). *The effects of paper recycling and its environmental impact.* Environmental Management in Practice; doi:10.5772/23110.2011

Caberara, G. L., Rodriguez, D. M. G., & Maruri, A. B. (1999). Genotoxicity of the extracts from the compost of organic and the total municipal garbage using three plant bioassays. *Mutation Research*, *426*(2), 201–206. doi:10.1016/S0027-5107(99)00068-8 PMID:10350598

Cadoche, L., & López, G. D. (1989). Assessment of size reduction as a preliminary step in the production of ethanol from lignocellulosic wastes. *Biological Wastes*, *30*(2), 153–157. doi:10.1016/0269-7483(89)90069-4

Camacho, S. L. (1995). The Plasma Arc Torch, Its Electrical and Thermal Characteristics. In *Proceedings of the International Symposium on Environmental Technologies: Plasma Systems and Applications*. Georgia Institute of Technology Research Corporation.

Camill, P. (2010). Global Change. *Nature Education Knowledge*, *3*(10), 49.

Canesi, L., Ciacci, C., & Balbi, T. (2015). Interactive effects of nanoparticles with other contaminants in aquatic organisms: Friend or foe? *Marine Environmental Research*, *111*, 128–134. doi:10.1016/j.marenvres.2015.03.010 PMID:25842999

Capalbo, D.M.F., & Valicente, F.H., Oliveira Moraes, I., & Pelizer, L.H. (2001). Solid-statefermentation of Bacillus thuringiensis tolworthi to control fall armyworm inmaize. *Electronic Journal of Biotechnology*, 4.

Capoor, M. R., & Bhowmik, K. T. (2017). Current perspectives on biomedical waste management: Rules, conventions and treatment technologies. *Indian Journal of Medical Microbiology*, *35*(2), 157. PMID:28681801

Carpenter, E. J., & Smith, K. L. Jr. (1972). Plastic on the Sargasso Sea surface. *Science*, *175*(4027), 1240–1241. doi:10.1126cience.175.4027.1240 PMID:5061243

Ceballos, D., Gong, W., & Page, E. (2015). A pilot assessment of occupational health hazards in the U.S. electronic scrap recycling industry. *Journal of Occupational and Environmental Hygiene*, *12*(7), 482–488. doi:10.1080/15459624.2015.1018516 PMID:25738822

CEET. (2008). *Biomedical Waste Management- Burgeoning issue*. CEET.

CEFC. (2015). *Clean Energy Finance Corporation: The Australian bioenergy and energy from waste market November 2015*. Sydney: CEFC.

Census. (2011). Provisional Population Totals, India.

Central Pollution Control Board (CPCB). (2007). *Bio-methanation potential of solid wastes from agro-based industries.* Delhi, India: Ministry of Environment and Forests.

Central Pollution Control Board. (2000). *Management of Municipal Solid Waste*. Delhi Centre Pollution Control Board. Available online: www.cpcb.nic.in

Central Pollution Control Board. (2005). *Management of Municipal Solid Waste*. Delhi, India: Ministry of Environment and Forests.

Central Pollution Control Board. (2007). *Bio-methanation potential of solid wastes from Agro- based Industries*. Delhi, India: Ministry of Environment and Forests.

Central Pollution Control Board. (2016-17). *Consolidated annual report (For the year 2016-2017)*. Author.

Cerdá, E., Caparrós, A., & Ovando, P. (2017). *Bioenergía en la Unión Europea*. Madrid: Fedea. Disponível em: https://www.researchgate.net/profile/Alejandro_Caparros/publication/277259293_Bioenergia_en_la_Union_Europea/links/0f3175358c6e9ac1d8000000/Bioenergia-en-la-Union-Europea.pdf

Cerezal, P., & Duarte, G. (2005). Utilización de cáscaras en la elaboración de productosconcentrados de tuna (Opuntiaficus-índica (L.)Miller). *Journal of the Professional Association for Cactus Development*, 61-83.

Cesaro, A., & Belgiorno, V. (2015). Combined Biogas and Bioethanol Production: Opportunities and Challenges for Industrial Application. *Energies*, *8*(8), 8121–8144. doi:10.3390/en8088121

Chacko, C. M., & Estherlydia, D. (2014). Antimicrobial evaluation of jams made from indigenous fruit peels. *International Journal of Advanced Research*, *2*(1), 202–207.

Chakrabarty, D. (2008). *Vermicompost and Organic Pisciculture*. West Bengal: Akshay Krishi Vikash.

Chakraborty, S., Veeregowda, B., Gowda, L., Sannegowda, S. N., Tiwari, R., Dhama, K., & Singh, S. V. (2013). Biomedical waste management. *Interaction*.

Chambers, J. C., & Siddle, R. C. (1991). Fate of heavy metals in abandoned lead zinc tailing ponds: Vegetation. *Journal of Environmental Quality*, *20*(4), 745. doi:10.2134/jeq1991.00472425002000040005x

Chander, G., Wani, S. P., Gopalakrishnan, S., Mahapatra, A., Chaudhury, S., Pawar, C. S., ... Rao, A. K. (2018). Microbial consortium culture and vermi-composting technologies for recycling on-farm wastes and food production. *International Journal of Recycling of Organic Waste in Agriculture*, *7*(2), 99–108. doi:10.100740093-018-0195-9

Chandler, A. J., Eighmy, T. T., Hjelmar, O., Kosson, D. S., Sawell, S. E., Vehlow, J., ... Hartlén, J. (1997). *Municipal solid waste incinerator residues* (Vol. 67). Elsevier.

Chaney, R. L. (1973). Crop and food chain effects of toxic elements in sludge and effluents: Recycling municipal sludges and effects on land. Academic Press.

Chang, N. B., Pires, A., & Martinho, G. (2011). Empowering systems analysis for solid waste management: Challenges, trends, and perspectives. *Critical Reviews in Environmental Science and Technology*, *41*(16), 1449–1530.

Chaoui, I., Zibiliske, M., & Ohno, T. (2003). Effects of earthworm casts and compost on soil microbial activity and plant nutrient availability. *Soil Biology & Biochemistry*, *35*(2), 295–302. doi:10.1016/S0038-0717(02)00279-1

Chao, W. L., Lin, C. M., Shiung, I. I., & Kuo, Y. L. (2006). Degradation of di-butyl-phthalate by soil bacteria. *Chemosphere*, *63*(8), 1377–1383. doi:10.1016/j.chemosphere.2005.09.021 PMID:16289698

Chartier, Y. (Ed.). (2014). *Safe management of wastes from health-care activities*. World Health Organization.

Chatterjee, R. (2010). Municipal solid waste management in Kohima city— India. *Iranian Journal of Environmental Health Sciences & Engineering*, *7*(2), 173–180.

Chatterjee, S. (2012). Sustainable electronic waste management and recycling process. *American Journal of Environmental Engineering*, *2*(1), 23–33. doi:10.5923/j.ajee.20120201.05

Chattopadhyay, G. (2012). Use of vermicomposting biotechnology for recycling organic wastes in agriculture. *International Journal of Recycling of Organic Waste in Agriculture*, *1*(1), 1–6. doi:10.1186/2251-7715-1-8

Chaudhuri, P. S., Pal, T. K., Bhattacharjee, G., & Dey, S. K. (2000). Chemical changes during vermin- composting (Perionyxexcavatus) of kitchen wastes. *Tropical Ecology*, *41*, 107–110.

Chavan, A. (2013). Use of plastic waste in flexible pavements. *IJAIEM*, *2*(4), 540–553.

Chen, D., Yina, L., Wanga, H., & He, P. (2014). Pyrolysis technologies for municipal solid waste. *Waste Management*, 1-21.

Cheng, H., & Hu, Y. (2010). Municipal solid waste (MSW) as a renewable source of energy: Current and future practices in China. *Bioresource Technology*, *101*(11), 3816–3824. doi:10.1016/j.biortech.2010.01.040 PMID:20137912

Cheng, Y. W., Sung, F. C., Yang, Y., Lo, Y. H., Chung, Y. T., & Li, K. C. (2009). Medical waste production at hospitals and associated factors. *Waste Management (New York, N.Y.)*, *29*(1), 440–444. doi:10.1016/j.wasman.2008.01.014 PMID:18359619

Chen, J. (2007). Rapid urbanization in China: A real challenge to soil protection and food security. *Catena*, *69*(1), 1–15.

Cheremisinoff, N. P. (2003). *Handbook of solid waste management and waste minimization technologies*. Elsevier Science.

Chertow, M. R. (2000). Industrial symbiosis: Literature and taxonomy. *Annual Review of Energy and the Environment*, *25*(1), 313–337. doi:10.1146/annurev.energy.25.1.313

Cherubini, F., Bargigli, S., & Ulgiati, S. (2009). Life cycle assessment (LCA) of waste management strategies: Landfilling, sorting plant and incineration. *Energy*, *34*(12), 2116–2123. doi:10.1016/j.energy.2008.08.023

Cheung, W. M., & Pachisia, V. (2015). Facilitating waste paper recycling and repurposing via cost modelling of machine failure, labour availability and waste quantity. *Resources, Conservation and Recycling*, *101*, 34–41. doi:10.1016/j.resconrec.2015.05.011

Chibuike, G. U., & Obiora, S. C. (2014). Heavy Metal Polluted Soils:Effect on Plants and Bioremediation Methods. *Applied and Environmental Soil Science*, *2014*, 1–12. doi:10.1155/2014/752708

Chiemchaisri, C., Juanga, J. P., & Visvanathan, C. (2007). Municipal solid waste management in Thailand and disposal emission inventory. *Environmental Monitoring and Assessment*, *135*, 13–20. PMID:17492361

Chinwan, D., & Pant, S. (2014). Waste to Energy in India and its Management. *Journal of Basic and Applied Engineering Research*, *1*(10), 89–94.

Chirashree, G. (2004). Integrated Vermi-Pisciculture – an alternative option for recycling of solid municipal waste in rural India. *Bioresource Technology*, *93*(1), 71–75. doi:10.1016/j.biortech.2003.09.014 PMID:14987723

Chowdhary, A. (2018). Study of knowledge, behaviour and practice of biomedical waste among health personnel. *International Journal of Community Medicine and Public Health*, *5*(8), 3330–3334. doi:10.18203/2394-6040.ijcmph20183056

Chowdhury, A. K. M. M. B., Konstantinou, F., Damati, A., Akratos, C. S., Vlastos, D., Tekerlekopoulou, A. G., & Vayenas, D. V. (2015). Is physicochemical evaluation enough to characterize olive mill waste compost as soil amendment? The case of genotoxicity and cytotoxicity evaluation. *Journal of Cleaner Production*, *93*, 94–102. doi:10.1016/j.jclepro.2015.01.029

Christensen, T. (2011). *Solid waste technology & management. In International Journal of Advanced Science and Research Chichester* (p. 108). West Sussex, UK: Wiley.

Christos, C., & Lasaridi, K. (2009). A GIS based model for the optimization of municipal solid waste collection: The case study of Nikea, Athens, Greece. *WSEAS Transactions on Environment and Development*, *5*(10), 640–650.

Chronopoulos, J., Haidouti, C., Chronopoulousereli, A., & Massas, I. (1997). Variations in plant and soil lead and cadmium content in urban parks in Athens. Greece. *The Science of the Total Environment*, *196*(1), 91–98. doi:10.1016/S0048-9697(96)05415-0

Chung, S. S., & Lo Carlos, W. H. (2008). Local waste management constraints and waste administrators in China. *Waste Management (New York, N.Y.)*, *28*(2), 272–281. doi:10.1016/j.wasman.2006.11.013 PMID:17280826

City of Toronto. (2012). *How We Calculate Waste Diversion Rates*. Available: http://www.toronto.ca/environment/initiatives/wdt.htm

Clapp, J. (2002). The distancing of waste: Overconsumption in a global economy. *Confronting Consumption*, 155-176.

Coker, A., Sangodoyin, A., Sridhar, M., Booth, C., Olomolaiye, P., & Hammond, F. (2009). Medical waste management in Ibadan, Nigeria: Obstacles and prospects. *Waste Management (New York, N.Y.)*, *29*(2), 804–811. doi:10.1016/j.wasman.2008.06.040 PMID:18835151

Cole, M., & Galloway, T. S. (2015). Ingestion of nanoplastics andmicroplastics by Pacific oyster larvae. *Environmental Science & Technology*, *49*(24), 14625–14632. doi:10.1021/acs.est.5b04099 PMID:26580574

Cole, M., Lindeque, P., Fileman, E., Halsband, C., & Galloway, T. S. (2015). The impact of polystyrene microplastics on feeding, function and fecundity in the marine copepod *Calanushel golandicus*. *Environmental Science & Technology*, *49*(2), 1130–1137. doi:10.1021/es504525u PMID:25563688

Colon, M., & Fawcett, B. (2006). Community-based household waste management: Lessons learnt from EXNORA's 'zero waste management' scheme in two South Indian cities. *Habitat International*, *30*(4), 916–931. doi:10.1016/j.habitatint.2005.04.006

Comisión Federal de Electricidad; Instituto de Investigaciones Eléctricas. (2012). *Guía de usuario: generación de electricidad mediante residuos sólidos urbanos*. Cuauhtémoc, MX: CFE. Disponível em: http://www.olade.org/realc/docs/doc_106229_20170612115908.pdf

Communication from the Commission Europe. (2010). *A strategy for smart, sustainable and inclusive growth*. Author.

Cook, J. (1977). Environmental pollution by heavy metals. *International Journal of Environmental Qulaity*, *9*, 253–266.

Coombs, J. (1996). *Bioconversion Assessment Study*. CPL Scientific Ltd.

Coronel, B., Duroselle, P., Behr, H., Moskovtchenko, J. F., & Freney, J. (2002). In situ decontamination of medical wastes using oxidative agents: A 16-month study in a polyvalent intensive care unit. *The Journal of Hospital Infection*, *50*(3), 207–212. doi:10.1053/jhin.2002.1188 PMID:11886197

Couth, R., & Trois, C. (2012). Cost effective waste management through composting in Africa. *Waste Management (New York, N.Y.)*, *32*, 2518–2525. PMID:22857939

Cox, J., Giorgi, S., Sharp, V., Strange, K., Wilson, D. C., & Blakey, N. (2010). Household waste prevention—A review of evidence. *Waste Management & Research*, *28*(3), 193–219. doi:10.1177/0734242X10361506 PMID:20215491

Cozar, A., Echevarría, F., Gonzalez-Gordillo, J. I., Irigoien, X., Úbeda, B., Hernandez-Leon, S., ... Duarte, C. M. (2014). Plastic debris in the open ocean. *Proceedings of the National Academy of Sciences of the United States of America*, *111*(28), 10239–10244. doi:10.1073/pnas.1314705111 PMID:24982135

CPCB. (2000a). *Status of Solid Waste Generation, Collection, Treatment and Disposal in Metro cities, Series: CUPS/46/1999–2000*. Author.

CPCB. (2000b). *Status of Municipal Solid waste Generation, Collection, Treatment and Disposal in Class I Cities, Series: ADSORBS/31/1999–2000*. Author.

CPCB. (2000c). *Management of municipal solid waste in Delhi*. Retrieved from http://www.cpcb.nic.in/divisionsofheadoffice/pcp/MSW_Report

CPCB. (2000d). *Management of municipal solid waste Delhi*. Retrieved from http://www.cpcb.nic.in/divisionsofheadoffice/pcp/ MSW_Report.pdf

CPCB. (2004b). Management of Municipal Solid Waste. Ministry of Environment and Forests.

CPHEEO — Central Public Health and Environmental Engineering Organization. (2000). Manual on Municipal Solid Waste Management. Ministry of Urban Development, Govt. of India.

Cresser, M., Killham, K., & Edwards, T. (1993). *Soil chemistry and its applications*. Cambridge, UK: Cambridge University Press. doi:10.1017/CBO9780511622939

Crowley, D., Staines, A., Collins, C., Bracken, J., Bruen, M., Fry, J., … Thunhurst, C. (2003). *Health and environmental effects of landfilling and incineration of waste-A literature review*. Academic Press.

Cunningham, W., Cunningham, M. A., & Saigo, B. W. (2007). Environmental Science, A global concern. *Mc. Graw Hill. Higher Education.*

Cuny, D., Haluwyn, C. V., & Pesch, R. (2001). Biomonitoring of trace elements in air and soil compartments along the major motorway in France. *Water, Air, and Soil Pollution*, *125*(1/4), 273–289. doi:10.1023/A:1005278900969

Curran, T., & Williams, I. D. (2012). A zero waste vision for industrial networks in Europe. *Journal of Hazardous Materials*, *207-208*, 3–7. doi:10.1016/j.jhazmat.2011.07.122 PMID:21899950

Da Silva, C. E., Hoppe, A. E., Ravanello, M. M., & Mello, N. (2005). Medical wastes management in the south of Brazil. *Waste Management (New York, N.Y.)*, *25*(6), 600–605. doi:10.1016/j.wasman.2004.03.002 PMID:15993344

da-Costa, J. P., Santos, P. S. M., Duarte, A. C., & Rocha-Santos, T. (2016). (Nanoplastics in the environment sources, fates and effects. *The Science of the Total Environment*, *566–567*, 15–26. doi:10.1016/j.scitotenv.2016.05.041 PMID:27213666

Dalentoft, E., & Thulin, P. (1997). The use of aerobic selectors in activated sludge systems for treatment of wastewater from the pulp and paper industry. *Water Science and Technology*, *35*(2-3), 181–188. doi:10.2166/wst.1997.0513

Daly, M. J. (2009). A new perspective on radiation resistance based on Deinococcus radiodurans. *Nature Reviews. Microbiology*, *7*(3), 237–245. doi:10.1038/nrmicro2073 PMID:19172147

Damgaard, A., Manfredi, S., Merrild, H., Stensøe, S., & Christensen, T. H. (2011). LCA and economic evaluation of landfill leachate and gas technologies. *Waste Management (New York, N.Y.)*, *31*(7), 1532–1541. doi:10.1016/j.wasman.2011.02.027 PMID:21435856

Dandin, S. B., Das, P. K., & Bhogesha, K. (2006). *National Seminar on soil health and water management of sustainable sericulture*. Central Silk Board.

Danthurebandara, M., Van Passel, S., Nelen, D., Tielemans, Y., & Van Acker, K. (2012). Environmental and socioeconomic impacts of landfills. *Linnaeus Eco-Tech*, *2012*, 40–52.

Darwin, C. (1981). *The formation of vegetable mould through the action of worms, with observations on their habitats*. London: Murray.

Dasgupta, B., Yadav, V.L., & Mondal, M.K. (2013). Seasonal characterization and present status of municipal solid management in Varanasi, India. *Adv. Environ. Res., 2*, 51–60. doi:.1.051 doi:10.12989/aer.2013.2

Dasgupta, B., & Mondal, M. (2012). Bio Energy Conversion of Organic Fraction of Varanasi's Municipal Solid Waste. *Energy Procedia, 14*, 1931–1938.

Dash, M. C. (1978). *Role of Earthworms in the Decomposer System*. New Delhi, India: India International Scientific Publication.

Das, M., Uppal, H. S., Singh, R., Beri, S., Mohan, K. S., Gupta, V. C., & Adholeya, A. (2011). Co-composting of physic nut (Jatrophacurcas) deoiled cake with rice straw and different animal dung. *Bioresource Technology, 102*(11), 6541–6546. doi:10.1016/j.biortech.2011.03.058 PMID:21489784

Das, P. K. (2002). *Advancein Indian Sericultural Research*. Mysore: CSRTI.

Das, S., & Bhattarcharya, B. K. (2014). Estimation of municipal solid waste generation and future trends in greater metropolitan regions of Kolkata, India. *Journal of Industrial Engineering and Management Innovation, 1*(1), 31–38.

Datar, M. T., & More, A. B. (2006). *Vermi-composting – A Technological Option for Solid Waste Management*. Available at:http://www. bvucoepune.edu.in/pdf's/Research%20and%20Publication Research%20Publications_2006-07/International%20Conference_2006-07/Vermi-composting%20Mr%20AB% 20More.pdf

Datta, P., Mohi, G. K., & Chander, J. (2018). Biomedical waste management in India: Critical appraisal. *Journal of Laboratory Physicians, 10*(1), 6. doi:10.4103/JLP.JLP_89_17 PMID:29403196

Datta, S., Singh, J., Singh, S., & Singh, J. (2016). Earthworms, pesticides and sustainable agriculture: A review. *Environmental Science and Pollution Research International, 23*(9), 8227–8243. doi:10.100711356-016-6375-0 PMID:26951221

Daud, W. M. A. W., & Ali, W. S. W. (2004). Comparison on pore development of activated carbon produced from palm shell and coconut shell. *Bioresource Technology, 93*(1), 63–69. doi:10.1016/j.biortech.2003.09.015 PMID:14987722

Dave, P. N., & Joshi, A. K. (2010). Plasma pyrolysis and gasification of plastics waste – a review. *Journal of Scientific and Industrial Research, 177*(69), 177–179.

Davorin, K. (2011). Innovative systemic approach for promoting sustainable innovation for zero construction waste. *Kybernetes, 40*(1/2), 275–289. doi:10.1108/03684921111118040

De Araujo, A. S. F., de Melo, W. J., & Singh, R. P. (2010). Municipal solid waste compost amendment in agricultural soil: Changes in soil microbial biomass. *Reviews in Environmental Science and Biotechnology, 9*(1), 41–49. doi:10.100711157-009-9179-6

De Brito, M. P., & Dekker, R. (2004). A framework for reverse logistics. In *Reverse logistics* (pp. 3–27). Berlin: Springer. doi:10.1007/978-3-540-24803-3_1

De Guardia, A., Petiot, C., Benoist, J. C., & Druilhe, C. (2012). Characterization and modelling of the heat transfers in a pilot-scale reactor during composting under forced aeration. *Waste Management (New York, N.Y.), 32*(6), 1091–1105. doi:10.1016/j.wasman.2011.12.028 PMID:22301461

Deanin, R. D. (1975). Additives in plastics. *Environmental Health Perspectives, 11*, 35–39. doi:10.1289/ehp.751135 PMID:1175566

DEFRA (Department for Environment, Food and Rural Affairs). (2007). *Incineration of municipal solid waste*. Waste Management Technology Brief, the New Technologies Work Stream of the DEFRA Waste Implementation Programme.

Defra, W. S., & Hannan, M. (2006). Review of England's waste strategy. In Environmental report under the 'SEA' directive, (p. 96). London, UK: DEFRA.

DEFRA. (2007). Incineration of Municipal Solid Waste. Report of Department for Environment, Food & Rural Affairs.

Defra. (2007). *Waste strategy for England.* Norwich, UK: Department of Environment food and Rural Affairs, HMSO.

Dekker, R., Fleischmann, M., Inderfurth, K., & Wassenhove, V. (2003). *Reverse Logistics – Quantitative Models for Closed-Loop Supply Chains.* Berlin: Springer.

Demirbas, A. (2010). Waste management, waste resources facilities and waste conversion process. *Energy Conversion and Management, 52*(2), 1280–1287. doi:10.1016/j.enconman.2010.09.025

Dene, L. O. (2013). *Solid Waste Collection Vehicle Route Optimization for the City of Redlands.* University of Redlands.

Dennis, P. (2017). *The Roles of Strategic Environmental Assessment and Learning in Planning for Successful Community-Based Solid Waste Management in Kenya.* Academic Press.

Deo, B. (2004). Heavy metal accumulation by plant species from a coal mining area in Orissa. *Journal of Environmental Biology, 25,* 163–166. PMID:15529873

Department of Environment and Energy. (2015). *Australian National Greenhouse Gas Accounts – State and Territory Greenhouse Gas Inventories: Table 3.* Author.

Department of Environment and Energy. (2016). *Quarterly Update of Australia's National Greenhouse Gas Inventory: June 2016.* Commonwealth of Australia.

Department of Environment, Land, Water, and Planning. (2016). *Victoria's Climate Change Framework.* Retrieved from:https://www.climatechange.vic.gov.au/__data/assets/pdf_file/0021/552 54/DELWPClimateChange_Framework.pdf

Devine, T. J., & Clark, B. T. (2006). *Mobile apparatus and process for treating infectious waste.* U.S. Patent 7,144,550.

Dhameja, S. K. (2009). *Environmental Engineering and management.* Delhi: S.K. Kataria and Sons.

Dharini, K., Cynthia, J. B., Kamalambikai, B., Celestina, J. P. A. S., & Muthu, D. (2017). Hazardous E-waste and its impact on soil structure. *Earth and Environmental Science, 80,* 1–5.

Dhawan, R., Bisht, B.M.S., Kumar, R., Kumari, S., & Dhawan, S.K. (2019). Recycling of plastic waste into tiles with reduced flammability and improved tensile strength. *Process Safety and Environmental Protection.* . doi:10.1016/j.psep.2019.02.018

Dias, B. O., Silva, C. A., Higashikawa, F. S., Roig, A., & Sánchez-Monedero, M. A. (2010). Use of biochar as bulking agent for the composting of poultry manure: Effect on organic matter degradation and humification. *Bioresource Technology, 101*(4), 1239–1246. doi:10.1016/j.biortech.2009.09.024 PMID:19796932

Dias-Ferreira, C., Valente, S., & Vaz, J. (2016). Practices of pharmaceutical waste generation and discarding in households across Portugal. *Waste Management & Research, 34*(10), 1006–1013. doi:10.1177/0734242X16639388 PMID:27026517

Diaz, L. F., Savage, G. M., Eggerth, L. L., & Chiumenti, A. (2007). Systems used in composting. In Compost science and technology (pp. 67-87). Elsevier.

Diaz, L. F., Savage, G. M., & Eggerth, L. L. (2005). Alternatives for the treatment and disposal of healthcare wastes in developing countries. *Waste Management (New York, N.Y.), 25*(6), 626–637. doi:10.1016/j.wasman.2005.01.005 PMID:15993347

Dionne, A., Tweddell, R. J., Antoun, H., & Avis, T. J. (2012). Effect of non-aerated compost teas on damping-off pathogens of tomato. *Canadian Journal of Plant Pathology*, *34*(1), 51–57. doi:10.1080/07060661.2012.660195

Directive, E. C. (2008). 105/EC of the European Parliament and of the Council of 16 December 2008 on environmental quality standards in the field of water policy, amending and subsequently repealing Council Directives 82/176. *EEC*, *83*(513), 84–97.

DIT. (2003). *Environmental management for Information Technology industry in India*. Department of Information Technology, Government of India.

Divya, K., Singh, T., & Kumar, A. (2012). E-waste management: An overview. *International Journal of Research in IT & Management*, *2*, 763–770.

Divya, U. K. (2001). *Relevance of vermiculture in sustainable agriculture*. Agriculture General World.

Dobre, P., Nicolae, F., & Matei, F. (2014). Main factors affecting biogas production – an overview. *Romanian Biotechnological Letters*, *19*, 9283–9296.

Domínguez, J., & Edwards, C. A. (1997). Effects of stocking rate and moisture content on the growth and maturation of *Eiseniaandrei* (Oligochaeta) in pig manure. *Soil Biology & Biochemistry*, *29*(3-4), 743–746. doi:10.1016/S0038-0717(96)00276-3

Dominguez, J., & Edwards, C. A. (2004). *Vermicomposting organic wastes: Soil Zoology for Sustainable Development in the 21st Century*. Cairo, Egypt: MikhaTl.

Dominguez, J., Edwards, C., & Subler, S. (1997). Comparison of vermicomposting and composting *BioCycle. Journal of Composting and Organics Recycling*, *38*, 57–59.

Dominguez, J., Edwards, C., & Subler, S. (1997). Comparison of vermicomposting and composting. *BioCycle*, *4*, 57–59.

Doubleday, G. P. (1974). The reclamation of land after coal mining. *Outlook on Agriculture*, *8*(3), 156–162. doi:10.1177/003072707400800307

Drumright, R. E., Gruber, P. R., & Henton, D. E. (2000). Polylactic acid technology. *Advanced Materials*, *12*(23), 1841–1846. doi:10.1002/1521-4095(200012)12:23<1841::AID-ADMA1841>3.0.CO;2-E

DTIE, Division of Technology, Industry and Economics. International Environmental Technology Centre. (2007). E-*waste: Volume 1: Inventory assessment manual (Osaka/Shiga)*. Available at: http://www.unep.or.jp/ietc/Publications/spc/EWasteManual_Vol1.pdf

Duhan, J. S., Kumar, A., & Tanwar, S. K. (2013). Bioethanol production from starchy part of tuberous plant (potato) using Saccharomyces cerevisiae MTCC-170. *African Journal of Microbiological Research*, *7*(46), 5253–5260. doi:10.5897/AJMR2013.6122

Dutta, S. K., Upadhyay, V. P., & Sridharan, U. (2006). Environmental management of industrial hazardous wastes in India. *Journal of Environmental Science & Engineering*, *48*(2), 143. PMID:17913193

E&P Forum/UNEP. (1997). *Environmental management in oil and gas exploration and production-An over view of issues and management approaches*. Joint E&P Forum/UNEP Technical Report 37.

Edge, M., Hayes, M., Mohammadian, M., Allen, N. S., Jewitt, T. S., Brems, K., & Jones, K. (1991). Aspects of polyethylene terephthalate degradation for archival life and environmental degradation. *Polymer Degradation & Stability*, *32*(2), 131–153. doi:10.1016/0141-3910(91)90047-U

Edward, C. A. (1988). *Breakdown of animal, vegetable and industrial organic wastes by earthworms*. The Hague: SPB.

Edwards, C. A. (1998). The use of earthworms in the breakdown and management of organic wastes. In C. A. Edwards (Ed.), *Earthworm ecology* (pp. 327–354). Boca Roton, FL: CRC press.

Edwards, C. A., & Bohlen, P. J. (1996). *Biology and Ecology of Earthworms (3rd ed.)*. London: Chapman and Hall.

Edwards, C. A., & Fletcher, K. E. (1988). Interaction between Earthworms and Micro-Organisms in Organic Matter Breakdown. *Agriculture, Ecosystems & Environment*, *24*(1-3), 235–247. doi:10.1016/0167-8809(88)90069-2

Edwards, C. A., & Lofty, J. R. (1977). *Biology of Earthworms*. London: Chapman and Hall. doi:10.1007/978-1-4613-3382-1

Elahi, H., Rehman, W., Swati, R. F., Khan, M. H., Mushtaq, T., Haider, Z., ... Riaz, S. (2017). Automatic sorting, counting and bottle filling robotic statio. *International Journal of Scientific Engineering and Applied Science*, *3*, 5.

El-Haggar, S. M. (2007). *Sustainable industrial design and waste management: Cradle-to-cradle for sustainable development*. Oxford, UK: Elsevier/Academic Press.

Ellabban, O., Abu-Rub, H., & Blaabjerg, F. (2014). Renewable energy resources: Current status, future prospects and their enabling technology. *Renewable & Sustainable Energy Reviews*, *39*, 748–764. doi:10.1016/j.rser.2014.07.113

ElShafei, G. M. S., ElSherbiny, I. M. A., & Philip, C. A. (2013). Silkworms' feces-based activated carbons as cheap adsorbents for removal of Cadmium and Methylene Blue from aqueous solutions. *Chemical Engineering Research & Design*. doi:10.1016/j.cherd.2013.09.004

Emery, A., Davies, A., Griffiths, A., & Williams, K. (2007). Environmental and economic modelling: A case study of municipal solid waste management scenarios in Wales. *Resources, Conservation and Recycling*, *49*(3), 244–263. doi:10.1016/j.resconrec.2006.03.016

Enayetullah, I., Sinha, A. H. M. M., & Khan, S. S. A. (2005). *Urban solid waste management scenario of Bangladesh: problems and prospects*. Waste Concern Technical Documentation. Retrieved from http://www.wasteconcern.org/Publication/Waste%20Survey_05.pdf

EnCams. (2006). Litter segmentation 2006. Wigan, UK: Environmental Campaigns Limited (ENCAMS).

Engeli, H., Edelmann, W., Fuchs, J., & Rottermann, K. (1993). Survival of plant pathogens and weed seeds during anaerobic digestion. *Water Science and Technology*, *27*(2), 69–76. doi:10.2166/wst.1993.0079

Enger, E., Smith, B., & Bockarie, A. T. (2006). Environmental Science. A study ofinterrelationships.*McGraw Hill. Higher Education.*

EPA, Environmental Protection Agency. (1997). *Test Methods for Evaluating Solid Waste. n-Hexane Extractable Material (HEM) for Sludge, Sediment and Solid Samples, Method 9071B, Revision 2. Environmental Protection Agency, OSW 486*. Office of Solid Waste.

Epstein, E. (1974). Effect of sewage sludge on some soil physical properties. *Journal of Environmental Quality*, *4*(1), 139–142. doi:10.2134/jeq1975.00472425000400010033x

Epstein, E., Taylor, J. M., & Chaney, R. L. (1976). Effects of Sewage Sludge and Sludge Compost Applied to Soil on some Soil Physical and Chemical Properties1. *Journal of Environmental Quality*, *5*(4), 422. doi:10.2134/jeq1976.00472425000500040021x

Epstein, E., Willson, G. B., Burge, W. D., Mullen, D. C., & Enkiri, N. K. (1976). A forced aeration system for composting wastewater sludge. *Journal - Water Pollution Control Federation*, *48*, 688–694.

Erdogan, N., & Baris, E. (2007). Environmental protection programs and conservation practices of hotels in Ankara, Turkey. *Tourism Management*, *28*(2), 604–614.

Ericson, S. (2010). No waste allowed. *Threads, 149*, 60–63.

Eriksen, M., Lebreton, L. C., Carson, H. S., Thiel, M., Moore, C. J., Borerro, J. C., ... Reisser, J. (2014). Plastic pollution in the world's oceans:more than 5 trillion plastic pieces weighing over 250,000 tons afloat at sea. *PLoS One*, *9*(12), e111913. doi:10.1371/journal.pone.0111913 PMID:25494041

ESCAP. (1997). *Guidelines and manual on land use planning and practices in watershed management and disaster reduction. Economic and social commission for Asia and Pacific*. Bangkok: United Nations.

ESCAP. (1997). *Road transport and the environment: Areas of concern for the Asia and Pacific region* (ST/ESCAP/1840). ESCAP.

EU. (1999). Biomass Conversion Technologies. EUR 18029 EN.

Europe, P. (2017). *Plastics—The facts*. Plastic Europe. Retrieved from https://www.plasticseurope.org/en/resources/publications/274-plastics-facts-2017

Europe, P. (2018). *Plastics—the Facts 2017*. Retrieved from www.plasticseurope.org/application/files/5715/1717/4180/Plastics_the_facts_2017_FINAL_for_website_one_page.pdf

European Association of Plastics Recycling and Recovery. (2017). Available online: http://www.eproplasticsrecycling.org/pages/75/epro_statistics

European Union. (2012). Directive 2012/19/EU of the European Parliament and of the Council of 4 July 2012 on Waste Electrical and Electronic Equipment (WEEE) recast. *Official Journal of the European Union*, *55*, 197.

Europe, P. (2012). *Plastics—the Facts 2012. An Analysis of European Plastics Production, Demand and Waste Data for 2011*. Brussels: Plastic Eur.

Europe, P. (2015). *Plastics the Facts 2015: An Analysis of European Plastics Production, Demand and Waste Data, 30*. Brussels: Plastics Europe.

Evans Steve, D. (2009). *Plasma Gasification Plant Benefits for Municipal Waste Management*. Available at http://www.articlesbase.com/literaturearticles/plasma-gasificationplant-benefits-for-municipal-waste-management-850915

Ezechi, E. H., Nwabuko, C. G., Enyinnaya, O. C., & Babington, C. J. (2017). Municipal solid waste management in Aba, Nigeria. *Challenges and Prospects Environmental Engineering and Resources, 22*(3), 231-236.

Ezechi, E. H., Isa, M. H., Kutty, S. R., & Sapari, N. B. (2011). Boron recovery, application and economic significance: A review. In *National Postgraduate Conference (NPC)*. Kuala Lumpur, Malaysia. Institute of Electrical and Electronic Engineering.

Fang, M., Wong, J., Ma, K., & Wong, M. (1999). Co-composting of sewage sludge and coal fly ash: Nutrient transformations. *Bioresource Technology*, *67*, 19–24.

Farid, G., & Sarwar, N. (2015). Heavy metals (Cd, Ni & Pb) contamination of soils, plants and waters in Madina town of Faisalabad Metropolitan and preparation of GIS based maps. *Advances in Crop Science and Technology*, *4*(01), 1–7. doi:10.4172/2329-8863.1000199

Fatoki, O. S. (1996). Trace zinc and copper concentrations in roadside surface soils and vegetation measurement of local atmospheric pollution in Alice, South Africa. *Environment International*, *22*(6), 759–762. doi:10.1016/S0160-4120(96)00068-2

Favero, A., & Mendelsohn, R. (2014). Using markets for woody biomass energy to sequester carbon in forests. *Journal of the Association of Environmental and Resource Economists*, *1*(1-2), 75-95. Disponível em: http://www.jstor.org/stable/10.1086/676033

Federico, A., Borraz, J. S., Molina, J. A. N., Nafate, C. C., Archila, M. A., Oliva, L. M., & … . (2007). Vermicompost as a soil supplement to improve growth, yield and fruit quality of tomato (*Lycopersicon esculentum*). *Bioresource Technology*, *98*(15), 2781–2786. doi:10.1016/j.biortech.2006.02.032 PMID:17400447

Fei, F., Wen, Z., Huang, S., & Clercq, D. (2018). Mechanical biological treatment of municipal solid waste: Energy efficiency, environmental impact and economic feasibility analysis. *Journal of Cleaner Production*, *178*, 731–739.

Feigin, A., & Kipnis, T. (1980). Agrochemicals in soils. Pergamon Press.

Felipe, L., de Oliveira, A. M., & Bicas, J. L. (2017). Bioaromas–Perspectives for sustainable development. *Trends in Food Science & Technology*, *62*, 141–153. doi:10.1016/j.tifs.2017.02.005

Feng Li, R. (2010). The composting treatment of residue from biological municipal waste anaerobic digestion. *Advanced Materials Research*, *129*(131), 703–707.

Feng, S., Wang, X., Wei, G., Peng, P., Yang, Y., & Cao, Z. (2006). Leachates of municipal solid waste incineration bottom ash from Macao: Heavy metal concentrations and genotoxicity. *Chemosphere*, *67*(6), 1133–1137. doi:10.1016/j.chemosphere.2006.11.030 PMID:17217988

Ferrer, Y., & Pérez, H. (2010). Los microorganismos en la digestión anaerobia y la producción de biogás. Consideraciones en la elección del inóculo para el mejoramiento de la calidad y el rendimiento. *ICIDCA sobre los derivados de la caña de azúcar*, *43*(1), 9-20. Disponível em: http://www.redalyc.org/pdf/2231/223120681002.pdf

Ficeriova, J., Balaz, P., Dutkova, E., & Gock, E. (2008). Leaching of gold and silver from crushed Au-Ag wastes. *The Open Chemical Engineering Journal*, *2*(1), 6–9. doi:10.2174/1874123100802010006

Flotats & Gibert. (2002). Mas El Cros Biogas Plant. Evaluation of 18 years in Operation. *7th FAO/SREN Workshop. Anaerobic digestion for sustainability in waste (water) treatment and re-use*, 1-8.

Follet, R., Danahue, R., & Murphy, L. (1981). *Soil and Soil Amendments*. Prentice-Hall, Inc.

Francis, R. (2016). *Recycling of polymers: methods, characterization and applications*. Hoboken, NJ: John Wiley & Sons.

Franz, R., & Welle, F. (2003). Recycling packaging materials A2—Ahvenainen, Raija. In *Novel Food Packaging Techniques* (pp. 497–518). Cambridge, UK: Woodhead Publishing. doi:10.1533/9781855737020.4.497

French, K. J., Siruton, D. A., & Anderson, M. R. (1999). Influence of physical and chemical characteristics on mercury in aquatic sediments. *Water, Air, and Soil Pollution*, *110*(3/4), 347–362. doi:10.1023/A:1005007402287

Friedman, M. S., Powell, K. E., Hutwagner, L., Graham, L. M., & Teague, W. G. (2001). Impact of changes in transportation and commuting behaviors during the 1996 Summer Olympic Games in Atlanta on air quality and childhood asthma. *Journal of the American Medical Association*, *285*(7), 897–905. doi:10.1001/jama.285.7.897 PMID:11180733

Friends of the Earth. (2008). Brief Autoclaving. Author.

Fu, J., Zhou, Q., Liu, J., Liu, W., Wang, T., Zhang, Q., & Jiang, G. (2008). High levels of heavy metals in rice (*Oryza sativa* L.) from a typical E-waste recycling area in southeast China and its potential risk to human health. *Chemosphere*, *71*(7), 1269–1275. doi:10.1016/j.chemosphere.2007.11.065 PMID:18289635

Funk, K., Milfold, J., & Simpkins, T. (2013). *Technology for Site-Specific Optimization of Renewable Energy Options*. Technical Report. NREL/TP-6A50-52829. Retrieved from http://www.osti.gov/bridge

Gabhane, J., William, S. P., Bidyadhar, R., Bhilawe, P., Anand, D., Vaidya, A. N., & Wate, S. R. (2012). Additives aided composting of green waste: Effects on organic matter degradation, compost maturity, and quality of the finished compost. *Bioresource Technology*, *114*, 382–388. doi:10.1016/j.biortech.2012.02.040 PMID:22444633

Gadde, B., Menke, C., & Wassmann, R. (2009). Rice straw as a renewable energy source in India, Thailand, and the Philippines: Overall potential and limitations for energy contribution and greenhouse gas mitigation. *Biomass and Bioenergy*, *33*(11), 1532–1546. doi:10.1016/j.biombioe.2009.07.018

GAIA. (2013). *Individual Zero Waste Case Studies*. Available at http://no-burn.org/-1-18

Gajalakshmi, V., & Abbasi, S. A. (2003). High rate vermicomposting system for recycling paper waste. *Indian Journal of Biotechnology*, *2*, 613–615.

Galgani, F., Fleet, D., van Franeker, J., Katsanevakis, S., Maes, T., Mouat, J., ... Janssen, C. (2010). Marine Strategy Framework Directive, Task Group 10 Report: Marine Litter. In N. Zampoukas (Ed.), *JRC Scientific and Technical Reports*. Ispra: European Comission Joint Research Centre.

Gallardo, A., Carlos, M., Bovea, M. D., Colomer, F. J., & Albarrán, F. (2014). Analysis of refuse-derived fuel from the municipal solid waste reject fraction and its compliance with quality standards. *Journal of Cleaner Production*, *83*, 118–125.

Gallop, W. A., Evans, M.G., & Mithal, A.K. (2009). *Knife Having Superior Functionality and Appeal*. Academic Press.

Galloway, T.S., Lewis, C.N. (2016). *Marine microplastics spell big problems for future generations*. Academic Press.

Gall, S. C., & Thompson, R. C. (2015). The impact of debris on marine life. *Marine Pollution Bulletin*, *92*(1-2), 170–179. doi:10.1016/j.marpolbul.2014.12.041 PMID:25680883

Gan, S., Lau, E. V., & Ng, H. K. (2009). Remediation of soils contaminated with polycyclic aromatic hydrocarbons (PAHs). *Journal of Hazardous Materials*, *172*(2-3), 532–549. doi:10.1016/j.jhazmat.2009.07.118 PMID:19700241

Garbisu, C., & Alkorta, I. (2001). Phytoextraction: A cost-effective plant based technology for the removal of metals from the environment. *Bioresource Technology*, *77*(3), 229–236. doi:10.1016/S0960-8524(00)00108-5 PMID:11272009

Garcia, M. J. M., Moreno-Grau, S., Garcia, J. J. M., Moreno, J., Bayo, J., Perez, J. J. G., & Moreno-clavel, J. (2001). Distribution of the metals lead, cadmium, copper and zinc in the top soil of cartagena, spain. *Water, Air, and Soil Pollution*, *131*(1/4), 329–347. doi:10.1023/A:1011916218755

Garcia, M., Scheffczyk, A., Garcia, T., & Römbke, J. (2011). The effects of the insecticide lambda-Cyhalothrin on the earthworm *Eisenia fetida* under experimental conditions of tropical and temperate regions. *Environmental Pollution*, *159*(2), 398–400. doi:10.1016/j.envpol.2010.10.038 PMID:21106283

Garg, P., Gupta, A., & Satya, S. (2006). Vermicomposting of different types of waste using *Eiseniafetida*: A comparative study. *Bioresource Technology*, *97*(3), 391–395. doi:10.1016/j.biortech.2005.03.009 PMID:16168639

Garg, V. K., Chand, S. A., & Yadav, A. (2005). Growth and reproduction of *Eiseniafetida* in various animal wastes during vermicomposting. *Applied Ecology and Environmental Research*, *3*(2), 51–59. doi:10.15666/aeer/0302_051059

Garg, V. K., & Gupta, R. (2011). Effect of temperature variations on vermicomposting of household solid waste and fecundity of *Eiseniafetida*. *Bioremediation Journal*, *15*(3), 165–172. doi:10.1080/10889868.2011.598487

Garg, V. K., Suthar, S., & Yadav, A. (2011). Management of food industry waste employing vermicompost technology. *Bioresource Technology*, *10*, 1016. PMID:22197330

Gaustad, G., Olivetti, E., & Kirchain, R. (2012). Improving aluminum recycling: A survey of sorting and impurity removal technologies. *Resources, Conservation and Recycling*, *58*, 79–87.

Gayen, S., & Ghosh, U. (2011). Pectinmethylesterase Production from mixed agro- wastes by Penicilliumnotatum NCIM. 923 in Solid-State fermentation. *Journal of Bioremediation & Biodegradation*, *2*(02), 119. doi:10.4172/2155-6199.1000119

Gellynck, X., Jacobsen, R., & Verhelst, P. (2011). Identifying the key factors in increasing recycling and reducing residual household waste: A case study of the Flemish region of Belgium. *Journal of Environmental Management*, *92*(10), 2683–2690. doi:10.1016/j.jenvman.2011.06.006 PMID:21704444

Gencer, Y. G. (2016). *Mystery of Recycling: Glass and Aluminum Examples*. Doi:10.4018/978-1-4666-9723-2.ch009

George. (2001). *Vermicomposting*. Retrieved at July 22th, 2010, from http://aces.nmsu.edu/pubs/_h/h-164.pdf

Gershon, R. R., Vlahov, D., Felknor, S. A., Vesley, D., Johnson, P. C., Delcios, G. L., & Murphy, L. R. (1995). Compliance with universal precautions among health care workers at three regional hospitals. *American Journal of Infection Control*, *23*(4), 225–236. doi:10.1016/0196-6553(95)90067-5 PMID:7503434

Gesamp. (2016). Sources, Fate and Effects of Microplastics in the Marine Environment – Part Two of a Global Assessment. In *IMO/FAO/UNESCO-IOC/UNIDO/WO/IAEA/UN/UNEP/UNDP Joint Group of Experts on the Scientific Aspects of Marine Environmental Protection*. Academic Press.

Geschwind, S. A., Stolwijk, J. A. J., Bracken, M., Fitzgeralg, E., Stark, A., Oslen, C., & Melius, J. (1992). Risk of congenital malformations associated with proximity to hazardous waste sites. *American Journal of Epidemiology*, *135*(11), 1197–1207. doi:10.1093/oxfordjournals.aje.a116226 PMID:1626538

Geyer, R., Jambeck, J. R., & Law, K. L. (2017). Production, use, and fate of all plastics ever made. *Science Advances*, *3*(7), e1700782. doi:10.1126ciadv.1700782 PMID:28776036

Ghasemi, M. K., & Yusuff, R. B. (2016). Advantages and Disadvantages of Healthcare Waste Treatment and Disposal Alternatives: Malaysian Scenario. *Polish Journal of Environmental Studies*, *25*(1).

Ghatnekar, S. D., Mahavash, F. K., & Ghatnegar, G. S. (1998). Management of solid waste through vermiculture biotechnology. In R. K. Trivedy & A. Kumar (Eds.), *Ecotechnology for pollution control and environmental management* (pp. 58–67). EnviromediaKarad.

Ghosh, R., & Kansal, A. (2014). Urban challenges in India and the mission for a sustainable habitat. *Interdisciplina*, *2*(2), 281–304.

Giang, N. V., Kochanek, K., Vu, N. T., & Duan, N. B. (2018). Landfill leachate assessment by hydrological and geophysical data: Case study NamSon, Hanoi, Vietnam. *Journal of Material Cycles and Waste Management*, *20*(3), 1648–1662. doi:10.100710163-018-0732-7

Giboz, J., Copponnex, T., & Mélé, P. (2007). Microinjection molding of thermoplastic polymers: A review. *Journal of Micromechanics and Microengineering*, *17*(6), R96–R109. doi:10.1088/0960-1317/17/6/R02

Gidde, M. R., Todkar, V. V., & Kokate, K. K. (2008). Municipal solid waste management in emerging mega cities: a case study of Pune city. *Proceedings of Indo Italian Conference on Green and Clean Environment*.

Gilberto, T. Z. Z., Maria, V. S., & Carvalho, D. G. (2008). A case study of fuel savings through optimization of MSW transportation routes. *Management of Environmental Quality*, *19*(4), 444–454.

Gillilan, S., Werner, C. M., Olson, L., & Adams, D. (1996). Teaching the concept of precycling: A campaign and evaluation. *The Journal of Environmental Education*, *28*(1), 11–18. doi:10.1080/00958964.1996.9942810

Giusti, L. (2009). A review of waste management practices and their impact on human health. *Waste Management (New York, N.Y.)*, *29*(8), 2227–2239. doi:10.1016/j.wasman.2009.03.028 PMID:19401266

Glavic, P., & Lukman, R. (2007). Review of sustainability terms and their definitions. *Journal of Cleaner Production, 15*(18), 1875–1885.

Global Methane Inititive. (2018). *Mexico Landfill Gas Model 2.0* [Modelo de Biogas Mexicano 2.0]. Disponível em https://www.globalmethane.org/partners/country.aspx?c=mexico

Gomez, E., Rani, D. A., Cheeseman, C. R., Deegan, D., Wise, M., & Boccaccini, A. R. (2009). Thermal plasma technology for the treatment of wastes: A critical review. *Journal of Hazardous Materials, 161*(2-3), 614–626. doi:10.1016/j.jhazmat.2008.04.017 PMID:18499345

Gonawala, S. S., & Jardosh, H. (2018). Organic Waste in Composting: A brief review. *International Journal of Current Engineering and Technology, 8*(1), 36–38. doi:10.14741/ijcet.v8i01.10884

Gong, J., Yao, K., Liu, J., Wen, X., Chen, X., Jiang, Z., Mijowska, E., & Tang, T. (2013). Catalytic conversion of linear low density polyethylene into carbon nanomaterials under the combined catalysis of Ni2O3 and poly(vinyl chloride). *Chemical Engineering Journal, 215-216*, 339-347.

Gopal, M., Gupta, A., Palaniswami, C., Dhanapal, R., & Thomas, G. V. (2010). Coconut leaf vermiwash: A bio-liquid from coconut leaf vermicompost for improving the crop production capacities of soil. *Current Science, 98*(9), 1202–1210.

Goswami, K. B. (2018). Challenges and opportunities associated with waste management in India. *International Journal of Research in Social Sciences, 8*(2).

Goto, M., Sasaki, M., & Hirose, T. (2006). Reactions of polymers in supercritical fluids for chemical recycling of waste plastics. *Journal of Materials Science, 41*, 1509–1515.

Government of India. (2009). *The solid waste management sector in India*. Department of Economic Affairs Ministry of Finance.

Gowe, C. (2015). Review on potential use of fruit and vegetables by-products as a valuable source of natural food additives. *Food Science and Quality Management, 45*, 47–61.

Grant, T., James, K.L., & Partl, H. (2003). *Life Cycle Assessment Of Waste Andresource Recovery*. Academic Press.

Grapplli, A., Galli, E., & Tomati, U. (1987). Earthworm casting effect on *Agaricusbisporus* fructification. *Agrochimica, 21*, 457–462.

Grau, A., & Farre, O. (2011). *Situation and Potential of the DirectWaste Energy Recovery. Technical study. PER 2011-2020*. IDAE.

Green, D. S. (2016). Effects of Microplastics on European Flat Oysters, Ostrea Edulis and their Associated Benthic Communities. *Environmental Pollution, 216*, 95–103. doi:10.1016/j.envpol.2016.05.043 PMID:27239693

Green, D. S., Boots, B., Blockley, D. J., Rocha, C., & Thompson, R. (2015). Impacts of Discarded Plastic Bags on Marine Assemblages and Ecosystem Functioning. *Environmental Science & Technology, 49*(9), 5380–5389. doi:10.1021/acs.est.5b00277 PMID:25822754

Green, D. S., Boots, B., O'Connor, N., & Thompson, R. C. (2017). Microplastics Affect the Ecological Functioning of an Important Biogenic Habitat. *Environmental Science & Technology, 51*(1), 68–77. doi:10.1021/acs.est.6b04496 PMID:27936642

Gregory, M. R. (2009). Environmental Implications of Plastic Debris in Marine Settings – Entanglement, Ingestion, Smothering, Hangers-on, Hitch-hiking, and Alien Invasions. *Philosophical Transactions of the Royal Society of London. Series B, Biological Sciences, 364*(1526), 2013–2026. doi:10.1098/rstb.2008.0265 PMID:19528053

Greyson, J. (2007). An economic instrument for zero waste, economic growth and sustainability. *Journal of Cleaner Production*, *15*(13-14), 1382–1390. doi:10.1016/j.jclepro.2006.07.019

Grigore, M. E. (2017). Methods of recycling, properties and applications of recycled thermoplastic polymers. *Recycling*, *2*, 24.

Grósz, Z. (2003). Az ABV Védelemalapjai. Budapest: Zrínyi Miklós Nemeztvédelmi Egyetem, (Tankönyv).

Guerrero, L. A., Maas, G., & Hogland, W. (2013). Solid waste management challenges for cities in developing countries. *Waste Management*, *33*(1), 220-232.

Guerrero, L. A., Maas, G., & Hogland, W. (2013). Solid waste management challenges for cities in developing countries. *Waste Management (New York, N.Y.)*, *33*(1), 220–232. doi:10.1016/j.wasman.2012.09.008 PMID:23098815

Guidoni, L. L. C., Marques, R. V., Moncks, R. B., Botelho, F. T., da Paz, M. F., Corrêa, L. B., & Corrêa, É. K. (2018). Home composting using different ratios of bulking agent to food waste. *Journal of Environmental Management*, *207*, 141–150. doi:10.1016/j.jenvman.2017.11.031 PMID:29156437

Gulyurt, M. (2012). Biomedical instrument application: Medical waste treatment technologies. In *A Roadmap of Biomedical Engineers and Milestones*. IntechOpen. doi:10.5772/39276

Gume, B., Muleta, D., & Abate, D. (2013). Evaluation of locally available substrates for cultivation of oyster mushroom (Pleurotus ostreatus) in Jimma, Ethiopia. *African Journal of Microbiological Research*, *7*(20), 2228–2237. doi:10.5897/AJMR12.895

Gundupalli, S. P., Haitb, S., & Thakur, A. (2017). A review on automated sorting of source-separated municipal solid waste for recycling. *Waste Management*, 1-39.

Gupta, A. (2014). Human hair "waste" and its utilization: gaps and possibilities. *Journal of Waste Management*.

Gupta, Sangita, & Kaur. (2009). Electronic waste: A case study. *Research Journal of Chemical Sciences*, *1*, 49–56.

Gupta, Yadav, Kumar, Gupta, Yadav, & Kumar. (2015). A review on current status of municipal solid waste management in India. *Journal of Environmental Sciences*, *37*, 206–217. Retrieved from www.journals.elsevier.com/journal-of-environmental-sciences

Gupta, N., Yadav, K. K., & Kumar, V. (2015). A review on current status of municipal solid waste management in India. *Journal of Environmental Sciences (China)*, *37*, 206–217. PMID:26574106

Gupta, P. K. (2004). Vermicomposting for sustainable agriculture, *Agrobios*, India. *Indian Journal of Biotechnology*, *3*, 486–494.

Gupta, R., & Garg, V. (2011). Potential and possibilities of vermi-composting in sustainable solid waste management: A review. *International Journal of Environment and Waste Management*, *7*(3/4), 210.

Gupta, R., Mutiyar, P. K., Rawat, N. K., Saini, M. S., & Garg, V. K. (2007). development of a water hyacinth based vermireactor using an epigeic earthworm Eiseniafoetida. *Bioresource Technology*, *98*(13), 2605–2610. doi:10.1016/j.biortech.2006.09.007 PMID:17071077

Gupta, S. (2017). Biomedical waste in India a review. *Journal of Advanced Medical and Dental Sciences Research*, *5*(4), 99.

Gupta, S. K., Pandey, S. K., Singh, N. B., Singh, A., Singh, V. K., & Verma, T. (2013). Municipal solid waste characterizations and management strategies for the Lucknow city, India. *International Journal of Engine Research*, *8*(17), 2031–2036.

Gupta, S., Boojh, R., Mishra, A., & Chandra, H. (2009). Rules and management of biomedical waste at Vivekananda Polyclinic: A case study. *Waste Management (New York, N.Y.)*, *29*(2), 812–819. doi:10.1016/j.wasman.2008.06.009 PMID:18684608

Gupta, S., Mohan, K., Prasad, R., Gupta, S., & Kansal, A. (1998). Solid waste management in India: Options and opportunities. *Resources, Conservation and Recycling*, *24*(2), 137–154. doi:10.1016/S0921-3449(98)00033-0

Gupta, V. K., Ali, I., Saleh, T. A., Nayak, A., & Agarwal, S. (2012). Chemical treatment technologies for waste-water recycling—an overview. *RSC Advances*, *2*(16), 6380–6388.

Gurav, M. V., & Pathade, G. R. (2011). Production of vermicompost from temple waste (Nirmalya): A case study. *Universal Journal of Environmental Research and Technology*, *1*(2), 182–192.

Gustafsson, J., Cederberg, C., Sonesson, U., & Emanuelsson, A. (2013). *The methodology of the FAO study: Global Food Losses and Food Waste-extent, causes and prevention.* SIK Institutetförlivsmedelochbioteknik.

Gutberlet, J. (2010). Waste, poverty and recycling. *Waste Management (New York, N.Y.)*, *30*(2), 171–173. PMID:19962085

Haberle, S. G., & Maslin, M. A. (1999). Late Quaternary Vegetation and Climate Change in the Amazon Basin Based on a 50,000 Year Pollen Record from the Amazon Fan, ODP Site 932. *Quaternary Research*, *51*(1), 27–38. doi:10.1006/qres.1998.2020

Haimour, N. M., & Emeish, S. (2006). Utilization of date stones for production of activated carbon using phosphoric acid. *Waste Management (New York, N.Y.)*, *26*(6), 651–660. doi:10.1016/j.wasman.2005.08.004 PMID:16256326

Hajinezhad, A., Halimehjani, E. Z., & Tahani, M. (2016). Utilization of Refuse-Derived Fuel (RDF) from Urban Waste as an Alternative Fuel for Cement Factory: A Case Study. *International Journal of Renewable Energy Research.*, *6*(2), 702–714.

Halász, L., & Nagy, K. (2001). Mérgezőanyagokkémiája. Budapest: Zrínyi Miklós Nemeztvédelmi Egyetem, (Egyetemijegyzet)

Hall, R., Belson, D., Murali, P., & Dessouky, M. (2006). Modeling patient flows through the healthcare system. In Patient flow: Reducing delay in healthcare delivery (pp. 1-44). Academic Press. doi:10.1007/978-0-387-33636-7_1

Hamad, K., Kaseem, M., & Deri, F. (2013). Recycling of waste from polymer materials: An overview of the recent works. *Polymer Degradation & Stability*, *98*(12), 2801–2812. doi:10.1016/j.polymdegradstab.2013.09.025

Hand, P., Hayes, W. A., Satchell, J. E., Frankland, J. C., Edwards, C. A., & Neuhauser, E. F. (2004). The vermicomposting of cow slurry. *Earthworms in Waste and Environmental Management*, *31*, 49–63.

Hansen, M. S., & Gilberg, U. M. (2003). *Cost-benefit analysis and LCA in environmental decision-making – potentials and limitations* (Master thesis). E&R Technical University of Denmark, Kongens Lyngby.

Harbec, S. (2009). *The Uptake of Sustainable Wastes Management: the Case of Electronic Medical Equipment* (Doctoral dissertation). École Polytechnique de Montréal.

Hargreaves, J. C., Adl, M. S., & Warman, P. R. (2008). A review of the use of composted municipal solid waste in agriculture. *Agriculture, Ecosystems & Environment*, *123*(1-3), 1–14.

Haro, P., Ollero, P., & Villanueva, A. L. (2013). Thermochemical biorefinery based on dimethyl ether as intermediate: Technoeconomic assessment. *Applied Energy*, *102*, 950–961. doi:10.1016/j.apenergy.2012.09.051

Hartenstein, R., & Bisesi, M. S. (1988). Use of earthworm biotechnology for the management of effluents from intensively housed livestock. *Outlook on Agriculture*, *18*(2), 72–76. doi:10.1177/003072708901800205

Hartenstein, R., Neuhauser, E. F., & Kaplan, D. L. (1979). Reproductive potential of the earthworm *Eisenia foetida. Oecologia*, *43*(3), 329–340. doi:10.1007/BF00344959 PMID:28309889

Hartley, B. L., Pahl, S., & Thompson, R. C. (2013). Baseline Evaluation of Stakeholder Perceptions and Attitudes Towards Issues Surrounding Marine Litter, Deliverable D2.1 Report hazardous. *Nature*, *494*, 169–171. PMID:23407523

Hasanbeigi, A., Lu, H., Williams, C., & Price, L. (2012). International Best Practices for Pre-Processing and Co Processing Municipal Solid Waste and Sewage Sludge in the Cement Industry. China Energy Group Energy Analysis and Environmental Impacts Department Environmental Energy Technologies Division Lawrence Berkeley National Laboratory. *LBNL*, *5581E*, 1–115.

Hashiguchi, T., Toya, M., Uemura, K., Fuchigami, T., & Ito, K. (1976). Detection of heavy metals in vegetables grown in the Sukeye area. *Kageshima Ken Kegai Kenkyushe Ho*, *12*, 55–56.

Hassan, A., Tudor, T., & Vaccari, M. (2018). Healthcare waste management: A case study from Sudan. *Environments*, *5*(8), 89. doi:10.3390/environments5080089

Hastik, R., Basso, S., Geitner, C., Haida, C., Poljanec, A., Portaccio, A., ... Walzer, C. (2015). Renewable energies and ecosystem service impacts. *Renewable & Sustainable Energy Reviews*, *48*, 608–623. doi:10.1016/j.rser.2015.04.004

Heap, B. (2009). Preface. *Philosophical Transactions of the Royal Society of London. Series B, Biological Sciences*, *364*(1526), 1971–1971. doi:10.1098/rstb.2009.0030 PMID:19528048

Heberlein, J., & Murphy, A. B. (2008). Thermal plasma waste treatment. *Journal of Physics. D, Applied Physics*, *41*(5), 053001. doi:10.1088/0022-3727/41/5/053001

Heffner, H. A. (2004). *Analysis of post-use hypodermic needle medical waste disposal* (Doctoral dissertation). Georgia Institute of Technology.

Hegde, V., Kulkarni, R. D., & Ajantha, G. S. (2007). Biomedical waste management. *Journal of Oral and Maxillofacial Pathology*, *11*(1), 5. doi:10.4103/0973-029X.33955

Held, J. S., & Sharp, J. W. (1996). *Apparatus and method for processing medical waste*. U.S. Patent 5, 508, 004.

Henry, J. G., & Heinke, G. W. (1989). *Environmental Science and Engineering*. Prentice Hall.

He, R., Liu, X., Zhang, Z., & Shen, D. (2007). Characteristics of the bioreactor landfill system using an anaerobic–aerobic process for nitrogen removal. *Bioresource Technology*, *98*(13), 2526–2532. PMID:17071082

Hermann, B. G., Debeer, L., De Wilde, B., Blok, K., & Patel, M. K. (2011). To compost or not to compost: Carbon and energy footprints of biodegradable materials' waste treatment. *Polymer Degradation & Stability*, *96*(6), 1159–1171. doi:10.1016/j.polymdegradstab.2010.12.026

Hernandez-Atonal, F., Ryu, C., Sharifi, V., & Swithenbank, J. (2007). Combustion of refuse-derived fuel in a fluidised bed. *Chemical Engineering Science*, *62*(1-2), 627–635.

Herwaldt, L. A., & Rutala, W. A. (1996). Disinfection and sterilization of patient-care items. *Infection Control and Hospital Epidemiology*, *17*(6), 377–384. doi:10.2307/30141141 PMID:8805073

Heskett, M., Takada, H., Yamashita, R., Yuyama, M., Ito, M., Geok, Y. B., ... Mermoz, J. (2012). Measurement of persistent organic pollutants (POPs) in plastic resin pellets from remote islands: Toward establishment of background concentrations for International Pellet Watch. *Marine Pollution Bulletin*, *64*(2), 445–448. doi:10.1016/j.marpolbul.2011.11.004 PMID:22137935

Higham, I. (1998). Advanced thermal conversion technologies for energy from solid waste. IEA Caddett Centre for Renewable Energy, ETCU.

Hill, M. H., & Bucknell, P. A. (1967). *The evolution of fashion: Pattern and cut from 1066 to 1930*. London: Batsford.

Hiraki, A., Ikuno, Y., Kim, J., & Ueda, S. (1996). Suppression of enveloped virus production with a substance from silk worm fecal matter cell. *Struc. Func.*, *21*(6), 501–514. doi:10.1247/csf.21.501 PMID:9078408

Hirani, D. P., Villaitramani, K. R., & Kumbhar, S. J. (2014). Biomedical waste: An introduction to its management. *International Journal of Innovative Research in Advanced Engineering*, *1*(8), 82–87.

Hoitink, H. A., & Fahy, P. (1986). Basis for the control of soil borne plant pathogens with composts. *Annual Review of Phytopathology*, *24*(1), 93–114. doi:10.1146/annurev.py.24.090186.000521

Holmberg, J., & Robert, K. H. (2000). Backcasting from non-overlapping sustainability principles: A framework for strategic planning. *International Journal of Sustainable Development and World Ecology*, *7*, 291–308. doi:10.1080/13504500009470049

Hoornweg, D. (1999). *What a Waste: Solid Waste Management in Asia*. The International Bank for Reconstruction and Development/The World Bank. Retrieved from http://technogreen.net.in/downloads/nisargruna-biogas-technology_concept-note.pdf[REMOVED HYPERLINK FIELD]

Hoornweg, D. (1999). *What a Waste: Solid Waste Management in Asia*. Washington, DC: The International Bank for Reconstruction and Development/The World Bank.

Hoornweg, D., & Bhada-Tata, P. (2012). *What a waste: a global review of solid waste management*. Washington, DC: World Bank.

Hopewell, J., Dvorak, R., & Kosior, E. (2009). Plastics recycling: Challenges and opportunities. *Philosophical Transactions of the Royal Society of London. Series B, Biological Sciences*, *364*(1526), 2115–2126. doi:10.1098/rstb.2008.0311 PMID:19528059

Hossain, M. S., Balakrishnan, V., Rahman, N. N. N. A., Sarker, M. Z. I., & Kadir, M. O. A. (2012). Treatment of clinical solid waste using a steam autoclave as a possible alternative technology to incineration. *International Journal of Environmental Research and Public Health*, *9*(3), 855–867. PMID:22690168

Hossain, M. S., Santhanam, A., Norulaini, N. N., & Omar, A. M. (2011). Clinical solid waste management practices and its impact on human health and environment–A review. *Waste Management (New York, N.Y.)*, *31*(4), 754–766. doi:10.1016/j.wasman.2010.11.008 PMID:21186116

Ho, Y. B., & Tai, K. M. (1988). Elevated levels of level and other metals in roadside soil and grass and their use to monitor aerial metal deposition in Hong Kong. *Environ. Pollut.*, *49*(1), 37–51. doi:10.1016/0269-7491(88)90012-7 PMID:15092672

Huang, H., & Tang, L. (2007). Treatment of organic waste using thermal plasma pyrolysis technology. *Energy Conversion and Management*, *48*(4), 1331–1337. doi:10.1016/j.enconman.2006.08.013

Huang, Y., Bird, R. N., & Heidrich, O. (2007). A review of the use of recycled solid waste materials in asphalt pavements. *Resources, Conservation and Recycling*, *52*(1), 58–73. doi:10.1016/j.resconrec.2007.02.002

Hu, B., Jia, X., Hu, J., Xu, D., Xia, F., & Li, Y. (2017). Assessment of heavy metal pollution and Health risks in the Soil-Plant-Human System in the Yangtze River Delta, China. *International Journal of Environmental Research and Public Health*, *14*(9), 1–18. doi:10.3390/ijerph14091042 PMID:28891954

Hunt, C. (1996). Child Waste Pickers in India; The Occupation and its Health Risks. *Environment and Urbanization, 8*(2), 111–118. doi:10.1177/095624789600800209

Hurst, C., Longhurst, P., Pollard, S., Smith, R., Jefferson, B., & Gronow, J. (2005). Assesment of municipal waste compost as daily cover material for odour at landfill sites. *Environmental Pollution, 135*(1), 171–177. doi:10.1016/j.envpol.2004.09.004 PMID:15701404

Hussain, N., Abbasi, T., & Abbasi, S.A. (2005). *Vermicomposting eliminates the toxicity of Lantana (*Lantana camara*) and turns it into a plant friendly organic fertilizer*. Academic Press.

Idowu, A., Adagunodo, E., Esimai, O., & Olapade, T. (2012). Development of a web based gis waste disposal management system for Nigeria. *International Journal of Information Engineering and Electronic Business, 4*(3), 40–48.

Imam, A., Mohammed, B., Wilson, D. C., & Cheeseman, C. R. (2008). Solid waste management in Abuja, Nigeria. *Waste Management (New York, N.Y.), 28*(2), 468–472. doi:10.1016/j.wasman.2007.01.006 PMID:17379496

Indrajeet, S. N., & Singh, J. (2010). Vermicomposting of farm garbage in different combination. *Journal of Recent Advances in Applied Science, 25*, 15–18.

Indris, A., Inane, B., & Hassan, M. N. (2004). Overview of waste disposal and landfills/dumps in Asian countries. *Journal of Material Cycles and Waste Management, 6*, 104–110.

Ingale, S., Joshi, S. J., & Gupte, A. (2014). Production of bioethanol using agricultural waste: Banana pseudo stem. *Brazilian Journal of Microbiology, 45*(3), 885–892. doi:10.1590/S1517-83822014000300018 PMID:25477922

Insam, H., & de Bertoldi, M. (2007). Microbiology of the composting process. In L. F. Diaz, M. de Bertoldi, W. Bidlingmaier, & E. Stentiford (Eds.), *Compost science and technology. Waste management series 8*. Amsterdam: Elsevier.

Intergovernmental Panel on Climate Change. (2006). *IPCC guidelines for national greenhouse gas inventories*. Hayama, JP: IPCC. Disponível em: https://www.ipcc-nggip.iges.or.jp/public/2006gl/index.html

International Engineering Consortium. (2015). Retrieved from http://www.uky.edu/~jclark/mas355/GSM.PDF

Ioakeimidis, C., Fotopoulou, K. N., Karapanagioti, H. K., Geraga, M., Zeri, C., Papathanassiou, E., ... Papatheodorou, G. (2016). The degradation potential of PET bottles in the marine environment: An ATR-FTIR based approach. *Scientific Reports, 6*(1), 1–8. doi:10.1038rep23501 PMID:27000994

Ioakeimidis, C., Zeri, C., Kaberi, H., Galatchi, M., Antoniadis, K., Streftaris, N., ... Papatheodorou, G. (2014). A comparative study of marine litter on the seafloor of coastal areas in the Eastern Mediterranean and Black Seas. *Marine Pollution Bulletin, 89*(1-2), 296–304. doi:10.1016/j.marpolbul.2014.09.044 PMID:25440189

Ioannidou, O., & Zabaniotou, A. (2007). Agricultural residues as precursors for activated carbon production—A review. *Renewable & Sustainable Energy Reviews, 11*(9), 1966–2005. doi:10.1016/j.rser.2006.03.013

Iqbal, H. M. N., & Asgher, M. (2013). Characterization and decolorization applicability of xerogel matrix immobilized manganese peroxidase produced from Trametes versicolor IBL-04. *Protein and Peptide Letters, 20*(5), 591–600. doi:10.2174/0929866511320050013 PMID:23016633

Iqbal, H. M. N., Kyazze, G., & Keshavarz, T. (2013). Advances in the valorization of lignocellulosic materials by biotechnology: An overview. *BioResources, 8*(2), 3157–3176. doi:10.15376/biores.8.2.3157-3176

Isacsson, U., & Lu, X. (1995). Testing and Appraisal of Polymer Modified Road Bitumen. *Materials and Structures, 28*(3), 139–159. doi:10.1007/BF02473221

Ismail, Z. Z., & Hashmi, E. A. A. L. (2008). Use of waste plastic in concrete mixture as aggregate replacement. *Waste Management (New York, N.Y.)*, *28*(11), 2041–2047. doi:10.1016/j.wasman.2007.08.023 PMID:17931848

ISWA (International Solid Waste Association). (2012). *Globalization and waste management final report from the ISWA task force*. Retrieved from http://www.iswa. org/knowledgebase/tfgfinal

Jacko, R. B., & Neurendrof, D. W. (1977). Trace metal particulate emission test results from a number of industrial and municipal point sources. *Journal of the Air Pollution Control Association*, *27*(10), 989–994. doi:10.1080/00022470.1977.10470522

Jackson, T. (2005). *Motivating Sustainable Consumption: A Review of Evidence on Consumer Behaviour and Behavioural Change* (T. Jackson, Ed.). Surrey, UK: Sustainable Development Research Network.

Jafari, A., Godini, H., & Mirhousaini, S. H. J. (2010). Municipal solid waste management in Khoram Abad city and experiences. *World Academy of Science, Engineering and Technology*, *4*, 163–168.

Jafari, A., Godini, H., & Mirhousaini, S. H. J. (2010). Municipal solid waste management in KhoramAbad city and experiences. *World Academy of Science, Engineering and Technology*, *4*, 163–168.

Jaffery, S. (2013). *Medical waste illegally sold off from Pakistan hospital*. Academic Press.

Jain, P., Rani, B. M., & Bhawsar, J. (2009). Study recycling of plastic and CRTs from electronic waste. *Oriental Journal of Computer Science & Technology*, *2*, 51–56.

Jain, S., Agarwal, A., Jani, V., Singhal, S., Sharma, P., & Jalan, R. (2017). Assessment of carbon neutrality and sustainability in educational campuses (CaNSEC): A general framework. *Ecological Indicators*, *76*, 131–143. doi:10.1016/j.ecolind.2017.01.012

Jambeck, J. R., Geyer, R., Wilcox, C., Siegler, T. R., Perryman, M., Andrady, A., ... Law, K. L. (2015). Plastic Waste Inputs from Land into the Ocean. *Science*, *347*(6223), 768–771. doi:10.1126cience.1260352 PMID:25678662

Jang, Y. C., Lee, C., Yoon, O. S., & Kim, H. (2006). Medical waste management in Korea. *Journal of Environmental Management*, *80*(2), 107–115. doi:10.1016/j.jenvman.2005.08.018 PMID:16338054

Jayathilakan, K., Sultana, K., Radhakrishna, K., & Bawa, A. S. (2012). Utilization of byproducts and waste materials from meat, poultry and fish processing industries: A review. *Journal of Food Science and Technology*, *49*(3), 278–293. doi:10.100713197-011-0290-7 PMID:23729848

Jecu, L. (2000). Solid state fermentation of agricultural wastes for endoglucanase production. *Industrial Crops and Products*, *11*(1), 1–5. doi:10.1016/S0926-6690(99)00022-9

Jenkins, R. R., Martinez, S. A., & Podolsky, P. K. (2013). The determinant of household recycling; A material-specific analysis of recycling program features and unit recycling. *Journal of Environmental Economics and Management*, *45*(2), 294–318.

Jenne, E. A., & Louma, S. N. (1997). Forms of trace elements in soils and associated waters: an overview of their determination and biological availability. In Biological implications of metals in the environment. Conf. 750929.

Jensen, P. (2014). Behavior genetics and the domestication of animals. *Annual Review of Animal Biosciences*, *2*(1), 85–104. doi:10.1146/annurev-animal-022513-114135 PMID:25384136

Jha, M. K., Sondhi, O. A. K., & Pansare, M. (2003). Solid waste management- a case study. *Indian Journal of Environment Protection*, *23*, 1153–1160.

Jiang, T., Ma, X., Tang, Q., Yang, J., Li, G., & Schuchardt, F. (2016). Combined use of nitrification inhibitor and struvite crystallization to reduce the NH_3 and N_2O emissions during composting. *Bioresource Technology*, *217*, 210–218. doi:10.1016/j.biortech.2016.01.089 PMID:26865057

Jicong, H., Yagnum, Q., Guangging, I., & Dong, R. (2005). The influence of temperature, pH, and C/N ratio on the growth and survival of earthworms in Municipal Solid Waste. CIGR Ejournal, 7.

Jinhui, L., Huabo, D., & Pixing, S. (2011). Heavy metal contamination of surface soil in electronic waste dismantling area: Site investigation and source-apportionment analysis. *Waste Management & Research*, *29*(7), 727–738. doi:10.1177/0734242X10397580 PMID:21406506

Joardar, S. D. (2000). Urban residential solid waste management in India. *Public Works Management & Policy*, *4*(4), 319–330.

Joardar, S. D. (2000). Urban residential solid waste management in India: Issues related to institutional arrangements. *Public Works Management & Policy*, *4*(4), 319–330. doi:10.1177/1087724X0044006

Jobling, S., Reynolds, T., White, R., Parker, M. G., & Sumpter, J. P. (1995). A Variety of Environmentally Persistent Chemicals, Including Some Phthalate Plasticizers, are Weakly Estrogenic. *Environmental Health Perspectives*, *10*(6), 582–587. doi:10.1289/ehp.95103582 PMID:7556011

Joint Group of Experts on the Scientific Aspects of Marine Environmental Protection (GESAMP). (Ed.). (2015). *Sources, Fate and Effects of Microplastics in the Marine Environment: A Global Assessment*. London: Int. Mar. Org.

Joint Group of Experts on the Scientific Aspects of Marine Environmental Protection (GESAMP). (Ed.). (2016). *Sources, Fate and Effects of Microplastics in the Marine Environment: Part Two of a Global Assessment*. London: Int. Mar. Org.

Jones, A., & Harrison, R. (2016). Emission of ultrafine particles from the incineration of municipal solid waste: A review. *Atmospheric Environment*, *140*, 519–528.

Jones, D. L., & Healey, J. R. (2010). Organic amendments for remediation: Putting waste to good use. *Elements*, *6*(6), 369–374. doi:10.2113/gselements.6.6.369

Jones, G. E. (1981). Sludge Characteristics. In G. E. Jones & R. T. Sprague (Eds.), *Sludge and Its Ultimate Disposal*. Ann Arbor Science.

Joshi, R., & Ahmed, S. (2016). Status and challenges of municipal solid waste management in India: A review. *Cogent Environmental Science*, *2*(1), 1139434. doi:10.1080/23311843.2016.1139434

JRC-IES. (2010). *International Reference Life Cycle Data System (ILCD) Handbook: General Guide for Life Cycle Assessment – Detailed Guidance (Plus 6 Supporting Documents)*. Author.

Kaewlue, W. (2012). *Solid Waste Management of Tha Khon Yang Municipality, Kantharawichai District, Maha Sarakham Province* (Master's Thesis). Mahasarakham University, Talat, Thailand.

Kaisarevic, S., Andric, N., Bobic, S., Trickovic, J., Teodorovic, I., Miloradov, M. V., & Kovacevic, R. Z. (2007). Detection of Dioxin like Contaminants in Soil from the Area of oil Refineries in Vojvodina Region of Serbia. Springer Science Business Media, 79, 422-426.

Kakuta, Y., Hirano, K., Sugano, M., & Mashimo, K. (2008). Study on chlorine removal from mixture of waste plastics. *Waste Management (New York, N.Y.)*, *28*(3), 615–621. doi:10.1016/j.wasman.2006.12.023 PMID:17482803

Kalargaris, Tian, & Gu. (2017). The utilisation of oils produced from plastic waste at different pyrolysis temperatures in a DI diesel engine. *Energy*, *131*, 179-185. Retrieved from http://www.ipcc.ch/pdf/assessmentreport/ar4/syr/ar4_syr.pdf

Kale, R. D. (1998). *Earthworm- Cinderella of Organic Farming*. Bangalore, India: Publishers- Prism Book Pvt. Ltd.

Kale, R. D., & Bano, K. (1988). Earthworm cultivation and culturing techniques for production of Vee COMP83E UAS, Vee MEAL 83P UAS. *Mysore Journal of Agricultural Sciences*, *22*, 339–344.

Kale, R. D., Bano, K., & Krishnamoorthy, R. V. (1982). Potential of *Perionyx excavatus* for utilizing organic wastes. *Pedobiologia*, *23*, 419–425.

Kale, R. D., Mallesh, B. C., Kubra, B., & Bhagyaraj, D. J. (1992). Influence of vermicompost application on available micronutrients and selected microbial populations in paddy field. *Soil Biology & Biochemistry*, *24*(12), 1317–1320. doi:10.1016/0038-0717(92)90111-A

Kalinauskas, Z. (2010). Interdependence between the Advance in Science and the Advance in Technology. *Evolution of Science and Technology*, *1*(2), 157–166.

Kalra, V. (2004). E-*waste in India: System failure imminent: Take action now! (Delhi, Toxics Link)*. Available at: http://www.toxicslink.org/pub-view.php?pubnum=40.pdf

Kanat, G. (2010). Municipal solid-waste management in Istanbul. *Waste Management (New York, N.Y.)*, *30*(8-9), 1737–1745. PMID:20185290

Kang, H. Y., & Schoenung, J. M. (2005). Electronic waste recycling: A review of US infrastructure and technology options. *Resources, Conservation and Recycling*, *45*(4), 368–400.

Kansal, A. (2002). Solid Waste management Strategies for India. *Indian Journal of Environment Protection*, *22*(4), 444–448.

Kansal, A. (2002). Solid waste management strategies for India. *Indian Journal of Environment Protection*, *22*, 444–448.

Kansal, A. (2002). Solid waste management strategies for India. *Indian Journal of Environmental Protection*, *22*(4), 444–448.

Karakashev D. & Angelidaki I. (2011). Emerging biological technologies: biofuels and biochemicals. *Solid Waste Technology & Management.*

Karayannidis, G. P., & Achilias, D. S. (2007). Chemical recycling of poly (ethylene terephthalate). *Macromolecular Materials and Engineering*, *292*(2), 128–146.

Karlsson, H., Ahlgren, S., Strid, I., & Hansson, P. A. (2015). Faba beans for biorefinery feedstock or feed? Greenhouse gas and energy balances of different applications. *Agricultural Systems*, *141*, 138–148. doi:10.1016/j.agsy.2015.10.004

Karthikeyan, V., Sathymoorthy, G. L., & Murugesan, R. (2007).Vermicomposting of market waste in Salem, Tamil Nadu, India. *Proceedings of the international conference on sustainable solid waste management*, 276-281.

Karthikeyan, A., & Sivakumar, N. (2007). Sericulture pupal waste – a new productionmedium for mass cultivation of Bacillus thuringiensis. *Indian Journal of Biotechnology*, *6*, 557–559.

Karthikeyan, K., & Balasubramanian, S. (2010). Studies on the characterization and possibilities of reutilization of solid wastes from a waste paper based paper industry. *Global Journal of Environmental Research*, *4*(1), 18–22.

Kashem, M. A., & Singh, B. (1999). Heavy metal contamination of soil and vegetation in the vicinity of industries in Bangladesh. *Water, Air, and Soil Pollution*, *115*(1/4), 347–361. doi:10.1023/A:1005193207319

Kathiravale, S., & Muhd Yunus, M. N. (2008). Waste to wealth. *Asia Europe Journal*, *6*(2), 359–371. doi:10.100710308-008-0179-x

Katnoria, J. K., Bhardwaj, R., & Nagpal, A. (2011). Evaluation of genotoxic potential of industrial waste contaminated soil extracts of Amritsar, India. *Journal of Environmental Biology*, *32*, 363–367. PMID:22167950

Kaushal, R. K., Varghese, G. K., & Chabukdhara, M. (2012). Municipal solid waste management in India—current state and future challenges: A review. *International Journal of Environmental Science and Technology*, *4*, 1473–1489.

Kaviraj & Sharma. (2003). Municipal solid waste management through vermicomposting employing exotic and local species of earthworms. *Bioresour Technol*, *90*, 169–173.

Kaviraj & Sharma. (2003). Municipal solid waste management through vermicomposting employing exotic and local species of earthworms. *Biosource Tech*nology, *90*(2), 169-173.

Kaviraj, & Sharma, S. (2003). Municipal solid waste management through vermicomposting employing exotic and local species of earthworms. *Bioresource Technology*, *90*(2), 169–173. doi:10.1016/S0960-8524(03)00123-8 PMID:12895560

Kawai, F., Watanabe, M., Shibata, M., Yokoyama, S., Sudate, Y., & Hayashi, S. (2004). Comparative study on biodegradability of polyethylene wax by bacteria and fungi. *Polymer Degradation & Stability*, *86*(1), 105–114. doi:10.1016/j.polymdegradstab.2004.03.015

Kedzierski, M., D'Almeida, M., Magueresse, A., Le Grand, A., Duval, H., César, G., ... Le Tilly, V. (2018). Threat of plastic ageing in marine environment. Adsorption/desorption of micro pollutants. *Marine Pollution Bulletin*, *127*, 684–694. doi:10.1016/j.marpolbul.2017.12.059 PMID:29475712

Keeny, D. R., & Wildung, R. E. (1972). Chemical properties of soils. In Soils for management of organic wastes and waste waters. American Soc. Agron.

Khajuria, A., Matsui, T., Machimura, T., & Morioka, T. (2010). Assessment of the challenge of sustainable recycling of municipal solid waste management in India. *International Journal of Environmental Technology and Management*, *13*, 171–187.

Khalid, A., Arshad, M., Anjum, M., Mahmood, T., & Dawson, L. (2011). The anaerobic digestion of solid organic waste. *Waste Management*, *31*(8), 1737-1744.

Khan, A. G. (2001). Relationships between chromium biomagnification ratio, accumulation factor and mycorrhizae in plants growing on tannery effluent-polluted soil. *Environment International*, *26*(5-6), 417–423. doi:10.1016/S0160-4120(01)00022-8 PMID:11392761

Khan, N. A. (2014). Perspectives of transport and disposal of municipal solid waste in Srinagar city. *International Journal of Engineering Research and General Science*, *2*(4), 238–251.

Khan, R. R. (1994). Environmental management of municipal solid wastes. *Indian J. Environ. Prot.*, *14*(1), 26–30.

Khan, R. R. (1994). Environmental management of municipal solid wastes. *Indian Journal of Environment Protection*, *14*(1), 26–30.

Khan, S., & Singh, J. (1999). Mobility of heavy metals as affected by some insecticides through soils, saturated with different cations. *Pollution Research*, *18*, 133–136.

Kim, K. T., Graves, P. B., Safadi, G. S., Alhadeff, G., & Metcalfe, J. (1998). Implementation recommendations for making health care facilities latex safe. *AORN Journal*, *67*(3), 615–632. doi:10.1016/S0001-2092(06)62830-5 PMID:9541704

Kim, K. Y., Kang, P. D., Lee, K. G., Hyung, K., Kim, M. J., Kim, K. H., & ... . (2010). Microsatellite analysis of the silkworm strains (Bombyx mori): High variability and potential markers for strain identification. *Genes & Genomics*, *32*(6), 532–543. doi:10.100713258-010-0066-x

King, G. N., & King, H. W. (1986). Polymer Modified Asphalts: An Overview. American Society of Civil Engineers.

King, S. (2004). *Strategic Framework—The Foundation of an Effective Planning Process*. Available online: http://www.infomanagementcenter.com/enewsletter/200411/third.htm

King, A. M., Burgess, S. C., Ijomah, W., & McMahon, C. A. (2006). Reducing waste: Repair, recondition, remanufacture or recycle. *Sustainable Development*, *14*(4), 257–267. doi:10.1002d.271

Kiran, E. U., Trzcinski, A., Ng, W., & Liu, Y. (2014). Bioconversion of food waste to energy: A review. *Fuel*, *134*, 389–399.

Kirkeby, J. T., Birgisdottir, H., Hansen, T. L., Christensen, T. H., Bhander, G. S., & Hauschild, M. (2006). Environmental assessment of solid waste systems and technologies: EASEWASTE. *Waste Management & Research*, *24*(1), 3–15. doi:10.1177/0734242X06062580 PMID:16496866

Kiser, D.M.R.E., Johnson, N.E., Russell, J..T.G., & MacDonald. (1998). *Low level class A/C waste, Near Surface Land Disposal Level Feasibility Design Description*. INEEL/EXT-98-00051, Lock head Martin Idaho Technologies Company.

Kjeldsen, P., Barlaz, M. A., Rooker, A. P., Baun, A., Ledin, A., & Christensen, T. H. (2002). Present and long-term composition of MSW landfill leachate: A review. *Critical Reviews in Environmental Science and Technology*, *32*(4), 297–336. doi:10.1080/10643380290813462

Klein, D. H., & Russel, P. (1973). Heavy metals: Fall out around a power plant. *Environmental Science & Technology*, *7*(4), 357–358. doi:10.1021/es60076a004

Klein, L. A., Aang, M., Nash, N., & Krishchner, S. L. (1974). Sources of metals in New York city waste water. *Journal - Water Pollution Control Federation*, *46*, 26–53.

Klundert, A. V. D., & Anschutz, J. (2001). *Integrated Sustainable Waste Management—The Concept: Tools for Decision-Makers, Experiences from the Urban Waste Expertise Programme (1995–2001)*. Gouda, The Netherlands: WASTE.

Knasmuller, S., Gottmann, E., Steinkellner, H., Fomin, A., Pickl, C., Paschki, A., ... Kundi, M. (1998). Detection of genotoxic effects of heavy metal contaminated soils with plant bioassays. *Mutation Research*, *420*(1-3), 37–48. doi:10.1016/S1383-5718(98)00145-4 PMID:9838036

Koch, H. M., & Calafat, A. M. (2009). Human body burdens of chemicals used in plastic manufacture. *Philosophical Transactions of the Royal Society of London. Series B, Biological Sciences*, *364*(1526), 2063–2078. doi:10.1098/rstb.2008.0208 PMID:19528056

Koe, L. C. C., & Aziz, M. I. A. (1995). *Suggested approach to control marine population from land-based sources*. Paper presented at Asia Pacific Regional conference on Prevention of Marine Population From Land Sources and Sea Sources, Singapore.

Kolek, Z. (2001). Recycled polymers from food packaging in relation to environmental protection. *Polish Journal of Environmental Studies*, *10*, 73–76.

Kong, Z., Wang, X., Liu, Q., Li, T., Chen, X., Chai, L., ... Shen, Q. (2018). Evolution of various fractions during the windrow composting of chicken manure with rice chaff. *Journal of Environmental Management*, *207*, 366–377. doi:10.1016/j.jenvman.2017.11.023 PMID:29182981

Koniecki, D., Wang, R., Moody, R. P., & Zhu, J. (2011). Phthalates in cosmetics and personal care products:concentrations and possible dermal exposure. *Environmental Research*, *111*(3), 329–336. doi:10.1016/j.envres.2011.01.013 PMID:21315328

Kosky, J. P., Jones, J. F., & Corp, M. E. I. (1995). *System and method for the treatment of hazardous waste material*. U.S. Patent 5, 471,937.

Kothari, R., Tyagi, V., & Pathak, A. (2010). Waste-to-energy: A way from renewable energy sources to sustainable development. *Renewable & Sustainable Energy Reviews*, *14*(9), 3164–3170.

Kowalska, H., Czajkowska, K., Cichowska, J., & Lenart, A. (2017). What's new in biopotential of fruit and vegetable by-products applied in the food processing industry. *Trends in Food Science & Technology*, *67*, 150–159. doi:10.1016/j.tifs.2017.06.016

Krishna, C. (2005). Solid-state fermentation systems—An overview. *Critical Reviews in Biotechnology*, *25*(1-2), 1–30. doi:10.1080/07388550590925383 PMID:15999850

Krishnamoorthy, R. V., & Vajrabhiah, S. N. (1986). Biological activity of earthworm casts: An assessment of plant growth promoter levels in casts. *Proceedings of the Indiana Academy of Sciences*, *95*(3), 341–351. doi:10.1007/BF03179368

Kudyakov, R., Baibergenova, A., Zdeb, M., & Carpenter, D. O. (2004). Respiratory disease in relation to patient residence near to hazardous waste sites. *Environmental Toxicology and Pharmacology*, *18*(3), 249–257. doi:10.1016/j.etap.2004.06.005 PMID:21782756

Kuhn, S. B., Rebolledo, E.L., & van Franeker, J.A. (2015). Deleterious Effects of Litter on Marine Life. In Marine Anthropogenic Litter (pp. 75–116). Heidelberg, Germany: Springer.

Kulkarni, N. P., & Agrawal, Y. K. (2003). Physico-chemical characteristics of the topsoil of a hospital. *Indian Journal of Environmental Protection*, *23*, 503–507.

Kum, V., Sharp, A., & Harnpornchai, N. (2005). Improving the solid waste management in Phnom Penh city: a strategic approach. *Waste Management, 25*(1), 101-109.

Kumar, A. (1998). Effect of mixed industrial effluents on properties of ground water and irrigated soils. *Journal of Indian Soiety for Soil Sciences*, *46*, 427–429.

Kumar, A., Duhan, J. S., Gahlawat, S., & Gahlawat, S. K. (2014). Production of ethanol from tuberous plant (sweet potato) using *Saccharomyces cerevisiae* MTCC-170. *African Journal of Biotechnology*, *13*(28).

Kumari, M., Kumar, S., Chauhan, R. S., & Ravikanth, K. (2011). Bioconversion of herbal industry waste into vermicompost using an epigeic earthworm *EudrilusEugeniae*. *Waste Management & Research*, *29*(11), 1205–1212. doi:10.1177/0734242X10384432 PMID:20952444

Kumar, J. S., Subbaiah, K. V., & Rao, P. V. P. (2014). Municipal solid waste management scenario in India. *Australian Journal of Civil Engineering Research*, *2*, 1–8.

Kumar, J. S., Subbaiah, K. V., & Rao, P. V. V. P. (2010). Waste to Energy: A Case Study of Eluru, A.P, India. *International Journal of Environmental Sciences and Development*, *1*(3), 238–243.

Kumar, K. N., & Goel, S. (2009). Characterization of municipal solid waste (MSW) and a proposed management plan for Kharagpur,West Bengal, India. *Resources, Conservation and Recycling*, *53*, 166–174.

Kumar, R., Khan, E. A., Ahmed, J., Khan, Z., Magan, M., Nousheen, N., & Mughal, M. I. (2010). Healthcare waste management (HCWM) in Pakistan: Current situation and training options. *Journal of Ayub Medical College, Abbottabad*, *22*(4), 101–106. PMID:22455273

Kumar, R., Shaikh, B. T., Somrongthong, R., & Chapman, R. S. (2015). Practices and challenges of infectious waste management: A qualitative descriptive study from tertiary care hospitals in Pakistan. *Pakistan Journal of Medical Sciences*, *31*(4), 795. doi:10.12669/pjms.314.7988 PMID:26430405

Kumar, R., Somrongthong, R., & Shaikh, B. T. (2015). Effectiveness of intensive healthcare waste management training model among health professionals at teaching hospitals of Pakistan: A quasi-experimental study. *BMC Health Services Research*, *15*(1), 81. doi:10.118612913-015-0758-7 PMID:25889451

Kumar, S. (2011). Composting of municipal solid waste. *Critical Reviews in Biotechnology*, *31*(2), 112–136. doi:10.3109/07388551.2010.492207 PMID:20854128

Kumar, S., Bhattacharya, J. K., Vaidya, A. N., Chakrabarti, T., Devotta, S., & Akolkar, A. B. (2009). Assessment of the status of municipal solid waste management in metro cities, state capitals, class I cities, and class II towns in India: An insight. *Journal of Waste Management*, *29*, 883–895. PMID:18595684

Kumar, S., Bhattacharyya, J. K., Vaidya, A. N., Chakrabarti, T., Devotta, S., & Akolkar, A. B. (2009). Assessment of the status of municipal solid waste management in metro cities, state capitals, class I cities, and class II towns in India: An insight. *Waste Management (New York, N.Y.)*, *29*(2), 883–895. doi:10.1016/j.wasman.2008.04.011 PMID:18595684

Kumar, S., Dhar, H., Nair, V. V., Bhattacharyya, J. K., Vaidya, A. N., & Akolkar, A. B. (2016). Characterization of municipal solid waste in high-altitude sub-tropical regions. *Environmental Technology*, *37*(20), 2627–2637. doi:10.1080/09593330.2016.1158322 PMID:26915419

Kumar, S., Negi, S., Mandpe, A., Singh, R. V., & Hussain, A. (2018). Rapid composting techniques in Indian context and utilization of black soldier fly for enhanced decomposition of biodegradable wastes-A comprehensive review. *Journal of Environmental Management*, *227*, 189–199. doi:10.1016/j.jenvman.2018.08.096 PMID:30193208

Kumar, S., Pandaa, A. K., & Singha, R. K. (2011). A review on tertiary recycling of high-density polyethylene to fuel. *Resources, Conservation and Recycling*, *55*, 893–910.

Kumar, S., & Singh, V. (2019). E-Waste: Generation, Environmental and Health Impacts, Recycling and Status of E-Waste Legislation. *Journal of Emerging Technologies and Innovative Research*, *6*(1), 592–600.

Kumar, S., Smith, S. R., Fowler, G., Velis, C., Kumar, S. J., Arya, S., ... Cheeseman, C. (2017). Challenges and opportunities associated with waste management in India. *Royal Society Open Science*, *4*(3), 160764. doi:10.1098/rsos.160764 PMID:28405362

LaBrecque, S. (2015). What does zero waste mean to you? *The Guardian*.

Lai, C. M., Ke, G. R., & Chung, M. Y. (2009). Potential of food wastes for power generation and energy conservation in Taiwan. *Renewable Energy*, *34*(8), 1913–1915. doi:10.1016/j.renene.2008.12.007

Laist, D. W. (1997). Impacts of marine debris: entanglement of marine life in marine debris including a comprehensive list of species with entanglement and ingestion records. In J. M. Coe & D. B. Rogers (Eds.), *Marine Debris: Sources, Impacts, and Solutions* (pp. 99–140). New York: Springer-Verlag. doi:10.1007/978-1-4613-8486-1_10

Lal, P. M., & Singh, C. J. (2012). Plasma Gasification: A Sustainable Solution for the Municipal Solid Waste Management in the State of Madhya Pradesh, India. *International Journal of Environmental Sciences*, *3*(1), 306.

Lambert, D. M. (2004). *Supply Chain Management: Processes, Partnerships, Performance* (1st ed.). Sarasota, FL: Supply Chain Management Institute.

Lamb, J. B., Willis, B. L., Fiorenza, E. A., Couch, C. S., Howard, R., Rader, D. N., ... Harvell, C. D. (2018). Plastic waste associated with disease on coral reefs. *Science*, *359*(6374), 460–462. doi:10.1126cience.aar3320 PMID:29371469

Lam, C. H., Ip, A. W., Barford, J. P., & McKay, G. (2010). Use of incineration MSW ash: A review. *Sustainability*, *2*(7), 1943–1968.

Landrigan, P. J., Schechter, C. B., Lipton, J. M., Fahs, M. C., & Schwartz, J. (2002). Environmental pollutants and disease in American children: Estimates of morbidity, mortality, and costs for lead poisoning, asthma, cancer, and developmental disabilities. *Environmental Health Perspectives, 110*(7), 721–728. doi:10.1289/ehp.02110721 PMID:12117650

Langeveld, H., Sanders, J., & Meeusen, M. (2012). *The Biobased Economy: Biofuels*. Materials, and Chemicals in the Post-oil Era, Earthscan.

Lang, I. A., Galloway, T. S., Scarlett, A., Henley, W. E., Depledge, M., Wallace, R. B., & Melzer, D. (2008). Association of urinary bisphenol A concentration with medical disorders and laboratory abnormalities in adults. *Journal of the American Medical Association, 300*(11), 1303–1310. doi:10.1001/jama.300.11.1303 PMID:18799442

Lansforsakringar Miljo. (2002). *Recycling Insurance*. Retrieved from http://www.lansforsakringar.se

Laroche, M., Bergeron, J., & Barbaro-Forleo, G. (2001). Targeting consumers who are willing to pay more for environmentally friendly products. *Journal of Consumer Marketing, 18*(6), 503–520.

Larsen, T. A., Lienert, J., Joss, A., & Siegrist, H. (2004). How to avoid pharmaceuticals in the aquatic environment. *Journal of Biotechnology, 113*(1-3), 295–304. doi:10.1016/j.jbiotec.2004.03.033 PMID:15380662

Larson, G., & Fuller, D. Q. (2014). The evolution of animal domestication. *Annual Review of Ecology Evolution and Systematics, 45*(1), 115–136. doi:10.1146/annurev-ecolsys-110512-135813

Late, A., & Mule, M. B. (2013). Composition and characterization study of solid waste from Aurangabad city. *Universal Journal of Environmental Research and Technology, 3*(1), 55–60.

Latifah, A. M., Mohd, L. C., Mohd, K. Y., Tengku, H., Tengku, I., Rosta, H., ... Kamaruzaman, J. (2009). Influences of bedding material in vermicomposting process. *International Journal of Biology, 1*(1), 81–91.

Latifah, A., & Mohd, A. A., & NurIlyana, M. (2009). Municipal solid waste management in Malaysia: Practices and challenges. *Waste Management (New York, N.Y.), 29*, 2902–2906. PMID:19540745

Law, K. L., Morét-Ferguson, S. E., Goodwin, D. S., Zettler, E. R., DeForce, E., Kukulka, T., & Proskurowski, G. (2014). Distribution of Surface Plastic Debris in the Eastern Pacific Ocean from an 11-Year Data Set. *Environmental Science & Technology, 48*(9), 4732–4738. doi:10.1021/es4053076 PMID:24708264

Law, K. L., Morét-Ferguson, S., Maximenko, N. A., Proskurowski, G., Peacock, E. E., Hafner, J., & Reddy, C. M. (2010). Plastic Accumulation in the North Atlantic Subtropical Gyre. *Science, 329*(5996), 1185–1188. doi:10.1126cience.1192321 PMID:20724586

Lebreton, L. C. M., van der Zwet, J., Damsteeg, J. W., Slat, B., Andrady, A., & Reisser, J. (2017). River plastic emissions to the world's oceans. *Nature Communications, 8*(1), 15611. doi:10.1038/ncomms15611 PMID:28589961

Lee, B., Pometto, A. L., Fratzke, A., & Bailey, T. B. (1991). Biodegradation of degradable plastic polyethylene by Phanerochaete and *Streptomyces* species. *Applied and Environmental Microbiology, 57*, 678–685. PMID:16348434

Lee, C. C., Huffman, G., & Nalesnik, R. (1991). Medical waste management. *Environmental Science & Technology, 25*(3), 360–363. doi:10.1021/es00015a607

Lee, L. H., Wu, T. Y., Shak, K. P. Y., Lim, S. L., Ng, K. Y., Nguyen, M. N., & Teoh, W. H. (2018). Sustainable approach to biotransform industrial sludge into organic fertilizer via vermicomposting: A mini-review. *Journal of Chemical Technology and Biotechnology (Oxford, Oxfordshire), 93*(4), 925–935. doi:10.1002/jctb.5490

Lemmens, B., Elslander, H., Vanderreydt, I., Peys, K., Diels, L., Oosterlinck, M., & Joos, M. (2007). Assessment of plasma gasification of high caloric waste streams. *Waste Management (New York, N.Y.), 27*, 1562–1569. PMID:17134888

Leonelli, C., & Mason, T. J. (2010). Microwave and ultrasonic processing: now a realistic option for industry. *Chemical Engineering and Processing: Process Intensification, 49*(9), 885-900.

Liang, C., Das, K. C., & Clendon, R. (2003). The influence of temperature and moisture contents regimes on the aerobic microbial activity of a biosolids composting blend. *Bioresource Technology*, *86*, 131–137. PMID:12653277

Liao, C. H., & Chiu, A. S. F. (2011). Evaluate municipal solid waste management problems using hierarchical framework. *Procedia: Social and Behavioral Sciences*, *25*, 353–362. doi:10.1016/j.sbspro.2011.10.554

Li, C. S., & Jenq, F. T. (1993). Physical and chemical composition of hospital waste. *Infection Control and Hospital Epidemiology*, *14*(3), 145–150. doi:10.1086/646700 PMID:8478527

Lieberman, M. A., & Lichtenberg, A. J. (2005). *Principles of plasma discharges and materials processing* (Vol. 2). New York: Wiley-interscience. doi:10.1002/0471724254

Li, G., Yun, Y., Li, H., & Sang, N. (2008). Effect of landfill leachate on cell cycle, micronucleus, and sister chromatid exchange in Triticumaestivum. *Journal of Hazardous Materials*, *155*(1–2), 10–16. doi:10.1016/j.jhazmat.2007.10.106 PMID:18068897

Li, J., Shang, X., Zhao, Z., Tanguay, R. L., Dong, Q., & Huang, C. (2010). Polycyclic aromatic hydrocarbons in water, sediment, soil, and plants of the Aojiang River waterway in Wenzhou, China. *Journal of Hazardous Materials*, *173*(1-3), 75–81. doi:10.1016/j.jhazmat.2009.08.050 PMID:19726127

Li, J., Wang, C., Du, L., Lv, Z., Li, X., Hu, X., ... Zhang, Y. (2017). Did municipal solid waste landfill have obvious influence on polychlorinated dibenzo-p-dioxins and polychlorinated dibenzofurans (PCDD/Fs) in ambient air: A case study in East China. *Waste Management (New York, N.Y.)*, *62*, 169–176. doi:10.1016/j.wasman.2017.02.014 PMID:28223077

Likens, G. E., & Bormann, F. H. (2004). Acid Rain: A Serious Regional Environmental Problem. *Science*, *184*(4142), 1176–1179. doi:10.1126cience.184.4142.1176 PMID:17756304

Limayem, A., & Ricke, S. C. (2012). Lignocellulosic biomass for bioethanol production: Current perspectives, potential issues and future prospects. *Progress in Energy and Combustion Science*, *38*(4), 449–467. doi:10.1016/j.pecs.2012.03.002

Lim, P. N., Wu, T. Y., Shyang Sim, E. Y., & Lim, S. L. (2011). The potential reuse of soybean husk as feedstock of Eudriluseugeniae in vermicomposting. *Journal of the Science of Food and Agriculture*, *91*(14), 2637–2642. doi:10.1002/jsfa.4504 PMID:21725978

Lim, S. L., Lee, L. H., & Wu, T. Y. (2016). Sustainability of using composting and vermicomposting technologies for organic solid waste biotransformation: Recent overview, greenhouse gases emissions and economic analysis. *Journal of Cleaner Production*, *111*, 262–278. doi:10.1016/j.jclepro.2015.08.083

Lim, S. L., Wu, T. Y., Sim, E. Y. S., Lim, P. N., & Clarke, C. (2012). Biotransformation of rice husk into organic fertilizer through vermicomposting. *Ecological Engineering*, *41*, 60–64. doi:10.1016/j.ecoleng.2012.01.011

Linak, W. P., & Wendt, J. O. (1993). Toxic metal emissions from incineration: Mechanisms and control. *Progress in Energy and Combustion Science*, *19*(2), 145–185.

Lithner, D., Larsson, A., & Dave, G. (2011). Environmental and health hazard ranking and assessment of plastic polymers based on chemical composition. *Science Total Environment, 409*(18).

Lithner, D., Larsson, A., & Dave, G. (2011). Environmental and health hazard ranking and assessment of plastic polymers based on chemical composition. *The Science of the Total Environment*, *409*(18), 3309–3324. doi:10.1016/j.scitotenv.2011.04.038 PMID:21663944

Liu, C., & Wu, X. W. (2010). Factors influencing municipal solid waste generation in China: A multiple statistical analysis study. *Waste Management & Research*, *29*, 371–378. PMID:20699292

Liu, D. (2008). *Bio-hydrogen production by dark fermentation from organic wastes and residues. Kgs.* Lyngby: DTU Environment.

Liu, G., Niu, Z., Van Niekerk, D., Xue, J., & Zheng, L. (2008). Polycyclic aromatic hydrocarbons (PAHs) from coal combustion: Emissions, analysis, and toxicology. *Reviews of Environmental Contamination and Toxicology*, *192*, 1–28. doi:10.1007/978-0-387-71724-1_1 PMID:18020302

Liu, K., & Price, G. W. (2011). Evaluation of three composting systems for the management of spent coffee grounds. *Bioresource Technology*, *102*(17), 7966–7947. doi:10.1016/j.biortech.2011.05.073 PMID:21704514

Liu, Y. Y., Wang, Y., Walsh, T. R., Yi, L. X., Zhang, R., Spencer, J., ... Yu, L. F. (2016). Emergence of plasmid-mediated colistin resistance mechanism MCR-1 in animals and human beings in China: A microbiological and molecular biological study. *The Lancet Infectious Diseases*, *16*(2), 161–168. doi:10.1016/S1473-3099(15)00424-7 PMID:26603172

Liu, Y., Ma, L., Li, Y., & Zheng, L. (2006). Evolution of heavy metal speciation during the aerobic composting process of sewage sludge. *Chemosphere*, *67*(5), 1025–1032. doi:10.1016/j.chemosphere.2006.10.056 PMID:17184823

Li, Z., Huang, G., Yu, H., Zhou, Y., & Huang, W. (2015). Critical factors and their effects on product maturity in food waste composting. *Environmental Monitoring and Assessment*, *187*(4), 217. doi:10.100710661-015-4430-9 PMID:25822329

Lobelle, D., & Cunliffe, M. (2011). Early microbial biofilm formation on marine plastic debris. *Marine Pollution Bulletin*, *62*(1), 197–200. doi:10.1016/j.marpolbul.2010.10.013 PMID:21093883

Łochyńska, M., & Frankowski, J. (2018). The biogas production potential from silkworm waste. *Waste Management (New York, N.Y.)*, *79*, 564–570. doi:10.1016/j.wasman.2018.08.019 PMID:30343788

Lodha, B. (2007). Effect of maturity and nutritional parameters during vermicomposting of municipal solid waste. *International Journal of Environmental Technology and Management*, *7*(3/4), 454–463. doi:10.1504/IJETM.2007.015157

Lonnstedt, O. M., & Eklov, P. (2016). Environmentally relevant concentrations of microplastic particles influence larval fish ecology. *Science*, *352*(6290), 1213–1216. doi:10.1126cience.aad8828 PMID:27257256

Lopez-Real, J. M. (1996). Composting of Agricultural Wastes. In M. de Bertoldi, P. Sequi, B. Lemmes, & T. Papi (Eds.), *The Science of Composting* (pp. 542–550). Dordrecht, The Netherlands: Springer. doi:10.1007/978-94-009-1569-5_51

Loureiro, S., Santos, C., Pinto, G., Costa, A., Monteiro, M., Nogueira, A. J., & Soares, A. M. (2006). Toxicity assessment of two soils from Jales mine (Portugal) using plants: Growth and biochemical parameters. *Archives of Environmental Contamination and Toxicology*, *50*(2), 182–190. doi:10.100700244-004-0261-3 PMID:16307215

Lucier, C. A., & Gareau, B. J. (2009). *Solving the E-Waste Problem. Recycling - From E-Waste to Resources: Sustainable Innovation and Technology Transfer Industrial Sector Studies* Available from: http:// www.unep.fr/shared/publications/pdf/ DTIx1192xPA-Recycling%20from%20 ewaste%20to%20Resources.pdf

Luıs, L. G., Ferreira, P., Fonte, E., Oliveira, M., & Guilhermino, L. (2015). Does the presence of microplastics influence the acute toxicity of chromium (VI) to early juveniles of the common goby (*Pomato schistusmicrops*)? A study with juveniles from two wild estuarine populations. *Aquatic Toxicology (Amsterdam, Netherlands)*, *164*, 163–174. doi:10.1016/j.aquatox.2015.04.018 PMID:26004740

Lund, L. J., Page, A. L., & Nelon, C. O. (1968). Movement of heavy metals below sewage disposal pond. *Journal of Environmental Quality*, *5*(3), 330–334. doi:10.2134/jeq1976.00472425000500030023x

Lyytimäki, J. (2018). Renewable energy in the news: Environmental, economic, policy and technology discussion of biogas. *Sustain. Prod. Consum.*, *15*, 65–73. doi:10.1016/j.spc.2018.04.004

Maamari, O., Mouaffak, L., Kamel, R., Brandam, C., Lteif, R., & Salameh, D. (2016). Comparison of steam sterilization conditions efficiency in the treatment of Infectious Health Care Waste. *Waste Management (New York, N.Y.)*, *49*, 462–468. doi:10.1016/j.wasman.2016.01.014 PMID:26803472

Ma, C., Ye, S., Lin, T., Ding, X., Yuan, H., & Guo, Z. (2014). Sorce apportionment of polycyclic aromatic hydrocarbons in soils of wetlands in the Liao River Delta, Northeast China. *Marine Pollution Bulletin*, *80*(1-2), 160–167. doi:10.1016/j.marpolbul.2014.01.019 PMID:24461691

Macfadyen, G., Huntington, T., & Cappell, R. (2009). Abandoned, Lost or Otherwise Discarded Fishing Gear. In *UNEP Regional Seas Reports and Studies, 115*. Rome: UNEP/FAO.

Maczulak, A. E. (2010). *Waste treatment: reducing global waste* (Vol. 2). Infobase Publishing.

Madalina, E. G. R. (2017). Methods of Recycling. *Properties and Applications of Recycled Thermoplastic Polymers Recycling*, *2*, 24. doi:10.3390/recycling2040024

Maiti, S., Sarma, S. J., Brar, S. K., Le Bihan, Y., Drogui, P., Buelna, G., & Verma, M. (2016). Agro-industrial wastes as feedstock for sustainable bio-production of butanol by *Clostridium beijerinckii. Food and Bioproducts Processing*, *98*, 217–226. doi:10.1016/j.fbp.2016.01.002

Majumder, S. K. (1997). Scope for new commercial products from sericulture. *Indian Silk.*, *35*(12), 13–18.

Malińska, K., Zabochnicka-Świątek, M., & Dach, J. (2014). Effects of biochar amendment on ammonia emission during composting of sewage sludge. *Ecological Engineering*, *71*, 474–478. doi:10.1016/j.ecoleng.2014.07.012

Malviya, R., Chaudhary, R., & Buddhi, D. (2002). Study on solid waste assessment and management -Indore city. *Indian Journal of Environmental Protection*, *22*(8), 841–846.

Manaf, L.A., Samah, M. A. A., & Zukki, N. I. M. (2009). Municipal solid waste management in Malaysia: Practices and challenges. *Waste Management, 29*, 2902–2906.

Mandal, S. K., & Dutta, J. (2009). Integrated bio-medical waste management plan for Patna city. Institute of Town Planners. *India Journal*, *6*(2), 1–25.

Manju, R., Sathya, S., & Sheema, K. (2017). Use of Plastic Waste in Bituminous Pavement. *International Journal of Chemtech Research*, *8*, 804–811.

Manna, M. C., Singh, M., Kundu, S., Tripathi, A. K., & Takkar, P. N. (1997). Growth and reproduction of the vermicomposting earthworm *Perionyx excavatus* as influenced by food materials. *Biology and Fertility of Soils*, *24*(1), 129–132. doi:10.1007/BF01420233

Mantzouridou, F. T., Paraskevopoulou, A., & Lalou, S. (2015). Yeast flavour production by solid state fermentation of orange peel waste. *Biochemical Engineering Journal*, *101*, 1–8. doi:10.1016/j.bej.2015.04.013

Manyapu, V., Mandpe, A., & Kumar, S. (2018). Synergistic effect of fly ash in in-vessel composting of biomass and kitchen waste. *Bioresource Technology*, *251*, 114–120. doi:10.1016/j.biortech.2017.12.039 PMID:29274517

Manyuchi, M. M., & Phiri, A. (2013). Vermi-composting in solid waste management: A review. *International Journal of Engineering Science and Technology*, *2*(12), 1234–1242.

Manyuchi, M. M., & Phiri, A. (2013a). Vermi-composting in solid waste management: A review. *Int. J. Sci. Eng. Technol.*, *2*(12), 1234–1242.

Manyuchi, M. M., Phiri, A., & Muredzi, P. (2013). Effect of Vermi-compost, Vermi-wash and Application Time on Soil Micronutrients Composition. *International Journal of Engineering and Advanced Technology.*, *2*(5), 215–218.

Manyuchi, M. M., Phiri, A., Muredzi, P., & Chirinda, N. (2013). Effect of drying on vermicompost macronutrient composition. *Internatioanl Journal of Engineering Science and Invention*, *1*(10), 1–3.

Mao, C., Feng, Y., Wang, X., & Ren, G. (2015). Review on research achievements of biogas from anaerobic digestion. *Renewable & Sustainable Energy Reviews*, *45*, 540–555. doi:10.1016/j.rser.2015.02.032

Marculescu, C. (2011). Thermal-chemical treatment of solid waste mixtures. *Energy Procedia*, *6*, 558–564.

Marinkovic, N., Vitale, K., Holcer, N. J., Džakula, A., & Pavic, T. (2008). Management of hazardous medical waste in Croatia. *Waste Management, 28*(6), 1049-1056.

Marshall, R. E., & Farahbakhsh, K. (2013). Systems approaches to integrated solid waste management in developing countries. *Waste Management (New York, N.Y.)*, *33*(4), 988–1003. PMID:23360772

Masera Cerutti, O. (2011). *La bioenergía en México: situación actual y perspectivas*. Red Mexicana de Bioenergía. Disponível em: http://rembio.org.mx/wp-content/uploads/2014/12/CT4.pdf

Masood, M., Barlow, C. Y., & Wilson, D. C. (2014). An assessment of the current municipal solid waste management system in Lahore, Pakistan. *Waste Management & Research*, *32*, 834–847. PMID:25106534

Masters, G. M., & Ela, W. P. (2008). *Environmental engineering and science* (3rd ed.). Prentice Hall.

Mastorakis, N. E., Bulucea, C. A., Oprea, T. A., Bulucea, C. A., & Dondon, P. (2011). Holistic approach of biomedical waste management system with regard to health and environmental risks. Development, Energy, Environment. *Economics*, *5*(3), 287–295.

Materazzi, M., Lettieri, P., Mazzei, L., Taylor, R., & Chapman, C. (2014). Tar evolution in a two stage fluid bed–plasma gasification process for waste valorization. *Fuel Processing Technology*, *128*, 146–157. doi:10.1016/j.fuproc.2014.06.028

Matete, N., & Trois, C. (2008). Towards zero waste in emerging countrie Sea South African experience. *Waste Management (New York, N.Y.)*, *28*(8), 1480–1492. doi:10.1016/j.wasman.2007.06.006 PMID:17714928

Mathur, P., Patan, S., & Shobhawat, S. (2012). Need of biomedical waste management system in hospitals-An emerging issue-a review. *Current World Environment*, *7*(1), 117–124. doi:10.12944/CWE.7.1.18

Mato, Y., Isobe, T., Takada, H., Kanehiro, H., Ohtake, C., & Kaminuma, T. (2001). Plastic resin pellets as a transport medium for toxic chemicals in the marine environment. *Environmental Science & Technology*, *35*(2), 318–324. doi:10.1021/es0010498 PMID:11347604

Mattsson, K., Elyse, V., Malmendal, J. A., Linse, S., Hansson, L., & Cedervall, T. (2017). Brain damage and behavioural disorders in fish induced by plastic nanoparticles delivered through the food chain. *Scientific Reports*, *7*(1), 11452. doi:10.103841598-017-10813-0 PMID:28904346

Maudgal, S. (1995). Waste management in India. *J. India Assoc. Environ. Manag.*, *22*(3), 203–208.

Ma, Y., Wang, L., & Sun, X. (2011). Deinking technology and deinking agent of waste paper. *Advanced Materials Research*, *335-336*, 1340–1343.

Maya, D. M. Y., Sarmiento, A. L. E., Oliveira, C. A. V. B. S., Lora, E. E. S., & Andrade, R. V. (2016). Gasification of municipal solid waste for power generation in Brazil, a review of available technologies and their environmental benefits. *Journal of Chemistry and Chemical Engineering*, *10*, 249–255.

Mbongwe, B., Mmereki, B. T., & Magashula, A. (2008). Healthcare waste management: Current practices in selected healthcare facilities, Botswana. *Waste Management (New York, N.Y.)*, *28*(1), 226–233. doi:10.1016/j.wasman.2006.12.019 PMID:17350817

McCalla, T. M., Paterson, J. R., & Lue-Hing, C. (1977). *Properties of agricultural and municipal wastes. In soils for management of organic waste and wastewater* (R. J. Stevenson, Ed.). Madison, WI: Amer. Soc. Agron.

Mccormick, K., & Kautto, N. (2017). The bioeconomy in Europe: an overview. *Sustainability*, *5*(6), 2589-2608. Disponível em: http://www.mdpi.com/2071-1050/5/6/2589/html

McCormick, K., & Kautto, N. (2013). The bioeconomy in Europe: An overview. *Sustainability*, *5*(6), 2589–2608. doi:10.3390u5062589

McDermid, K. J., & McMullen, T. L. (2004). Quantitative analysis of small-plastic debris on beaches in the Hawaiian archipelago. *Marine Pollution Bulletin*, *48*(7-8), 790–794. doi:10.1016/j.marpolbul.2003.10.017 PMID:15041436

McDougall, F. R., White, P. R., Franke, M., & Hindle, P. (2001). *Integrated solid waste management: a life cycle inventory* (2nd ed.). Blackwell. doi:10.1002/9780470999677

McDougall, F. R., White, P., Francke, M., & Hindle, P. (2001). *Integrated Solid Waste Management – A Lifecycle Inventory 2*. London: Blackwell.

McKendry, P. (2002). Energy production from biomass (part 2): Conversion technologies. *Bioresource Technology*, *83*(1), 47–54. doi:10.1016/S0960-8524(01)00119-5 PMID:12058830

McQuillan, H. (2011). Zero-waste design practice: Strategies and risk taking for garment design. In A. Gwilt & T. Rissanen (Eds.), Shaping sustainable fashion: Changing the way we make and use clothes. London: Earthscan.

McQuillan, H., & Rissanen, T. (2011). *YIELD exhibition catalogue*. Brooklyn, NY: Textile Arts Center. Retrieved August 20, 2012, from http://yieldexhibition.com/yieldexhibitioncatalogue.pdf

Mead, D. (2007). Forest for energy and the role of planted trees. *Critical Reviews in Plant Sciences*, *24*(5-6), 407-421, 2007. Disponível em: http://www.tandfonline.com/doi/abs/10.1080/07352680500316391

Meadows, D. (1999). *Leverage points: places to intervene in a system*. Sustainability Institute. Retrieved from http://www.sustainabilityinstitute.org

Medina, M. (2010). *Solid wastes, poverty and the environment in developing country cities: challenges and opportunities*. Helsinki: WIDER. Disponível em: https://www.wider.unu.edu/sites/default/files/wp2010-23.pdf

Medipally, S. R., Yusoff, F. M., Banerjee, S., & Shariff, M. (2015). Microalgae as sustainable renewable energy feedstock for biofuel production. *BioMed Research International*. PMID:25874216

Meeker, J. D., Sathyanarayana, S., & Swan, S. H. (2009). Phthalates and other additives in plastics: Human exposure and associated health outcomes. *Philosophical Transactions of the Royal Society of London. Series B, Biological Sciences*, *364*(1526), 2097–2113. doi:10.1098/rstb.2008.0268 PMID:19528058

Mehta, K., & Duhan, J. S. (2014). Production of invertase from Aspergillusniger using fruit peel waste as a substrate. *International Journal of Pharma and Bio Sciences*, *5*(2), 353–360.

Mehta, P. (2019). E-Waste, Chemical Toxicity, and Legislation in India. In *Advanced Methodologies and Technologies in Engineering and Environmental Science* (pp. 144–156). IGI Global.

Meidl, J. A. (1997). Responding to changing conditions: How powdered activated carbon systems can provide the operational flexibility necessary to treat contaminated groundwater and industrial wastes. *Carbon*, *35*(9), 1207–1216. doi:10.1016/S0008-6223(97)00093-6

Melanie, C., & Victoria, K. (2014). Expanding zero-waste design practices: A discussion paper, International Journal of Fashion Design. *Technology and Education*, *7*(1), 58–64. doi:10.1080/17543266.2013.837967

Mesina, M. B., De Jong, T. P. R., & Dalmijn, W. L. (2003). Improvements in separation of non-ferrous scrap metals using an electromagnetic sensor. *Physical Separation in Science and Engineering*, *12*(2), 87–101.

Messerle, V. E., Mosse, A. L., & Ustimenko, A. B. (2018). Processing of biomedical waste in plasma gasifier. *Waste Management (New York, N.Y.)*, *79*, 791–799. doi:10.1016/j.wasman.2018.08.048 PMID:30343813

Metcalfe, A., Riley, M., Barr, S., Tudor, T., Robinson, G., & Guilbert, S. (2012). Food waste bins: Bridging infrastructures and practices. *The Sociological Review*, *60*, 135–155.

Miandad, R., Rehan, M., Nizami, A., El-FetouhBarakat, M. A., & Ismail, I. M. (2016). The Energy and Value-Added Products from Pyrolysis of Waste Plastics. In Recycling of Solid Waste for Biofuels and Bio-chemicals. Environmental Footprints and Eco-design of Products and Processes, (pp. 333–355). Academic Press.

Michler, G. H., & Balta-Calleja, F. J. (2016). *Mechanical Properties of Polymers Based on Nanostructure and Morphology* (Vol. 71). Roca Raton, FL: CRC Press. doi:10.1201/9781420027136

Mickael, D. (2016). Categorization and Sorting for Waste Management. *International Journal of Waste Resources*, *6*(2).

Milinković, M., Lalević, B., Jovičić-Petrović, J., Golubović-Ćurguz, V., Kljujev, I., & Raičević, V. (2019). Biopotential of compost and compost products derived from horticultural waste—Effect on plant growth and plant pathogens' suppression. *Process Safety and Environmental Protection*, *121*, 299–306. doi:10.1016/j.psep.2018.09.024

Min, B., Cheng, S., & Logan, B. E. (2005). Electricity generation using membrane and salt bridge microbial fuel cells. *Water Research*, *39*(9), 1675–1686. doi:10.1016/j.watres.2005.02.002 PMID:15899266

Ministry of Environment and Forests (MoEF). (2015). Municipal solid waste (Management and Handling) rules. The Gazette of India.

Ministry of New and Renewable Energy (MNRE). (2016-17). *Annual report*. Author.

Mirabella, N., Castellani, V., & Sala, S. (2014). Current options for the valorization of food manufacturing waste: A review. *Journal of Cleaner Production*, *65*, 28–41.

Mishra, R. K., Singh, B. K., Upadhyay, R. K., & Singh, S. (2009). Technology for vermicompost production. *Indian Farming*.

Misra, V., & Pandey, S. D. (2005). Hazardous waste, impact on health and environment for development of better waste management strategies in future in India. *Environment International*, *31*(3), 417–431. doi:10.1016/j.envint.2004.08.005 PMID:15734194

Miyazaki, M., & Une, H. (2005). Infectious waste management in Japan: A revised regulation and a management process in medical institutions. *Waste Management*, *25*(6), 616-621.

Modak, P., Jiemian, Y., Hongyuan, Yu., & Mohanty, C. R. (2010). Municipal solid waste management: turning waste into resources. In Shanghai manual: a guide for sustainable urban development in the 21st century, (pp. 1–36). Academic Press.

Mohammad, N., Alam, M. Z., Kabbashi, N. A., & Ahsan, A. (2012). Effective composting of oil palm industrial waste by filamentous fungi: A review. *Resources, Conservation and Recycling*, *58*, 69–78. doi:10.1016/j.resconrec.2011.10.009

Mohanty, C. R., Mishra, U., & Beuria, P. R. (2014). Municipal solid waste management in Bhubaneswar, India — a review. *International Journal of Latest Trends in Engineering & Technology, 3*(3), 303–312.

Mohanty, C. R., Mishra, U., & Beuria, P. R. (2014). Municipal solid waste management in Bhubaneswar, India— a review. *Int. J. Latest Trends Eng. Technol., 3*(3), 303–312.

Mohee, R., & Mudhoo, A. (2005). Analysis of the physical properties of an in-vessel composting matrix. *Powder Technology, 155*(1), 92–99. doi:10.1016/j.powtec.2005.05.051

Mohite, B. J. (2013). Issues and strategies in managing E-waste in India. *Indian Journal of Research in Management. Business and Social Sciences, 1*, 46–51.

Moh, Y. C., & Manaf, L. A. (2017). Solid waste management transformation and future challenges of source separation and recycling practice in Malaysia. *Resources, Conservation and Recycling, 116*, 1–14. doi:10.1016/j.resconrec.2016.09.012

Moldoveanu, A. M. (2014). Assessment of Soil Pollution with Heavy Metals in Romania: Environmental Risk Assessment of Soil Contamination. INTECH.

Monreal, C. (2008). *La utilización de los residuos agrícolas y otros desechos para la producción de bioenergía, biocombustibles y bioproductos*. Ottawa: ECORC.

Montaño Arango, O., Corona Armenta, J. R., & Montelongo Reyes, M. (2009). Metodología sistémica para el desarrollo de un proyecto de Biogás. *Congreso Internacional de Investigación en Ciencias Administrativas, 13.* Disponível em: https://www.uaeh.edu.mx/investigacion/productos/5173/prod_12.pdf

Moore, C. J. (2008). Synthetic polymers in the marine environment: A rapidly increasing, long-term threat. *Environmental Research, 108*(2), 131–139. doi:10.1016/j.envres.2008.07.025 PMID:18949831

Moore, C. J., Moore, S. L., Leecaster, M. K., & Weisberg, S. B. (2001). A comparison of plastic and plankton in the North Pacific central gyre. *Marine Pollution Bulletin, 42*(12), 1297–1300. doi:10.1016/S0025-326X(01)00114-X PMID:11827116

Moore, S. A. (2012). Garbage matters: Concepts in new geographies of waste. *Progress in Human Geography, 36*(6), 780–799. doi:10.1177/0309132512437077

Moreno, E. L., Bazoglu, N., & Mboup, G. (2008). State of the World's Cities 2008/2009: Harmonious Cities. Earthscan.

Moreno, E. L., Bazoglu, N., Mboup, G., & Warah, R. (2008). *State of the World's Cities 2008/2009: Harmonious Cities.* London: Earthscan.

Morgan, M., & Burrows, I. (1982). *Earthworms/Microorganisms Interactions.* Rothamsted Experimental Station.

Morita, M., & Sasaki, K. (2012). Factors influencing the degradation of garbage in methanogenic bioreactors and impacts on biogas formation. Applied Microbiology and Biotechnology, 94(3), 575-582.

Mor, S., Ravindra, K., Dahiya, R. P., & Chandra, A. (2006). Leachate characterization and assessment of groundwater pollution near municipal solid waste landfill site. *Environmental Monitoring and Assessment, 118*(1-3), 435–456. PMID:16897556

Mostafa, G. M., Shazly, M. M., & Sherief, W. I. (2009). Development of a waste management protocol based on assessment of knowledge and practice of healthcare personnel in surgical departments. *Waste Management, 29*(1), 430-439.

Mouat, T., Lopez-Lozano, R., & Bateson, H. (2010). *Economic Impacts of Marine Litter.* KIMO.

Moustakas, K., & Loizidou, M. (2010). Solid waste management through the application of thermal methods. *Waste Manage*. Retrieved from http://www. intechopen.com/books/waste-management/solid-wastemanagement-through-the-application-of-thermal-methods

Munch, D. (1993). Concentration profiles of arsenic, cadmium, chromium, copper, lead, mercury, nickel, zinc, vanadium and polynucleated aromatic hydrocarbons in forest soil beside an urban road. *The Science of the Total Environment*, *138*(1-3), 47–55. doi:10.1016/0048-9697(93)90404-T

Munnoli, P. M. (2007). *Management of industrial organic solid wastes through vermiculture biotechnology with special reference to microorganisms*. Goa University.

Munroe, G. (2004). *Manual of on-farm vermicomposting and vermiculture*. Academic Press.

Murphy, J. D., & McKeogh, E. (2004). Technical, economic and environmental analysis of energy production from municipal solid waste. *Renewable Energy*, *29*(7), 1043–1057. doi:10.1016/j.renene.2003.12.002

Murphy, S., & Pincetl, S. (2013). Zero waste in Los Angeles: Is the emperor wearing any clothes? *Resources, Conservation and Recycling*, *81*, 40–51. doi:10.1016/j.resconrec.2013.09.012

Mushimiyimana, I., &Tallapragada, P. (2016). *Bioethanol production from agro wastes by acid hydrolysis and fermentation process*. Academic Press.

Mutha, N. H., Patel, M., & Premnath, V. (2006). Plastics materials flow analysis for India. *Resources, Conservation and Recycling*, *47*(3), 222–244. doi:10.1016/j.resconrec.2005.09.003

Muthukumaravel, K., Amsath, A., & Sukumaran, M. (2008). Vermicomposting of vegetable wastes using cow dung. *E-Journal of Chemistry*, *5*(4), 810–813. doi:10.1155/2008/572431

Nagarajan, R., & Thirumalaisaumalaisamy, S., Lakshumanan, & Eakshumanan, E. (2012). Impact of leachate on groundwater pollution due to non-engineered municipal solid waste landfill sites of erode city, Tamil Nadu, India. *Iranian Journal of Environmental Health Sciences & Engineering*, *9*(1), 1–12. PMID:23369323

Nagavallemma, K. P., Wani, S. P., Lacroix, S., Padmaja, V. V., Vineela, C., Rao, M. B., & Sahrawat, K. L. (2004). *Vermi-composting: Recycling wastes into valuable organic fertilizer*. Global Theme on Agroecosystems Report no. 8.

Nagy, A., & Kuti, R. (2016).. . *The Environmental Impact of Plastic Waste Incineration*, *15*(3), 231–237.

Najafi, G., Ghobadian, B., Tavakoli, T., & Yusaf, T. (2009). Potential of bioethanol production from agricultural wastes in Iran. *Renewable & Sustainable Energy Reviews*, *13*(6-7), 1418–1427. doi:10.1016/j.rser.2008.08.010

Nambiar, K. K. M. (1990). Long-term fertility effects on wheat productivity. In . *International Conference on Wheat for the Nontraditional Warm Areas*. CIMMYT.

Nandan, A., Yadav, B. P., Baksi, S., & Bose, D. (2017). Recent Scenario of Solid Waste Management in India. *World Scientific News*, *66*, 56–74.

Narayana, T. (2009). Municipal solid waste management in India: From waste disposal to recovery of resources. *Waste Management (New York, N.Y.)*, *29*(3), 1163–1166. PMID:18829290

Narkhede, S. D., Attarde, S. B., & Ingle, S. T. (2011). Study on effect of chemical fertilizer and vermicompost on growth of chilli pepper plant (*Capsicum annum*). *Journal of Applied Science in Environmental Sanitation*, *6*(3), 327–332.

Nasir, I., Ghazi, I., & Omar, R. (2012). Production of biogas from solid organic wastes through anaerobic digestion: a review. Applied Microbiology and Biotechnology, 95(2), 321-329.

Nath, G., Singh, K., & Singh, D. K. (2009). Chemical analysis of vermicomposts/vermiwash of different combinations of animal, agro and kitchen wastes. *Australian Journal of Basic and Applied Sciences*, *3*(4), 3671–3676.

Naveen, B. P., Sivapullaiah, P. V., & Sitharam, T. G. (2013). Disposal options for solid waste of Bangalore city base on its characteristics. *International Journal of Environment and Waste Management*, *12*(1), 77–88.

Ndegwa, P. M., & Thompson, S. A. (2001). Integrating composting and vermicomposting in the treatment and bioconversion of bio solids. *Bioresource Technology*, *76*(2), 107–112. doi:10.1016/S0960-8524(00)00104-8 PMID:11131792

Ndegwa, P. M., Thompson, S. A., & Das, K. C. (2000). Effect of stocking density and feeding rate on vermicomposting of bio solids. *Bioresource Technology*, *7*(1), 5–12. doi:10.1016/S0960-8524(99)00055-3

Neelagund, S. E., Ingalhalli, S. S., Savanurmath, C. J., Hinchigeri, S. B., & Hiremath, M. B. (2007). Purification and characterization of antiviral protein from fecal matter of silkworm. *Journal of Environmental Sciences (China)*, *5*, 77–85.

Negi, S., Dhar, H., Hussain, A., & Kumar, S. (2018). Biomethanation potential for co-digestion of municipal solid waste and rice straw: A batch study. *Bioresource Technology*, *254*, 139–144. doi:10.1016/j.biortech.2018.01.070 PMID:29413914

Neklyudov, A. D., Fedotov, G. N., & Ivankin, A. N. (2006). Aerobic processing of organic waste into composts. *Applied Biochemistry and Microbiology*, *42*(4), 341–353. doi:10.1134/S0003683806040016

Neklyudov, A., Fedotov, G., & Ivankin, A. (2008). Intensification of composting processes by aerobic microorganisms: A review. *Applied Biochemistry and Microbiology*, *44*(1), 6–18. doi:10.1134/S000368380801002X

Nelms, S. E., Coombes, C., Foster, L. C., Galloway, T. S., Godley, B. J., Lindeque, P. K., & Witt, M. J. (2017). Marine Anthropogenic Litter on British beaches: A 10-year Nationwide Assessment using Citizen Science Data. *The Science of the Total Environment*, *579*, 1399–1409. doi:10.1016/j.scitotenv.2016.11.137 PMID:27913017

Nema, A. K. (2004). *Collection and transport of municipal solid waste. Training Program on Solid Waste Management*. Delhi, India: Springer.

Nema, A., Pathak, A., Bajaj, P., Singh, H., & Kumar, S. (2011). A case study: Biomedical waste management practices at city hospital in Himachal Pradesh. *Waste Management & Research*, *29*(6), 669–673. doi:10.1177/0734242X10396753 PMID:21382874

Nema, S. K., & Ganeshprasad, K. S. (2002). Plasma pyrolysis of medical waste. *Current Science*, 271–278.

Nemathaga, F., Maringa, S., & Chimuka, L. (2008). Hospital solid waste management practices in Limpopo Province, South Africa: A case study of two hospitals. *Waste Management (New York, N.Y.)*, *28*(7), 1236–1245. doi:10.1016/j.wasman.2007.03.033 PMID:18060760

Ngoc, H. N., Agusa, T., Ramu, K., Phuc, C. T. N., Murata, S., & Bulbule, K. A. (2009). Contamination by trace elements at E-waste recycling sites in Bangalore, India. *Chemosphere*, *76*(1), 9–15. doi:10.1016/j.chemosphere.2009.02.056 PMID:19345395

Ngoc, U. N., & Schnitzer, H. (2009). Sustainable solutions for solid waste management in Southeast Asian countries. *Waste Management (New York, N.Y.)*, *29*(6), 1982–1995. doi:10.1016/j.wasman.2008.08.031 PMID:19285384

Ngwuluka, N., Ochekpe, N., Odumosu, P., & John, S. A. (2009). Waste management in healthcare establishments within Jos Metropolis, Nigeria. *African Journal of Environmental Science and Technology*, *3*(12).

Nic, M., Jirat, J., & Kosata, B. (Eds.). (2006). esters. IUPAC Compendium of Chemical Terminology. doi:10.1351/goldbook.E02219

Nicholson, J. (2017). *The Chemistry of Polymers*. London, UK: Royal Society of Chemistry.

Nicolau, L., Marcalo, A., Ferreira, M., Sa, S., Vingada, J., & Eira, C. (2016). Ingestion of marine litter by loggerhead sea turtles, Carettacaretta, in Portuguese continental waters. *Marine Pollution Bulletin*, *103*(1-2), 179–185. doi:10.1016/j.marpolbul.2015.12.021 PMID:26763321

Nidoni, P. G. (2017). Incineration Process for Solid Waste Management and Effective Utilization of By Products. *International Research Journal of Engineering and Technology*, *4*(12), 378–382.

Nielsen, M. C., DePuy, R. A., Kumar, A., Lyons, J. P. F., Lissianski, V., Shi, R., . . . Zamansky, V. (2015). *Plasma-assisted waste gasification system*. U.S. Patent 9,074,152.

Ni, H. G., & Zeng, E. Y. (2009). Law enforcement and global collaboration are the keys to containing e-waste tsunami in China. *Environmental Science & Technology*, *43*(11), 3991–3994. doi:10.1021/es802725m PMID:19569320

Nkwachukwu, O. I., Chidi, N. I., & Charles, K. O. (2010). Issues of roadside disposal habit of municipal solid waste, environmental impacts and implementation of sound management practices in developing country "Nigeria." *International Journal of Environmental Sciences and Development*, *5*, 409–418.

Noorjahan, A., Dhanakumar, S., Mohanraj, R., & Ravichandran, M. (2012). Status of heavy metals distribution in municipal solid waste in Tiruchirappalli City, India. *International Journal of Applied Biology and Pharmaceutical Technology*, *3*(3), 252–258.

Nording, M., Nichkova, M., Spinnel, E., Persson, Y., Gee, S. J., Hammock, B. D., & Haglund, P. (2006). Rapid screening of dioxin-contaminated soil by accelerated solvent extraction/purification followed by immunochemical detection. *Analytical and Bioanalytical Chemistry*, *385*(2), 357–366. doi:10.100700216-006-0402-9 PMID:16598456

Norman, Q. (2004). Effects of vermicompost's produced from food waste on the growth and yields of greenhouse peppers. *Bioresource Technology*, *93*(2), 139–144. doi:10.1016/j.biortech.2003.10.015 PMID:15051075

North, D. C. (1990). *Institutions, institutional change and economic performance*. Cambridge, UK: Cambridge University Press. doi:10.1017/CBO9780511808678

Nwachukwu, N. C., Orji, F. A., & Ugbogu, O. C. (2013). Health care waste management–public health benefits, and the need for effective environmental regulatory surveillance in federal Republic of Nigeria. In Current topics in public health. InTechOpen.

O'Brine, T., & Thompson, R. C. (2010). Degradation of plastic carrier bags in the marine environment. *Marine Pollution Bulletin*, *60*(12), 2279–2283. doi:10.1016/j.marpolbul.2010.08.005 PMID:20961585

O'Rorke, M. (1988). *Public information campaign on precycling*. Prepared for City of Berkeley.

OECD. (2001). Extended Producer Responsibility: A Guidance Manual for Governments. Organization for Economic Cooperation and Development.

Oehlmann, J., Schulte-Oehlmann, U., Kloas, W., Jagnytsch, O., Lutz, I., Kusk, K. O., ... Tyler, C. R. (2009). A critical analysis of the biological impacts of plasticizers on wildlife. *Philosophical Transactions of the Royal Society of London. Series B, Biological Sciences*, *364*(1526), 2047–2062. doi:10.1098/rstb.2008.0242 PMID:19528055

Ogawa, H. (2000). Sustainable solid waste management in developing countries. In *7th ISWA International Congress and Exhibition*. World health Organization.

Ogwueleka, T. C. (2009). Municipal solid waste characteristics and management in Nigeria. *Iranian Journal of Environmental Health Sciences & Engineering*, *6*(3).

Olah, G. A., Goeppert, A., & Prakash, G. S. (2008). Chemical recycling of carbon dioxide to methanol and dimethyl ether: From greenhouse gas to renewable, environmentally carbon neutral fuels and synthetic hydrocarbons. *The Journal of Organic Chemistry*, *74*, 487–498. PMID:19063591

Olajire, A. A., & Ayodele, E. T. (1997). Contamination of roadside soil and grass with heavy metals. *Environment International*, *23*(1), 91–101. doi:10.1016/S0160-4120(96)00080-3

Oli, A. N., Ekejindu, C. C., Adje, D. U., Ezeobi, I., Ejiofor, O. S., Ibeh, C. C., & Ubajaka, C. F. (2016). Healthcare waste management in selected government and private hospitals in Southeast Nigeria. *Asian Pacific Journal of Tropical Biomedicine*, *6*(1), 84–89. doi:10.1016/j.apjtb.2015.09.019

Omar, D., Nazli, S. N., Subramaniam, A., & Karuppannan, L. (2012). Clinical waste management in district hospitals of Tumpat, Batu Pahat and Taiping. *Procedia: Social and Behavioral Sciences*, *68*, 134–145. doi:10.1016/j.sbspro.2012.12.213

Omran, A., & Mahmood, A., AbdulAziz, H., & Robinson, G.M. (2009). Investigatinghouseholds attitude towards recycling of solid waste in Malaysia:a case study. *International Journal of Environmental of Research*, *3*, 275–288.

Onanuga, M. Y., & Odunsi, O. (2018). Health is wealth: Concern for households' solid waste self-disposal practices. *Environmental Quality Management*, 1–9.

Onifade, A. A., & Babatunde, G. M. (1998). Comparison of theutilization of palm kernel meal, brewer's dried grain andmaize offals by broiler chicks. *British Poultry Science*, *39*(2), 245–250. doi:10.1080/00071669889204 PMID:9649879

Onyari, J. M., Wandiga, S. O., & Njentede, J. O. (1991). Lead contamination in street soils of Nairoubi city, Mombassa Island, Kenya. *Bulletin of Environmental Contamination and Toxicology*, *46*(5), 782–789. doi:10.1007/BF01689968 PMID:2039875

Orozco, F. H., Cegarra, J., Trvjillo, L. M., & Roig, A. (1996). Vermicomposting of coffee pulp using the earthworm *Eisenia foetida*: Effects on C and N contents and the availability of nutrients. *Biology and Fertility of Soils*, *22*(1-2), 162–166. doi:10.1007/BF00384449

Ostrander, R., & Clark, S. C. (1991). Soil heavy metal concentrations and erosion damage in upland Grasslands in the Pennines, England. *Water, Air, and Soil Pollution*, *59*(1-2), 81. doi:10.1007/BF00283173

Othman, I., Al-Qudat, M., & Al-Masri, M. S. (1997). Lead levels in roadside soils and vegetation of Damascus city. *The Science of the Total Environment*, *207*(1), 43–48. doi:10.1016/S0048-9697(97)00243-X PMID:9397598

Othman, N., Irwan, J. M., & Roslan, M. A. (2012). Vermicomposting of food waste. *InternationalJournal of Integrated Engineering*, *4*(2), 39–48.

Oweis, R., Al-Widyan, M., & Ohood Al-Limoon, O. (2005). *Medical waste management in Jordan: a study at the King Hussein*. Academic Press.

Pacey, J., Augenstein, D., Morck, R., Reinhart, D., & Yazdani, R. (1999). The bioreactor landfill-an innovation in solid waste management. *Municipal Solid Waste Management*, 53-60.

Paepatung, N., Nopharatana, A., & Songkasiri, W. (2009). Bio-methane potential of biological solid materials and agricultural wastes. *Asian Journal on Energy and Environment*, *10*(1), 19–27.

Palmer, P. (2004). *Getting to Zero Waste*. Sebastopol, CA: Purple Sky Press.

Palsania, J., Sharma, R., Srivastava, J. K., & Sharma, D. (2008). Effect of moisture content variation over kinetic reaction rate during vermicomposting process. *Applied Ecology and Environmental Research*, *6*(2), 49–61. doi:10.15666/aeer/0602_049061

Pande, K. S., & Sharma, S. O. (1999). Distribution of organic matter and toxic metals in the sediments of Ramganga river at Moradabad. *Pollution Research, 18*, 43–47.

Pandey, P. K., Cao, W., Biswas, S., & Vaddella, V. (2016). A new closed loop heating system for composting of green and food wastes. *Journal of Cleaner Production, 133*, 1252–1259. doi:10.1016/j.jclepro.2016.05.114

Pang, L. I., Scorpio, D. E., & Kibert, C. J. (2010). Strategies for successful construction and demolition waste recycling operations. *Journal of Construction Management and Economics, 15*, 49–50.

Paolini, V. (2018). Environmental impact of biogas: A short review of current knowledge. *J. Environ. Sci. Heal. Part A*, 1–8, doi:10.1080/10934529.2018.1459076

Papadimitriou, E. K., Barton, J. R., & Stentiford, E. I. (2008). Sources and levels of potentially toxic elements in the biodegradable fraction of autoclaved non-segregated household waste and its compost/digestate. *Waste Management & Research, 26*(5), 419. PMID:18927061

Parajuli, R., Dalgaard, T., Jørgensen, U., Adamsen, A. P. S., Knudsen, M. T., Birkved, M., & Schjørring, J. K. (2015). Biorefining in the prevailing energy and materials crisis: A review of sustainable pathways for biorefinery value chains and sustainability assessment methodologies. *Renewable & Sustainable Energy Reviews, 43*, 244–263. doi:10.1016/j.rser.2014.11.041

Parihar, D. K. (2012). Production of lipase utilizing linseed oilcake as fermentation substrate. *Int. J. Sci. Environ. Technol, 1*(3), 135–143.

Park, J. H., Lee, D. Y., Yun, P., Yeon, S. W., Ko, J. H., Kim, Y. S., & Baek, N. I. (2011). Flavonoids from silkworm droppingsand their promotional activities on hemeoxygenase-1. *Journal of Asian Natural Products Research, 13*(4), 377–382. doi:10.1080/10286020.2011.555331 PMID:21462044

Park, S. W., Lee, J. S., Yang, W. S., Kang, J. J., Sung, J. H., Alam, M. T., ... Oh, J. H. (2016). For Waste to Energy, Assessment of Fluff Type Solid Refuse Fuel by Thermal Characteristics Analyses International Conference on Solid Waste Management, 5IconSWM 2015. *Procedia Environmental Sciences, 35*, 498–505.

Park, Y. J., Kim, W. S., Ko, S. H., Lim, D. S., Lee, H. J., Lee, W. Y., & Lee, D. W. (2003). Separation and characterization ofchlorophyll degradation products insilkworm using HPLC-UV-APCI-MS. *Journal of Liquid Chromatography & Related Technologies, 26*(19), 3183–3197. doi:10.1081/JLC-120025517

Pasupathi, P., Sindhu, S., & Ponnusha, B. S. (2011). Biomedical waste management for health care industry: A review. *International Journal of Biological & Medical Research, 2*(2), 472–486.

Pathma, J., & Sakthivel, N. (2012). Microbial diversity of vermicompost bacteria that exhibit useful agricultural traits and waste management potential. *SpringerPlus, 1*(26), 2–20. PMID:23961356

Pathma, J., & Sakthivel, N. (2012). Microbialdiversity of vermicompost bacteria that exhibit useful agricultural traits and waste management potential. *SpringerPlus*, 1. PMID:23961356

Patil, B. R., Singh, K. K., Pawar, S. E., Maarse, L., & Otte, J. (2009). *Sericulture: an alternative source of income to enhance the livelihoods of small-scale farmers and tribal communities.* Pro-Poor Livestock Policy Initiative. A Living from Livestock Research Report 09-03.

Patil, A. D., & Shekdar, A. V. (2001). Health-care waste management in India. *Journal of Environmental Management, 63*(2), 211–220. doi:10.1006/jema.2001.0453 PMID:11721600

Patil, G. V., & Pokhrel, K. (2005). Biomedical solid waste management in an Indian hospital: A case study. *Waste Management (New York, N.Y.), 25*(6), 592–599. doi:10.1016/j.wasman.2004.07.011 PMID:15993343

Patil, S. R., Amena, S., Vikas, A., Rahul, P., Jagadeesh, K., & Praveen, K. (2013). Utilization of silkworm litter and pupal waste-an eco-friendly approach for mass production of Bacillus thuringiensis. *Bioresource Technology*, *131*, 545–547. doi:10.1016/j.biortech.2012.12.153 PMID:23403062

Pati, R. K., Vrat, P., & Kumar, P. (2008). A goal programming model for paper recycling system. The international journal of management science. *Omega*, *36*, 405–417.

Patle, A. V., Williams, S. P. M. P., Gabhane, J., Dhar, H., & Nagamaik, P. B. (2014). Microbial assisted rapid composting of agriculture residues. *International Journal of Scientific and Engineering Research*, *5*(5).

Pattnaik, S., & Reddy, M. V. (2009). Assesment of municipal solid waste management in Puducherry (Pondicherry), India. *Resources, Conservation and Recycling*, *54*(8), 512–520. doi:10.1016/j.resconrec.2009.10.008

Patwary, M. A., O'Hare, W. T., & Sarker, M. H. (2011). Assessment of occupational and environmental safety associated with medical waste disposal in developing countries: A qualitative approach. *Safety Science*, *49*(8-9), 1200–1207. doi:10.1016/j.ssci.2011.04.001

Pavani & Rajeswari. (2014). Impact of Plastics On Environmental Pollution. *Journal of Chemical and Pharmaceutical Sciences.*

Pehren, H. R., & Clark, C. S. (1987). Microorganisms in municipal solid waste and public health implications. *Critical Reviews in Environmental Control*, *17*(3), 187–228. doi:10.1080/10643388709388334

Peng, M. (2012). *Global business*. Dallas, TX: Cengage Learning.

Perrin, D., & Barton, J. (2001). Issues associated with transforming household attitudes and opinions into materials recovery: A review of two kerbside recycling schemes. *Resources, Conservation and Recycling*, *33*(1), 61–74. doi:10.1016/S0921-3449(01)00075-1

Peterson, C. (2013). *Assessment of solid waste management practices and its vulnerability to climate risks in Maldives Tourism Sector*. Retrieved from http://tourism.gov.mv/downloads/tap/2014/ Solid_Waste.pdf

Phillips, P. S., Tudor, T., Bird, H., & Bates, M. (2011). A critical review of a key waste strategy initiative in England: Zero waste places projects 2008-2009. *Resources, Conservation and Recycling*, *55*(3), 335–343. doi:10.1016/j.resconrec.2010.10.006

Piccone, G., Biosoil, B., Deluca, G., & Minelli, L. (1986). Vermicomposting of different organic wastes. Udine, Italy: Academic Press.

Pietrzak, W., Kawa-Rygielska, J., Król, B., Lennartsson, P. R., & Taherzadeh, M. J. (2016). Ethanol, feed components and fungal biomass production from field bean (Vicia faba var. equina) seeds in an integrated process. *Bioresource Technology*, *216*, 69–76. doi:10.1016/j.biortech.2016.05.055 PMID:27233099

Pino Pérez, F. (2013). *Asociación público-privada entre Bioeléctrica S. A. de C. V. y el gobierno del Estado de Nuevo León Bioenergía de Nuevo León S. A. de C. V.* Monterrey: Tecnológico de Monterrey. Disponível em: http://www.cca.org.mx/ps/funcionarios/muniapp/descargas/Documentos_de_apoyo/informaciontematica/capp/APP_Bioelectrica.pdf

Pires, A., Martinho, G., & Chang, N.-B. (2010). Solid waste management in European countries: A review of systems analysis techniques. *Journal of Environmental Management*, *92*(4), 1033–1050. doi:10.1016/j.jenvman.2010.11.024 PMID:21194829

Pittet, D., Allegranzi, B., Storr, J., & Donaldson, L. (2006). 'Clean care is safer care': The global patient safety challenge 2005–2006. *International Journal of Infectious Diseases*, *10*(6), 419–424. doi:10.1016/j.ijid.2006.06.001 PMID:16914344

Planning Commission, Government of India. (2014). *Report of the task Force on waste to energy (Volume I) in the context of integrated municipal solid waste management*. Retrieved from http://planningcommission.nic.in/reports/genrep/rep_ wte1205.pdf

Plastic Coding Guidelines in the United States. (2017). Available online: http://www.natureworksllc.com/~/media/The_Ingeo_Journey/EndofLife_Options/plastic_codes/2008_04_10_plastic_code_guidelines_pdf.pdf

Plastics Europe. (2016). *Plastics - the Facts 2016. An analysis of European plastics production, demand and waste data*. Brussels, Belg.: Plast. Eur.

Plavac, B., Filipan, V., Sutlovic, I., & Sveticic, J. (2017). Sustainable waste management with mechanical biological treatment and energy utilization. *Technical Gazette*, *24*(4), 1283–1289.

Poon, C. S., Yu, A. T., & Jaillon, L. (2004). Reducing building waste at construction sites in Hong Kong. *Construction Management and Economics*, *22*(5), 461–470. doi:10.1080/0144619042000202816

Population Reference Bureau. (2017). *World Population Data Sheet*. PRB Publ. 1–22.

Porta, D., Milani, S., Lazzarino, A. I., Perucci, C. A., & Forastiere, F. (2009). Systematic review of epidemiological studies on health effects associated with management of solid waste. *Environmental Health*, *8*(1), 60. doi:10.1186/1476-069X-8-60 PMID:20030820

Porter, M. E. (1991). *La ventaja competitiva de las naciones*. Barcelona: Plaza & Janés.

Porter, M. E., & Kramer, M. R. (2006). Strategy and society: The link between competitive advantage and corporate social responsibility. *Harvard Business Review*, *84*(12), 78–92. PMID:17183795

Porter, M. E., & Van der Linde, C. (1995). Toward a new conception of the environment-competitiveness relationship. *The Journal of Economic Perspectives*, *9*(4), 97–118. doi:10.1257/jep.9.4.97

Poskrobko, B., & Poskrobko, T. (2012). *Environment management in Poland*. Bialystok: Bialystok University Press.

PPCB (Punjab Pollution Control Board). (2010). *Status report on municipal solid waste in Punjab*. Punjab Pollution Control Board, Patiala. Retrieved from http://www. ppcb.gov.in/Attachments/Annual%20Reports/AR.pdf

Pramanik, P., Ghosh, G. K., Ghosal, P. K., & Banik, P. (2007). Changes in organic-C, N, P and K and enzyme activities in vermicompost of biodegradable organic wastes under liming and microbial inoculants. *Bioresource Technology*, *98*(13), 2485–2494. doi:10.1016/j.biortech.2006.09.017 PMID:17081750

Price, A. (2005). UK power: An ever changing challenge for civil engineers. *Proceedings of the Institution of Civil Engineers. Civil Engineering*, *158*(6), 4–11.

Pruss, A., & Townend, W. K. (1998). *Teacher's guide: management of wastes from health-care activities* (No. WHO/EOS/98.6). Geneva: World Health Organization.

Pruss-Ustün, A., Rapiti, E., & Hutin, Y. (2005). Estimation of the global burden of disease attributable to contaminated sharps injuries among health-care workers. *American Journal of Industrial Medicine*, *48*(6), 482–490. doi:10.1002/ajim.20230 PMID:16299710

Prusty, B. G., Sahu, K. C., & Godgul, G. (1994). Metal contamination due to mixing and milling activities at Zamar zinc mine, Rajasthan, India (Part I). *Chemical Geology*, *112*(3-4), 275–291. doi:10.1016/0009-2541(94)90029-9

Przybylinska, P. A., & Wyszkowski, M. (2016). Environmental contamination with phthalates and its impacts on living organisms. *Ecological Chemistry and Engineering. S*, *23*(2), 347–356. doi:10.1515/eces-2016-0024

Punjab State Council for Science and Technology Chandigarh. (2010). Retrieved on September 30, 2010 from http://agri.and.nic.in/vermi_culture.htm

Qian, X., Shen, G., Wang, Z., Guo, C., Liu, Y., Lei, Z., & Zhang, Z. (2014). Co-composting of livestock manure with rice straw: Characterization and establishment of maturity evaluation system. *Waste Management (New York, N.Y.)*, *34*(2), 530–535. doi:10.1016/j.wasman.2013.10.007 PMID:24188923

Quaik, S., Embrandiri, A., Rupani, P. F., Singh, R. P., & Ibrahim, M. H. (2012). Effect of vermiwash and vermicomposting leachate in hydroponics culture of Indian Borage (Plectranthus ambionicus) plantlets. In *UMT 11th International Annual Symposium on Sustainability Science and Management* (pp. 210-214). Academic Press.

Quaik, S., & Ibrahim, M. H. (2013). A review on potential of vermicomposting derived liquids in agricultural use. *International Journal of Scientific and Research Publications*, *3*(3), 1–6.

Rachel, O. A., Komine, H., Yasuhara, K., & Murakami, S. (2009). *Municipal solid waste management in developed and developing countries-Japan and Nigeria as case studies.* Available at: http://wwwgeo.civil.ibaraki.ac.jp/komine/mypapers/JGSPaper/JGS2009(973)Rachel.pdf

Rada, E., Ragazzi, M., & Fedrizzi, P. (2013). Web-GIS oriented systems viability for municipal solid waste selective collection optimization in developed and transient economies. *Waste Management (New York, N.Y.)*, *33*(4), 785–792. PMID:23402896

Radha, G. (2002). *A Study of the Performance of the Indian IT Sector*. Available at: www.nautilus.org

Radha, K. V., Kalaivani, K., & Lavanya, R. (2009). A case study of biomedical waste management in hospitals. *Global Journal of Health Science*, *1*(1), 82–88.

Rafati, L. (2016). Waste-to-energy: Challenges and opportunities in Iran. *Journal of Environmental Health and Sustainable Development, Yazd, IR*, *1*(3), 163–172.

Rahimi, A., & García, J.M. (2017). Chemical recycling of waste plastics for new materials production. *Nat. Rev. Chem.*, 1.

Rai, R. D., & Ahlawat, O. P. (2002). Edible fungi: Biotechnological approaches. In *Applied Mycology and Biotechnology* (Vol. 2, pp. 87–121). Elsevier.

Rajan, R., Robin, D. T., & Vandanarani, M. (2018). Biomedical waste management in Ayurveda hospitals e current practices & future prospective. *Journal of Ayurveda and Integrative Medicine*, 1–8. PMID:29555257

Rajiv, K., Sinha, Agarwal, A., Chauhan, K., Chandran, V., & Brijal Kiranbhai Soni, B.K. (2010). Vermiculture Technology: Reviving the Dreams of Sir Charles Darwin for Scientific Use of Earthworms in Sustainable Development Programs. *Technology and Investment*, 155-172.

Rajkumar, N., Subramani, T., & Elango, L. (2010). Groundwater contamination due to municipal solid waste diposal — a GIS based study in Erode City. *International Journal of Environmental Science*, *1*(1), 39–55.

Rajput,, R., & Prasad,, G., & Chorea. (2009). Scenario of solid waste management in present Indian context. *Caspian Journal of Environment Science*, *7*(1), 45-53.

Rajput, R., Prasad, G., & Chopra, A. K. (2009). Scenario of solid waste management in present Indian context. *Journal of Environmental Sciences (China)*, *7*(1), 45–53.

Raju, S. (1996). Utilization of sericultural byproducts: A Chinese example. *IndianSilk*, *35*, 19–20.

Ramachandran, S., Patel, A. K., Nampoothiri, K. M., Francis, F., Nagy, V., Szakacs, G., & Pandey, A. (2004). Coconut oil cake—a potential raw material for the production of α-amylase. *Bioresource Technology*, *93*(2), 169–174. doi:10.1016/j.biortech.2003.10.021 PMID:15051078

Rampal, R. K., Kour, J., & Jamwal, R. (2002). Soild waste generation in government hospitals of Jammu city, India. *Pollution Research*, *21*, 39–43.

Rana, P. R., Yadav, D., Ayub, S., & Siddiqui, A. A. (2014). Status and challenges in solid waste management: A case study of Aligarh city. *Journal of Civil Engineering & Environmental Technology*, *1*, 19–24.

Rana, R., Ganguly, R., & Gupta, A. K. (2015). An assessment of solid waste management system in Chandigarh city India. *The Electronic Journal of Geotechnical Engineering*, *20*, 1547–1572.

Rao, R. K., & Sharm, V. R. (2004). Biomedical Waste Management: An Infrastructural Survey of Hospitals. *MJAFI*, *60*, 379-382.

Rao, A. S., & Rao, P. R. (2001). Heavy metals concentrations in the sediments from Kolleru lake, India. *Indian Journal of Environmental Health*, *43*, 148–153. PMID:12395518

Rao, K. J., & Shantaram, M. V. (1996). Soil pollution due to disposal of urban solid wastes at landfill site, Hyderabad. *Indian Journal of Environment Protection*, *16*(5), 373–385.

Rao, S. K. M., Ranyal, R. K., Bhatia, S. S., & Sharma, V. R. (2004). Biomedical waste management: An infrastructural survey of hospitals. *Medical Journal, Armed Forces India*, *60*(4), 379–382. doi:10.1016/S0377-1237(04)80016-9 PMID:27407678

Rao, S. L. N., & Sarma, D. R. R. (1998). Trace metals in soils near an industrial belt in Visakhapatnam, India. *Indian Journal of Environmental Protection*, *17*, 377–380.

Ras, A. V., & Rintala, J. (2007). Trace compounds of biogas from different biogas production plants. *Journal of Energy*, *32*(8), 1375–1380. doi:10.1016/j.energy.2006.10.018

Rathi, S. (2006). Alternative approaches for better municipal solid waste management in Mumbai, India. *J. Waste Manag.*, *26*(10), 1192–1200. doi:10.1016/j.wasman.2005.09.006 PMID:16288861

Rathi, S. (2006). Alternative approaches for better municipal solid waste management in Mumbai, India. *Waste Management (New York, N.Y.)*, *26*, 1192–1200. PMID:16288861

Rathi, S. (2007). Optimization model for integrated municipal solid waste management in Mumbai, India. *Environment and Development Economics*, *12*, 105–121.

Rauch, Hrbek, & Hofbauer. (2013). Biomass gasification for synthesis gas production and applications of the syngas. *WIREs Energy Environ*. doi:10.1002/wene.97

Raut, J. S., & Karuppayil, S. M. (2014). A status review on the medicinal properties of essential oils. *Industrial Crops and Products*, *62*, 250–264. doi:10.1016/j.indcrop.2014.05.055

Raut, M. P., William, S. P., Bhattacharyya, J. K., Chakrabarti, T., & Devotta, S. (2008). Microbial dynamics and enzyme activities during rapid composting of municipal solid waste– a compost maturity analysis perspective. *Bioresource Technology*, *99*(14), 6512–6519. doi:10.1016/j.biortech.2007.11.030 PMID:18155903

Razo, C. (2017). *Producción de biomasa para biocombustibles liquidos: el potencial de América Latina y el Caribe*. Santiago: CEPAL. Disponível em: http://www20.iadb.org/intal/catalogo/PE/2008/02231.pdf

Rebitanim, N. Z., Karim, G. W. A., Rebitanim, N. A., & Amran, M. S. M. (2013). Potential applications of wastes from energy generation particularly biochar in Malaysia. *Renewable & Sustainable Energy Reviews, 21*, 694–702.

Rechberger, H. (2011). *Incineration: Co-combustion in Solid waste technology & management.* Academic Press.

Rekha, K. S. S., Lakshmi, M. V. C., Devi, V. S., & Siddartha Kumar, M. (2012). Production and optimization of lipase from *Candida rugosa* using groundnut oilcake under solid state fermentation. *Biosensors, 27*, 31.

Rex, E., & Baumann, H. (2008). Implications of an interpretive understanding of LCA practice. *Business Strategy and the Environment, 17*(7), 420–430. doi:10.1002/bse.633

Rezaiyan, J., & Cheremisinoff, N. P. (2005). *Gasification technologies: a primer for engineers and scientists.* CRC Press.

Ribeiro, M. C. S., Fiúza, A., Ferreira, A., Dinis, M. D. L., Castro, M. A. C., Meixedo, J. P., & Alvim, M. R. (2016). Recycling approach towards sustainability advance of composite materials industry. *Recycling, 1*, 178.

Rillig, M. C. (2012). Microplastic in terrestrial ecosystems and the soil. *Environmental Science & Technology, 46*(12), 6453–6454. doi:10.1021/es302011r PMID:22676039

Rissanen, T. (2011). *Zero waste fashion repatterned exhibition catalogue*. Chicago, IL: Columbia College Chicago. Retrieved July 5, 2013, fromhttp://www.colum.edu/ADGallery/ Exhibits/2010_zerowaste/zerowastecatalog_final.pdf

Rivas, B., Torrado, A., Torre, P., Converti, A., & Domínguez, J. M. (2008). Submerged citric acid fermentation on orange peel autohydrolysate. *Journal of Agricultural and Food Chemistry, 56*(7), 2380–2387. doi:10.1021/jf073388r PMID:18321055

Robinson, J. B. (1990, October). Future under glassda recipe for people who hate to predict. *Futures.*

Rodriguez, A. (2015). Management of municipal solid waste in Mexico. *International Waste Management and Landfill Symposium, 15.* Disponível em: https://www.researchgate.net/profile/Sanchez_Salinas_Enrique/publication/292326943_management_of_municipal_solid_waste_in_mexico/links/56ad1a7a08ae19a3851408ed/management-of-municipal-solid-waste-in-mexico.pdf?origin=publication_detail

Rogers, D. S., & Tibben-Lembke, R. S. (2001). An examination of reverse logistics practices. *Journal of Business Logistics, 22*(2), 129–148. doi:10.1002/j.2158-1592.2001.tb00007.x

Romera-Castillo, C., Pinto, M., Langer, T. M., Álvarez-Salgado, X. A., & Herndl, G. J. (2018). Dissolved organic carbon leaching from plastics stimulates microbial activity in the ocean. *Nature Communication, 9*, 1430. Retrieved from rural.nic.in/sites/downloads/pura/ZWM_models

Roseland, M. (2012). *Toward sustainable communities: Solutions for citizens and their governments.* New Society Publishers.

Rossem, C. V. (2008). *Individual producer responsibility in the WEEE Directive: from theory to practice?* (Ph.D. Dissertation). IIIEE, Lund University, Sweden.

Rotter, S. (2011). *Incineration: RDF and SRF - Solid waste from fuels in Solid waste technology & management.* Academic Press.

Rudel, R. A., Brody, J. G., Spengler, J. C., Vallarino, J., Geno, P. W., Sun, G., & Yau, A. (2001). Identification of selected hormonally active agents and animal mammary carcinogens in commercial and residential air and dust samples. *Journal of the Air & Waste Management Association, 51*(4), 499–513. doi:10.1080/10473289.2001.10464292 PMID:11321907

Rudel, R. A., Camann, D. E., Spengler, J. D., Korn, L. R., & Brody, J. G. (2003). Phthalates, alkylphenols, pesticides, polybrominated diphenyl ethers, and other endocrine-disrupting compounds in indoor air and dust. *Environmental Science & Technology*, *37*(20), 4543–4553. doi:10.1021/es0264596 PMID:14594359

Ruj, B., & Ghosh, S. (2014). Technological aspects for thermal plasma treatment of municipal solid waste—A review. *Fuel Processing Technology*, *126*, 298–308. doi:10.1016/j.fuproc.2014.05.011

Rupani, P. F., Mohd, F., Md, D., & Zelam, D., & Kanhere. (2013). Sustainable management of different organic waste by vermicomposting technology. *Agricultural Technology (Thailand)*, *9*(3), 529–539.

Rutala, W. A., & Mayhall, C. G. (1992). Medical waste. *Infection Control and Hospital Epidemiology*, *13*(1), 38–48. doi:10.2307/30146966 PMID:1545111

Rutala, W. A., & Weber, D. J. (2007). How to assess risk of disease transmission to patients when there is a failure to follow recommended disinfection and sterilization guidelines. *Infection Control and Hospital Epidemiology*, *28*(2), 146–155. doi:10.1086/511700 PMID:17265395

Rutala, W. A., & Weber, D. J. (2010). Guideline for disinfection and sterilization of prion-contaminated medical instruments. *Infection Control and Hospital Epidemiology*, *31*(2), 107–117. doi:10.1086/650197 PMID:20055640

Ryckeboer, J., Mergaert, J., Vaes, K., Klammer, S., De Clercq, D., Coosemans, J., ... Swings, J. (2003). A survey of bacteria and fungi occurring during composting and self-heating processes. *Annals of Microbiology*, *53*(4), 349–410.

Saad, S. A. G. (2013). Management of hospitals solid waste in Khartoum State. *Environmental Monitoring and Assessment*, *185*(10), 8567–8582. doi:10.100710661-013-3196-1 PMID:23644667

Sabbas, T., Polettini, A., & Pomib, R. (2003). Management of municipal solid waste incineration residues. *Waste Management (New York, N.Y.)*, *23*, 61–88. PMID:12623102

Sadh, P. K., Duhan, S., & Duhan, J. S. (2018). Agro-industrial wastes and their utilization using solid state fermentation: A review. *Bioresources and Bioprocessing*, *5*(1), 1. doi:10.118640643-017-0187-z

Sadiq, M., Alam, I., El-Mubarek, A., & Al-Mohdhar, H. M. (1989). Preliminary evaluation of metal pollution from wear auto tires. *Bulletin of Environmental Contamination and Toxicology*, *42*(5), 743–749. doi:10.1007/BF01700397 PMID:2743004

Safavi, S. M., Masoumi, H., Mirian, S. S., & Tabrizchi, M. (2010). Sorting of polypropylene resins by color in MSW using visible reflectance spectroscopy. *Waste Management (New York, N.Y.)*, *30*, 2216–2222. PMID:20674323

Sahu, Nair, & Sharma. (2014). Review on Solid Waste Management Practice in India: A State of Art. *International Journal of Innovative Research and Development*, *3*(3).

SAICM, Strategic Approach to International Chemicals Management. (2009). *Background information in relation to the emerging policy issue of electronic waste*. Paper presented at the International Conference on Chemicals Management, Geneva.

Saikia, N., & de Brito, J. (2012). Use of plastic waste as aggregate in cement mortar and concrete preparation: A review. *Construction & Building Materials*, *34*, 385–401. doi:10.1016/j.conbuildmat.2012.02.066

Saini, J. K., Saini, R., & Tewari, L. (2014). Lignocellulosic agriculture wastes as biomass feedstocks for second-generation bioethanol production: Concepts and recent developments. 3. *Biotechnology (Faisalabad)*, *5*(4), 337–353. doi:10.100713205-014-0246-5 PMID:28324547

Saison, C., Degrange, V., Oliver, R., Millard, P., Commeaux, C., Montange, D., & Le Roux, X. (2006). Alteration and resilience of the soil microbial community following compost amendment: Effects of compost level and compost-borne microbial community. *Environmental Microbiology*, *8*(2), 247–257. doi:10.1111/j.1462-2920.2005.00892.x PMID:16423013

Sakal, R., Prasad, A. K., & Singh, A. P. (1992). Depth wise distribution of trace metals in soils receiving sewage and sludge. *Journal of the Indian Society of Soil Science*, *40*, 732–737.

Sakamoto, T., Hasunuma, T., Hori, Y., Yamada, R., & Kondo, A. (2012). Direct ethanol production from hemicellulosic materials of rice straw by use of an engineered yeast strain codisplaying three types of hemicellulolytic enzyme on the surface of xylose-utilizing Saccharomyces cerevisiae cells. *Journal of Biotechnology*, *158*(4), 4203–4210. doi:10.1016/j.jbiotec.2011.06.025 PMID:21741417

Saleem, W., Zulfiqar, A., Tahir, M., Asif, F., & Yaqub, G. (2016). Latest technologies of municipal solid waste management in developed and developing countries: A review. *International Journal of Advanced Science and Research*, *1*(10), 22–29.

Salim, A. A., Grbavčić, S., Šekuljica, N., Stefanović, A., Tanasković, S. J., Luković, N., & Knežević-Jugović, Z. (2017). Production of enzymes by a newly isolated Bacillus sp. TMF-1 in solid state fermentation on agricultural by-products: The evaluation of substrate pretreatment methods. *Bioresource Technology*, *228*, 193–200. doi:10.1016/j.biortech.2016.12.081 PMID:28063362

Salinas Callejas, E., & Gasca Quezada, V. (2009). Los biocombustibles. *El cotidiano*, *157*, 75-82. Disponível em: http://www.redalyc.org/pdf/325/32512739009.pdf

Sanasam, S. D., & Talukdar, N. C. (2017). Quality compost production from municipality biowaste in mix with rice straw, cow dung, and earthworm EiseniaFetida. *Compost Science & Utilization*, *25*(3), 141–151. doi:10.1080/1065657X.2016.1249312

Sanchez-Martin, M. J., Sanchez-Camazano, M., & Lorenzo, L. F. (2000). Chromium and lead contents in suburban and urban soils of medium-sized cities of spain: Influence of traffic. *Environmental Contamination and Toxicology*, *64*(2), 250–257. doi:10.1007001289910037 PMID:10656892

Sánchez, Ó. J., Ospina, D. A., & Montoya, S. (2017). Compost supplementation with nutrients and microorganisms in composting process. *Waste Management (New York, N.Y.)*, *69*, 136–153. doi:10.1016/j.wasman.2017.08.012 PMID:28823698

Sangawar & Deshmukh. (2012). a short overview on development of the plastic waste management: environmental issues and challenges. *Sci. Revs. Chem. Commun.*, *2*(3), 349-354.

Sankhla, M. S., Kumari, M., Nandan, M., Mohril, S., Singh, G., Chaturvedi, B. & Kumar, R. (2016). Effect of Electronic waste on Environmental & Human health-A Review. *IOSR Journal of Environmental Science, Toxicology and Food Technology, 10*, 2319-2399. doi:10.9790/2402-10090198104

Sankoh, F. P., Yan, X., & Tran, Q. (2013). Environmental and health impact of solid waste disposal in developing cities: A case study of granville brook dumpsite, Freetown, Sierra Leone. *Journal of Environmental Protection*, *4*(07), 665–670. doi:10.4236/jep.2013.47076

Sarc, R., & Lorber, K. (2013). Production, quality and quality assurance of Refuse Derived Fuels (RDFs). *Waste Management (New York, N.Y.)*, *33*(9), 1825–1834. PMID:23746983

Sarkar, A., & Kaul, P. (2014). Evaluation of Tomato Processing By-Products: A Comparative Study in a Pilot Scale Setup. *Journal of Food Process Engineering*, *37*(3), 299–307. doi:10.1111/jfpe.12086

Sarkar, N., Ghosh, S. K., Bannerjee, S., & Aikat, K. (2012). Bioethanol production from agricultural wastes: An overview. *Renewable Energy*, *37*(1), 19–27.

Sastraatmadja, D. D., Tomita, F., & Ksai, T. (2001). Production of high-quality oncom, a traditional Indonesian fermented food, by the inoculation with selected mold strains in the form of pure culture and solid inoculum. *Journal-Graduate School of Agriculture Hokkaido University*, *70*(2), 111–128.

Satchell, J. E. (1955). *Some aspects of earthworm ecology*. London: Butterworth's.

Savino, A. A. (2009). *Waste and Climate Change: ISWA White Paper*. International Solid Waste Association.

Saxena, S., Srivastava, R. K., & Samaddar, A. B. (2010). Sustainable waste management issues in India. *IUP J. Soil Water Sci.*, *3*(1), 72–90.

Saxena, S., Srivastava, R. K., & Samaddar, A. B. (2010). Sustainablewaste management issues in India. *Journal of Soil & Water Science*, *3*(1), 72–90.

Scarlat, N., Dallemand, J. F., & Fahl, F. (2018). Biogas: Developments and perspectives in Europe. *Renewable Energy*, *129*, 457–472. doi:10.1016/j.renene.2018.03.006

Scarlat, N., Motola, V., Dallemand, J. F., Monforti-Ferrario, F., & Mofor, L. (2015). Evaluation of energy potential of Municipal Solid Waste from African urban areas. *Renewable & Sustainable Energy Reviews*, *50*, 1269–1286. doi:10.1016/j.rser.2015.05.067

Scharfe, D. (2010). *Integrated Waste Management Plan*. Centre & South Hastings Waste Services Board/Waste Diversion Ontario and Stewardship Ontario.

Scheirs, J. (1998). *Polymer Recycling*. New York: Wiley.

Schlosser, E., Nass, B., & Wanzke, W. (2003). *Flame Retardant Combination for Thermoplastic Polymers L*. U.S. Patent 6,547,992.

Schluep, M., Hagelueken, C., Kuehr, R., Magalini, F., Maurer, C., Meskers, C., . . . Wang, F. (2009). *Recycling: From E-waste to resources, Sustainable Innovation and Technology Transfer Industrial Sector Studies (Nairobi and Bonn, UNEP and STeP)*. Available at: http://www.unep.org/PDF/PressReleases/EWaste_publication_screen_FINALVERSION sml.pdf

Schmidt, C., Krauth, T., & Wagner, S. (2017). Export of plastic debris by rivers into the sea. *Environmental Science & Technology*, *51*(21), 12246–12253. doi:10.1021/acs.est.7b02368 PMID:29019247

Schnitzer, M., & Hanson, E. H. (1970). Organo-metallic interactions in soils and evaluation of the methods for the determination of stability constants of metal-fulvic acid complexes. *Soil Science*, *109*(6), 333–339. doi:10.1097/00010694-197006000-00001

Scholey, D. V., Burton, E. J., & Williams, P. E. V. (2016). The bio refinery; producing feed and fuel from grain. *Food Chemistry*, *197*, 937–942. doi:10.1016/j.foodchem.2015.11.063 PMID:26617037

Schott, A. (2001). *Modern solid waste management in practice*. Academic Press.

Schweitzer, J. P., Gionfra, S., Pantzar, M., Mottershead, D., Watkins, E., Petsinaris, F., ... Janssens, C. (2018). *Unwrapped: How Throwaway Plastic is Failing to Solve Europe's Food Waste Problem (And What We Need To Do Instead)*. Institute for European Environmental Policy.

Scott, F., Venturini, F., Aroca, G., & Conejeros, R. (2013). Selection of process alternatives for lignocellulosic bioethanol production using a MILP approach. *Bioresource Technology*, *148*, 525–534. doi:10.1016/j.biortech.2013.09.008 PMID:24080291

Seadon, J. K. (2006). Integrated waste management--looking beyond the solid waste horizon. *Waste Management (New York, N.Y.)*, *26*(12), 1327–1336. doi:10.1016/j.wasman.2006.04.009 PMID:16765038

SEISA. (2006). *Our History*. Disponível em: https://seisa.com.mx/nosotros/

Seltenrich, N. (2017). *Incineration versus Recycling: In Europe, a Debate over Trash*. Available online:http://360.yale.edu/feature/incineration_versus_recycling__in_europe_a_debate_over_trash/2686/

Sembiring, E., & Nitivattananon, V. (2010). Sustainable solid waste management toward an inclusive society: Integration of the informal sector. *Resources, Conservation and Recycling*, *54*(11), 802–809.

Senapati, B. K. (1992). Biotic interactions between soil nematodes and earthworms. *Soil Biology & Biochemistry*, *24*(12), 1441–1444. doi:10.1016/0038-0717(92)90130-P

Seng, B., Kaneko, H., Hirayama, K., & Katayama-Hirayama, K. (2010). Municipal solid waste management in Phnom Penh, capital city of Cambodia. *Waste Management & Research*, *29*(5), 491–500. doi:10.1177/0734242X10380994 PMID:20813763

Sentime, K. (2013). The impact of legislative framework governing waste management and collection in South Africa. *Afr. Geogr. Rev.*, *33*, 81–93.

Sexton, D., & Spelman, D. (2003). Current best practices and guidelines. *Cardiology Clinics*, *21*(2), 273–282. PMID:12874898

Shah, A. A., Hasan, F., Hameed, A., & Ahmed, S. (2008). Biological degradation of plastics: A comprehensive review. *Biotechnology Advances*, *26*(3), 246–265. doi:10.1016/j.biotechadv.2007.12.005 PMID:18337047

Shahsavarani, H., Hasegawa, D., Yokota, D., Sugiyama, M., Kaneko, Y., Boonchird, C. H., & Harashima, S. (2013). Enhanced bio-ethanol production from cellulosic materials by semi-simultaneous saccharification and fermentation using high temperature resistant Saccharomyces cerevisiae TJ14. *Journal of Bioscience and Bioengineering*, *115*(1), 20–23. doi:10.1016/j.jbiosc.2012.07.018 PMID:22925900

Shammas, N. K., & Wang, L. K. (2009). Biosolids composting. In L. K. Wang, N. C. Pereira, Y. T. Hung, & N. K. Shammas (Eds.), *Biological Treatment Processes. Handbook of Environmental Engineering* (Vol. 8, pp. 669–713). Totowa, NJ: Humana Press. doi:10.1007/978-1-60327-156-1_16

Sharholy, M. (2005). Analysis of municipal solid waste management systems in Delhi—a review. *Proceedings for the second International Congress of Chemistry and Environment*, 773–777.

Sharholy, M., Ahmad, K., Mahmood, G., & Trivedi, R. C. (2005). Analysis of municipal solid waste management systems in Delhi – a review. *Proc. for the second International Congress of Chemistry and Environment*, 773–777.

Sharholy, M., Ahmad, K., Mahmood, G., & Trivedi, R. C. (2006). Development of prediction models for municipal solid waste generation for Delhi city. *Proceedings of National Conference of Advances in Mechanical Engineering*.

Sharholy, M., Ahmad, K., Mahmood, G., & Trivedi, R. C. (2008). Municipal solid waste management in Indian cities – A review. *Waste Management (New York, N.Y.)*, *28*(2), 459–467. doi:10.1016/j.wasman.2007.02.008 PMID:17433664

Sharholy, M., Ahmad, K., Vaishya, R. C., & Gupta, R. D. (2007). Municipal solid waste characteristics and management in Allahabad India. *Journal of Waste Management*, *27*, 490–496. PMID:16766176

Sharholy, M., Ahmad, K., Vaishya, R. C., & Gupta, R. D. (2007). Municipal solid waste characteristics and management in Allahabad, India. *J. Waste Manag.*, *27*(4), 490–496. doi:10.1016/j.wasman.2006.03.001 PMID:16766176

Sharma, K., Moser, B. R., Vermillion, K. E., Doll, K. M., & Rajagopalan, N. (2014). Production, characterization and fuel properties of alternative diesel fuel from pyrolysis of waste plastic grocery bags. *Fuel Processing Technology*, *122*, 79–90. doi:10.1016/j.fuproc.2014.01.019

Shaw, D. G., & Day, R. H. (1994). Colour- and form-dependent loss of plastic micro-debris from the North Pacific Ocean. *Marine Pollution Bulletin*, *28*(1), 39–43. doi:10.1016/0025-326X(94)90184-8

Shazwin, T. M., & Nakagoshi, N. (2010). Sustainable waste management through international cooperation: Review of comprehensive waste management technique and training course. *J. Int. Dev. Coop.*, *16*(1), 23–33.

Sheavly, S. B., & Register, K. M. (2007). Marine debris & plastics: Environmental concerns, sources, impacts and solutions. *Journal of Polymers and the Environment*, *15*(4), 301–305. doi:10.100710924-007-0074-3

Shekdar, A. V. (2009). Sustainable solid waste management: An integrated approach for Asian countries. *Waste Management (New York, N.Y.)*, *29*(4), 1438–1448. doi:10.1016/j.wasman.2008.08.025 PMID:19081236

Shrivastava, S., & Singh, K. (2013). Vermicompost to save our agricultural land. *Research Journal of Agriculture and Forestry Science*, *1*(4), 18–20.

Shrivastva, V. S., Rai, A. K., & Mehrotra, R. C. (1989). Rapid acid dissolution of wheat plant tissue for iron, copper and cadmium determination. *Indian Journal of Environmental Protection*, *10*, 198–202.

Shweta, R. K. (2011). Enhancement of wood waste decomposition by microbial inoculation prior to vermicomposting. *Bioresource Technology*, *102*(2), 1475–1480. doi:10.1016/j.biortech.2010.09.090 PMID:21036037

Siddique, R., Khatib, J., & Kaur, I. (2008). Use of recycled plastic in concrete: A review. *Waste Management (New York, N.Y.)*, *28*(10), 1835–1852. doi:10.1016/j.wasman.2007.09.011 PMID:17981022

Siddiqui, T. Z., Siddiqui, F. Z., & Khan, E. (2006). Sustainable development through integrated municipal solid waste management (MSWM) approach— a case study of Aligarh District. *Proceedings of National Conference of Advances in Mechanical Engineering.*

Siddiqui, T. Z., Siddiqui, F. Z., & Khan, E. (2006). Sustainable development through integrated municipal solid waste management (MSWM) approach — a case study of Aligarh District. *Proceedings of National Conference of Advances in Mechanical Engineering.*

Sims, J. T., & Kline, J. S. (1991). Chemical fraction and plant uptake of heavy metals in soils amended with co-composed sewage sludge. *Journal of Environmental Quality*, *20*(2), 387–395. doi:10.2134/jeq1991.00472425002000020009x

Sims, R. (2006). *Energy and fuelwood.* Palmerston North, NZ: Massey University.

Sindiri, M. K., Machavarapu, M., & Vangalapati, M. (2013). Alfa-amylase production and purification using fermented orange peel in solid state fermentation by *Aspergillusniger. Ind J Appl Res*, *3*(8), 49–5. doi:10.15373/2249555X/AUG2013/16

Singh, N., Hui, D., Singh, R., Ahuja, I. P. S., Feo, L., & Fraternali, F. (2016). *Recycling of plastic solid waste: A state of art review and future applications*. Academic Press.

Singh, R., Ahirwar, N.K., Tiwari, J., & Pathak, J. (2018). Review on sources and effect of heavy metal in soil: its bioremedition. *Impact Journals,* 1-22.

Singhal, B. K., Dhar, A., Khan, M. A., & Bindroo, B. B. (2005). Utilization of sericulturalbyproducts as urgent need for sustainable sericulture. In Progress of Research in Organic Sericulture and Seri byproduct Utilization. Seri Scientific Publishers.

Singhal, B. K., Dhar, A., Khan, M. A., Sengupta, D., & Dhar, S. L. (2005). Mulberry by-products utilization for sustenance of sericulture industry of Jammu and Kashmir. *Proc. The 20th Congress of the International Sericultural Commission*, 152-156.

Singhal, S., & Pandey, S. (2001). Solid waste management in India: Status and future directions. *TERI Information Monitoring on Environment Science*, *6*(1), 1–4.

Singh, D., & Suthar, S. (2012). Vermicomposting of herbal pharmaceutical industry solid wastes. *Ecological Engineering*, *39*, 1–6. doi:10.1016/j.ecoleng.2011.10.015

Singh, G. K., Gupta, K., & Chaudhary, S. (2014). Solid Waste Management: Its Sources, Collection, Transportation and Recycling. *International Journal of Environmental Sciences and Development*, *5*, 4.

Singh, J., & Kalamdhad, A. S. (2011). Effect of heavy metals on soil, plants, human health and aquatic life. *International Journal of Research in Chemistry and Environment*, *1*, 15–21.

Singh, N., Hui, D., Singh, R., Ahuja, I. P. S., Feo, L., & Fraternali, F. (2017). Recycling of plastic solid waste: A state of art review and future applications. *Composites. Part B, Engineering*, *115*, 409–422. doi:10.1016/j.compositesb.2016.09.013

Singh, N., Hui, D., Singh, R., Ahuja, I., Feo, L., & Fraternali, F. (2017). Recycling of plastic solid waste: A state of art review and future applications. *Composites Part B Engineering Journal*, *115*, 409–422.

Singh, R. P., Ibrahim, M. H., Esa, N., & Iliyana, M. (2010). Composting of waste from palm oil mill: A sustainable waste management practice. *Reviews in Environmental Science and Biotechnology*, *9*, 331–344.

Singh, R. P., Singh, P., Araujo, A. S., Ibrahim, M. H., & Sulaiman, O. (2011). Management of urban solid waste: Vermicomposting a sustainable option. *Resources, Conservation and Recycling*, *55*(7), 719–729. doi:10.1016/j.resconrec.2011.02.005

Singh, R. P., Tyagi, V., Allen, T., Ibrahim, M. H., & Kothari, R. (2011). An overview for exploring the possibilities of energy generation from municipal solid waste (MSW) in Indian scenario. *Renewable & Sustainable Energy Reviews*, *15*, 4797–4808.

Singh, R., Embrandiri, A., Ibrahim, M., & Esa, N. (2011). Management of biomass residues generated from palm oil mill: Vermicomposting a sustainable option. *Resources, Conservation and Recycling*, *55*(4), 423–434. doi:10.1016/j.resconrec.2010.11.005

Singh, S., Ram, L. C., Masto, R. E., & Verma, S. K. (2011). A comparative evaluation of minerals and trace elements in the ashes from lignite, coal refuse, and biomass fired power plants. *International Journal of Coal Geology*, *87*(2), 112–120. doi:10.1016/j.coal.2011.05.006

Singh, & Sharma, Teixeira da Silva, Satya, & Arora. (2011). Vermicomposting of municipal solid waste employing *Eiseniafetida* together with *Penicilliumspp.*and*AzotobacterBioinoculants. Dynamic Soil. Dynamic Plant*, *1*, 89–95.

Sinha, R. K., Agarwal, S., Chauhan, K., & Valani, D. (2010). The wonders of earthworms and its vermicompost in farm production: Charles Darwin's 'friends of farmers', with potential to replace destructive chemical fertilizers. *Agricultural Science*, *1*(2), 76–94. doi:10.4236/as.2010.12011

Sinha, R. K., Heart, A. S., Asadi, R., & Carretero, E. (2002). Vermiculture technology for environmental management: Study of action of earthworms *Eiseniafetida, Eudriluseuginiae* and *Perionyx excavates* on biodegradation of some community wastes in India and Australia. *The Environmentalist*, *22*(2), 261–268. doi:10.1023/A:1016583929723

Sinha, S. (1999). Accumulation of Cu, Cd, Cr, mn and Pb from artificially contaminated soil by *Bacopa monnieri. Environmental Monitoring and Assessment*, *57*(3), 253–264. doi:10.1023/A:1006091215529

Siracusa, V., Rocculi, P., Romani, S., & Dalla Rosa, M. (2008). Biodegradable polymers for food packaging: A review. *Trends in Food Science & Technology*, *19*(12), 634–643.

Sistema Nacional de información ambiental y de recursos naturales. (2018). *Sistema nacional de información ambiental y de recursos naturales: informe del medio ambiente*. Disponível em: http://apps1.semarnat.gob.mx/dgeia/informe15/tema/cap7.html#tema1

Sjollema, S. B. (2016). Do plastic particles affect microalgal photosynthesis and growth? *AquaticToxicology*, *170*, 259–261. PMID:26675372

Smahi, D., Hammoumi, O. E., & Fekri, A. (2013). Assessment of the impact of the landfill on groundwater quality: A case study of Mediouna Site, Casablanca, Morocco. *Journal of Water Resource and Protection*, *5*, 440–445.

Smith, C. A. (2002). Managing pharmaceutical waste. *The Journal of the Pharmacy Society of Wisconsin: JPSW*, *5*, 17–22.

Smith, T., Sonnenfeld, D. A., & Naguib, P. D. (2006). *Challenging the chip: Labor rights and environmental justice in the global electronics industry*. Philadelphia, PA: Temple University Press.

Sodhi, H. K., Sharma, K., Gupta, J. K., & Soni, S. K. (2005). Production of a thermostable α-amylase from Bacillus sp. PS-7 by solid state fermentation and its synergistic use in the hydrolysis of malt starch for alcohol production. *Process Biochemistry*, *40*(2), 525–534. doi:10.1016/j.procbio.2003.10.008

Somani, M., Datta, M., Ramana, G. V., & Sreekrishnan, T. R. (2018). Investigations on fine fraction of aged municipal solid waste recovered through landfill mining: Case study of three dumpsites from India. *Waste Management & Research*, *36*(8), 744–755. doi:10.1177/0734242X18782393 PMID:29939108

Song, J.H., Murphy, R.J., Narayan, R., & Davies, G.B.H. (2009). Biodegradable and compostable alternatives to conventional plastics. *Phil. Trans. R. Soc. B*, *364*, 2127–.2139.

Song, Q., Li, J., & Zeng, X. (2015). Minimizing the increasing solid waste through zero waste strategy. *Journal of Cleaner Production*, *104*, 199–210. doi:10.1016/j.jclepro.2014.08.027

Soobhany, N. (2018). Preliminary evaluation of pathogenic bacteria loading on organic Municipal Solid Waste compost and vermicompost. *Journal of Environmental Management*, *206*, 763–767. doi:10.1016/j.jenvman.2017.11.029 PMID:29161678

Sreekanth, K., Vishal, G. N., Raghunandan, H. V., & Nitin, K. U. (2014). A review on managing of pharmaceutical waste in industry. *International Journal of Pharm Tech Research*, *6*(3), 899–907.

Sridevi, P., Modi, M., Lakshmi, M. V. V. C., & Kesavarao, L. (2012). A review on integrated solid waste management. *Int. J. Eng. Sci. Adv. Technol.*, *2*, 1491–1499.

Srivastava, R., Krishna, V., & Sonkar, I. (2014). Characterization and management of municipal solid waste: A case study of Varanasi city, India. *International Journal of Current Research and Academic Reviews*, *2*(8), 10–16.

Standard, StE. P. W. P. (2014). *Solving the E-Waste Problem. Recommendations for Standards Development for the Collection, Storage, Transport and Treatment of E-Waste*. Available from: http://www.step-initiative.org/files/_documents/whitepapers/StEP_WP_Standard_20140602.pdf

Staniskis, J. (2005). Integrated waste management: Concept and implementation. *Environmental Research, Engineering and Management*, *33*(3), 40–46.

Steinkraus, K. H. (1983). *Handbook of indigenous fermented foods*. New York: M Dekker Inc.

Stone, R. (1978). Municipal solid wastes and their disposal. *Environmental Health Perspectives*, *27*, 239–244. PMID:738240

Sukan, A., Roy, I., & Keshavarz, T. (2014). Agro-industrial waste materials as substrates for the production of poly (3-hydroxybutyric acid). *Journal of Biomaterials and Nanobiotechnology*, *5*(4), 229–240. doi:10.4236/jbnb.2014.54027

Sukholthaman, P., & Sharp, A. (2016). A system dynamics model to evaluate effects of source separation of municipal solid waste management: A case of Bangkok, Thailand. *Waste Management (New York, N.Y.)*, *52*, 50–61. doi:10.1016/j.wasman.2016.03.026 PMID:27026497

Sun, S., Sidhu, V., Rong, Y., & Zheng, Y. (2018). Pesticide Pollution in Agricultural soils and Sustainable Remediation Methods: A Review. Springer International Publishing AG.

Sundaravadivelan, C., Isaiarasu, L., Manimuthu, M., Kumar, P., Kuberan, T., & Anburaj, J. (2011). Impact analysis and confirmative study of physicochemical, nutritional and biochemical parameters of vermiwash produced from different leaf liters by using two earthworm species. *Agricultural Technology (Thailand)*, *7*(5), 1443–1457.

Sunitha, N. D., Giraddi, R. S., Kulkarni, K. A., & Lingappa, S. (1997). Evaluation methods of vermicomposting under open field conditions. *Karnataka Journal of Agricultural Sciences*, *10*(4), 987–990.

Supreme Court Committee Report. (1999). *Report of the committee on Solid Waste Management in Class I cities in India.*

Surek, E., & Nilufer-Erdil, D. (2016). Phenolic contents, antioxidant activities and potential bioaccessibilities of industrial pomegranate nectar processing wastes. *International Journal of Food Science & Technology*, *51*(1), 231–239. doi:10.1111/ijfs.13000

Suriyanarayanan, S., Mailappa, A. S., Jayakumar, D., Nanthakumar, K., Karthikeyan, K., & Balasubramania, S. (2010). Studies on the characterization and possibilities of reutilization of solid wastes from a waste paper based paper industry. *Global Journal Environmental Research*, *4*(1), 18–22.

Sussarellu, R., Suquet, M., Thomas, Y., Lambert, C., Fabioux, C., Pernet, M. E. J., ... Huvet, A. (2016). Oyster Reproduction is Affected by Exposure to Polystyrene Microplastics. *Proceedings of the National Academy of Sciences of the United States of America*, *113*(9), 2430–2435. doi:10.1073/pnas.1519019113 PMID:26831072

Suthar, S. (2007). Nutrients changes and biodynamic of epigeic earthworms Perionyx excavatus during recycling of some agricultural wastes. *Bioresource Technology.*

Suthar, S. (2009). Potential of Allolobophora parva (Oligochaeta) in vermicomposting. *Bioresource Technology*, *100*(24), 6422–6427. doi:10.1016/j.biortech.2009.07.004 PMID:19648000

Suthar, S. (2009). Vermicomposting of vegetable-market solid waste using *Eiseniafetida*: Impact of bulking material on earthworm growth and decomposition rate. *Ecological Engineering*, *35*(5), 914–920. doi:10.1016/j.ecoleng.2008.12.019

Suthar, S., & Singh, S. (2007). Vermicomposting of domestic waste by using two epigeic earthworms. *International Journal of Environmental Science and Technology*, *5*(1), 99–106. doi:10.1007/BF03326002

Suthar, S., & Singh, S. (2008). Feasibility of vermicomposting in bio stabilization of sludge from a distillery industry. *The Science of the Total Environment*, *394*(2-3), 237–243. doi:10.1016/j.scitotenv.2008.02.005 PMID:18313726

Sutherland, R. A., & Tack, F. M. (2000). Metal phase association in an urban watershed. Honolulu. Hawaii. *The Science of the Total Environment*, *4*(2-3), 103–113. doi:10.1016/S0048-9697(00)00458-7 PMID:10902838

Suttibak, S., & Nitivattananon, V. (2008). Resources, Conservation and Recycling Assessment of factors influencing the performance of solid waste recycling programs. *Conservation And Recycling*, *53*(1-2), 45–56. doi:10.1016/j.resconrec.2008.09.004

Svoboda, J. (2004). *Magnetic Techniques for the Treatment of Materials.* Springer.

Swaileh, K. M., Hussein, R. M., & Elhaj, S. A. (2004). Assessment of heavy metal contamination in roadside surface soil and vegetation from the west bank. *Archives of Environmental Contamination and Toxicology*, *47*(1), 23–30. doi:10.100700244-003-3045-2 PMID:15346775

Swan, S. H. (2008). Environmental phthalate exposure in relation to reproductive outcomes and other health endpoints in humans. *Environmental Research*, *108*(2), 177–184. doi:10.1016/j.envres.2008.08.007 PMID:18949837

Swan, S. H., Main, K. M., Liu, F., Stewart, S. L., Kruse, R. L., Calafat, A. M., ... Teague, J. L. (2005). Decrease in anogenital distance among male infants with prenatal phthalate exposure. *Environmental Health Perspectives*, *113*(8), 1056–1061. doi:10.1289/ehp.8100 PMID:16079079

Swinwood, J. F., Waite, T. D., Kruger, P., & Rao, S. M. (1994). Radiation technologies for waste treatment: A global perspective. *International Atomic Energy Agency Bulletin*, *36*(1), 11–15.

Szycher, M. (2017). Medical pharmaceutical markets for medical plastics. In High Performance Biomaterials (pp. 3-30). Academic Press. doi:10.1201/9780203752029-1

Tabata, T., Hishinuma, T., Ihara, T., & Genchi, Y. (2011). Life cycle assessment of integrated municipal solid waste management systems, taking account of climate change and landfill shortage trade-off problems. *Waste Management & Research*, *29*(4), 423–432. doi:10.1177/0734242X10379493 PMID:20699290

Talashikar, S. C., & Powar, A. G. (1998). Vermibiotechnology for eco-friendly disposal of wastes. Ecotechnology for pollution control and environment management. *Indian Journalof EnvironmentalEcoplanning*, *7*, 535–538.

Talsness, C. E., Andrade, A. J. M., Kuriyama, S. N., Taylor, J. A., & vom Saal, F. S. (2009). Components of plastic: Experimental studies in animals and relevance for human health. *Philosophical Transactions of the Royal Society of London. Series B, Biological Sciences*, *364*(1526), 2079–2096. doi:10.1098/rstb.2008.0281 PMID:19528057

Talyan, V., Dahiya, R. P., & Sreekrishnan, T. R. (2008). State of municipal solid waste management in Delhi, the capital of India. *Waste Management (New York, N.Y.)*, *28*, 1276–1287. PMID:17692510

Tangahu, B. V., Abdullah, S. R. S., Basri, H., Idris, M., Anuar, N., & Mukhlisin, M. (2011). A review on heavy metals (As, Pb, & Hg) uptake by plants through Phytoremediation. *International Journal of Chemical Engineering*, *2011*, 1–31. doi:10.1155/2011/939161

Tan, S. T., Ho, W. S., Hashim, H., Lee, C. T., Taib, M. R., & Ho, C. S. (2015). Energy, economic and environmental (3E) analysis of waste-to-energy (WTE) strategies for municipal solid waste (MSW) management in Malaysia. *Energy Conversion and Management*, *102*, 111–120.

Tatasumi, Y., Ikeda, A., & Yoda, K. (1983). Soil pollution by heavy metals in Sakai city, Japan. *Journal of Ecology*, *32*, 241–249.

Tchobanoglous, G., & Kreith, F. (2002). *Handbook of Solid Waste Management* (2nd ed.). McGraw-Hill.

Tchobanolglous, G., Theisen, H., & Vigil, S. (1993). *Integrated Solid Waste Management*. McGraw Hill, Inc.

Ter Halle, A., Jeannau, L., Martignac, M., Jarde, E., Pedrono, B., Brach, L., & Gigault, J. (2017). Nanoplastic in the North atlantic subtropical gyre. *Environmental Science & Technology*, *51*(23), 13689–13697. doi:10.1021/acs.est.7b03667 PMID:29161030

Ter Halle, A., Ladirat, L., Gendre, X., Goudouneche, D., Pusineri, C., Routaboul, C., ... Perez, E. (2016). Understanding the fragmentation pattern of marine plastic debris. *Environmental Science & Technology*, *50*(11), 5668–5675. doi:10.1021/acs.est.6b00594 PMID:27138466

Terazono, A., Murakami, S., Abe, N., Inanc, B., Moriguchi, Y., Sakai, S., ... Williams, E. (2006). Current status and research on E-waste issues in Asia. *Journal of Material Cycles and Waste Management*, *8*(1), 1–12. doi:10.100710163-005-0147-0

TERI Energy Data Directory & Yearbook (TEDDY). (2010). New Delhi, India: The Energy and Resource Institute.

Ternes, T. A., Joss, A., & Siegrist, H. (2004). *Peer reviewed: scrutinizing pharmaceuticals and personal care products in wastewater treatment*. Academic Press.

Teuten, E. L. (2009). Transport and release of chemicals from plastics to the environment and to wildlife. *Philosophical Transactions of the Royal Sociey B*, *364*, 2027–2045. PMID:19528054

Teuten, E. L., Saquing, J. M., Knappe, D. R. U., Barlaz, M. A., Jonsson, S., Björn, A., ... Takada, H. (2009). Transport and release of chemicals from plastics to the environment and to wildlife. *Philosophical Transactions of the Royal Society of London. Series B, Biological Sciences*, *364*(1526), 2027–2045. doi:10.1098/rstb.2008.0284 PMID:19528054

Teuten, E., Rowland, S., Galloway, T., & Thompson, R. (2007). Potential for plastics to transport hydrophobic contaminants. *Environmental Science & Technology*, *41*(22), 7759–7764. doi:10.1021/es071737s PMID:18075085

Textiles Environment Design. (2012). *TED's Ten: 1 – design to minimise waste*. Retrieved October 26, 2012, from http://www.tedresearch.net/1-minimise-waste

Thamaraj, K., Ganesh, P., Kolanjinathan, K., Suresh, K. R., & Anandan, A. (2011). Influence of vermicompost and vermiwash on physicochemical properties of rice cultivated soil. *Current Biotica*, *2*(3), 18–21.

Tharmaraj, K., Ganesh, P., Kolanjinathan, K., Suresh Kumar, R., & Anandan, A. (2011). Influence of vermicompost and vermiwash on physico chemical properties of rice cultivated soil. *Current Biotica*, *2*, 18–21.

The Carbon Trust. (2006). *Carbon Footprint in the Supply Chain: The Next Step for Business*. Author.

The Ecotourism and Conservation Society of Sikkim (ECOSS). (2014). *Plastics and the Environment Assessing the Impact of the Complete Ban on Plastic Carry Bag*. Retrieved from http://www.toxicslink.org

The New Plastics Economy: Rethinking the Future of Plastics. (n.d.). Retrieved from https://www. ellenmacarthurfoundation.org/publications/the-new-plastics-economy-rethinking-the-future-ofplastics

The State of Victoria Department of Environment, Land, Water, and Planning. (2017). *Turning waste into energy*. Retrieved from http://creativecommons.org/licenses/by/4.0/

Thomas, K. V., Dye, C., Schlabach, M., & Langford, K. H. (2007). Source to sink tracking of selected human pharmaceuticals from two Oslo city hospitals and a wastewater treatment works. *Journal of Environmental Monitoring*, *9*(12), 1410–1418. doi:10.1039/b709745j PMID:18049781

Thomas, S. (2008). From 'Green Blur' to ecofashion: Fashioning an eco-lexicon. *Fashion Theory*, *12*(4), 525–539. doi:10.2752/175174108X346977

Thomas, S., & Trivedy, R. K. (2002). *Earthworm biotechnology for waste management and crop improvement a review of research*. Bangalore, India: Allied Publication Pvt. Ltd.

Thompson, R. C., Moore, C. J., vomSaal, S. F., & Swan, S. H. (2009). Plastics, the environment and human health: Current consensus and future trends. *Philosophical Transactions of the Royal Society of London. Series B, Biological Sciences*, *364*(1526), 2153–2166. doi:10.1098/rstb.2009.0053 PMID:19528062

Thompson, R. C., Olsen, Y., Mitchell, R. P., Davis, A., Rowland, S. J., John, A. W. G., ... Russell, A. E. (2004). Lost at sea: Where is all the plastic? *Science*, *304*(5672), 838–838. doi:10.1126cience.1094559 PMID:15131299

Thompson, R. C., Swan, S. H., Moore, C. J., & vom Saal, F. S. (2009). Our plastic age. *Philosophical Transactions of the Royal Society of London. Series B, Biological Sciences*, *364*(1526), 1973–1976. doi:10.1098/rstb.2009.0054 PMID:19528049

Tirado-Montiel, M. I., Tyagi, R. D., & Valero, J. R. (2001). Wastewater treatment sludge asa raw material for the production of Bacillus thuringiensis based biopesticides. *Water Res. N.Y.*, *35*(16), 3807–3816. doi:10.1016/S0043-1354(01)00103-8

Tiwari, A. V., & Kadu, P. A. (2013). Biomedical Waste Management Practices in India-A Review. *International Journal of Current Engineering and Technology*, *3*(5), 2030–2033.

Tognetti, C., Mazzarino, M. J., & Laos, F. (2007). Cocomposting biosolids and municipal organic waste: Effects of process management on stabilization and quality. *Biology and Fertility of Soils*, *43*(4), 387–397. doi:10.100700374-006-0164-8

Tognetti, C., Mazzarino, M. J., & Laos, F. (2007). Improving the quality of municipal organic waste compost. *Bioresource Technology*, *98*(5), 1067–1076. doi:10.1016/j.biortech.2006.04.025 PMID:16822669

Tojo, N. (2004). *Extended producer responsibility as a driver for design change – utopia or reality?* (Ph.D. Dissertation). IIIEE, Lund University, Sweden.

Toles, C. A., Marshall, W. E., & Johns, M. M. (1997). Granular activated carbons from nutshells for the uptake of metals and organic compounds. *Carbon*, *35*(9), 1407–1414. doi:10.1016/S0008-6223(97)00073-0

Topçu, E. N., Tonay, A. M., Dede, A., Öztürk, A. A., & Öztürk, B. (2013). Origin and abundance of marine litter along sandy beaches of the Turkish Western Black Sea Coast. *Marine Environmental Research*, *85*, 21–28. doi:10.1016/j.marenvres.2012.12.006 PMID:23290790

Torbert, H. A., Gebhart, D. L., & Busby, R. R. (2011). New municipal solid waste processing technology reduces volume and provides beneficial reuse applications for soil improvement and dust control. In Integrated Waste Management (vol. 1). In Tech.

Townsend, K., & Mills, F. (2013). Mastering zero: How the pursuit of less waste leads to more creative pattern cutting. *International Journal of Fashion Design, Technology, and Education*, *6*(2), 104–111. doi:10.1080/17543266.2013.793746

Toxic Link Organisation (T.L.O). (2014). Available at: http://www.toxicslink.org

Toxic Link Organisation (T.L.O). (2015). Available at: http://www.toxicslink.org

Tozlu, A., Özahi, E., & Abuşoglu, A. (2016). Waste to energy technologies for municipal solid waste management in Gaziantep. *Renewable & Sustainable Energy Reviews*, *54*, 809–815.

Traversa, A., Loffredo, F., Gattullo, C. E., & Senesi, N. (2010). Water-extractable organic matter of different composts: A comparative study of properties and allelochemical effects on horticultural plants. *Geoderma*, *156*(3-4), 287–292. doi:10.1016/j.geoderma.2010.02.028

Trigo, M., & Rodarte, A. B. (2017). Biofuels: socio-environmental violence in rural Mexico. *Latin American Perspectives*, *42*(5). Disponível em: http://journals.sagepub.com/doi/abs/10.1177/0094582X15594398

Tripathi, R. M., Khandekar, R. N., & Mishra, U. C. (1990). Toxic trace metals in the atmosphere of Mordabad (India). *Indian Journal of Environmental Health*, *32*, 140–147.

Troschinetz, A. M., & Mihelcic, J. R. (2009). Sustainable recycling of municipalsolid waste in developing countries. *Waste Management (New York, N.Y.)*, *29*(2), 915–923. doi:10.1016/j.wasman.2008.04.016 PMID:18657963

Tsakona, M., Anagnostopoulou, E., & Gidarakos, E. (2007). Hospital waste management and toxicity evaluation: A case study. *Waste Management (New York, N.Y.)*, *27*(7), 912–920. doi:10.1016/j.wasman.2006.04.019 PMID:16872822

Tsang, C., & Apps, J. (2005). *Underground injection science and technology*. Amsterdam: Elsevier.

Tudor, T. L., Noonan, C. L., & Jenkin, L. E. (2005). Healthcare waste management: A case study from the National Health Service in Cornwall, United Kingdom. *Waste Management (New York, N.Y.)*, *25*, 606–615. doi:10.1016/j.wasman.2004.10.004 PMID:15993345

Tudor, T., Robinson, G., Riley, M., Guilbert, S., & Barr, S. (2011). Challenges facing the sustainable consumption and waste management agendas: Perspectives on UK households. *Local Environment*, *16*(1), 51–66. doi:10.1080/13549839.2010.548372

Tulp, M., & Bohlin, L. (2004). Unconventionalnatural sources for future drug discovery. *Drug Discovery Today*, *9*(10), 450–458. doi:10.1016/S1359-6446(04)03066-1 PMID:15109950

Turan, N. G., Çoruh, S., Akdemir, A., & Ergun, O. N. (2009). Municipal solid waste management strategies in Turkey. *Waste Management (New York, N.Y.)*, *29*(1), 465–469. doi:10.1016/j.wasman.2008.06.004 PMID:18644708

Tweib, S. A., Rahman, R. A., & Kalil, M. S. (2011). A Literature Review on the Composting. *Proceedings of International Conference on Environment and Industrial Innovation (ICEII 2011)*.

U.S Environmental Protection Agency. (2010). Reduce, Reuse, Recycle. *Wastes - Resource Conservation*. Retrieved January 5, 2011, from http://www.epa.gov/wastes/conserve/rrr/

Uddin, A., Koizumi, K., Murata, K., & Sakata, Y. (1997). Thermal and catalytic degradation of structurally different types of polyethylene into fuel oil. *Polymer Degradation & Stability*, *56*(1), 37–44. doi:10.1016/S0141-3910(96)00191-7

Umicore Precious Metals Refining. (2007). *Metals Recovery from e-scrap in a global environment*. Available at: http://archive.basel.int/industry/sideevent030907/umicore.pdf

UNEP (United Nations Environment Programme). (2005). *Solid waste management*. Retrieved from http://www.unep.org/ietc/Portals/136/SWM-Vol1-Part1-Chapters1to3.pdf

UNEP, United Nations Environment Programme. (1989). *Basel convention on the control of transboundary movements of hazardous waste and their disposal*. http://www.basel.int/Portals/4/Basel%20Convention/docs/text/BaselConventionText-e. pdf

UNEP. (2016). *Marine Plastic Debris and Microplastics – Global Lessons and Research to Inspire Action and Guide Policy Change, 192*. Nairobi: United Nations Environment Programme.

UN-HABITAT. (2010). *Solid Waste Management in the World's Cities. Water and Sanitation in the World's Cities Report 2010*. Earthscan.

United Nations Environment Programme (UNEP). (2009). *Converting Waste Plastics into Fuel: Report on Situation Analysis of Existing Solid Waste Management System for Bangkok Metropolitan Administration* (p. 40). Bangkok, Thailand: International Environmental Technology Centre.

United Nations Environmental Program. Converting Waste Plastic into a Resource. (2009). Retrieved from http://www.unep.or.jp/

United Nations Environmental Programme. (2009). *Developing integrated solid waste management plan: Training manual: Assessment of current waste management systems and gaps therein* (Vol. 2). Osaka, Japan: Author. Retrieved from http://www.unep.or.jp/ietc/Publications/spc/ISWMPlan_Vol2.pdf

United Nations Newscentre. (2017). *UN declares war on ocean plastic*. Retrieved from http://web.unep.org/newscentre/un-declares-war-ocean-plastic

United States Department of Energy. (2013). *Energy Efficiency and Renewable Energy*. DOE/EE-0948. Bioenergy Energy Gov.

United States Environmental Protection Agency. (2018). *Report on environment: municipal solid waste*. Disponível em: https://cfpub.epa.gov/roe/indicator_pdf.cfm?i=53

Upadhyay, V., Jethoo, A. S., & Poonia, M. P. (2012). Solid waste collection and segregation: A case study of MNIT campus, Jaipur. *International Journal of Engineering and Innovative Technology*, *1*(3), 144–149.

US EPA. (2008). *Introduction to the Concept of Green Chemistry*. Retrieved from http://www.epa.gov/oppt/greenchemistry/pubs/aboutgc.html

USEPA (United States Environmental Protection Agency). (2002). *Solid waste management: a local challenge with global impacts*. Retrieved from http://www. epa.gov/osw/nonhaz/municipal/pubs/ghg/f02026.pdf

USEPA (United States Environmental Protection Agency). (2014). *Municipal solid waste generation, recycling, and disposal in the United States: Facts and figures for 2012*. Washington, DC: United States Environmental Protection Agency.

Van Gestel, C. A. M., Dirven-van Breemen, E. M., & Baerselman, R. (1992). Influence of environmental conditions on the growth and reproduction of the earthworm *Eiseniaandrei* in an artificial soil substrate. *Pedobiologia*, *36*, 109–120.

Van Hamme, J. D., Singh, A., & Ward, O. P. (2003). Recent advances in petroleum microbiology. *Microbiology and Molecular Biology Reviews*, *67*(4), 503–549. doi:10.1128/MMBR.67.4.503-549.2003 PMID:14665675

van Sebille, E., Wilcox, C., Lebreton, L., Maximenko, N., Hardesty, B. D., van Franeker, J. A., ... Law, K. L. (2015). A global inventory of small floating plastic debris. *Environmental Research Letters*, *10*(12), 124006. doi:10.1088/1748-9326/10/12/124006

VanEwijk, S., & Stegemann, J. A. (2014). Limitations of the WH for achieving absolute reductions in material throughput. *Journal of Cleaner Production*.

VanVeen, A. G., Graham, D. C. W., & Steinkraus, K. H. (1968). Fermented peanut press cake. *Cereal Sci Today*, *13*, 96–99.

Vasanthi, K., Chairman, K., & Ranjit Singh, A. J. A. (2013). Influence of animal wastes on growth and reproduction of the African earthworm species *Eudriluseugeniae* (Kinberg). *Pollution Research*, *32*(2), 337–342.

Vasistha, P., Ganguly, R., & Gupta, A. K. (2018). Biomedical Waste Generation and Management in Public Sector Hospital in Shimla City. In *Environmental Pollution* (pp. 225–232). Singapore: Springer. doi:10.1007/978-981-10-5792-2_19

Vaverková, M., Toman, F., Adamcová, D., Kotovicová, J., Vaverková, M., Stejskal, B., ... Yapsakli, K. (2013). Verifying research of waste landfill environmental impact using bioindicators. *Journal of Environmental Studies*, *4*(7), 8.

Vehlow, J. (1996). Municipal solid waste management in Germany. *Waste Management (New York, N.Y.)*, *16*(5-6), 367–374.

Velis, C. A., Longhurst, P. J., Drew, G. H., Smith, R., & Pollard, S. J. (2010). Production and quality assurance of solid recovered fuels using mechanical—biological treatment (MBT) of waste: A comprehensive assessment. *Critical Reviews in Environmental Science and Technology*, *40*(12), 979–1105.

Venglovsky, J., Martinez, J., & Placha, I. (2006). Hygienic and ecological risks connected with utilization of animal manures and biosolids in agriculture. *Livestock Science*, *102*(3), 197–203. doi:10.1016/j.livsci.2006.03.017

Venkatachalaiah, B. K. (1996). The role of Karnataka compost development corporation Ltd., In: Organic farming. Abstract. *National Seminar on Organic Farming and Sustainable Agriculture*, 24-25.

Verdone, N., & De Filippis, P. (2004). Thermodynamic behaviour of sodium and calcium based sorbents in the emission control of waste incinerators. *Chemosphere*, *54*(7), 975–985. doi:10.1016/j.chemosphere.2003.09.041 PMID:14637355

Vergara, S. E., & Tchobanoglous, G. (2012). Municipal Solid Waste and the Environment: A Global Perspective. *Annual Review of Environment and Resources*, *37*(1), 277–309. doi:10.1146/annurev-environ-050511-122532

Verma, L. K., Mani, S., Sinha, N., & Rana, S. (2008). Biomedical waste management in nursing homes and smaller hospitals in Delhi. *Waste Management (New York, N.Y.)*, *28*(12), 2723–2734. doi:10.1016/j.wasman.2007.12.013 PMID:18358710

Vermeulen, I., Van Caneghem, J., Block, C., Baeyens, J., & Vandecasteele, C. (2011). 'Automotive Shredder Residue (ASR): Reviewing Its Productions from End-of-Life Vehicles (ELVs) and Its Recycling, Energy and Chemicals Valorization. *Journal of Hazardous Materials*, *190*(1-3), 8–27. doi:10.1016/j.jhazmat.2011.02.088 PMID:21440364

Victoria E. P. A. (2017). *Environment Protection Authority Victoria (2017b), Your environment – Waste webpage*. EPA Victoria. Retrieved from http://www.epa.vic.gov.au/your-environment/waste

Victoria, E. P. A. (2013). *Energy from waste*. EPA Victoria. Retrieved from http://www.sustainability.vic.gov.au/

Victoria, S. (2017). *Statewide Waste and Resource Recovery Infrastructure plan Victoria 2017-2046 (amendment consultation draft)*. Sustainability Victoria.

Vij, D. (2012). Urbanization and solid waste management in India: Present practices and future challenges. *Procedia: Social and Behavioral Sciences*, *37*, 437–447. doi:10.1016/j.sbspro.2012.03.309

Vikram Reddy, Gath, & Rossi. (2009). Fertilizers from vermiculture as an option for organic wastes recovery. *Agrochemical*, *27*(3), 244–251.

Villarrubia-Gómez, P., Cornell, S. E., & Fabres, J. (2018). Marine plastic pollution as a planetary boundary threat–The drifting piece in the sustainability. *Marine Policy*, *96*, 213–220. doi:10.1016/j.marpol.2017.11.035

Virani, T. (2011). *The economic feasibility of bionergy production from Miscanthus for the Ontario greenhouse industry* (Master's thesis). The Faculty of Graduate Studies, University of Guelph, Ontario, Canada.

Visvanathan, C., Trankler, J., Jospeh K., & Nagendran, R. (2005). *Vermicomposting as an Eco-Tool in Sustainable Solid Waste Management*. Asian Institute of Technology, Annamalai University.

Viswanath, P., & Nand, K. (1994). Anaerobic digestion of silk industry wastes. *Bioresource Technology*, *49*(3), 273–276. doi:10.1016/0960-8524(94)90052-3

von Moos, N., Burkhardt-Holm, P., & Köhler, A. (2012). Uptake and effects of microplastics on cells and tissue of the blue mussel *Mytilusedulis* L. after an experimental exposure. *Environmental Science & Technology*, *46*(20), 11327–11335. doi:10.1021/es302332w PMID:22963286

Vu, K. D., Tyagi, R. D., Surampalli, R. Y., & Valéro, J. R. (2012). Mathematical relationshipsbetween spore concentrations, delta-endotoxin levels, and entomotoxicity ofBacillus thuringiensis preparations produced in different fermentation media. *Bioresource Technology*, *123*, 303–311. doi:10.1016/j.biortech.2012.07.110 PMID:22940334

Wade, M. J., Davis, B. K., Carlisle, J. S., Klein, A. K., & Valoppi, L. M. (1993). Environmental transformation of toxic metals. *Occupational Medicine*, *8*, 575–601. PMID:8272980

Waggoner, Miller, De Roo, & Henry. (2006). Plastic mulching: Principles and benefits. *Bulletin of the Connecticut Agricultural Experiment Station*, *634*, 1–44.

Wagner, M., & Oehlmann, J. (2009). Endocrine disruptors in bottled mineral water: Total estrogenic burden and migration from plastic bottles. *Environmental Science and Pollution Research International*, *16*(3), 278–286. doi:10.100711356-009-0107-7 PMID:19274472

Wagner, T. P., & Raymond, T. (2015). Landfill mining: Case study of a successful metals recovery project. *Waste Management (New York, N.Y.)*, *45*, 448–457. doi:10.1016/j.wasman.2015.06.034 PMID:26152366

Wahab, D. A., Hussain, A., Scavino, E., Mustafa, M. M., & Basri, H. (2006). Development of a Prototype Automated Sorting System for Plastic Recycling. *American Journal of Applied Sciences*, *3*(7), 1924–1928.

Wang, K., Li, L., Tan, F., & Wu, D. (2018). Treatment of Landfill Leachate Using Activated Sludge Technology: A Review. *Archaea (Vancouver, B.C.)*. PMID:30254508

Wang, Y., Tian, Z., Zhu, H., Cheng, Z., Kang, M., Luo, C., ... Zhang, G. (2012). Polycyclic aromatic hydrocarbons (PAHs) in soils and vegetation near an e-waste recycling site in South China: Concentration, distribution, source, and risk assessment. *The Science of the Total Environment*, *439*, 187–193. doi:10.1016/j.scitotenv.2012.08.018 PMID:23063924

Wangyao, K., Towprayoon, S., Chiemchaisri, C., Gheewala, S., & Nopharatana, A. (2009). Application of the IPCC Waste Model to solid waste disposal sites in tropical countries: Case study of Thailand. *Environmental Monitoring and Assessment*, *164*(1-4), 249–261. PMID:19415519

Wani, S. P. (2002). *Improving the livelihoods: New partnerships for win-win solutions for natural resource management.* Paper submitted in the 2nd International Agronomy Congress, New Delhi, India.

Watanabe, T., & Hirayama, T. (2001). Genotoxicity of soil. *Journal of Health Science*, *47*(5), 433–438. doi:10.1248/jhs.47.433

Waters, C. N., Zalasiewicz, J., Summerhayes, C., Barnosky, A. D., Poirier, C., Ga uszka, A., ... Wolfe, A. P. (2016). The Anthropocene is functionally and stratigraphically distinct from the Holocene. *Science*, *351*(6269), 137. doi:10.1126cience.aad2622 PMID:26744408

Wath, S. B., Dutt, P. S., & Chakrabarti, T. (2011). E-Waste scenario in India, its management and implications. *Environmental Monitoring and Assessment*, *172*(1-4), 249–262. doi:10.100710661-010-1331-9 PMID:20151189

Watts, A. (2015). *Are we really "choking the ocean with plastic"? Tracing the creation of an eco-myth.* Available at: https://wattsupwiththat.com/2015/12/24/are-we-really-choking-theocean- with-plastic-tracing-the-creation-of-an-eco-myth/

Webb, H., Arnott, J., Crawford, R., & Ivanova, E. (2013). Plastic Degradation and Its Environmental Implications with Special Reference to Polyethylene terephthalate. *Polymers*, *5*(1), 1–18. doi:10.3390/polym5010001

WEC. (2013). *World Energy Resources: Waste to Energy*. Retrieved from https://www.worldenergy.org/wp

Wenhua, L. (2001). *Agro-ecological Farming Systems in China (Man and the Biosphere Series). Chinese Academy of Sciences*. Beijing, China: UNESCO.

Werner, S., Budziak, A., van Franeker, J., Galgani, F., Hanke, G., Maes, T., ... Vlachogianni, T. (2016). *Harm Caused by Marine Litter. MSFD GES TG Marine Litter–Thematic Report. JRC Technical Report.* Luxembourg: European Union.

Westinghouse Plasma Corporation (WPC). (2010). *Technology and solutions-Plasma Gasification Vitrification Reactor.* Retrieved 22 July 2010, from http://www.westinghouse-plasma.com/

Wilburn, S. Q., & Eijkemans, G. (2004). Preventing needle stick injuries among healthcare workers: A WHO-ICN collaboration. *International Journal of Occupational and Environmental Health*, *10*(4), 451–456. doi:10.1179/oeh.2004.10.4.451 PMID:15702761

Wilcke, W. (2007). Global patterns of polycyclic aromatic hydrocarbons (PAHs) in soil. *Science Direct*, *141*, 157–166.

Wilcox, C., Hardesty, B. D., Sharples, R., Griffin, D. A., Lawson, T. J., & Gunn, R. (2013). Ghost net Impacts on Globally Threatened Turtles, a Spatial Risk Analysis for Northern Australia. *Conservation Letters*, *6*(4), 247–254. doi:10.1111/conl.12001

Wildung, R. E., Garland, T. R., & Drucker, H. (1976). *Transformation of trace metals by soil microorganisms*. Agronomy abstracts, No. 28.

Willaman, J., & Eskew, R. (1943). *Uses for vegetable wastes*. Academic Press.

William, P. T. (2005). Water treatment and disposal. John Wiley.

Williams, M. (2017). *Waste-to-energy success factors in Sweden and the United States: analyzing the transferability of the swedish waste-to-energy model to the United States*. Disponível em: https://pt.scribd.com/document/122133655/Waste-to-energy-success-factors-in-Sweden-and-the-United-States

Wilson, D. C., Rodic, L., Modak, P., Soos, R., Carpintero, A., Velis, K., ... Simonett, O. (2015). *Global waste management outlook*. UNEP.

Wilson, D. C., Velis, C., & Cheeseman, C. (2006). Role of informal sector recycling in waste management in developing countries. *Habitat International*, *30*(4), 797–808. doi:10.1016/j.habitatint.2005.09.005

Winchester, N., & Reilly, J. (2015). The feasibility, costs, and environmental implications of large-scale biomass energy. *Energy Economics*, *51*, 188-203. doi:10.1016/j.eneco.2015.06.016

Wittmann, M. (2010). *Sustainable healthcare design*. In *Evidence-based Design for Healthcare Facilities*. Sigma Theta Tau International.

Wohlleben, W., & Neubauer, N. (2016). Quantitative rates of release from weathered nanocomposites are determined across 5 orders of magnitude by the matrix, modulated by the embedded nanomaterial. *Nano Impact*, *1*, 39–45.

Wolny-Koładka, K., & Malinowski, M. (2015). Assessment of the microbiological contamination of air in a municipal solid waste treatment company. *Ecological Chemistry and Engineering A*, *22*(2), 175–183.

Wong, C. S., Duzgoren-Aydin, N. S., Aydin, A., & Wong, M. H. (2007). Evidence of excessive releases of metals from primitive e-waste processing in Guiyu, China. *Environ Pollut*, *148*(1), 62–72. doi:10.1016/j.envpol.2006.11.006 PMID:17240013

Wong, J. W. C., Fung, S. O., & Selvam, A. (2009). Coal fly ash and lime addition enhances the rate and efficiency of decomposition of food waste during composting. *Bioresource Technology*, *100*(13), 3324–3331. doi:10.1016/j.biortech.2009.01.063 PMID:19268581

Wong, M. H., & Wong, J. W. C. (1986). Effects of fly ash on soil microbial activity. *Environmental Pollution. Series A. Ecological and Biological*, *40*(2), 127–144. doi:10.1016/0143-1471(86)90080-2

Wong, S. L., Ngadi, N., Abdullah, T. A. T., & Inuw, I. M. (2015). Current State and Future Prospects of Plastic Waste as Source of Fuel: A Review'. *Renewable & Sustainable Energy Reviews*, *50*, 1167–1180. doi:10.1016/j.rser.2015.04.063

Wong, S. S., Teng, T. T., Ahmad, A. L., Zuhairi, A., & Najafpour, G. (2006). Treatment of pulp and paper mill wastewater by polyacrylamide (PAM) in polymer induced flocculation. *Journal of Hazardous Materials*, *135*(1-3), 378–388. doi:10.1016/j.jhazmat.2005.11.076 PMID:16431022

Woodall, L. C., Sanchez-Vidal, A., Canals, M., Paterson, G. L., Coppock, R., Sleight, V., ... Thompson, R. C. (2014). The deep sea is a major sink for microplastic debris. *Royal Society Open Science*, *1*(4), 140317. doi:10.1098/rsos.140317 PMID:26064573

Woolridge, A. C., Ward, G. D., Phillips, P. S., Collins, M., & Gandy, S. (2006). Life cycle assessment for reuse/recycling of donated waste textiles compared to use of virgin material: An UK energy saving perspective. *Resources, Conservation and Recycling*, *46*(1), 94–103. doi:10.1016/j.resconrec.2005.06.006

World Bank. (1995). Lebanese republic solid waste/environmental management project, Draft Report. Washington, DC: Author.

World Bank. (1998). Sectoral Analysis in Bangladesh, Draft Report. Washington, DC: Author.

World Bank. (1999). *What a waste: solid waste management in Asia*. Retrieved from http://web.mit.edu/urbanupgrading/urbanenvironment/resources/references/pdfs/WhatAWasteAsia.pdf

World Bank. (2012). *What a Waste: Global Review of Solid Waste Management*. Author.

World Bank. (2017). *Brief: Solid waste management*. Washington, DC: Author. Retrieved from https://www.worldbank.org/en/topic/urban development/brief/solid-waste-management

World Bioenergy Association (WBA). (2015). Thermochemical Gasification of Biomass. *Holländargatan, 17*, 1-8.

World Bioenergy Association (WBA). (2015). Thermochemical gasification of Biomass. *Holländargatan, 17*. Retrieved from www.worldbioenergy.org

World Economic Forum. (2016). *The new plastics economy. Rethinking the future of plastics*. World Economic Forum.

World Energy Resources. (2013). *Waste to Energy*. World Energy Council.

World Health Organization. (2015). *Guidance on regulations for the transport of infectious substances 2015–2016: Applicable as of 1 January 2015*. Author.

World Resources Institute; World Businesses Council for Sustainable Development. (2004). *The greenhouse gas protocol: a corporate accounting and reporting standard* (Revised edition). Washington, DC: WRI.

Worrell, W. A., & Vesilind, P. A. (2011). *Solid waste engineering*. Cengage Learning.

Wrap. (2006). *Environmental benefits of recycling: an international review of life cycle comparisons for key materials in the UK recycling sector*. Banbury, UK: WRAP.

Wrap. (2008). *The carbon impact of bottling Australian wine in the UK: PET and glass bottles*. Banbury, UK: WRAP.

Wright, B. (1982). *The British Fire Mark 1680-1879*. Cambridge, UK: WoodheadFaulkner. Retrieved from http://www.firemarkcircle.fsnet.co.uk/history3.htm

Wright, S. L., Rowe, D., Reid, M. J., Thomas, K. V., & Galloway, T. S. (2015). Bioaccumulation and biological effects of cigarette litter in marine worms. *Scientific Reports*, *5*(1), 14119. doi:10.1038rep14119 PMID:26369692

Wright, S. L., Rowe, D., Thompson, R. C., & Galloway, T. S. (2013). Microplastic Ingestion Decreases Energy Reserves in Marine Worms. *Current Biology*, *23*(23), 1031–1033. doi:10.1016/j.cub.2013.10.068 PMID:24309274

Wright, S. L., Thompson, R. C., & Galloway, T. S. (2013). The physical impacts of microplastics on marine organisms: A review. *Environmental Pollution*, *178*, 483–492. doi:10.1016/j.envpol.2013.02.031 PMID:23545014

WRI-WBCSD. (2007). *The GHG Protocol for Project Accounting*. Author.

Wu, J., Xiao, Y. Z., & Yu, H. Q. (2005). Degradation of lignin in pulp mill wastewaters by white-rot fungi on biofilm. *Bioresource Technology, 96*(12), 1357–1363. doi:10.1016/j.biortech.2004.11.019 PMID:15792583

Wu, L., & Bradshaw, A. D. (1972). Actual pollution and the rapid evolution of copper tolerance. *Nature, 238*(5360), 167–169. doi:10.1038/238167a0

Wu, L., Bradshaw, A. D., & Thurraan, D. A. (1975). The potential for evolution of heavy metal tolerance in plants. III. The rapid evolution of copper tolerance in *Agrostis stolonifera. Hereditas, 34*(2), 165–187. doi:10.1038/hdy.1975.21

Wyszkowski, M., & Sivitskaya, V. (2012). Changes in the content of organic carbon and available forms of macronutrients in soil under the influence of soil contamination with fuel oil and application of different substances. *Journal of Elementology, 17*, 139–148.

Xavier, R., Nagarathinam, P., Gopalakrishnan, M. V., & Jayaraman, K. (2007). Isolation ofLepidopteran active native Bacillus thuringiensis strains through PCR Panning. *Asia-Pacific Journal of Molecular Biology and Biotechnology, 15*, 61–67.

Yadav, S. K., Babu, S., Yadav, M. K., Singh, K., Yadav, G. S., & Pal, S. (2013). A review of organic farming for sustainable agriculture in Northern India. International Journal of Agronomy.

Yadav, I. C., & Devi, N. L. (2009). Studies on municipal solid waste management in Mysore city — a case study. *Report and Opinion, 1*(3), 15–21.

Yamashita, R., & Tanimura, A. (2007). Floating plastic in the Kuroshio Current area, western North Pacific Ocean. *Marine Pollution Bulletin, 54*(4), 485–488. doi:10.1016/j.marpolbul.2006.11.012 PMID:17275038

Yang, Y., Boom, R., Irion, B., van Heerden, D. J., Kuiper, P., & de Wit, H. (2012). Recycling of composite materials. Chem. Eng. Process. *Process Intensif., 51*, 53–68. doi:10.1016/j.cep.2011.09.007

Yan, Y. W. (2013). Comparison of plant nutrient contents in vermicompost from selected plant residues. *African Journal of Biotechnology, 12*(17), 2207–2214. doi:10.5897/AJB11.3164

Yarsley, V. E., & Couzens, E. G. (1945). *Plastics*. Middlesex, UK: Penguin Books Limited.

Yassoglou, N., Kosmas, C., Asimakopoulos, J., & Kallinou, C. (1987). Heavy metal contamination of roadside soils in the greater Athens area. *Environmental Pollution, 47*(4), 293–304. doi:10.1016/0269-7491(87)90149-7 PMID:15092703

Yokel, J., & Delistraty, D. A. (2003). *Arsenic, lead and other trace elements in soils contaminated with pesticide residues at the Hansford site*. Wiley Periodicals.

Yong, Z., Gang, X., Guanxing, W., Tao, Z. & Dawei, J. (2009). Medical waste management in China: a case study of Nanjing. *Waste Management, 29*(4), 1376-1382.

Yoshida, S., Hiraga, K., Takehana, T., Taniguchi, I., Yamaji, H., Maeda, Y., ... Oda, K. (2017). A bacterium that degrades and assimilates polyethylene terephthalate. *Science, 351*(6278), 1196–1199. doi:10.1126cience.aad6359 PMID:26965627

Young, G. C. (2010). *Municipal Solid Waste to Energy Conversion Processes: Economic, Technical and Renewable Comparison*. John Wiley & Sons, Inc.

Yuan, H., & Zhu, N. (2016). Progress in inhibition mechanisms and process control of intermediates and by-products in sewage sludge anaerobic digestion. *Renewable & Sustainable Energy Reviews, 58*, 429–438.

Yusuf, M. (2017). Agro-Industrial Waste Materials and their Recycled Value-Added Applications. Handbook of Ecomaterials, 1-11.

Yu, X. Z., Gao, Y., Wu, S. C., Zhang, H. B., Cheung, K. C., & Wong, M. H. (2006). Distribution of polycyclic aromatic hydrocarbons in soils at Guiyu area of China, affected by recycling of electronic waste using primitive technologies. *Chemosphere*, *65*(9), 1500–1509. doi:10.1016/j.chemosphere.2006.04.006 PMID:16713614

Zaman, A. U. (2009). Life Cycle Environmental Assessment of Municipal Solid Waste to Energy Technologies. EESI, School of Architecture and the Built Environment, KTH Royal Institute of Technology.

Zaman, A. U. (2009). Life Cycle Environmental Assessment of Municipal Solid Waste to Energy Technologies. Global Journal of Environmental Research, 3(3), 155-163.

Zaman, A. U. (2015). A Strategic Waste Management Framework and Tool for the Development of Zero Waste Cities. School of Art, Architecture and Design; University of South Australia.

Zaman, A. U. (2010). Comparative study of municipal solid waste treatment technologies using life cycle assessment method. *International Journal of Environmental Science and Technology*, *7*(2), 225–234.

Zaman, A. U. (2015). A comprehensive review of the development of zero waste management: Lessons learned and guidelines. *Journal of Cleaner Production*, *91*, 12–25. doi:10.1016/j.jclepro.2014.12.013

Zaman, A. U., & Lehmann, S. (2011). Challenges and Opportunities in Transforming a City into a "Zero Waste City". *Challenges*, *2*(4), 73–93. doi:10.3390/challe2040073

Zaman, A. U., & Lehmann, S. (2013). The zero waste index: A performance measurement tool for waste management systems in a'zero waste city'. *Journal of Cleaner Production*, *50*, 123–132. doi:10.1016/j.jclepro.2012.11.041

Zare, Y. (2017). *3Recycled Polymers: Properties and Applications*. Available online: https://www.smithersrapra.com/SmithersRapra/media/Sample-Chapters/Recycled-Polymers-Properties-and-Applications,-Volume-2.pdf

Zeng, F., Cui, K., Xie, Z., Wu, L., Luo, D., Chen, L., ... Sun, G. (2009). Distribution of phthalate esters in urban soils of subtropical city, Guangzhou, China. *Journal of Hazardous Materials*, *164*(2-3), 1171–1178. doi:10.1016/j.jhazmat.2008.09.029 PMID:18963455

Zero Waste International Alliance. (2004). *Zero Waste Definition Adopted by Zero Waste Planning Group*. Retrieved from http://www.zwia.org/main/index.php?option=com_content&view=article&id=49&Itemid=37

Zero Waste SA. (2013). *South Australia's Waste Strategy 2011-2015*. Available at Error! Hyperlink reference not valid.

Zero Waste South Australia. (2013). *South Australia's Kerbside Three-Bin System Waste Report*. Retrieved from http://www.zerowaste.sa.gov.au/upload/resource-centre/publications/localgovernment/ZWSA%20Kerbside%20report%20 2015%20DE_02.pdf

Zettler, E. R., Mincer, T. J., & Amaral-Zettler, L. A. (2013). Life in the "plastisphere": Microbial communities on plastic marine debris. *Environmental Science & Technology*, *47*(13), 7137–7146. doi:10.1021/es401288x PMID:23745679

Zhang, H. J., Zhang, Y. H., Wang, Y., Yang, Y. H., Zhang, J., Wang, Y. L., & Wang, J. L. (2013). Investigation of medical waste management in Gansu Province, China. *Waste Management & Research*, *31*(6), 655–659. doi:10.1177/0734242X13482161 PMID:23567844

Zhang, S., & Herbell, J. D., Gaye, & Haake, B. (2004). Biodegradable organic matter in municipal solid waste incineration bottom ash. *Waste Management (New York, N.Y.)*, *24*, 673–679. PMID:15288299

Zhou, B. W., Shin, S. G., Hwang, K., Ahn, J. H., & Hwang, S. (2010). Effect of microwave irradiation on cellular disintegration of Gram positive and negative cells. *Applied Microbiology and Biotechnology*, *87*(2), 765–770. doi:10.100700253-010-2574-7 PMID:20393700

Zhuang, S., Fu, J., Powell, C., Huang, J., Xia, Y., & Yan, R. (2015). Production of medium-chain volatile flavour esters in Pichiapastoris whole-cell biocatalysts with extracellular expression of Saccharomyces cerevisiae acyl-CoA: Ethanol O-acyltransferase Eht1 or Eeb1. *SpringerPlus*, *4*(1), 467. doi:10.118640064-015-1195-0 PMID:26357598

Zhu, D., Asnani, P. U., Zurbrugg, C., Anapolsky, S., & Mani, S. (2008). *Improving solid waste management in India. A Source book for Policy Makers and Practitioners, World Bank Institute, WBI Development Studies*. The World Bank.

Ziadat, A., & Mott, H. (2005). Assessing solid waste recycling opportunities for closed campuses. *Management of Environmental Quality*, *16*(3), 250–256.

Zia, H., & Devadas, V. (2008). Urban solid waste management in Kanpur: Opportunities and perspectives. *Habitat International*, *32*(1), 58–73.

Ziolkowska, A., & Wyszkowski, M. (2010). Toxicity of petroleum substances to microorganisms and plants. *Ecological Chemistry and Engineering. S*, *17*, 73–82.

Žmak & Hartmann. (2017). Current State Of The Plastic Waste Recycling System In The European Union And In Germany. *Technical Journal*, *11*(3), 138–142.

Zorpas, A. A., & Loizidou, M. (2008). Sawdust and natural zeolite as a bulking agent for improving quality of a composting product from anaerobically stabilized sewage sludge. *Bioresource Technology*, *99*(16), 7545–7552. doi:10.1016/j.biortech.2008.02.014 PMID:18439819

Zorrob, S. E., & Suparama, L. B. (2004). Laboratory Design and Investigation of Proportion of Bituminous Composite Containing Waste Recycled Plastics Aggregate Replacement (Plastiphalt). *Cement and Concrete Composites*, *22*(4), 233–242. doi:10.1016/S0958-9465(00)00026-3

Zubris, K. A. V., & Richards, B. K. (2005). Synthetic fibers as an indicator of land application of sludge. *Environ. Pollut.*, *138*(2), 201–211. doi:10.1016/j.envpol.2005.04.013 PMID:15967553

Zurbrugg, C., Gfrerer, M., Ashadi, H., Brenner, W., & Küper, D. (2012). Determinants of sustainability in solid waste management–The Gianyar Waste Recovery Project in Indonesia. *Waste Management (New York, N.Y.)*, *32*(11), 2126–2133. doi:10.1016/j.wasman.2012.01.011 PMID:22330265

ZWSA. (2011). *Zero Waste SA. About Us*. Available at http://www.zerowaste.sa.gov.au/about-us

About the Contributors

Rouf Ahmad Bhat has his specialisation in Limnology, Toxicology, Phytoremediation and waste management techniques. He has been teaching graduate and post graduate students for the past one year. He has published more than 40 papers in referred international journals and a number of books with international publishers.

Khursheed Ahmad Wani is Lecturer in the Department of Environmental Science, Government Degree College Bijbehara, J&K. Dr. Wani served at ITM University Gwalior as Assistant Professor in the Department of Environmental Science from 2011 to 2017. He has completed his M. Sc. and Ph. D. in Environmental Science from Jiwaji University, Gwalior, Madhya Pradesh. He has also qualified Jammu and Kashmir State Level Eligibility Test (JKSET). His areas of interest include Environmental management, Environmental Impact Assessment and Occupational Health. He has published research papers in various national and international journals and is the author of three books on Environmental Science. He is a member of several national and international organizations and is in the editorial board of various journals. He has organized different programmes on Environmental Awareness and has completed research project sponsored by Madhya Pradesh Council of Science and Technology, Bhopal, India. Dr. Wani has organized various national and international conferences/seminars/ workshops as well.

* * *

Ifra Ashraf is currently doing PHD in Soil and Water Engineering from College of Agricultural Engineering and Technology, SKUAST-K in 2018. She has received her M. Tech. Soil and Water Engineering from College of Agricultural Engineering and Technology, SKUAST-K in 2018. She is the author of one research article, two review articles and one conference paper.

Fatima Buchh obtained M. Sc, M. Phil and a Doctorate degree in Home Science Extension Education from University of Kashmir. She has contributed about twenty research papers on different topics in various National and International journals. She has been actively involved in research and teaching at post graduate and undergraduate levels for at least eight years. At present she is working as a Research Officer in Centre for women's studies and research, University of Kashmir and has experience of guiding research work at various levels, besides that she had written a book on "Impact of domestic violence on Kashmiri women".

Ashaq Ahmad Dar is a Doctoral Student in the Department of Ecology and Environmental Sciences, Pondicherry University, Puducherry – India. He has completed his masters in Environmental Science and has qualified various national and state level examinations including NET-JRF, ASRB-NET, and JKSET. He is also the author of various international and national publications.

Moonisa Aslam Dervash is currently Working as Assistant Professor in Higher Education Department and she has published more than 17 research articles in standard National and International Journals.

Syed Maqbool Geelani is Currently working as Assistant Professor in Higher Education Department Jammu and Kashmir. Dr. Geelani has Published more 40 research articles in repute Journals.

Azad Gull is Senior Field Assistant at P4, Basic Seed Farm, Central Silk Board, Manasbal, Ganderbal, Jammu & Kashmir (India).

Naila Irshad is presently working as a Senior Assistant professor in the Institute of Home science, University of Kashmir. She has completed her P. G and Ph. D in Food and Nutrition (Home science) from the same University, has been contributing to the field of teaching and research since 2004. She has a teaching experience of more than 15 years. Dr. Naila has been a gold medalist right from class 11th till completing her P.G. She has 4 gold medals to her credit besides being adjudged as the best graduate of the year in the field of science. In addition to this she has also won several prizes for being a meritorious student at U.G and P.G. levels.

Asima Jillani did B. Tech from Skuast - k. Then did M. Tech (with JRF) and Ph. D (with JRF & SRF) fom GBPUAT Pantnagar. Qualified ASRB-ICAR NET in 2014, 2015, 2017. I have more than 15 publication's in national and international journals. Presently, working as SRF under NICRA project at SKUAST-K.

G. Hassan Lone has completed his Masters degree in Social Work (MSW) from University of Kashmir and M.Phil in Social Work from IGNOU. Presently he is pursuing Ph.D in Social Work at the Department of Social Work, University of Kashmir. He has more than 13 years of work experience in different sectors like NGOs, Academics, Emergency relief and response, Development Induced Displacement and Resettlement (DIDR). The author has a specialized Post Graduate Diploma in Resettlement and Rehabilitation and a certificate course in Financing for Development from the World Bank. He has many national and international publications to his credit.

Ikhlaq A. Mir is JRF in Climate Change and Mountain Agriculture (CCCMA), Sher-e-Kashmir University of Agricultural Sciences and Technology of Kashmir. He has received his M.Sc. in Geoinformatics (Remote sensing & GIS) in 2013 from University of Kashmir and Postgraduation in Economics in 2012 from IGNOU. He is the author of one research publication and 11 conference publications.

Shubeena Nazir is working Department of Education, Government of Jammu and Kashmir, India.

Shazia Ramzan is the post doc fellow (DST-SERB) University of Kashmir, Deptt. of Geography and regional development. She is the author of twenty research publications, four book chapters, five review articles, one book and one technical report. She has received her PhD in soil science from Skuast-K in 2016.

Nowsheeba Rashid has recieved her M.tech in food technology from Amity University in 2018.She is the author of one research article, one review article and one conference paper.

José Vargas-Hernández, M.B.A.; Ph.D., Member of the National System of Researchers of Mexico and a research professor at University Center for Economic and Managerial Sciences, University of Guadalajara. Professor Vargas-Hernández has a Ph. D. in Public Administration and a Ph.D. in Organizational Economics. He has undertaken studies in Organisational Behaviour and has a Master of Business Administration, published four books and more than 200 papers in international journals and reviews (some translated to English, French, German, Portuguese, Farsi, Chinese, etc.) and more than 300 essays in national journals and reviews. He has obtained several international Awards and recognition.

Akhlaq Amin Wani is Currently working as Associate Professor in Sher-e-Kashmir University of Agricultural Sciences and Technology, Faculty of Forestry Benhama Ganderbal. Dr. Wani has Published more 50 research articles in repute Journals.

Index

Z

Purchase Print, E-Book, or Print + E-Book

IGI Global's reference books are available in three unique pricing formats:
Print Only, E-Book Only, or Print + E-Book.
Shipping fees may apply.

www.igi-global.com

Recommended Reference Books

ISBN: 978-1-5225-7727-0
© 2019; 335 pp.
List Price: $195

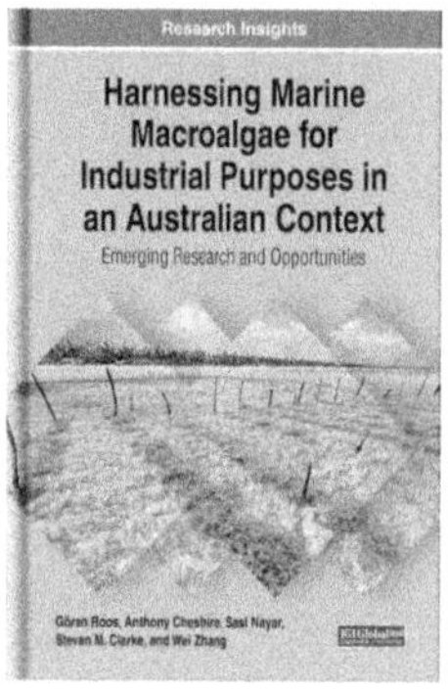

ISBN: 978-1-5225-5577-3
© 2019; 291 pp.
List Price: $195

ISBN: 978-1-5225-5291-8
© 2019; 252 pp.
List Price: $175

ISBN: 978-1-5225-5766-1
© 2019; 501 pp.
List Price: $195

ISBN: 978-1-5225-5909-2
© 2019; 308 pp.
List Price: $195

ISBN: 978-1-5225-8027-0
© 2019; 337 pp.
List Price: $205

Do you want to stay current on the latest research trends, product announcements, news and special offers?
Join IGI Global's mailing list today and start enjoying exclusive perks sent only to IGI Global members.
Add your name to the list at **www.igi-global.com/newsletters.**

Publisher of Peer-Reviewed, Timely, and Innovative Academic Research

IGI Global
DISSEMINATOR OF KNOWLEDGE

www.ingramcontent.com/pod-product-compliance
Lightning Source LLC
LaVergne TN
LVHW062119300926
849161LV00010B/2493